ACIDIC DEPOSITION: SULPHUR AND NITROGEN OXIDES

THE ALBERTA GOVERNMENT/INDUSTRY ACID DEPOSITION RESEARCH PROGRAM (ADRP)

EDITORS

A. H. LEGGE

KANANASKIS CENTRE FOR ENVIRONMENTAL RESEARCH
UNIVERSITY OF CALGARY
CALGARY, ALBERTA, CANADA

AND

S. V. KRUPA

DEPARTMENT OF PLANT PATHOLOGY
UNIVERSITY OF MINNESOTA
ST. PAUL, MINNESOTA, USA

Library of Congress Cataloging-in-Publication Data

Acidic deposition.

Bibliography: p.
Includes index
1. Alberta Government/Industry Acid Deposition Research Program. 2. Acid deposition—Environmental aspects—Alberta. 3. Sulphur oxides—Environmental aspects—Alberta. 4. Nitrogen oxides—Environmental aspects—Alberta. 5. Air quality—Alberta. I. Legge, Allan H. II. Krupa, Sagar V.
TD195.54C22A432 1990 628.5'32 89-12319
ISBN 0-87371-190-4

LEWIS PUBLISHERS, INC.
121 South Main Street, Chelsea, Michigan 48118

PRINTED IN THE UNITED STATES OF AMERICA

PREFACE

The Alberta Government/Industry Acid Deposition Research Program (ADRP) was jointly sponsored by the public and private sectors in Alberta, Canada. The Program began in 1983 with the objective of addressing rising public concerns on the environmental consequences of acidic and acidifying air pollutants on the environment in Alberta. This book presents the results of the Biophysics Research of the ADRP which began in 1985. It represents the culmination of three and one-half years of intensive study, including characterization of the background air quality of Alberta and determination of how this air quality is modified by emissions of oxides of sulphur and nitrogen within Alberta. The measured air quality was integrated with modelling to evaluate present and potential regional-scale responses of selected environmental receptors.

The ADRP was unique in its organization. A Members Committee, consisting of representatives of the Alberta Department of the Environment, Canadian Petroleum Association, Energy Resources Conservation Board, and the Electric Power Generating Utilities, set the overall objectives of the Program. These organizations sponsored the research. A Science Advisory Board, consisting of ten internationally recognized scientists in the field of air pollution and its related environmental effects, were involved in the initial research design, provided advice and evaluated the research on a regular basis, through peer review. Approval of the scientific aspects of ADRP rested solely with the Science Advisory Board. A Public Advisory Board was also formed in recognition of the value and need for public views and to facilitate involvement from and communication with the general public and environmental interest groups. The Public Advisory Board was represented on the Members Committee through a non-voting participant.

The ADRP Biophysics Research represents an integrated holistic approach to the evaluation of the chemistry of multiphase atmospheric systems and their possible effects on the environment. While no environmental response studies *per se* were conducted in the field, the information contained in the book serves to identify areas which require attention.

In addition to the joint public/private sector sponsorship of the ADRP, the Biophysics Research was unique in that it was a "joint venture" between the University of Calgary and the private sector. The Prime Contractor for the Biophysics Research was the Kananaskis Centre for Environmental Research, the University of Calgary, with the private sector represented by Western Research of Calgary as the primary partner.

The Biophysics Research team was international in scope and included the following groups and individuals.

- Kananaskis Centre for Environmental Research, the University of Calgary, Calgary, Alberta

  A. H. Legge, Principal Investigator
  E. Peake, Project Leader, atmospheric deposition and chemical analysis
  M. T. Strosher, Project Leader, atmospheric sulphur gases and organic compounds
  S. A. Telang, surface water acidification, literature review
  R. M. Danielson, and
  S. Visser, soil microbiology, literature review

- Western Research, Calgary, Alberta

  D. M. Leahey, Project Leader, meteorology and climatology
  M. C. Hansen, Project Leader, air quality and meteorology data quality control and quality assurance
  D. J. Picard, Project Leader, sulphur oxide and nitrogen oxide emissions inventory
  D. G. Colley, emissions inventory, compilation, and assessment

- A.S.L. and Associates, Helena, Montana

  A. S. Lefohn, international air quality comparisons

- Alberta Research Council, Terrain Science Department, Edmonton, Alberta

  S. A. Abboud, Project Leader, soil chemistry modelling

L. W. Turchenek, soil acidification, literature review and soil chemistry modelling

- Aquatic Resource Management Limited, Calgary, Alberta

  R. A. Crowther, water chemistry modelling

- Carnegie-Mellon University, Department of Civil Engineering, Pittsburgh, Pennsylvania

  C. I. Davidson, deposition velocity studies

- Ecosat Geobotanical Surveys, Inc.

  D. R. Jaques, biophysical description and inventory of Alberta

- Element Analysis Corporation, Tallahassee, Florida

  S. Bauman, elemental analysis of coarse and fine particles

- Emporia State University, Department of Biology, Emporia, Kansas

  J. M. Mayo, acidic deposition and forests, literature review

- Florida State University, Department of Oceanography, Tallahassee, Florida

  J. W. Winchester, analysis and interpretation of elemental composition of coarse and fine particles

- Illinois State Water Survey, University of Illinois, Champaign, Illinois

  E. C. Krug, Project Leader, surface water chemistry modelling

- Lawrence Livermore National Laboratory, Environmental Sciences Division, Livermore, California

  J. H. Shinn,
  M. S. Torn, and
  J. E. Degrange, acidic deposition and agriculture, literature review

- McVehil-Monnett Associates, Englewood, Colorado

  G. E. McVehil, Project Leader, atmospheric chemical modelling and transport

- Nuclear Environmental Analysis, Inc., Beaverton, Oregon

  J. A. Cooper, Project Leader, source apportionment

- Subsurface Technologies Ltd., Calgary, Alberta

  K. W. Campbell, geological and hydrological aspects of acidification, literature review

- The University of Calgary,
  Department of Biological Sciences

  E. J. Laishley, and
  R. D. Bryant, inorganic sulphur, soil microbiology, literature review

  Department of Geography

  E. C. Rhodes, meteorology and determination of background air quality trailer site location

  Department of Mathematics and Statistics

  M. Nosal, Project Leader, data management and statistical analysis

Department of Physics

H. R. Krouse, stable sulphur isotopes in the environment, literature review
T. Mathews, atmospheric physics and meteorology

- U.S. Department of Agriculture, Beltsville, Maryland

J. F. Parr, soil microbiology, literature review

It is hoped that the Biophysics Research carried out by the ADRP will serve as a model for investigations on the environmental consequences of acidic and acidifying pollutants and air pollution in general in the future.

A. H. Legge
Principal Investigator

S. V. Krupa
Chairman, Science Advisory Board and
Science Advisor, ADRP

ADRP Members Committee

Kenneth R. Smith (Co-Chairman)
Assistant Deputy Minister
Alberta Environment

Ronald L. Findlay (Co-Chairman)
Manager, Environmental Affairs and Safety
Amoco Canada Petroleum Co. Ltd.
Canadian Petroleum Association

P. Douglas Bruchet (Treasurer)
Manager, Environmental and Socio-Economic Development
Canadian Petroleum Association

Dr. Cornelius (Casey) G. Van Teeling (Secretary)
Alberta Environment

Edward R. Brushett
Manager, Environment Protection
Energy Resources Conservation Board

Dr. John B. Railton
Special Projects Manager
Monenco Consultants Limited
Utilities Group Representative

Carl L. Primus
Environmental Health
Alberta Community and Occupational Health

Dr. H. Percy Sims (Special Advisor)
Alberta Environment

Dr. Harby S. Sandhu (Special Advisor)
Alberta Environment

Susie Washington (Moderator, Public Advisory Board)
Western Environmental and Social Trends

Program Manager:
Dr. Ronald R. Wallace
Dominion Ecological Consulting, Ltd.

Communications Coordinator:
Jean L. Andryiszyn
Francis, Williams & Johnson, Ltd.

ADRP Science Advisory Board

Dr. S. V. Krupa (Chairman)
University of Minnesota
Department of Plant Pathology

Dr. C. M. Bhumralkar
National Oceanographic and Atmospheric Administration (NOAA)

Dr. H. C. Jones
Tennessee Valley Authority
Fisheries and Aquatic Ecology Branch

Dr. R. N. Kickert*
Consultant in modelling, Oregon

Dr. H. M. Liljestrand
University of Texas
Department of Civil Engineering

Dr. J. P. Lodge, Jr.
Consultant in Atmospheric Chemistry and Editor, Atmospheric Environment

Dr. D. P. Ormrod
University of Guelph
Department of Horticultural Science

Dr. C. L. Schofield
Cornell University
Department of Natural Resources

Mr. R. K. Stevens
U. S. Environmental Protection Agency

Dr. M. A. Tabatabai
Iowa State University
Department of Agronomy

Dr. T. C. Weidensaul
Ohio State University
Ohio Agricultural Research and Development Center

* Because of the program needs, the SAB member was requested to participate in the final data analyses. The SAB member did not participate as a peer reviewer of the appropriate chapter in which he was an author.

ADRP Public Advisory Board

Susie Washington (Moderator)
Western Environmental and Social Trends

Dr. Nicholas Bayliss
Alberta Health Care Association

Mr. Herman Bulten
Christian Farmers Federation and other
Agricultural Associations

Mr. Hank Gauthier
Energy and Chemical Workers Union

Mr. Dennis Gogal
Environmental and Outdoor Education Council,
Alberta Teachers Association

Dr. Lochan Bakshi
Athabasca University,
Public Advisory Committee, Environment Council of Alberta

Mr. Sam Sinclair
Alberta Aboriginal People

Dr. Dixon A. R. Thompson
University of Calgary,
Science Advisory Committee,
Environmental Council of Alberta,
Alberta Society of Professional Biologists,
Canadian Society of Environmental Biologists

Dr. Eric L. Tollefson
University of Calgary,
Association of Professional Engineers,
Geologists and Geophysicists of Alberta

ACKNOWLEDGMENTS

Financial support for the $5.3 million Biophysics Research was provided by the Alberta Government/Industry Acid Deposition Research Program (ADRP) and is gratefully acknowledged. The funding partners of the ADRP which made up the Members Committee were the Alberta Department of the Environment, Canadian Petroleum Association, Energy Resources Conservation Board, and the Electrical Power Generating Utilities of Alberta, who contributed 50%, 34.6%, 10%, and 5.4%, respectively. Appreciation is extended to the Co-Chairmen, Mr. R.L. Findlay (Industry) and Mr. K.R. Smith (Government) and the Members Committee of the ADRP for supporting this research and providing funds for the preparation of this book. The assistance of Dr. R.R. Wallace, ADRP Program Manager is noted with thanks. The assistance and skill of Ms. J. Andryiszyn of Francis, Williams, & Johnson Ltd. of Calgary in communicating the progress and final results of the ADRP to the public is most appreciated.

The guidance, assistance, and critical reviews of the Biophysics Research by the Members of the Science Advisory Board were extremely helpful, greatly contributed to the success of the research, and are gratefully acknowledged. We are highly appreciative of the support and encouragement of the Public Advisory Board.

Thanks are extended to Mr. J. Couillard, Manager, Fortress Mountain Ski Area, and to Kananaskis Country for permission to locate the ADRP background air quality and meteorology station at Fortress Mountain, and to Mr. and Mrs. A.L. Walroth and Mr. and Mrs. M.E. Forster, for their willingness to allow the ADRP to locate the remaining two ADRP air quality and meteorology stations on their farmland.

Thanks are extended to Mr. A. Schultz, Director of the Air Quality Branch, Pollution Control Division, Environment Protection Services, Alberta Environment, for providing the external audits of the three ADRP air quality monitoring trailers. The assistance of Dr. Y. Lau, E. Boyko, and D. Bensler in this matter is most appreciated.

Much of the success of a research program rests on the technical support staff. The dedication and enthusiasm of the following individuals is appreciated and recognized: Kelvin Adolph, John Bogner, Doug Connery, John Corbin, Gary Cross, Brian Fong,

Linda Fung, Diane Goode, Peter Graw, Indra Harry, Pavel Kubicek, Julie Lockhart, Alice MacLean, Joseph Mathews, Ron Minks, Ken Montgomery, Brenda Moddle, Tim Nugent, Dave Savage, Trent Schumann, Giovanna Stea, Stan Swacha, Doug Taylor, and Karel Traxler, Kananaskis Centre for Environmental Research, The University of Calgary; Bruce Granstrom, Pat Irwin, and Jesusa Pontoy, Department of Physics, The University of Calgary; and Murali Balachandran, Nelson Boychuk, Ben Kucewicz, Leslie Phillips, Craig Snider, Bob Spate, Kevin Warren, Laurie Whitely, Paul Won, Dean Wong, Paul Vetro, and Gary Vollo, Western Research, Calgary.

The many graphics in this book were skillfully prepared by Mr. B. Matheson, Matheson Graphic Services Limited, and reproduced by Riley's Reproduction of Calgary. The photographic skills of Mr. J. Peacock, Department of Communication Media, The University of Calgary, and Spectrum Photo Labs are also acknowledged.

Appreciation is extended to the staff of the Kananaskis Centre for Environmental Research, The University of Calgary, and the staff of Western Research of Calgary, for their positive support throughout the ADRP. Thanks are due to Lynn Ewing and Della Patton for typing the many documents prepared for the ADRP. A special note of thanks is extended to Linda Jones for her accounting and editorial skills.

At the University of Minnesota, Mark Jorgensen (Computer Applications Programmer), Rosemary Kumhera and Leslie Johnson (Word Processing Specialists) deserve recognition and gratitude for their many, long hours of hard work in helping to complete this effort.

List of Contributors

S. A. Abboud
Terrain Science Department, Alberta Research Council, Edmonton, Alberta.

D. G. Colley
Western Research, Calgary, Alberta.

J. A. Cooper
Nuclear Element Analysis (NEA), Inc., Beaverton, Oregon.

R. A. Crowther
Aquatic Resource Management, Ltd., Calgary, Alberta.

C. I. Davidson
Department of Civil Engineering, Carnegie-Mellon University, Pittsburgh, Pennsylvania.

M. C. Hansen
Western Research, Calgary, Alberta.

R. N. Kickert
Ecological Systems Scientist, Corvallis, Oregon.

E. C. Krug
Illinois State Water Survey, Champaign, Illinois.

A. S. Lefohn
A. S. L. & Associates, Helena, Montana.

A. H. Legge
Kananaskis Centre for Environmental Research, University of Calgary, Calgary, Alberta.

G. E. McVehil
McVehil-Monnett Associates, Inc., Englewood, Colorado.

M. Nosal
Department of Mathematics and Statistics, University of Calgary, Calgary, Alberta.

E. Peake
Kananaskis Centre for Environmental Research, University of Calgary, Calgary, Alberta.

D. J. Picard
Western Research, Calgary, Alberta.

M. T. Strosher
Kananaskis Centre for Environmental Research, University of Calgary, Calgary, Alberta.

L. W. Turchenek
Terrain Science Department, Alberta Research Council, Edmonton, Alberta.

Contents

CHAPTER 1
INTRODUCTION

A. H. Legge

Atmospheric sulphur and nitrogen oxides, their conversion products including acidic precipitation and their present and potential effects on the environment have been of significant public concern. While the Alberta (Canada) Government regulations regarding air pollutant emissions allowed from the provincial industries are considered to be very stringent, in a proactive step, the Alberta Government and industry jointly established as a case study, the Acid Deposition Research Program (ADRP) in 1983 with the following goals:

a. To provide a comprehensive understanding of the effects and consequences of acid-forming gases (sulphur and nitrogen oxides) and acidic deposition on the environment;
b. To provide a scientific basis for sound, long-term environmental management and regulatory control with respect to acid-forming gases;
c. To disseminate such information among program members, to the public, and to government bodies;
d. To undertake any research deemed appropriate; and
e. To encourage and include opportunities for public representation in the ADRP.

Within the ADRP framework, the biophysics research effort was initiated in 1985 with the work planned in two phases; they were designated by their terminal points, Critical Point I. and Critical Point II. Results and conclusions to Critical Point I. would in turn shape the second phase of the investigation.

The objectives of Critical Point I. were:

a. Are there at present observed or measurable adverse effects of acidic or acidifying (sulphur and nitrogen oxides) pollutants on Alberta's vegetation, soils, and surface waters?
b. If the answer to objective (a) is negative, when, where, and under what circumstances could such adverse effects occur in Alberta?

In addressing these two questions, the ADRP Biophysics Research Program was designed to characterize the Alberta air quality (background and polluted) and to apply the resulting data in the assessment and prediction of regional-scale effects of various air pollutant exposure scenarios on sensitive environmental components. It is critical to note that the ADRP was not designed to address the effects of specific point sources but rather to develop a regional-scale assessment of environmental effects within the Province of Alberta.

The research to Critical Point I. was planned as a 3.5 year study. The following steps were involved in this effort:

a. Critical review of world literature on cause-and-effect relationships of acidic and acidifying (sulphur and nitrogen oxides) air pollutants with reference to crops, forests, soils, and surface waters and their related components;
b. Comprehensive characterization of background air quality (incoming air) of Alberta;
c. Comprehensive characterization of an environment within Alberta influenced by area, line, and local point sources of pollutants;
d. Preparation of a detailed, provincial-scale emission inventory for the oxides of sulphur and nitrogen; and
e. Development of regional-scale air quality predictions and application of such data toward regional-scale environmental effects assessment (crops, soils, and surface waters). In the effects assessment, forest response evaluations were excluded; a literature review on this subject was completed, but no further work was mandated by the ADRP terms of agreement, since other agencies in Alberta were studying the subject and necessary data bases were not available during the course of the present program.

Over 30 professional scientists and their support personnel participated in the efforts, representing a variety of disciplines such as atmospheric and analytical chemistry, meteorology, statistics and computer applications, crop and tree biology, soil and aquatic chemistry, and ecological modelling.

The work relevant to steps (a) and (d) were completed and released to the public in 1987 (Appendix). Chapters 2 through 9 of this book provide the results of research relevant to steps (b), (c), and (e).

CHAPTER 2

SULPHUR AND NITROGEN IN THE ATMOSPHERE

A. H. Legge

2.1 INTRODUCTION

The chemical climatology of the earth is governed by natural processes and by man's intervention. It is the latter which is of much concern at the local, national, and international level.

Air pollutants occur as gases (e.g., ozone, O_3), vapours (e.g., nitric acid, HNO_3), and particles (e.g., dust). The particulate matter can be separated into two major classes (Whitby 1978): fine particles (size $<2.0\ \mu m$) and coarse particles (size $>2.0\ \mu m$). While the coarse particles are mechanically generated, the fine particles are mostly produced by chemical reactions in the atmosphere.

The natural sources of the chemical constituents in the atmosphere consist of volcanoes, the ocean floor, soils, and vegetation. On the other hand, man-made sources can be classified as regional (e.g., southern California); area (e.g., an urban centre such as Calgary); line (e.g., a highway); continuous, point (e.g., a gas plant or a coal burning power plant), and a single event, point (e.g., a chemical spill).

Once a pollutant is emitted into the atmosphere it is deposited continuously on to surfaces (e.g., vegetation, soils, surface waters, materials) by dry deposition (Figure 2.1). This dry deposition is mediated by diffusion, Brownian motion, interception, impaction, and sedimentation (Legge and Krupa 1986). At the same time, depending on the meteorological characteristics, pollutants are transported from a few to thousands of kilometres. Again, depending on the physical and chemical climatology, primary pollutants (e.g., nitrogen dioxide, NO_2) are transformed to secondary pollutants (e.g., O_3) before transfer from the atmosphere to surfaces (Figure 2.1). This transformation of primary to secondary pollutants varies significantly in time and space, but at some measurable rate (Finlayson-Pitts and Pitts 1986).

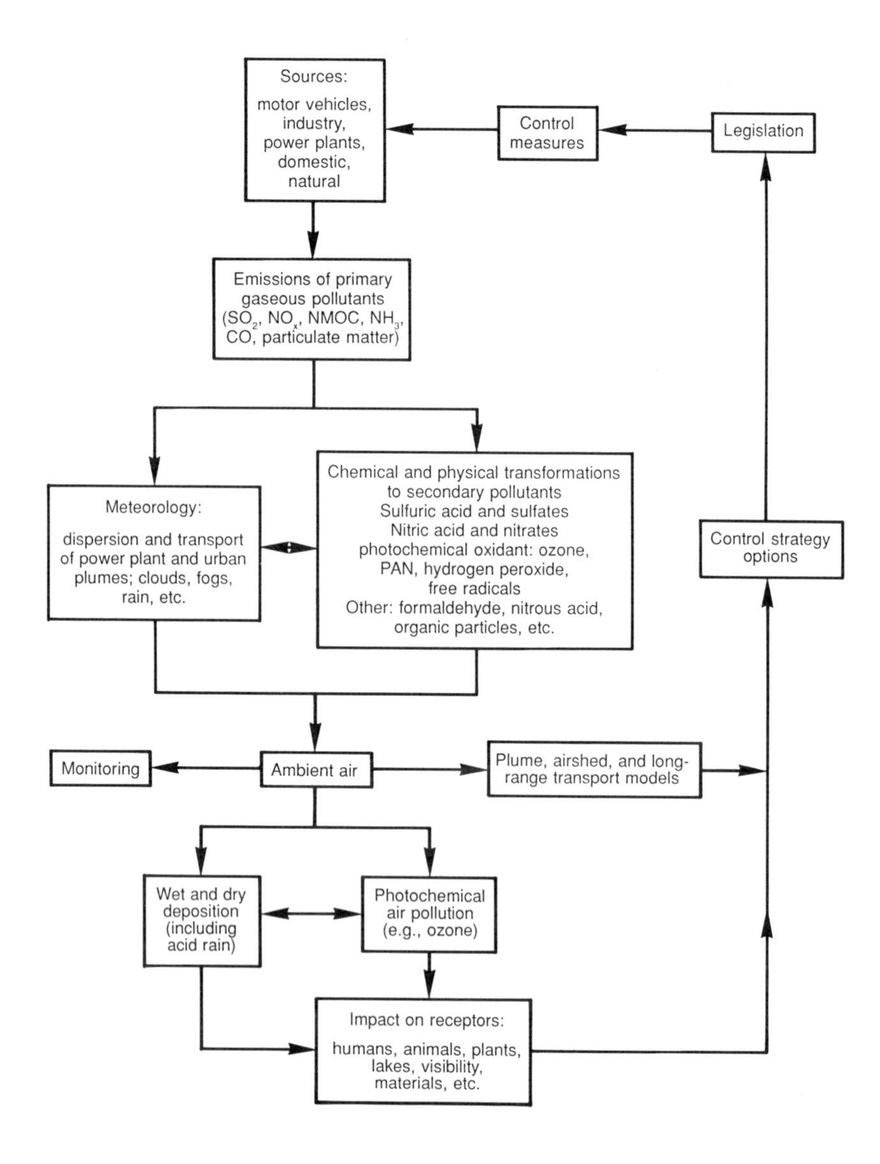

Figure 2.1 The air pollution system.
Source: Finlayson - Pitts and Pitts (1986). (Reprinted with permission, John Wiley & Sons, Inc. Copyright 1986).

In addition to the transport, transformation, and dry deposition, pollutants such as sulphur dioxide (SO_2) and fine particulate sulphate (SO_4^{2-}) are also incorporated into clouds. During precipitation events, these pollutants within the cloud (rainout), and to a degree what is in the air below the clouds (washout), are deposited on to surfaces by wet deposition. Of major concern in this context is the phenomenon of "acidic precipitation".

The relative magnitude and importance of dry versus wet deposition are known to vary in time and space (Pruppacher et al. 1983; U.S. NAPAP, Interim Assessment Report, Vol. III 1987). Thus, in evaluating the effects of air pollutants on environmental receptors, both atmospheric measurements and environmental response measurements must be performed on an integrated basis in common time and space.

Certain pollutants, such as O_3 and fine particulate SO_4^{2-}, are of regional-scale concern, while others such as SO_2 and hydrogen sulphide (H_2S) are of environmental concern on a local scale. Similarly, certain chemical constituents in the atmosphere occur as acids (e.g., sulphuric acid aerosol, H_2SO_4). Others such as SO_2 are acidifying in the sense that they form acids upon contact with wet surfaces and within plant cells upon uptake. Thus, within the direct context of "acidic" and "acidifying" pollutants, the emission of sulphur and nitrogen compounds is of main concern.

2.1.1 Sulphur Compounds

A number of gaseous sulphur compounds are emitted into the atmosphere through natural processes and/or by man's activities (Table 2.1). Of these SO_2 is the most important species of environmental concern. In this context the H_2S emitted into the atmosphere is also converted to SO_2 within hours to days (Meszaros 1981; Finlayson-Pitts and Pitts 1986). Table 2.2 provides a comparison of an estimate of the natural emission of sulphur compounds in the northern and southern hemispheres with man-made emissions. On a global scale, human activities accounted for 41% of the total emissions of sulphur in 1976. In the northern hemisphere where industrial activities are greatest, man-made emissions accounted for 56% of the total; in the southern hemisphere, they contributed only 8%.

Table 2.1 Some gaseous sulphur compounds identified in the atmosphere.

Sulphur dioxide (SO_2)	Hydrogen sulphide (H_2S)
Carbonyl sulphide (COS or OCS)	Carbon disulphide (CS_2)
Dimethyl sulphide (CH_3SCH_3)	Dimethyl disulphide (CH_3SSCH_3)
Methyl mercaptan (CH_3SH)	Unknowns (4 or more)

Once in the atmosphere, SO_2 is oxidized by homogeneous (gas to gas phase) (Figure 2.2) and heterogeneous (gas to particle phase) reactions resulting in the formation of SO_4^{2-}. For details of the SO_2 oxidation mechanisms, the reader is referred to Finlayson-Pitts and Pitts (1986); Hidy and Mueller (1986); and the U.S. National Research Council (1983).

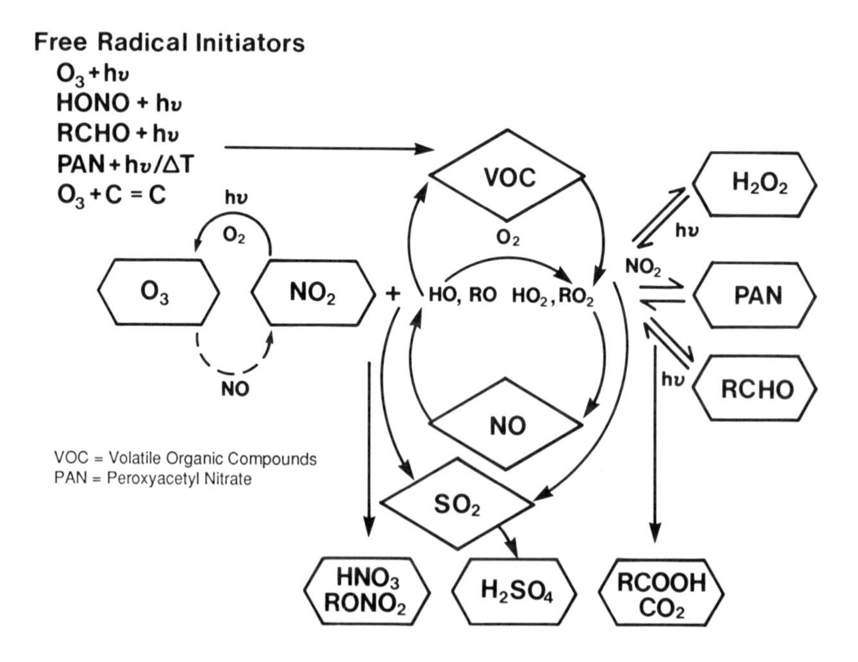

Figure 2.2 Schematic representation of the polluted atmosphere photo-oxidation cycle. Pollutants emitted from anthropogenic sources are shown in diamond-shaped boxes; pollutants formed in the atmosphere are shown in hexagonal boxes. The reaction scheme represents gas-phase chemistry only.
Source: U.S. NAPAP (1987).

Table 2.2 Estimated natural and anthropogenic global emissions of atmospheric sulphur ($Tg\ S\ y^{-1}$)[a].

Sources	1970 Northern and Southern Hemispheres	1976 Northern and Southern Hemispheres	1976 Northern Hemisphere[a]	1976 Southern Hemisphere[b]
Natural				
Volcanoes	5	5	3 (60)	2 (40)
Sea spray	44	44	19 (43)	25 (57)
Subtotal non-biogenic	49	49	22 (45)	27 (55)
Biogenic (land)	48	48	32 (67)	16 (33)
Biogenic (oceans)	50	50	22 (44)	28 (56)
Subtotal biogenic	98	98	54 (55)	44 (45)
Total Natural	147	147	76 (52)	72 (48)
Total Anthropogenic	86	104	98 (94)	6 (6)
Total (natural + anthropogenic)	233	251	174 (69)	77 (31)
Anthropogenic (% of total)	37	41	56	8

a $1\ Tg = 10^{12}$ g.
b Figures in parentheses represent the percentages in each hemisphere.

Source: Cullis and Hirschler 1980. (Reprinted with permission, Copyright 1980 Pergamon Press PLC).

In power plant plumes, SO_2 oxidation rates of up to 4% h^{-1} have been reported (Husar et al. 1978). Often such rates are much higher if the plume passes through clouds or fog banks (Eatough et al. 1984). Similarly, the rates of production of SO_4^{2-} from SO_2 are much higher during the summer compared with the winter (Richards et al. 1981). According to Gillani (1978) and Forrest et al. (1981), noontime SO_2 conversion rates in a power plant plume were 1-4% h^{-1} compared with the nighttime rates of <0.5% h^{-1}. However, significant SO_4^{2-} production (4.5-10.8% h^{-1}) can occur at nighttime if clouds were a contributing factor in the SO_2 conversion (Cass and Shair 1984). For details of the suggested mechanisms for the aqueous oxidation of SO_2, the reader is referred to Graedel and Goldberg (1983), Graedel et al. (1986), and Bielski et al. (1985).

In northeastern Alberta, Cheng et al. (1987) examined the transformation of SO_2 emitted by the oil sands extraction plants in the Fort McMurray area. In this study the rate of conversion of SO_2 to SO_4^{2-} was determined to be 0-1.8% h^{-1} during the winter and 0-8.6% h^{-1} during the summer. According to the authors, the production of a large concentration of the aerosol suggests that heterogeneous oxidation may be appreciable under certain meteorological conditions and close to the emission source.

2.1.2 Nitrogen Compounds

As with the gaseous sulphur compounds, a number of gaseous inorganic nitrogen species have been identified in the atmosphere (Table 2.3).

Table 2.3 Some gaseous nitrogen compounds identified in the atmosphere.

Nitric oxide (NO)	Nitrogen dioxide (NO_2)
Nitrous oxide (N_2O)	Nitrogen trioxide (NO_3)
Dinitrogen trioxide (N_2O_3)	Dinitrogen tetroxide (N_2O_4)
Dinitrogen pentoxide (N_2O_5)	Ammonia (NH_3)

At normal temperatures and small partial pressures, N_2O_3 and N_2O_4 decompose (Junge 1963) according to the following equations:

$$N_2O_3 \rightarrow NO + NO_2$$

$$N_2O_5 \rightarrow N_2O_3 + O_2$$

$$N_2O_4 \rightarrow 2NO_2$$

Similarly, under normal atmospheric conditions, NO_3 decomposes by photolysis or combines with NO (Crutzen 1974).

$$NO_3 + h\nu \rightarrow NO_2 + O$$

$$NO_3 + h\nu \rightarrow NO + O_2$$

$$NO_3 + NO \rightarrow 2NO_2$$

For these reasons, only N_2O and NO_x (NO or NO_2) can be identified in practice in the atmosphere, if we disregard HONO (HNO_2) and HNO_3 (formed by the interaction of NO_2 and H_2O vapour). In addition to the production of HNO_3, NO_2 is also converted to O_3 and peroxyacyl derivatives (e.g., peroxyacetyl nitrate, PAN) (Figure 2.2).

Table 2.4 provides a summary of an estimate of the global NO_x budget. Man-made sources, combustion of fossil fuels, and biomass fires associated with agriculture appear to exceed natural sources from lightning and soils by about a factor of 2.

Compared with SO_2, the oxidation of NO_x to NO_3^- in power plant plumes and in ambient air is less understood. In power plant plumes, rates of conversion of NO_x from roughly 0.2% up to 12% h^{-1} have been observed, with rates being much greater during midday than at night (Hegg and Hobbs 1979; Richards et al. 1981). The products of NO_x oxidation appear to be PAN and HNO_3, with a lesser amount of particulate NO_3^-. In urban plumes, rates of conversion of NO_x of <5% h^{-1} up to 24% h^{-1} have been reported (Chang et al. 1979; Spicer 1982a,b).

Table 2.4 Global budget for NO_x

	10^{12} gm N y^{-1}
Sources	
Fossil fuel combustion	21 (14-28)
Biomass burning	12 (4-24)
Lightning	8 (2-20)
Microbial activity in soils	8 (4-16)
Oxidation of ammonia[a]	1-10
Photolytic or biological processes in the ocean	>1
Input from the stratosphere	~0.5
Total	25-99
Sinks	
Precipitation	12-24
Dry deposition	12-22
Total	24-64

[a] Oxidation of ammonia may provide a sink for NO_x of similar magnitude.

Source: Logan (1983). (Reprinted with permission, Copyright 1983 American Geophysical Union).

2.1.3 H_2SO_4 and HNO_3

In the atmosphere, H_2SO_4 and HNO_3 differ in their physical and chemical behaviour. Nitric acid is more volatile and thus significant concentrations of that substance can exist in the gas phase. However, H_2SO_4 has a low vapour pressure under ambient conditions and thus exists in the fine particle (<2.0 μm size) phase (Whitby 1978; Roedel 1979). These particles, in addition to causing visibility degradation, can also act as cloud condensation nuclei (Husar et al. 1978).

Both H_2SO_4 and HNO_3 can react with bases present in the atmosphere to form salts. For example, H_2SO_4 can react rapidly with NH_3 in the atmosphere to form ammonium acid sulphate, NH_4HSO_4; letovicite, $(NH_4)_3H(SO_4)_2$; and ammonium sulphate, $(NH_4)_2SO_4$. Because of the characteristics of the equilibrium between HNO_3, NH_3 and NH_4NO_3, HNO_3 can revolatilize relatively easily even after forming the ammonium salt (Finlayson-Pitts and Pitts 1986). No analogous physical and chemical changes exist for H_2SO_4. In addition to the ammonium salts, H_2SO_4 and HNO_3 can readily form salts with other cations such as Ca^{2+} and Mg^{2+}.

2.2 DRY DEPOSITION OF SULPHUR AND NITROGEN COMPOUNDS

According to Garland (1978), deposition processes limit the lifetime of sulphur, nitrogen, and other pollutants in the atmosphere, control the distance travelled before deposition, and limit their atmospheric concentrations. Thus, understanding these deposition processes is essential for a proper assessment of the environmental significance of natural and man-made emissions of sulphur, nitrogen, and other pollutants.

Dry deposition leads to the direct collection of gases, vapours, and particles on land and water surfaces. Table 2.5 provides a summary, in the present context, of the distribution of chemical species of relevance to dry deposition, among trace gas, fine particle, and coarse particle fractions.

Table 2.5 The distribution of chemical species of relevance to acidic dry deposition, among trace gas, fine particle, and coarse particle fractions.

	Trace Gases	Fine Particles	Coarse Particles
Sulphur species	SO_2	Most SO_4^{2-}	Some SO_4^{2-}
Nitrogen species	HNO_3		
	NH_3	Most NH_4^+	Some NH_4^+
	NO_x	Most NO_3^-	Some NO_3^-
Oxidants	O_3		
	H_2O_2*		
Carbon species	VOC's*	Organic acids	Graphitic materials
Trace metals		Dominant	Some
Crustal materials		Some	Dominant
Oceanic materials		Some NaCl	Most NaCl

* H_2O_2 = hydrogen peroxide; VOC = volatile organic carbon.

Source: U.S. NAPAP (1987).

Dry deposition begins with turbulent transport and/or sedimentation to the near-surface layer which is then penetrated by convection, diffusion, or inertial processes, and ends with chemical or physical capture of the pollutants by the surface (Voldner et al. 1986). The rate of pollutant transfer by dry deposition is controlled by a wide range of chemical, physical, and biological factors which vary in their relative importance according to the nature and state of the surface, the characteristic of the pollutant, and the state of the atmosphere (U.S. National Research Council 1983). The complexity of the individual processes involved and the variety of possible interactions among them combine to prohibit easy generalization; nevertheless, a "deposition velocity" (V_d), analogous to a gravitational falling speed, is of considerable use. In practice, knowledge of V_d enables fluxes (F) to be estimated from air concentrations (C) as the simple product, $V_d \bullet C$ (U.S. National Research Council 1983).

Tables 2.6 through 2.8 provide summaries of the results obtained in studies on the dry deposition of SO_2, NO_x, and SO_4^{2-} size particles on to natural surfaces. For more information on the theory and results of dry deposition studies, the reader is referred to Davidson and Wu (1988), Garland (1978), Hicks (1984), Hosker and Lindberg (1982), International Electric Research Exchange (1981), U.S. National Research Council (1983), Sehmel (1980), and Voldner et al. (1986).

The importance of dry deposition processes in transferring sulphur and nitrogen oxides from the atmosphere on to surfaces is clearly demonstrated by the results of regional sulphur and nitrogen budgets in North America (Galloway and Whelpdale 1980; Voldner and Shannon 1983; Logan 1983). Such regional budgets show that about half the total pollutant deposition is through dry removal processes.

In Alberta, Kociuba (1984) has shown the importance of dry deposition of sulphate through modelling (Table 2.9). The results show that, on an average, the dry deposition of sulphate is 1.86 times greater than the corresponding wet deposition. These results can be compared with the values obtained by Lovett and Lindberg (1984) at Oak Ridge, Tennessee (Table 2.10). In the former case dry deposition appears to be the dominant process; in the latter wet deposition dominates.

Table 2.6 Dry deposition observations for SO_2 to natural surfaces.

Reference	V_z (cm s^{-1})	r_a	r_c (s cm^{-1})	REF HT Z(m)	Met. condition: Stab.	Met. condition: REF. HT Wind (ms^{-1})	Season	Time of day	Comment
Water									
Owers and Powell, 1974	0.5 ± 0.1			0.2	N	1.8	F	day	$S^{35}O_2$ field experiment, England, 2 × 10 m artificial lake of seawater, found no evidence contradicting assumption $r_s = 0$.
Shepherd, 1974	0.024 U* 0.001 U_{2m}			0.2	N			day	gradient method, small lake, pH 7.5
Whelpdale and Shaw, 1974	4.0 2.2 0.16			10	U N S		S. F.	day	gradient method, L. Ontario water pH 8.0–8.3
Garland, 1977	0.41	1.86†	0.56	3	$U_* = 0.13–0.30$		Sp	day	gradient method, reservoir water, pH 8, found two groups of results: 8 cases with $r_s = 0$ 6 cases with $r_s = 0.45$ cm^{-1} latter possibly due to organic film.
Snow									
Whelpdale and Shaw, 1974	0.17 0.05 0.005			10	U N S		F	day	gradient method, fresh to 12 day old snow, S. Ontario.
Dovland and Eliassen, 1976	0.10			2	S	Low	W		snowpack mass balance, old snow, S Norway.
Barrie and Walmsley, 1978	0.25 ± 0.20			3	U		W	aft.	snowpack mass balance, new snow, Alberta.
Granat and Johansson, 1983			> 10						chamber study, frozen snow, r_s much lower for melting snow.
Forest									
Garland and Branson, 1977			10‡ 1.5–5‡	0			S S	night day	$S^{35}O_2$ outdoor chamber study, Scots pine, young dry needles, England, r_s is controlling resistance, r_s to bark > 1000 s cm^{-1}
Galbally *et al.*, 1979			as above						eddy correlation to pine, England, results agree with Garland and Branson, 1977.
Dollard, 1980			> 5‡ 1.4‡	0 0				night day	$S^{35}O_2$ wind tunnel uptake, seedings of spruce, pine, birch, estimates assumes 50% spruce, 30% pine and 20% birch as mixed forest, Norway.

Continued...

Table 2.6 (Continued).

Reference	V_z (cm s^{-1})	r_a	r_s (s cm^{-1})	REF HT Z(m)	Met. condition: Stab.	Met. condition: REF. HT Wind (ms^{-1})	Season	Time of day	Comment
Hicks and Wesely, 1980		1–10 0.1	3–6 1.4				S	night day	eddy correlation S uptake to Loblolly pine, NC, strong diurnal cycle, max. resistance early evening.
Hallgren *et al.*, 1982			4.8§ 1.6§	0			Sp, S, F	nigh day	SO_2 uptake outdoor chamber, Scotch pine branches 1 year old, dry, Sweden.
Fowler and Cape, 1983	wet 0.3 (0.1–0.4) dry 0.1 (0.01–0.2) wet 0.3 (0.1–0.6) dry 0.5 (0.2–1.0)	0.05–2 0.05–2 0.05–0.2 0.05–0.2	2.5–10 5–100 1.5–10 1–5	11		> 2	Sp	night day	eddy correlation, SO_2 uptake to Scots pine, Scotland, mesophyll resistance observed during dry daytime conditions, deposition to wet canopy controlled by chemistry in liquid film, after rain during evaporation SO_2 is released, average values given.
Granat, 1983			2.5‡ > 10‡	0			Sp, S, F W	24-h mean	SO_2 uptake outdoor chamber, pine Sweden, leaf area index 3, takes wet canopy into account, dry deposition to wet canopy is small due to acidic water, daytime values higher than night-time.
Johansson *et al.*, 1983			7‡ 2‡ 8‡ 2‡ > 10 > 10				Sp F W	night day night day night day	SO_2 uptake outdoor chamber (see Granat, 1983), Scots pine, lower branches, Sweden, leaf area index 3, indicates stomatal and cuticular transport, saturation and higher resistance for high SO_2 concentrations, during wet conditions increased resistance in liquid film.
Grennfelt *et al.*, 1984	dry 0.8 wet 1.5						W, S		SO_2 coniferous forest, assumed values, no snow, stomata open
<u>Grass and crops</u>									
Hill, 1971	2.3		0.4	0			S	day	field deposition, $S^{35}O_2$ with stable SO_2 carrier, western U.S., Alfalfa, mass 1.5 kg m^{-2}.

Continued...

Table 2.6 (Continued).

					Met. condition				
Reference	V_z (cm s^{-1})	r_a	r_s (s cm^{-1})	REF HT Z(m)	Stab.	REF. HT Wind (ms^{-1})	Season	Time of day	Comment
Owers and Powell,	0.7	0.65†	0.75‖		U		S		field deposition $S^{35}O_2$ England,
1974	2.6 ± 0.3	0.38†	0.01‖	0.2	N		F	day	dry grass 9–13 cm high, surface
	0.7 ± 0.1	0.70†	0.73‖		N		F		roughness 1 cm
Garland *et al.*, 1973	0.85		0.70	1	$U_* = 0.27$–0.63		S, F	day	$S^{35}O_2$ over pasture, England
	(0.36–2.2)		(0.11–2.3)						summer winter
	0.50		0.95		$U_* = 0.05$–30		W	day	r_s/r_t 47% 40%
	(0.25–1.3)		(0.14–3.3)						
Shepherd, 1974	1.3		0.8	0.2	S	$U_* = 0.23$–0.43	S	day	gradient, 19 profiles short,
	0.3		3.0				F		medium, long dump grass
Whelpdale and Shaw,	2.4				U			mostly	gradient, S Ontario,
1974	2.0			10	N		S	day	wet grass $r_s = 0$,
	0.5				S				roughness ~ 3 cm.
Dennevik *et al.*, 1976	0.4–0.5	0.28	1.4–2				W	day	gradient, St. Louis, wheat
	1.2–1.4	0.11	0.6–0.7	1.5	wide range		S		$r_s \sim 80\% \, r_t$.
Garland, 1977	0.85	1.2†	0.34‖		U	$U_* = 0.20$–0.30	Sp	day	gradient, short grass, England
	0.89	0.46†	0.66‖	1			W, S		gradient, medium grass
	1.19	0.38†	0.45‖				S		$S^{35}O_2$ tracer, medium grass.
Fowler, 1978	0.8	0.5†	0.7	1			S	day	gradient, England, long grass,
	0.4	0.5†	2.0					night	dry $r_s \sim 0.7 \, r_t$, wet and
	0.5	0.5†	1.5				W		pH > 3.5, $r_s = 0$.
Platt, 1978	1.68						S, F	daily	gradient, W Germany, mainly
	2.1			15			W	means	agriculture, some forest and
	0.70						Sp, S		buildings, large spatial average,
									$r_a = 0.1$ at 15 m.
Davies and Mitchell,	wet 1.0(0.94) ¶		0.30	1			S	night	gradient, grass, England
1983	dry 0.45(0.39) ¶		1.65						$r_s = 0.3$–$0.4 \, r_t$
	wet 1.28(0.43) ¶		0.08				S	day	r_b taken as 0.14 s cm^{-1}
	dry 0.74(0.38) ¶		0.70						$r_s = f$(LAI, growth activity)
	wet 0.41(0.27) ¶		1.10				W	night	V_z error ± 50%
	dry 0.42(0.26) ¶		1.30						
	wet 0.95(0.51) ¶		0.55				W	day	
	dry 0.54(0.38) ¶		1.65						
Dolske and Gatz,	< 3		0				S	day	several methods, grass, Illinois,
1985							F		preliminary results.
Hicks *et al.*, 1983	0.2	1.0	5	5–10	U			night	eddy correlation, grassy weeds,
	0.8	0.3	1		S		F	day	wheat stubble, U.S.A., roughness
									9.4 cm, few measurements,
									indicative of process.
Hicks *et al.*, 1986	0.3			1	U or		F	day	eddy correlation, grassland,
					N				S. Ohio, S, SO_2, sulphate
									particles, NO_x, measurements
									indicative rather than final.

Continued...

Table 2.6 (Concluded).

Reference	V_z (cm s^{-1})	r_a	r_s (s cm^{-1})	REF HT Z(m)	Met. condition: Stab.	Met. condition: REF. HT Wind (ms^{-1})	Season	Time of day	Comment
Neumann and den Hartog, 1985	0.21 ± 0.11			2	S	$U_* = 0.16–0.37$	S	night	eddy correlation, grass, Illinois.
	0.33 ± 0.30				U	$U_* = 0.11–0.26$		day	
Soil and urban									
Payrissant and Beilke, 1975			0.4–1.0						indoor chamber, European soils, values are for relative humidity of 37–52%. To obtain values for relative humidity of 70–80% multiply by a factor of 3, 2 and 1.5 for pH 7.5, 6.2 and 4.2, respectively.
			0.8–1.3	0					
			3.8						
Judeikis and Stewart, 1976	2.5		0.4	0	cement				laboratory, flow reactor, r_a eliminated, saturation after long exposure (regenerated), surface roughness 1 mm.
	2.0		0.5		cement				
	1.8		0.56		exterior stucco I				
	0.9		1.1		cement II				
	0.7		1.5		dry adobe soil				
	0.7		1.5		sandy loam soil				
	0.04		25		asphalt				
Garland, 1977	1.2	0.83	0.01	1	S	$U_* = 0.11–0.30$	W	day	gradient method, calcareous soil, England.
Husar *et al.*, 1978	0.7								cities, based on London data.
Milne *et al.*, 1979		25	0						outdoor chamber, dry soil, acid to neutral, N Australia.

† Includes r_b.
‡ r_s is estimated for a forest canopy from a single branch or seedling measurement.
§ Inferred by the reviewer using a leaf area index of 3 and results in Table 3 of paper.
‖ Surface resistance only.
¶ Standard deviation given in parentheses.

Source: Voldner et al. (1986). Refer to that document for complete references. (Reprinted with permission, Copyright 1986 Pergamon Press PLC).

Table 2.7 Dry deposition observations for NO_x and HNO_3 to natural surfaces.

Reference	V_z (cm s^{-1})	r_a	r_s (s cm^{-1})	REF HT Z(m)	Met. condition: Stab.	Met. condition: REF. HT Wind (ms^{-1})	Season	Time of day	Comment
Water									
Bottger *et al.*, 1980†	7.10^{-4}		10^5						NO, box method, water,
	0.015		70						NO_2, box method, sea water,
	0.01		100						NO_2, box method, water.
Snow									
Gravenhorst and Bottger, 1982			>67						NO_2 chamber study, frozen snow.
Grant and Johansson, 1983			>33						NO_2, melting to -14°C NO, chamber study.
Forest									
Grennfelt *et al.*, 1983	av 0.4–0.8		1.25–2.5‡				May–Oct	day	NO_x outdoor chamber, pine needles Sweden, leaf area, index = 5,
	0							night	year by year variation, uptake is f (stomata/needles, size of stomata),
	0						Nov–Apr		uptake generally follows transpiration rate, importance of long term estimates.
Hicks *et al.*, 1983	av 0						S	night	NO_x, eddy correlation, deciduous forest (oak, hickery, tulip poplar) 22 m
	0.3							2 pm	high, complex terain, strong diurnal
	0.8							10 am	variation, high values of V_z indicates influence of HNO_3.
Huebert, 1983b	$V_d \sim 1/r_a$		0						modified Bowen ratio, HNO_3 forest, preliminary results.
Grennfelt *et al.*, 1984	0.5						snow cover		NO_x mass balance, coniferous forest,
	0.8						no snow	day	Sweden, V_z data based on
	0.2						no snow	night	Grennfelt unpublished and Bengtson *et al.*, 1982
	av 0.55						May–Oct		
	0.36						Nov–Apr		

Continued...

Table 2.7 (Continued).

Reference	V_z (cm s^{-1})	r_a	r_s (s cm^{-1})	REF HT Z(m)	Met. condition: Stab.	Met. condition: REF. HT Wind (ms^{-1})	Season	Time of day	Comment
Crops and grass and									
Hill, 1971†	0.1 0.4–1.9								box method, alfalfa, NO box method, alfalfa NO_2, potential influence of reaction and deposition of O_3
Platt *et al.*, 1980a†	0.5								alfalfa, oats, NO_x
Platt *et al.*, 1980b†	< 0.02								grass, NO_x
van Aalst, 1982	neg.						F		concentration measurements, 3 levels, tower, crop, Netherland, measurements indicate upward flux, potential influence of feed lot.
Gravenhorst and Bottger, 1982			2.5–10 3.3–60						meadow, chamber experiments, NO_2, pasture
Wesely *et al.*, 1982	0.05 0.6	 0.5	15 1.3 1.6 ± 0.2 2.8 ± 1.6 15 ± 5	6	 A, B, C D E		 S	night day 7:30–16 16–18 19–3:30	eddy correlation, Pennsylvania, well watered soybean, NO_x, r_s > stomata resistance, possible NO_2 reaction, re-emission of NO, data may be influenced by HNO_3 (Huebert, 1983).
Delany and Davis, 1983			2.4 1–5.6	1		$U_* = 0.05$–0.26	Sp	day	NO_x, gradient, grass.
Duyzer *et al.*, 1983	0–1.5 av 0.6 0						F W	day day	gradient method, grass, NO_x (can not separate NO and NO_2 due to reaction rates, potential NO release from plants).

Continued...

Table 2.7 (Concluded).

Reference	V_z (cm s^{-1})	r_a	r_s (s cm^{-1})	REF HT Z(m)	Met. condition: Stab.	Met. condition: REF. HT Wind (ms^{-1})	Season	Time of day	Comment
Hicks *et al.*, 1986			0.82 ± 1.45§ 0.40 ± 0.58§			U near N	F	day	eddy correlation, grass, S Ohio, NO_x measurement highly scattered, potential influence of local source, supports Wesely *et al.*, 1982, r_s day ~ 1 s cm^{-1}.
Huebert and Robert, 1985	2.5 ± 0.9 1.0 min 4.7 max		~0	1			S	day	HNO_3, modified Bowen ratio, grass Illinois, diurnal variation, max at mid day, r_s ~ 0.
Soil and urban									
Judeikis and Wren, 1978									cylindrical flow reactor, r_a eliminated, found surface saturation, surface regenerated after precip.
	0.9		5.3						NO sandy loam soil
	0.13		7.7						NO adobe clay
	0.21		4.8						NO cement
	0.6		1.7						NO_2 sandy loam soil
	0.77		1.3						NO_2 adobe clay
	0.32		3.1						NO_2 cement
Bottger *et al.*, 1980†									box method
	$< 10^{-5}$		10^5						NO, forest soil, dry and moist sand
	0.3		3.3						NO, forest soil
	0.6		1.7						NO_2, sand dry
	0.28		3.6						NO_2, sand moist
Gravenhorst and Bottger, 1982			av 10 6–60						chamber experiments, NO_2, bare soil.

† Values as reported in literature review by Aalst, 1982.
‡ Surface resistance only.
§ Outlier removed.

Source: Voldner et al. (1986). Refer to that document for complete references. (Reprinted with permission, Copyright 1986 Pergamon Press PLC).

Table 2.8 Dry deposition observations for SO_4^{2-} size particles to natural surfaces.

Reference	V_z (cm s^{-1})	r_a	r_b (s cm^{-1})	REF HT Z(m)	Met. condition: Stab.	Met. condition: REF. HT Wind (ms^{-1})	Season	Time of day	Comment
Water									
Prahm *et al.*, 1976	0.4 ± 0.2	—	—	—	S	—	—	—	spatial/temporal average v_z N Atlantic comparison of observations with model predictions in U.K. and Faroes, SO_4^{2-} particles.
Delumyea and Petel, 1979	0.6	—	—	—	S	—	Sp, S, F	—	Lake Huron, mass difference technique, particles with bimodal size distribution, mode and medians at 0.5–0.9 μm and 2.9–4.6 μm.
Dolske and Sievering, 1979	0.5 av 0.15 0.72 0 1.2			5	 S	 0 7	S, F		diabatic drag coefficient method, Lake Michigan, particle diam. 0.1–2.0 μm, v_z dependent on wind speed and stability.
Sievering *et al.*, 1979	0.20 ± 0.16 0.13 0.65 0.55	SO_4^{2-} Pb Mn Fe	88% 82% 49% 47%	mass < 1μm mass < 1μm mass < 1μm mass < 1μm	VS		Sp	—	L. Michigan, mass differencing techniques, bulk Richardson No. 0.17 ± 0.10.
Wesely and Williams, 1980	—	—	< 1	—	—	—	—	—	eddy correlation, Lake Michigan particles 0.3–1 μm, downward flux, particles 0.03–0.1 μm upward flux.
Sievering, 1981	—	—	small	5	—	2.4–8.2	Sp, S, F	—	profile measurements, L. Michigan r_b small, deposition velocity close to that of momentum. Little particle size dependence, 0.11–2 μm.
Snow									
Dovland and Eliassen, 1976	0.16 (0.6–2.0)	—	—	2	VS	0.5	W	all day	natural Pb tracer, Norway.
Wesely and Hicks, 1979	< 0.2	—	—	10	—	—	W	—	US mid—west, eddy correlation, small particles 0.05–0.1 μm.
Ibrahim *et al.*, 1983	0.04 0.10	— —	25 10	0.1 0.1	S U	$U_* = 0.17$ $U_* = 0.17$	W W	morn. aft.	ammonium sulphate tagged with S^{35} tracer study, Chalk River, Canada.
Forests									
Wesely and Hicks, 1979	0.5	—	1.5 ± 0.2	10 above zero plane	—	—	S	aft.	loblolly pine, North Carolina, eddy correlation sulphate particles.
Russell *et al.*, 1981	—	—	5–8	top of canopy	—	—	Sp, S, F	—	long term average r_b, New England natural Pb^{212} tracer, r_b estimates using double-sided leaf area index of 4.

Continued...

Table 2.8 (Continued).

Reference	V_z (cm s^{-1})	r_a	r_b (s cm^{-1})	REF HT Z(m)	Met. condition Stab.	REF. HT Wind (ms^{-1})	Season	Time of day	Comment
Hicks *et al.*, 1982	0.7	—	1.44 ± 1.5 (0.6–2.2)	10 above	—	—	S	day	eddy correlation, loblolly pine, North Carolina diurnal variation, $V_z \sim 0$ or negative at night (i.e. upward flux). Small particles 0.1–0.3 μm detected possibly from gas to particle conversion.
Hofken *et al.*, 1983	1.1 ± 0.3	—	0.0	top of canopy	—	—	May–Oct	Beech	SO_4^{2-} particles mass median diameter 0.6 μm,
	0.5 ± 0.2	—	1.7				Feb–Apr		mass budget of canopy, West Germany
	1.45 ± 0.5	—	0.7	top of	—	—	May–Oct	Spruce	includes fog deposition
	1.3 ± 0.5	—	0.8	canopy	—	—	Feb–Apr		thus r_b may be low
Wesely *et al.*, 1983	0.76 ± 0.8	—	1.3	10 above zero plane	U	$U_* = 0.57$	S		pine drought conditions)eddy)correlation
	−0.05 ± 0.24	—	20		U	$U_* = 0.44$	W		deciduous no leaves)SO_4
Grass and crops									
Little and Wiffen, 1977	0.3	—	3.3	close to surface	—	1.0	Sp, S	—	polydisperse, auto-Pb, dry short grass. Garland (1983) notes that 10% of the mass which is in > 3 μm particles can yield 50% of deposition.
Wesely *et al.*, 1977	~1	—	—	5	U	< 2.5	Sp	day	eddy correlation, particles 0.05–0.1 μm, smaller than SO_4^{2-}, mixture green grass
	~0.1	—	—	5	S	< 2.5	Sp	day	short brown vegetation and bare soil, z_0 = 3.0 cm.
Garland, 1978	< 0.1 0.03–0.56	—	—	—	—	—	—	—	SO_4^{2-} particles 0.4–1 μm, gradient method, England, grass
Everett *et al.*, 1979	1.0	—	—	22	S	< 4.5	S	all	SO_4^{2-} particles, gradient
	1.4	—	—		S	> 4.5	S	day	method, grass, U.S. midwest,
	1.0	—	—		U	< 4.5	S		8 days of measurements
	2.3	—	—		U	> 4.5	S		
Wesely and Hicks, 1979	1.0	—	< 0.6 ± 0.1	5	—	$U_* > 0.11$	W		particle size 0.5–0.1 μm, smaller
	0.02, 0.1	—	—	5	—	$U_* > 0.11$	F		than SO_4^{2-}, eddy correlation, U.S. midwest, winter grass, fall senescent maize.
Droppo, 1980	0.10	2.3	7.7			1.1	W		sulphate, arid vegetation, gradient
	0.27	0.44	3.3			1.3			method, Hanford, U.S.A.
Davies and Nicholson, 1982	0.08*			1		Annual		24–h av.	aerosol sulphate, gradient method,
	0.08*					S		night	rural East England, semi-continuous
	0.2*					S		day	measurements June 1979–1980, varying
	0.05*					W		night	meteorological and surface
	0.08*					W		day	conditions (crop, pasture), no
	> 1*(extreme value)								sampling during rapidly varying conditions (dawn/dusk), scattered
	0.04*				N				V_z values indicate stability and
	0.03*				S				seasonal dependance.
	0.16*				U				

Continued...

Table 2.8 (Concluded).

Reference	V_z (cm s^{-1})	r_a	r_b (s cm^{-1})	REF HT Z(m)	Met. condition Stab.	REF. HT Wind (ms^{-1})	Season	Time of day	Comment
Garland, 1982	0.05–0.12								gradient method, grass, wind tunnel in field, iron oxide monodispered $D = 1.8\,\mu m$, states V_z in field experiments is 2–3 times higher than measured in wind tunnel.
Garland and Cox, 1982	0.06 ± 0.03			close to grass		< 3	W	day	grass, particles 0.05–0.2 μm, gradient method, upward flux found for particle < 0.05 μm, possible production of small particles at surface.
Sievering, 1982	0.38 ± 0.29	—	—	~ 10	U	$U_* < 0.3$	F	Usually aft.	rye and wheat, surface roughness 1–20 cm gradient method, particles 1.5–0.3 μm, unsteady state conditions may have interfered with measurements.
Doran and Droppo, 1983	< (0.3–0.4)	—	—	0.75	N	—	S	—	gradient, method, SO_4^{2-} particles, upper limit of V_z.
Hicks *et al.*, 1983	0.7	—	2.9	5–10	U	$U_* = 0.14$	F	early aft	grass, weeds, wheat stubble, roughness 9.4 cm, eddy correlation SO_4^{2-} particles.
	0.0	—	—	5–10	U	$U_* = 0.14$	F	late aft	
Sievering, 1983	1.19 ± 0.18	0.24	0.6	5.5	U	$U_* = 0.59$	Sp, S	aft.	eddy correlation, particles 0.09–2.5 μm, mass median diameter 0.25 μm, grand average V_z is 0.05 cm s^{-1}, V_z observed 2–6 times greater than predicted from Slinn, 1982.
	0.37 ± 0.04	0.60	2.1		S	$U_* = 0.49$		morn.	
Wesely *et al.*, 1983	0.18 ± 0.02	—	5.7		U	$U_* = 0.025$		day,	SO_4^{2-} particles, eddy correlation, when r_b is low sees dependence of V_z on wind speed and stability.
	0.18 ± 0.02		5.6	5–10	S	$U_* = 0.30$	S	short	
	0.53 ± 0.18	—	1.5		very U	$U_* = 0.18$		grass	
	0.41 ± 0.05	—	1.8	5–10	mod. U	$U_* = 0.30$	S	day lush	
	0.10 ± 0.07	—	5.7		slight U	$U_* = 0.22$		grass	
Hicks *et al.*, 1986	~ 0.4	—	2.9 ± 1	5	U, N	1.5 2.5	F	day	eddy correlation, SO_4^{2-} particles, limited measurements, indicative only. V_z late afternoon is zero or negative.

* Median values.

Source: Voldner et al. (1986). Refer to that document for complete references. (Reprinted with permission, Copyright 1986 Pergamon Press PLC).

Table 2.9 Modelled dry and dry-wet sulphate deposition ratios for Alberta sites in 1982.[a]

Location	Dry Deposition (kg ha^{-1} y^{-1})	Dry-Wet Ratio
Beaverlodge	4.8	0.94
Calgary	20.7	1.86
Coronation	18.6	2.16
Edmonton	21.0	1.94
Edson	18.6	1.94
Fort McMurray	21.0	3.18
Lethbridge	13.8	1.86
Red Deer	12.3	1.84
Rocky Mountain House	20.2	1.94
Suffield	7.2	1.85
Whitecourt	21.9	1.43
Average	16.4	1.86[b]

[a] Source: Kociuba (1984).
[b] The average dry-wet ratio was determined by finding the ratio of the average dry (16.4 kg ha^{-1} y^{-1}) and average wet (8.8 kg ha^{-1} y^{-1}) depositions.

Table 2.10 Dry and wet deposition of sulphate by season in a chestnut oak canopy.

	SO_4^{2-} kg ha^{-1}		
Season	Dry	Wet	Dry-Wet Ratio
Growing	9.7	22.9	0.42
Dormant	2.0	16.1	0.12
Total	11.7	49.0	0.24

Source: Lovett and Lindberg (1984).

The results of the two studies (Alberta and Tennessee) discussed here substantiate the temporal and spatial variability in the magnitude and importance of dry versus wet deposition of pollutants. As opposed to SO_4^{2-}, there are no comparable studies on the dry deposition of NO_3^- in Alberta. However, Lovett and Lindberg provided results of NO_3^- dry deposition in their studies in Tennessee (Table 2.11).

Table 2.11 Dry and wet deposition of nitrate by season in a chestnut oak canopy.

	NO_3^- kg ha^{-1}		
Season	Dry	Wet	Dry-Wet Ratio
Growing	4.0	9.2	0.43
Dormant	3.9	5.4	0.72
Total	7.9	14.6	0.54

Source: Lovett and Lindberg (1984).

2.3 WET DEPOSITION OF SULPHUR AND NITROGEN COMPOUNDS

Both SO_2 and SO_4^{2-} can contribute significantly to the dissolved sulphur in rain. The contribution of SO_4^{2-} appears inevitable, since SO_4^{2-} particles serve as cloud condensation nuclei. However, the incorporation of SO_2 may be suppressed if the condensation nuclei are initially acidic. Dana et al. (1975) imply no more than 3% deposition in rain from a power plant plume in the first 10 km. Larson et al. (1975) deduced that only 8% of the sulphur emitted from a smelter while rain was falling was deposited within 60 km. Garland (1978) has summarized the information on mechanisms contributing to sulphur in rainwater (Table 2.12).

In addition to the rainout of condensation nuclei, several other physical processes may contribute to sulphate in rain. Diffusophoresis and Brownian diffusion may result in collection of small particles to the cloud droplets and raindrops may further collect particles by impaction, interception, or diffusion. According to Garland (1978), only the rainout of condensation nuclei appears capable of explaining the concentrations of several mg L^{-1} of SO_4^{2-} observed in practice.

Table 2.12 Mechanisms contributing to sulphur in rainwater (Garland 1978).

	Pollutant Lifetime During Rain (h)		Sulphate Conc. in Rainwater (mg L^{-1})	
Mechanism	(a)*	(b)*	(a)*	(b)*
Particulate Sulphate				
Diffusophoresis	--	--	10^{-2}	10^{-3}

Continued...

Table 2.12 (Concluded).

Mechanism	Pollutant Lifetime During Rain (h)		Sulphate Conc. in Rainwater (mg L^{-1})	
	(a)*	(b)*	(a)*	(b)*
Brownian diffusion to cloud droplets	100	4000	0.2	3 x 10^{-3}
Brownian diffusion to raindrops	10^4	2 x 10^{-5}	10^{-3}	10^{-5}
Impaction and interception by raindrops	400	1	0.02	1.2
Rainout of condensation nuclei	--	--	3-10	3-10
SO_2				
Solution and oxidation in cloud droplets	--	--	3	3
Uptake by falling raindrops	10	1	1	1

* Assumptions: i) 10 μg m^{-3} of sulphate of Composition (a)--80% submicron particles of typical diameter 0.2 μm and Composition (b)--20% particles of typically 4 μm diameter; ii)10 μg m^{-3} of SO_2; and iii) rain falling at 1 mm h^{-1} as 1 mm drops from a cloud containing 100 drops cm^{-3}, each of 10 μm radius. (Reprinted with permission of Harwell Laboratory, UK Atomic Energy Authority and Copyright 1978 Pergamon Press PLC).

The washout of large particles by raindrops may make a significant contribution, but this fraction of the aerosol will be exhausted by the first few millimetres of rain and may, therefore, account for the enhancement in sulphate concentration observed at the beginning of some periods of rain (Meurrens 1974; Pratt et al. 1983).

Diffusion and interception may be of greater significance in snow because of the larger surface area of the precipitation elements. In addition, the concentration of condensation nuclei collected in precipitation may be much reduced if distillation from liquid to solid phase dominates the aggregation of cloud droplets in the growth of snowflakes.

In summary, Table 2.12 shows that the probable contribution of dissolved SO_2 is smaller than the contribution due to the rainout of SO_4^{2-}. However, oxidation of SO_2 in clouds can make a substantial contribution to the SO_4^{2-} in rain.

In contrast to SO_4^{2-}, much less information has been published regarding the removal mechanisms of nitrogen species by precipitation. There is some evidence for the formation of HNO_3 in clouds and rainwater. Recently, both theory and experimental evidence suggest that HNO_3 may be formed rapidly from a combined gas-phase/liquid-phase process (U.S. National Research Council 1983). Although significant uncertainty remains concerning the source of HNO_3 in clouds and rainwater, the limited evidence currently available favours the probable importance of the formation of dinitrogen pentoxide (N_2O_5) from nitrogen dioxide (NO_2), followed by its reaction with water droplets to form HNO_3. HNO_3 can be effectively scavenged by precipitation.

Because of the significant concern arising from the occurrence of "acidic precipitation", numerous investigators have examined the qualitative and quantitative aspects of precipitation chemistry in the last 15-20 years. For more details than what will be described in the following section, the reader is referred to Chamberlain et al. (1981), U.S. National Research Council (1983), U.S. National Academy Press (1986), Husar et al. (1978), Knapp et al. (1988), Krupa et al. (in press), the Series on Acid Precipitation edited by Teasley (1984), and the Interim Assessment Report of the U.S. National Acid Precipitation Assessment Program, U.S. NAPAP (1987).

In comparison to seawater (Whitfield 1979), precipitation can be considered as a highly unbuffered, dilute solution of organic and inorganic ions. Precipitation is also composed of an insoluble fraction consisting of organic (pollen, pesticides, etc.) and inorganic (e.g., crustal) coarse (>2.0 μm size) particles (Krupa et al. 1976). However, it is the soluble fraction which is of immediate concern in the context of "acidic precipitation". Table 2.13 lists some of the inorganic ions important in precipitation chemistry. Similarly, Table 2.14 shows a comparison of the median ion concentrations (mg L^{-1}) during 1979-84 for three National Atmospheric Deposition Program (NADP) sites. Numerous ecological effects scientists have utilized the concentrations of many of these precipitation ions at various locations together with precipitation depth to compute ion deposition (kg ha^{-1}) in evaluating potential environmental effects (U.S. Environmental Protection Agency 1983).

Table 2.13 Some inorganic ions important in precipitation chemistry.*

Cations	Anions
H^+	Cl^-
NH_4^+	NO_3^-
Na^+	SO_3^{2-}
K^+	SO_4^{2-}
Ca^{2+}	PO_4^{2-}
Mg^{2+}	CO_3^{2-}

* All ions are presented here in their completely dissociated states. The reader should note, however, that various states of partial dissociation are possible as well (e.g., HSO_3^-, HCO_3^-).

Source: U.S. National Research Council (1983).

Table 2.14 Median ion concentrations (mg L^{-1}) during 1979-84 for three National Atmospheric Deposition Program (NADP) sites.

Ion	Lamberton Minnesota		N. Atlantic Lab, Massachusetts		Kane Pennsylvania	
pH	6.00	(82)*	4.67	(41)	4.27	(1)
H^+	0.001	(82)	0.021	(41)	0.068	(1)
SO_4^{2-}	1.88	(38)	1.54	(47)	3.48	(4)
NO_3^-	1.74	(19)	0.73	(65)	2.08	(10)
NH_4^+	0.81	(1)	0.08	(72)	0.28	(25)
Ca^{2+}	0.49	(2)	0.13	(54)	0.16	(49)
Precip. (mm)	8.30	(65)	22.10	(11)	21.80	(12)

* Values in parenthesis indicate relative ranking among the 82 NADP sites examined.

Source: Knapp et al. (1988).

The pH of natural precipitation is often assumed to be regulated by the dissociation of dissolved carbon dioxide (CO_2), thus having a value of 5.6 and that precipitation pH values below 5.6 are due to the addition of acidic components (primarily related to SO_4^{2-} and NO_3^-) by human activity (Garrels and MacKenzie 1971; Galloway et al. 1976; and Likens and Bormann 1974). Table 2.15 provides a summary of the molar ratios of SO_4^{2-} to NO_3^- for several sites in the northeastern United States during 1977-81.

Table 2.15 Molar ratios of sulphate to nitrate in precipitation in the United States, 1977-81*.

Sampling Location	April-September		October-March		Annual Average
Whiteface Mt., NY	1.4	(28)	0.65	(26)	1.08
Ithaca, NY	1.4	(25)	0.62	(21)	1.07
University Park, PA	1.3	(27)	0.74	(29)	1.04
Charlottesville, VA	1.3	(27)	0.81	(21)	1.16
Urbana, IL	1.4	(15)	0.94	(13)	1.25
Brookhaven, NY	1.12	(17)	1.0	(17)	1.10
Lewes, DE	1.4	(18)	0.93	(18)	1.16
Oxford, OH	1.5	(16)	1.2	(19)	1.37
Average	1.4 ± 0.1		0.86 ± 0.19		

* Numbers in parenthesis are the number of months of data in each sample.

Source: Adapted from MAP3S/RAINE Research Community (1982).

According to some investigators, the acidity and concentrations of SO_4^{2-}, NO_3^-, and some other components in precipitation have increased in recent years in certain geographic locations as a result of man's activities (Cogbill and Likens 1974; Galloway et al. 1976; Martin and Barber 1977; and Likens and Butler 1981).

The data from Hubbard Brook (New Hampshire) reveal several trends (Likens et al. 1980), which are supported at least qualitatively by bulk deposition monitoring with relatively unreliable quality control from nine sites in New York State (Miles and Yost 1982; Peters et al. 1982):

a. There has been a decrease in SO_4^{2-} concentration since 1964 but an increase in NO_3^- concentration over the same time (Table 2.16);
b. The annual pH of precipitation showed no long-term significant change from 1964-77, although several short-term changes did occur;
c. A linear regression equation of data points from 1964-77 indicated no statistically significant trends in H^+ deposition; and
d. Recent changes in H^+ deposition correspond more with changes in NO_3^- deposition than with SO_4^{2-} deposition, even though H_2SO_4 is the dominant species at Hubbard

Table 2.16 Annual average concentrations (mg L^{-1}) of sulphate and nitrate from weekly bulk samples at Hubbard Brook weighted by the annual amount of precipitation.*

Year	SO_4^{2-}	NO_3^-
1964-65	3.16	0.70
1965-66	3.33	1.39
1966-67	3.13	1.49
1967-68	3.27	1.56
1968-69	2.42	1.18
1969-70	2.24	1.14
1970-71	2.75	1.71
1971-72	2.67	1.74
1972-73	2.87	1.74
1973-74	2.84	1.67
1974-75	2.54	1.52
1975-76	2.14	1.22
1976-77	2.20	1.66
1977-78	2.04	1.32
1978-79	2.55	1.69
1979-80	1.91	1.44
1980-81	2.36	1.66

* Data were recorded in the water year, from June 1 through May 31.

Source: G.E. Likens, Cornell University, refer to U.S. National Research Council 1983.

Brook. The contribution of NO_3^- to total acidity has been increasing whereas that of SO_4^{2-} has been decreasing (Galloway and Likens 1981). Year-to-year changes superimposed on the long-term trend may be related to climatological influences (U.S. National Research Council 1983).

According to Hansen et al. (1981), available data are not of sufficient quantity and quality to support any long-term trends in precipitation acidity change over the past 50 years in the eastern United States. However, the observations do show that precipitation is definitely acidic over this region and is probably more acidic than expected from natural baseline conditions.

Recently Schertz and Hirsch (1985) performed a trend analysis (1978-83) of data from 19 sites of the National Atmospheric Deposition Program (NADP). They concluded that 41% of the trends detected in the ion concentrations were downward trends, 4% were upward trends, and 55% showed no trends at $\alpha=0.2$. The authors also concluded that the two constituents of greatest interest in terms of man-generated emissions and environmental effects, SO_4^{2-} and NO_3^-, showed only downward trends, and SO_4^{2-} showed the largest decreases in concentrations per year of all the ions tested.

In contrast, Stensland et al. (1986) derived the following conclusions:

a. The eastern half of the United States experiences concentrations of SO_4^{2-} and NO_3^- in precipitation that are, in general, greater by at least a factor of five than those in the remote areas of the world, indicating that levels have increased by this amount in northeastern North America since sometime before the 1950's;
b. Data on the chemistry of precipitation before 1955 should not be used for trend analysis;
c. Precipitation is currently more acidic in parts of the eastern United States than it was in the mid-1950's or mid-1960's; however, the amount of change and its mechanism are in dispute;
d. Precipitation SO_4^{2-} concentrations and possibly acidity have increased in the southeastern United States since the mid-1950's; and
e. In general, individual sites or groups of a few neighbouring sites cannot be assumed *a priori* to provide regionally representative information; regional representativeness must be demonstrated on a site-by-site basis.

Assuming that acidity has increased (at least in certain locations) and that SO_4^{2-} and NO_3^- are responsible for most of the free acidity, a strong statistical relationship should be observed between H^+ and SO_4^{2-} and/or NO_3^-. Information to date suggests that the degree of association is site specific, and that proximity to sources of SO_4^{2-} and NO_3^- as well as sources of other substances may influence the chemical nature of acidic substances in the atmosphere (Lefohn and Krupa 1988). At a given site, differences in the meteorology between events may result in wide variations in the measured acidity, concentrations of SO_4^{2-} and NO_3^-, and correlations between these ions (Pratt et al. 1984; Pratt and Krupa 1985). Sequeira (1982) found correlation coefficients in excess of 0.8 between SO_4^{2-} and H^+ at Mauna Loa, Hawaii, somewhat lower values at Monte Cimone, Italy, and 0.01 at Alamosa, Colorado. Barrie (1981) examined summer data from eastern Canadian sites and found correlation coefficients between SO_4^{2-} and H^+ as high as

0.99 and as low as -0.39 at different sites. Similarly Pratt et al. (1983), examining four years of rainfall chemistry at seven sites in a 600 km^2 area in central Minnesota, found correlation coefficients between SO_4^{2-} and H^+ to vary from 0.15-0.42, and between NO_3^- and H^+ from 0.06-0.62.

Kasina (1980) found no significant correlation between acidity and SO_4^{2-} in southern Poland. However, Madsen (1981) found good correlations between H^+ and excess SO_4^{2-} on the east coast of Florida during most of the months from late 1977 to late 1979. McNaughton (1981) found correlation coefficients in excess of 0.7 for all MAP3S sites except Illinois, where the value was below 0.4.

The preceding discussion shows the complexity in generalizing the characteristics of precipitation chemistry due to its significant spatial variability. In addition, it is well known that precipitation chemistry exhibits distinct temporal variability including seasonality (Pratt and Krupa 1983; Dana and Easter 1987). For example, Bowersox and de Pena (1980) concluded that, by applying multiple linear regression analysis for a central Pennsylvania site, on the average, H_2SO_4 was the principal contributor to H^+ concentration in rain but the acidity in snow was principally from HNO_3.

The pH of atmospheric precipitation at a given location depends on the chemical nature and relative proportions of acids and bases in the solution. Sequeira concluded that a pH of 5.6 may not be a reasonable reference value for unpolluted precipitation. Charlson and Rodhe (1982) also questioned the validity of using pH 5.6 as the background reference point, citing naturally occurring acids as possibly responsible for low pH values of rain. They stated that consideration of the natural cycling of water and sulphate through the atmosphere, precipitation rates, and experimentally determined rates of SO_4^{2-} scavenging indicates that average pH values of approximately 5.0 would be expected in pristine locations in the absence of basic materials. This value will vary considerably due to the variability in scavenging efficiencies as well as geographic patchiness in the sulphur and hydrological cycles. Thus, precipitation pH values might range from 4.5-5.6 due to these variabilities alone (Charlson and Rodhe 1982). Lefohn and Krupa (1984) found that pH of precipitation with minimum concentrations of SO_4^{2-} and NO_3^- was in the range of 4.6-5.5 for the northeastern United States.

An aspect of precipitation chemistry which has been largely ignored until recently is the presence of organic acids (Krupa et al. 1976). Meyers and Hites (1982), Kawamura and Kaplan (1983),

Keene and Galloway (1984), Guiang et al. (1984), and Chapman et al. (1986) have all shown the presence of organic acids in precipitation. Keene and Galloway (1984) estimated that organic acids may contribute 16-35% of the volume weighted free acidity in precipitation of North America. Krupa et al. (1987) calculated theoretical precipitation H^+ concentrations for Minnesota, based on the assumption that all of the organic anions were present as the corresponding acids. A plot of the calculated versus the measured H^+ concentrations showed poor correlation, yet the mean calculated and the mean measured H^+ concentrations were nearly equal on a yearly basis. The weak organic acids could account for all of the deposited acidity.

Since the early 1970's, several investigators have studied the precipitation composition and wet deposition in Alberta (Nyborg et al. 1977; Caiazza et al. 1978; Klemm and Gray 1982; Lau 1985; and Lau and Das 1985). Based on a study during 1973-74, Nyborg et al. (1977) concluded that rain and snow in Alberta were seldom acidic. Klemm and Gray (1982), through their study during 1977-78, found that less than 20% of the pH values of precipitation in central Alberta were below 5.0 and none were below 4.0. According to Lau and Das (1985) volume weighted average pH values of composite monthly precipitation samples at 11 Alberta CANSAP (Canadian Network for Sampling Precipitation) stations during 1978-84 ranged between 5.17 and 6.06 (Table 2.17). Wet deposition of H^+ (kg ha^{-1} y^{-1}) in Alberta during 1978-82 was 24 times less compared with the values for Ontario, Quebec, and Nova Scotia (Table 2.18). In a similar comparison, SO_4^{2-} deposition in Alberta was 1.76-3.7 times less and NO_3^- deposition was 1.8-6.0 times less compared with the three eastern Canadian provinces (Table 2.18).

Table 2.17 Wet deposition in Alberta (1978-84).

	Average Annual Deposition (mole m^{-2} y^{-1})			
Location	Sulphate	Nitrate	Hydrogen	Average pH
Beaverlodge	6.8	4.7	3.2	5.17
Calgary	12.8	8.1	0.7	5.77
Coronation	7.4	6.3	1.3	5.46
Edmonton	8.1	6.8	1.5	5.45
Edson	9.0	4.8	3.1	5.26

Continued...

Table 2.17 (Concluded).

Location	Average Annual Deposition (mole m^{-2} y^{-1}) Sulphate	Nitrate	Hydrogen	Average pH
Ft. McMurray	8.1	4.8	1.0	5.61
Lethbridge	10.5	10.8	0.4	6.06
Red Deer	8.8	7.4	2.0	5.39
Rocky Mtn House	9.9	6.0	1.5	5.53
Suffield	8.8	6.9	0.8	5.60
Whitecourt	11.4	6.9	1.1	5.71
Alberta Average	9.2	6.7	1.5	5.48

Sulphate: 1 mole m^{-2} = 0.961 kg ha^{-1}
Nitrate: 1 mole m^{-2} = 0.620 kg ha^{-1}
Hydrogen: 1 mole m^{-2} = 0.010 kg ha^{-1}

Source: Lau and Das (1985).

Table 2.18 Wet deposition of H^+, SO_4^{2-}, and NO_3^- (kg ha^{-1} y^{-1}) in Alberta and at selected Canadian stations from 1978 to 1982.

Location	Ion H^+	SO_4^{2-}	NO_3^-
Beaverlodge (Alta.)	0.031	6.8	3.2
Calgary (Alta.)	0.004	15.5	5.8
Coronation (Alta.)	0.011	8.0	4.6
Edson (Alta.)	0.330	9.0	3.1
Fort McMurray (Alta.)	0.010	8.0	3.6
Lethbridge (Alta.)	0.003	11.5	7.6
Red Deer (Alta.)	0.022	9.4	5.2
Rocky Mtn. House (Alta.)	0.013	10.3	3.7
Suffield (Alta.)	0.009	9.9	4.6
Whitecourt (Alta.)	0.015	10.5	4.3
Prince George (B.C.)	0.009	10.6	3.1
Revelstoke (B.C.)	0.060	6.5	3.9
Cree Lake (Sask.)	0.036	4.4	2.0
Wynyard (Sask.)	0.001	8.2	5.7
The Pas (Man.)	0.003	7.1	3.6
Bissett (Man.)	0.046	7.6	4.1
Moosonee (Ont.)	0.073	11.6	5.6
Simcoe (Ont.)	0.804	52.5	35.0
Maniwalki (Que.)	0.515	32.2	20.3
Quebec City (Que.)	0.573	57.2	25.8
Truro (N.S.)	0.432	30.0	10.9

Source: Lau (1985).

2.4 CHARACTERISTICS OF AIR POLLUTANT EXPOSURE DYNAMICS

In evaluating the effects of pollutants under ambient conditions, one must define the spatial and temporal variability. In addition, because of the nature of atmospheric processes, two or more pollutants can exhibit in a 24-hour period, increases or decreases in their concentrations simultaneously, sequentially, inversely, or in some variable or poorly defined pattern relative to each other. For example, in a typical situation of photochemistry, high concentrations of nitric oxide (NO) and nitrogen dioxide (NO_2) precede or succeed high concentrations of O_3 (U.S. NAS 1977). While high concentrations of O_3 are generally observed during daylight hours (the exceptions being meteorological phenomena at high elevations and stratospheric intrusion of O_3), high concentrations of fine particle sulphate (SO_4^{2-}) are observed during nighttime hours (Stevens et al. 1978). Thus, under field conditions, for example, vegetation is first exposed to high concentrations of fine particle SO_4^{2-} during the early part of the day and subsequently to high concentrations of O_3 during the late afternoon hours. The significance of this has been demonstrated by Herzfeld (1982) and reviewed by Chevone et al. (1986). When plants were exposed to fine particle acid SO_4^{2-} followed by O_3 under controlled conditions, the injurious effects of O_3 were significantly increased, in comparison with the effects induced by O_3 alone. The aerosol by itself did not produce detectable negative effects.

Certain pollutants, such as O_3 and sulphur dioxide (SO_2), are known to be relatively more phytotoxic than others, and under appropriate conditions these pollutants exist at sufficient ambient concentrations to cause visible injury. Other pollutants, such as NO_2 and SO_4^{2-}, are known to be relatively less phytotoxic and do not appear to exist at ambient concentrations sufficient to cause visible injury. However, what is critical is the joint effects of two or more types of pollutants. In addressing this issue, not only must one define the comparative relationships of the occurrence patterns of pollutants of concern, but also provide correctly and satisfactorily an appropriate numerical description of frequency distributions and time series analysis of individual pollutant occurrences relative to effects.

Many scientists assume that the frequency distribution of pollutant concentrations follows a normal, "bell-shaped" distribution. The following is an example of normal distribution: X is said to have the normal distribution if the distribution function of X is given by

$$F(x) = P(X \leq x) = \int_{-\infty}^{x} \frac{1}{\sqrt{2\pi}\ \sigma} e^{-1/2[(y-\mu)/\sigma]^2} dy$$

where it can be shown that the parameters μ and σ are the mean and standard deviation of X.

The standard normal distribution is the normal distribution with μ equal to 0 and σ equal to 1 (Snedecor and Cochran 1978). Long-term pollutant averaging techniques assume this (e.g., Oshima et al. 1976; Heagle et al. 1986). In reality, frequency distributions of the occurrence of O_3 and oxides of nitrogen exhibit a distinct skewness with a long tail at high values. This is also true for SO_2 and H_2S (Tables 2.19 through 2.21). However, according to Fowler and Cape (1982), using data from U.K., if the frequency classes are plotted in terms of the logarithm of SO_2 concentrations, a more symmetrical distribution is obtained. A similar conclusion was also reached by Male (1982) in the U.S. A random variable X has a lognormal distribution if its density is given as:

$$P_X(x) = \left[(x-\theta)\sqrt{2\pi}\,\sigma\right]^{-1} e^{-1/2[\log(x-\theta) - \zeta]^2/\sigma^2} \quad (x>\theta)$$

where, ζ is the expected value and σ is the standard deviation of the random variable log (x-θ) for some suitable parameter.

Table 2.19 Frequency distribution of averaged hourly SO_2 concentrations downwind from a large point source in Minnesota. Data from seven air quality monitoring sites were examined in developing this summary statistic.

SO_2 Concentration (ppb)	Absolute Frequency	Relative Frequency (%)	Cumulative Frequency(%)
0	89,192	94.4	94.4
5	3,446	3.6	98.0
10	1,316	1.4	99.5
15	260	0.3	99.8
20	142	0.2	99.9

Continued...

Table 2.19 (Concluded).

SO_2 Concentration (ppb)	Absolute Frequency	Relative Frequency (%)	Cumulative Frequency(%)
25	37	0.0	99.9
30	24	0.0	100.0
35	8	0.0	100.0
40	9	0.0	100.0
45	3	0.0	100.0
50	4	0.0	100.0
70	2	0.0	100.0
130	1	0.0	100.0

Source: Pratt (1982).

Table 2.20 Distribution of SO_2 concentrations in ambient air during January through December 1980 in the vicinity of a gas plant in Alberta.

Concentration (ppm)*	Frequency
.01	682
.02	418
.03	218
.04	167
.05	79
.06	66
.07	42
.08	39
.09	27
.10	20
.11	9
.12	8
.13	15
.14	4
.15	7
.16	5
.17	3
.18	6
.19	2
.20	1
.21	5
.22	1
.24	1
.25	1
.27	1
.28	1
.29	1

Continued...

Table 2.20 (Concluded).

Concentration (ppm)*	Frequency
.31	1
.38	1
.56	1
.64	1
.70	1
.76	1
.83	1

* 1/2 hour average.

Table 2.21 Distribution of H_2S concentrations in ambient air during January through December 1980 in the vicinity of a gas plant in Alberta.

Concentration (ppb)*	Frequency
1.00	46
2.00	7
3.00	9
4.00	8
5.00	4
6.00	4
7.00	2
8.00	4
10.00	6
11.00	2
13.00	3
14.00	1
17.00	1
18.00	2
20.00	2
21.00	1
22.00	1
31.00	1
36.00	1
41.00	1
48.00	1
75.00	1

* 1/2 hour average.

The assumption of lognormality has been questioned by Berger et al. (1982) in Belgium. These authors used 24-hour averages of SO_2 concentrations over a two and one-half year period in the region of Ghent, Belgium. Lognormal distribution was fitted to the data with poor results. The differences between the empirical data and the lognormal curve were most pronounced at extreme values of the 95th and higher percentiles. For this reason, the authors used the two-parameter gamma distribution, with the distribution function:

$$F(x) = \frac{1}{\beta^{\alpha}\Gamma(\alpha)} \int_0^x t^{\alpha-1} \exp\left\{\frac{-t}{\beta}\right\} dt$$

where the mean of the distribution is $\alpha\beta$.

The gamma distribution provided a much better fit, and the goodness-of-fit was confirmed by several tests of significance.

Most recently, Buttazzoni et al. (1986), evaluating the sulphur dioxide concentrations in the area of Venice, Italy, concluded that:

a. Their results seem to suggest that 2, 3, or 4-parameter lognormal distributions do not hold characteristics of generality or universality for the interpretation of air quality data;
b. In fact, Weibull and gamma distributions yield superior simulations in Venice;
c. Bell-shaped distributions cannot fit L-shaped experimental data sets; and
d. Gamma distributions are less convenient than Weibull distributions.

According to the authors, as a final comment, "...the research for a universal form allowing for direct comparisons among distributions underlined by differing sites or time spans, should head toward a generalization of such distributions."

In addition to the aforementioned studies, Lefohn and Benedict (1982) and Nosal (1984) examined the frequency distribution mathematical functions of ambient O_3 concentration in parts of the U.S. They concluded that such distributions follow a Weibull function (Figures 2.3 and 2.4). An example of Weibull distribution:

$$P_x(x) = c\,\alpha^{-1} \left[(x-\xi_0)/\alpha\right]^{c-1} e^{-\left[(x-\xi_0)/\alpha\right]^{c}} \qquad (\xi_0 < x)$$

where C is the shape, α is the scale, and ξ_0 is the location of the distribution.

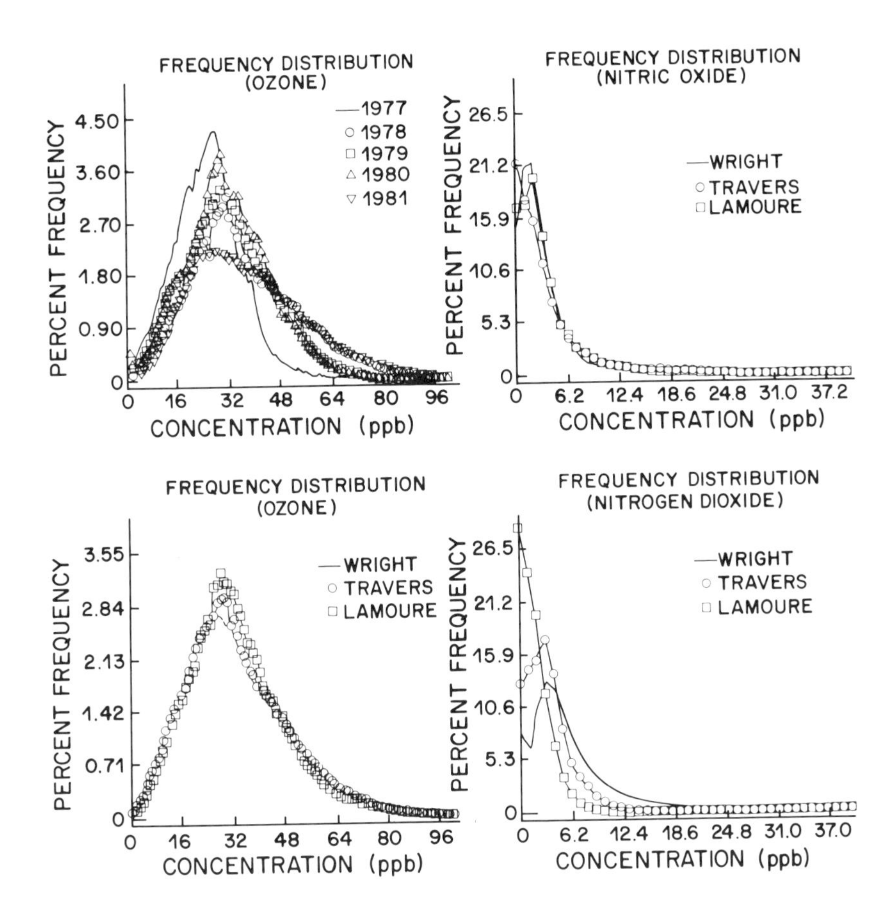

Figure 2.3 Frequency distributions for ambient ozone, nitric oxide, and nitrogen dioxide at three monitoring sites in the upper midwest USA.
Source: Pratt et al. (1983). (Reprinted with permission, Copyright 1983 Pergamon Press PLC).

In addition to the previous discussions on gaseous pollutants, the phenomenon of "acidic rain" is of major international, ecological concern. Our knowledge of the observed or potential effects of "simulated acidic rain" on crops and tree species have been summarized by Irving (1983); Jacobson (1984); and U.S. NAPAP (1987). There are no demonstrated cases of direct negative effects of "acidic rain" on vegetation under ambient conditions. Almost all studies relate to the use of "simulated rain". This consists of treating plant species with a "solution of constant chemical composition, applied artificially at constant or varying rates and amounts".

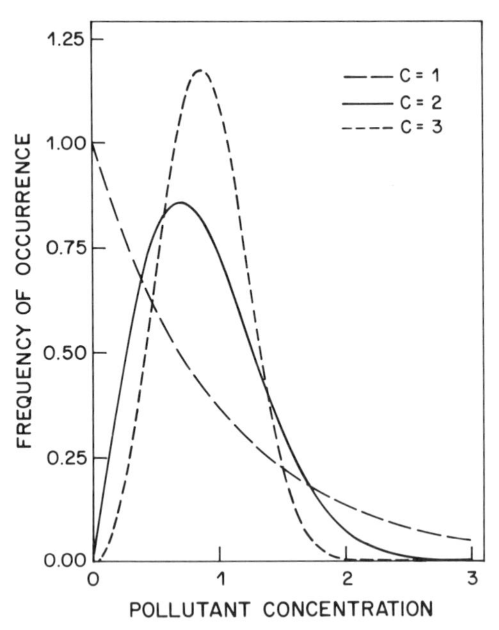

Figure 2.4 Theoretical distributions illustrating the Weibull family of frequency distributions.

This is inappropriate to the real world where the chemistry of precipitation varies significantly within (Table 2.22) and between individual rain events (Fowler and Cape 1984). The observed temporal and spatial variability of rainfall properties is regulated by: (a) regional versus local meteorology; (b) regional versus local distribution of pollutant sources and their characteristics; (c) the frequency of rainfall; (d) rainout versus washout; (e) the concentration of coarse particles in the air prior to the rain event; and (f) precipitation depth.

The literature contains numerous publications where "average" values of constituents in precipitation have been used to develop the "simulant" in plant exposure studies. As with the gaseous pollutants discussed previously, the frequency distributions of the concentrations of ions in precipitation also do not follow normal distribution. Such distributions are best described by a Weibull function (Nosal and Krupa 1986).

Table 2.22 Mean ion concentrations by 3.2 mm incremental fractions within rain events in Minnesota during 1980.

		Fraction Number[1]							
		1	2	3	4	5	6	7	8[2]
n =		85	48	33	27	19	10	4	2
pH	Weighted[3]	5.83	5.56	5.47	5.34	5.51	5.50	5.58	5.32
	S.E.	0.32	0.42	0.37	0.42	0.66	0.91	0.91	--
H^+	Weighted	3.47	5.54	6.20	6.85	7.05	5.58	3.96	7.08
	S.E.	0.69	1.39	1.22	1.28	2.11	2.22	2.16	--
NH_4^+	Weighted	51.36	37.00	34.77	29.80	31.14	28.49	14.48	30.56
	S.E.	4.00	3.81	3.88	3.48	6.46	5.75	4.88	--
K^+	Weighted	39.16	35.63	33.84	28.96	38.13	26.18	35.50	69.71
	S.E.	4.31	5.04	5.77	4.04	8.94	6.58	9.56	--
Ca^{2+}	Weighted	54.34	42.77	40.10	34.10	32.95	31.53	31.84	71.36
	S.E.	4.87	4.98	5.29	4.57	8.41	6.35	12.52	--
Mg^{2+}	Weighted	26.92	23.37	20.57	19.52	18.24	14.60	17.03	34.96
	S.E.	2.72	2.98	2.99	2.69	4.54	3.78	5.96	--
Al^{3+}	Weighted	21.82	23.59	21.23	23.01	19.61	18.65	24.77	38.36
	S.E.	2.19	3.17	3.13	3.27	4.64	4.27	6.20	--
Na^+	Weighted	38.95	34.97	31.74	32.18	31.00	33.33	50.01	71.12
	S.E.	4.63	5.50	5.95	5.22	8.45	8.63	15.63	--
SO_4^{2-}	Weighted	55.08	41.22	41.28	37.81	38.56	32.22	23.79	51.93
	S.E.	4.20	4.37	5.39	5.89	8.21	5.64	7.12	--
NO_3^-	Weighted	34.61	27.49	26.64	22.71	23.95	23.59	15.57	23.71
	S.E.	2.76	2.97	3.31	2.79	4.38	4.35	4.38	--

[1] Rain samples were collected as 3.2 mm sequential, incremental fractions within each rain event.
[2] Because the 2 samples of fraction 8 had equal volume, the volume weighted means were identical to the unweighted means.
[3] Weighted refers to volume weighted means.

Source: Pratt et al. (1983). (Reprinted with permission, Copyright 1983 Pergamon Press PLC).

Independent of the artifacts attached to pollutant averaging, Table 2.23 provides a summary of the exposure characteristics of major acidic and acidifying pollutants and O_3 in crop growing regions of the U.S.

Table 2.23 Concentrations and exposure characteristics of major acidic and acidifying pollutants and ozone in crop growing regions of the United States.

Pollutant	Average Growing Season Concentration	Exposure Characteristics
Acidic rain	Highest weighted average annual acidity, pH 4.1, in OH-PA-NY area, decreasing south and west to the Plains States (to pH 5.1); 2-4 mg L^{-1} SO_4^{2-} and 1-3 mg L^{-1} NO_3^- nationwide; SO_4^{2-}:NO_3^- ratio ~2 in east (NADP, 1983, 1984, 1985).	Fluctuations of H^+, SO_4^{2-}, and NO_3^- within and between events. Regional patterns in H^+ and SO_4^{2-}:NO_3^- ratio. Total deposition related to amount of rainfall. Concentrations of H^+ and SO_4^{2-} highest during growing season. Maximum recorded event acidity occurred in eastern U.S. rain (~pH 3.0) and Los Angeles fogs (~pH 1.7).
SO_2 (gas)	5-10 $\mu g\ m^{-3}$ (2-4 ppb); monthly ave. as high as 25-50 $\mu g\ m^{-3}$ (10-20 ppb) (U.S. EPA 1982). Hourly peaks of 100-3000 ppb near major sources.	In rural areas generally $\leq$20 $\mu g\ m^{-3}$. Peaks localized near sources and greatly affected by meteorological conditions; exposures often last only a few hours separated by days between fumigations. No strong diurnal variation on a regional scale.
NO_2 (gas)	3-25 $\mu g\ m^{-3}$ (2-13 ppb) in eastern U.S. <2 $\mu g\ m^{-3}$ in west; maximum hourly concentrations to 150 $\mu g\ m^{-3}$ (80 ppb) (U.S. EPA 1984b).	Short-duration peak exposures separated by days between fumigations. May accompany SO_2 peaks but co-occurrence is <50 h. during the growing season. Decreases with increasing O_3 diurnally.
O_3 (gas)	7-hour (0900-1600) ave. 88 $\mu g\ m^{-3}$ (45 ppb). Hourly peaks of 235 $\mu g\ m^{-3}$ (120 ppb).	Strong diurnal variation (highest in early afternoon, lowest in early morning). Peaks related to stagnant urban plumes. Generally unrelated to significant SO_2 or NO_x peaks.

Source: U.S. NAPAP (1987).

2.5 ENVIRONMENTAL RESPONSES TO THE DEPOSITION OF SULPHUR AND NITROGEN COMPOUNDS

2.5.1 Vegetation

2.5.1.1 Crops

While the subject of the present research is sulphur and nitrogen oxides and their products, one cannot, in isolation, examine their effects on crops under field conditions. The health of a crop is governed by its interaction with its physical (temperature, moisture, etc.), chemical (carbon dioxide, air pollutants, etc.), and biological (pathogens and pests) climatology. Further, these complex interactions are influenced by man's intervention (agricultural management practices).

Within the chemical climatology itself, acidic and acidifying pollutants cannot be viewed in isolation since ambient air is always composed of a mixture of chemical constituents. Table 2.5 provided a listing of pollutants relevant to dry acidic deposition. In addition, the previous section on "pollutant exposure dynamics" described some of the characteristics of pollutant occurrences in the atmosphere.

There are no demonstrated direct effects of ambient "acidic rain" *per se* on crops under field conditions. A number of investigators, however, have studied the effects of "artificial, simulated acidic rain" (SAR) on crops under laboratory, greenhouse, and controlled field type conditions (Irving 1983; U.S. NAPAP 1987). This subject is discussed in greater detail in a subsequent part of this section.

Under ambient conditions, crop responses to air pollution stress must be viewed as a product of both dry and wet deposition. As previously stated, dry deposition consists of the transfer of gases, vapours, and particles on to surfaces (e.g., crop canopies). Among the gaseous pollutants, at appropriate locations and at ambient concentrations, O_3, SO_2, PAN, and HF (hydrogen fluoride) are considered to be directly phytotoxic. It does not mean that other chemical constituents in the atmosphere do not have a role in producing jointly more than additive, additive, or less than additive effects on crops. Such joint effects are dependent on the crop species in question, sequence of exposure, and concentration of individual pollutants and their ratios in the air mixture, physiological growth stage of the crop and its sensitivity to stress, and physical and chemical environmental variables (Torn et al. 1987).

The all-pervasive O_3 is considered to be the most important phytotoxic gaseous air pollutant in North America (Krupa and Manning 1988). Its importance is being increasingly recognized in Europe and countries such as Australia, India, Japan, and Mexico. For a fundamental treatment of the formation of atmospheric O_3 and its effects on vegetation, the reader is referred to Krupa and Manning (1988) and Guderian (1985). While O_3 is neither "acidic" nor "acidifying", its importance cannot be ignored. In the atmosphere, O_3 is involved in the oxidation of SO_2 (Figure 2.2). Its joint effects with "acidic" or "acidifying" pollutants on vegetation have been reviewed by Chevone et al. (1986) and U.S. NAPAP (1987). In general, atmospheric or surface acidity increases the lifetime of O_3, its effective concentration, and results in certain areas in more than additive effects when plants are exposed sequentially to high concentrations of the two pollutants (acid aerosols or simulated acidic rain and ozone). This is consistent with atmospheric processes, where fine acid aerosols are known to accumulate during the nighttime hours (Stevens et al. 1978) and thus, plants are exposed to this pollutant during the early daytime hours prior to exposure to high concentrations of O_3 during the late afternoon hours (Krupa and Legge 1986; Krupa and Manning 1988).

Tables 2.24 through 2.27 provide a summary of the knowledge on visible injury, plant sensitivity, and growth and yield effects induced by O_3 under experimental conditions. For a more complete treatment of this subject, the reader is referred to U.S. National Academy of Sciences (1977), Guderian (1985), and U.S. EPA (1986).

Table 2.24 Sensitivity indices for agricultural crops under acute ozone exposures.

Crop	Sensitivity Index
Sensitive:	
Bean	127.57
Tomato	115.07
Grasses	83.72
Legumes	83.54
Oat	65.79

Continued...

Table 2.24 (Concluded).

Crop	Sensitivity Index
Intermediate:	
Vegetables	62.97
Wheat	52.45
Grasses	49.60
Clover	38.66
Legumes	38.94
Perennials	22.21
Resistant:	
Cucumber	22.90
Vegetables	16.98
Legumes	16.90
Grasses	9.92
Woody species	8.62

Source: U.S. Environmental Protection Agency (1978).

Table 2.25 Agricultural crops of Alberta known to be relatively sensitive to ozone.

Alfalfa (*Medicago sativa*)
Barley (*Hordeum vulgare*)
Bean (*Phaseolus vulgaris*)
Red clover (*Trifolium pratense*)
Corn, sweet (*Zea mays*)
Grass, bent (*Agrostis palustris*)
Grass, brome (*Bromus inermis*)
Grass, crab (*Digitaria sanguinalis*)
Grass, orchard (*Dactylis glomerata*)
Muskmelon (*Cucumis melo*)
Oat (*Avena sativa*)
Onion (*Allium cepa*)
Potato (*Solanum tuberosum*)
Radish (*Raphanus sativus*)
Rye (*Secale cereale*)
Spinach (*Spinacea oleracea*)
Tomato (*Lycopersicon esculentum*)
Wheat (*Triticum aestivum*)

Source: Hill et al. (1970).

Table 2.26 Effects of acute ozone exposure on growth and yield of agricultural crops.

Crop Species	Ozone Concentration (ppm)	Exposure Time (h)	Plant Response[1] Percent Reduction	Reference
Cucumber				
cv. Ohio Mosaic	1.0	1	19, top dry wt	1
	1.0	4	37, top dry wt	
Grapevine				
cv. Ives	0.08	4	60, shoot growth	2
cv. Delaware			33, shoot growth	
Pinto bean	0.05	24	Significant reduction in leaf growth	3
	0.10	12	Significant reduction in leaf growth	
Onion				
cv. Sparan Era	0.20	24	0, no effect	1
	1.0	1	19, plant dry wt	
	1.0	4	49, plant dry wt	
Potato				
cv. Norland	1.0	4	0, tuber dry wt	1
		4(3X)	30, tuber dry wt	
Radish				
cv. Cavalier	0.25	3	36, top dry wt	4
cv. Cherry Belle	0.25	3	38, root dry wt	
Radish	0.40	1.5(1X)	37, root dry wt	5
		1.5(2X)	63, root dry wt	
		1.5(3X)	75, root dry wt	
Snap bean	0.30	1.5(2X)	10, plant dry wt	6
			12, pod dry wt	
	0.60	1.5(2X)	25, plant dry wt	
			41, pod dry wt	
Soybean	0.30 to 0.45	1.5	Threshold for reduction of shoot growth	7
Tall fescue				
cv. Kentucky 3	0.30	2 (3X)	22, shoot dry wt	8
Tobacco				
cv. Bel W3	0.30	2	48, chlorophyll content	9
Tomato				
cv. Fireball	0.5	1	15, plant dry wt (grown in moist soil)	10
	1.0	1	20, plant dry wt (grown in moist soil)	

Continued...

Table 2.26 (Concluded).

Crop Species	Ozone Concentration (ppm)	Exposure Time (h)	Plant Response[1] Percent Reduction	Reference
Tomato (Continued) cv. Fireball	0.5	1	+15, plant dry wt (grown in dry soil)	10
	1.0	1	+25, plant dry wt (grown in dry soil)	
White clover cv. Tillman	0.30	2	17, shoot dry wt 33, root dry wt	8

[1] "+" = increase

References:
1. Ormrod et al. (1971)
2. Shertz et al. (1980)
3. Evans (1973)
4. Adedipe and Ormrod (1974)
5. Tingey et al. (1973a)
6. Blum and Heck (1980)
7. Heagle and Johnston (1979)
8. Kochhar et al. (1980), cited by Guderian (1985)
9. Adedipe et al. (1973)
10. Khatamian et al. (1973)

Source: Torn et al. (1987).

Table 2.27 Effects of long-term controlled ozone exposures on growth, yield, and foliar injury of various agricultural crops.

Crop Species	Ozone Concentration (ppm)	Exposure Time (h) (#/day)	Plant Response (% Reduction or Injury from Control)	Reference
Alfalfa	0.10	2/21 days	16, top dry wt	1
	0.15	2/21 days	26, top dry wt	
	0.20	2/21 days	39, top dry wt	
Alfalfa	0.05	7/68 days	30, shoot dry wt, 1st harvest 50, shoot dry wt, 2nd harvest	2
Bean, pinto	0.13	8/28 days	79, top dry wt 73, root fresh wt 70, height	3

Continued...

Table 2.27 (Continued).

Crop Species	Ozone Conc. (ppm)	Exposure Time (h) (#/day)	Plant Response (% Reduction or Injury from Control)	Reference
Bean, pinto	0.05	24/3-5 days	50, leaf chlorosis	4
	0.05	24/ 5 days	(fivefold increase in lateral bud elongation)	
Bean, pinto	0.15	2/63 days	33, plant dry wt 46, pod fresh wt	5
	0.25	2/63 days	95, plant dry wt 99, pod fresh wt	
	0.35	2/63 days	97, plant dry wt 100, pod fresh wt	
Bean, pinto	0.15	2/14 days	8, leaf dry wt	6
	0.15	3/14 days	8, leaf dry wt	
	0.15	4/14 days	23, leaf dry wt	
	0.15	6/14 days	49, leaf dry wt	
	0.225	2/14 days	44, leaf dry wt	
Bean, pinto	0.225	4/14 days	68, leaf dry wt	6
	0.30	1/14 days	40, leaf dry wt	
	0.30	3/14 days	76, leaf dry wt	
Bean, pinto	0.06	5 days/wk/ 40 days	48, shoot dry wt 50, root dry wt	7
Beet	0.20	3/38 days	50, top dry wt 40, storage root dry wt 67, fibrous root dry wt	8
Crimson clover	0.03	8/6 weeks	<10, dry wt	9
Corn, sweet cv. Golden:	0.20	3/3 days/wk/ until harvest	13, kernel dry wt 20, top dry wt 48, root dry wt	10
	0.35	3/3 days/wk/ until harvest	20, kernel dry wt 48, top dry wt 54, root dry wt	
	0.05	6/64 days	9, kernel dry wt 14, leaf injury	
	0.10	6-64 days	45, kernel dry wt 25, leaf injury	
Fescue, tall	0.09	6 weeks	17, leaf dry wt 15, shoot dry wt	12
Orchard grass	0.09	4-5 days/wk 5 weeks	14 to 21, shoot dry wt	13
Perennial ryegrass	0.09	4-5 days/wk 5 weeks	14 to 21, shoot dry wt	13

Continued...

Table 2.27 (Continued).

Crop Species	Ozone Conc. (ppm)	Exposure Time (h) (#/day)	Plant Response (% Reduction or Injury from Control)	Reference
Potato	0.20	3 h x 2(6)/week		14
2 seasons (Continued)				
cv. Norland:			30, tuber wt/19, tuber no.	
			20, tuber wt/21, tuber no.	
cv. Kennebec:			54, tuber wt/40, tuber no.	
			30, tuber wt/32, tuber no.	
Potato	0.05 (>or=)	326 to 533 total hours two years	34 to 50. tuber fresh wt	15
Radish	0.05	8/5 days/ wk/5 weeks	54, root fresh wt	16
Ryegrass, Italian	0.09	8/6 weeks	36, dry wt	
Soybean	0.05	6/133 days	3, seed yield	
			22, plant fresh wt	
			19, injury	
Spinach	0.06	7/days/ 37 days	18, fresh wt	18
	0.10	37, fresh wt		
	0.13	69, fresh wt		
Soybean	0.064	9/55 days	31, seed dry wt	18
	0.079		45, seed dry wt	
	0.094		56, seed dry wt	
Soybean	0.05 (>or=)	465/growing	28, seed wt season fresh wt	19
Tomato	0.20	2.5/3 days /wk/ 14 weeks	1, yield	10
			32, top dry wt	
			11, root dry wt	
	0.35	2.5/3 days /wk	45, yield; 72, top dry wt	
			59, root dry wt	
			8, tillering	
Wheat	0.20	4-7 days (anthesis)	30, yield	20
Wheat, winter	0.10	7-54 days	16, seed dry wt	21
	0.13	7-54 days	33, seed dry wt	

References:

1. Shinohara et al. (1974)
2. Neely et al. (1977)
3. Manning et al. (1971a)
4. Engle and Gabelman (1966,1967)
5. Hoffman et al. (1973)
6. Maas et al. (1973)
7. Manning (1978)
8. Ogata and Maas (1973)
9. Bennett and Runeckles (1977)
10. Oshima (1973)

Continued...

Table 2.27 (Concluded).

11. Heagle et al. (1972)
12. Johnston et al. (1980) (1966,1967)
13. Horsman et al. (1980)
14. Pell et al. (1980)
15. Heggestad (1973)
16. Tingey et al. (1973a)
17. Heagle et al. (1974)
18. Heagle et al. (1979a)
19. Kress and Miller (1981)
20. Shannon and Mulchi (1974)
21. Heagle et al. (1979b)

Source: Torn et al. (1987).

As opposed to O_3, SO_2 is a point source oriented problem. Depending on the source characteristics, meteorological features, and plume dispersion patterns, acute exposures and crop injury are observed under appropriate conditions in the vicinity of the source. Tables 2.28 through 2.30 provide a summary of the knowledge on plant sensitivity, growth, and yield effects induced by SO_2. For a more complete treatment of this subject, the reader is referred to Guderian (1977), Unsworth and Ormrod (1982), Winner et al. (1985), and International Electric Research Exchange (1981).

Table 2.28 Threshold sulphur dioxide concentrations (ppm) causing foliar injury to various agricultural crops.

A. Field Observations	1 h	2 h	3 h	4 h	8 h
Dreisinger and McGovern (1970) (Ni/Cu smelters - Sudbury, Canada)					
Sensitive crops	0.70	0.40	0.34	0.26	0.18
Intermediate	0.95	0.55	0.43	0.35	0.24
Resistant	1.88	1.1	0.86	0.70	0.49
Jones et al. (1979)	1 h		3 h		
(Power plants - Tennessee, U.S.)					
Sensitive	0.50 to 1.0		0.30 to 0.60		
Intermediate	1.0 to 2.0		0.60 to 0.80		
Resistant	2.0 +		0.80 +		
B. Controlled Environment Fumigations	1 h	2 h	3 h	4 h	8 h
Van Haut and Stratmann (1967)					
Sensitive (rye)	2.3	1.9	1.1	-	0.75
Katz and Ledingham (1939)					
Sensitive (alfalfa, barley)	1.5	1.0	0.89	-	0.55

Continued...

Table 2.28 (Concluded).

Thomas (1935)	1 h	2 h	3 h	4 h	8 h
Sensitive					
(alfalfa)	1.2	0.71	0.55	0.48	0.36
Fujiwara (1975)					
Sensitive	-	0.60	0.45	-	0.25
Zahn (1961)					
Sensitive	0.70	0.62	0.60	0.58	0.50
Intermediate	1.2	1.1	1.0	1.0	0.9
Resistant	1.8	1.7	1.6	1.6	1.4

Source: Original Table in International Electric Research Exchange (1981).

Table 2.29 Effects of ambient sulphur dioxide on yield of various agricultural crops.

Crop & Harvest Characteristics	Percentage of Control Value
Spring Canola	
(yield)	90.9
Alfalfa	
(yield)	81.0
Oats	
(yield)	76.1
Spring Wheat	
(yield)	73.4
Red Clover	
(yield)	63.6
Winter Rye	
(yield)	57.7
Winter Wheat	
(yield)	55.6

Exposure time: 4.3% of monitoring time.
Concentration: 0.44 ppm - during exposure time*.
0.083 ppm - average for monitoring time**.

* The exposure time was calculated by summing all time intervals, $\Delta t = 10$ minutes, with a mean SO_2 concentration greater than or equal to 0.10 ppm.

**The monitoring time is essentially equal to the exposure time of the test plants.

Source: Guderian and Stratmann (1968).

Table 2.30 Effects of SO_2 exposure on agricultural crops under field conditions.

Crop Species	Exposures	$\mu g\ m^{-3}$	(ppb)	% Yield Loss/Gain[a]	Reference
Corn-PAG 397	3 h x 12	186-1780 (11 concentrations used)	(70-670)	0	Miller et al. (1981)
Corn-Pioneer 3780	3 h x 12	186-1780 (11 concentrations used)	(70-670)	0	Miller et al. (1981)
Soybean-NK 1492	2.8 h x 10	399-638 (7 concentrations used)	(150-240)	0	Irving & Miller (1984)
Soybean-NK 1492	2.8 h x 10	798	(300)	6% loss	Irving & Miller, (1984)
Soybean-NK 1492	2.5 h x 18	240-940 (5 concentrations used)	(90-360)	5-19% loss	Sprugel et al. (1980)
Soybean-NK 1492	4.3 h x 1	3700, 4500, 5300	(1400, 1700, 2000)	5, 11, 15 % loss	Miller et al. (1979)
Wheat/Barley Alfalfa	3 h x 2	665-3190 (5 concentrations used)	(250-1200)	0	Wilhour et al. (1978)
Wheat/Barley	3 h x 10	665-3190 (5 concentrations used)	(250-1200)	0	Wilhour et al. (1978)
Kidney Bean	6 h x 56	260	(100)	0	Oshima (1978)
Winter Wheat/ Potato	10-13% growing season	475 (peaks at 700)	(190)	10% loss	Guderian and Stratmann (1981)

[a] Only statistically significant losses are reported.

Source: U.S. NAPAP (1987). Refer to that document for complete references.

In comparison to O_3 and SO_2, NO_2 is much less phytotoxic (Table 2.31). Only in rare cases (e.g., certain types of heated greenhouses in England), have ambient concentrations of NO_2 been shown to be directly phytotoxic (Law and Mansfield 1982).

As previously stated, a number of studies have examined the effects of simulated acidic rain (SAR) on crops (Table 2.32). One of the most complete dose-response studies illustrating the potential impact of SAR on crop yield was conducted on greenhouse-grown radish (Irving 1985). The results were derived from a collaborative study involving six laboratories using similar protocols. Radish plants were exposed to SAR over a complete growth cycle with six or eight rain treatment groups ranging in pH from 5.6-2.6. The experiment was repeated by four of the laboratories. The overall response functions from most of the laboratories were similar and indicated a threshold for significant yield decreases from simulated rain acidity between pH 3.0 and 3.4 over the growth season. Field grown radish plants appeared to be less sensitive compared with those grown in the greenhouse as no negative effects were reported from rain acidity levels as low as pH 3.0 in four field studies using simulated rain over a growth season (Irving 1985). The results of other studies on the effect of SAR on crops are summarized in Table 2.33.

There appear to be only two studies on the effects of fine particle aerosols on vegetation (Herzfeld 1982; Gmur et al. 1983). While Herzfeld was unable to show direct effects of submicron H_2SO_4 aerosols *per se* on soybean and bean plants, Gmur et al. observed chlorosis and necrosis on bean plants subjected to $(NH_4)_2SO_4$ aerosols. In both these studies, the aerosol concentrations used were orders of magnitude higher than the measured ambient concentrations.

As previously stated, under ambient conditions, vegetation is subjected to a pollutant mixture and not to single pollutants. Earlier experiments involving exposure of plants to O_3, SO_2, and NO_x at concentrations high enough to show readily measurable effects, generally indicated that the effect was roughly additive, although the results did not indicate consistent responses. At very high concentrations, injury from one gaseous pollutant may have been so extensive that further effects from a second pollutant were retarded, thus the result was less than additive. Recent well designed field studies have not indicated consistent effects from combinations of O_3 and SO_2.

Table 2.31 Comparison of yield effects from O_3, SO_2, and NO_2 on greenhouse-grown snap bean and radish.

	Concentration-ppm ($\mu g\ m^{-3}$)				Yield-% Difference From Control			
Species	O_3	SO_2	NO_2	Exposure Duration	O_3	SO_2	NO_2	Reference
Snap Bean	0.15 (294)	0.15 (393)	0.15 (281)	4 h x 3/wks for 4 wks	-27*	- 9	+12	Reinert & Heck, (1982)
Radish	0.30 (588)	0.30 (786)	0.30 (563)	3 h x 3/wks for 1 wk	-32*	+12	- 2.4	Sanders & Reinert, (1982)
Radish	0.2/0.4 (392/784)	0.2/0.4 (524/1048)	0.2/0.4 (375/750)	3 h/6 h for 1 time	-20*	- 2.6	+ 1	Reinert & Gray, (1981)

* Indicates statistical significance at $p \leq 0.05$.

Source: U.S. NAPAP (1987). Refer to that document for complete references.

Table 2.32 Summary of field studies to examine crop yield response to simulated acidic rain (SAR) dose.

Crop	Number of Cultivars Studied	Total Number of Studies	Studies Reporting Reduced Yield at Ambient pH[a]
Corn	2	8	0
Soybean	8	29	7[b]
Hay	2	2	0
Wheat	1	1	0
Tobacco	1	1	0
Potato	1	2	0
Oat	1	1	0
Bean (snap)	1	2	0
Total	17	46	7

[a] Yield reduction from simulated rain of pH 3.8 to 4.5 compared with pH 5.6.
[b] One study at University of Illinois; 6 studies at Brookhaven National Laboratory. There were no cultivars which consistently exhibited a negative response to simulated acidic rain.

Source: U.S. NAPAP (1987).

Reductions in crop yield have been observed to be caused by exposures to combinations of SO_2 and NO_2 that caused no effects when the pollutants were used singly. Forage grasses (timothy and rye) exhibited reduced leaf weight and area when exposed to SO_2 and NO_2. This effect was more than additive (Ashenden and Williams 1980). However, it is important to note that the co-occurrence of SO_2 and NO_2 was observed to be less than 50 h during the growing season in the U.S. (Table 2.23).

According to the U.S. NAPAP Interim Assessment Report (1987), although no significant negative interactions between acidic precipitation and gaseous pollutants have been observed, only a few experiments have been performed to understand this question; thus some uncertainty remains.

Table 2.33 Summary of most current replicated field studies of crop response to acidic deposition using simulated rain and ambient rain exclusion.

CROP	VALUE[a] 10^9s	AREA HARVESTED[b] 10^6 ha	CULTIVAR	EFFECT ON YIELD FROM[c]: AMBIENT ACIDITY[d]	EFFECT ON YIELD FROM[c]: HIGH ACIDITY[e]	cm RAIN	LOCATION[f]	REF	COMMENTS: EFFECTS ARE BASED ON COMPARISONS TO CONTROL RAIN WITH NO STRONG ACIDS (~ pH 5.6)
corn	20.5	29.1	Pioneer 3377	0	0	31,32	IL	h	no effect at pH 4.6, 4.2, 3.8, 3.4, 3.0; same results for 3 yrs.
			Pioneer 3377		0	32	IL	i	no effect at pH 3.0; same results using 3 different total rainfall amounts
			B73 × M017	0	0	31,32	IL	h	no effect at pH 4.6, 4.2, 3.8, 3.4, 3.0; same results for 3 yrs.
			B73 × M017		0	32	IL	i	no effect at pH 3.0 using normal rainfall amounts
			B73 × M017		–	32	IL	i	reduced yields using 50% or 75% of ave. rainfall amounts at pH 3.0
soybean	11.4	26.7	Williams 82	0	0	31,32	IL	h,i	no effect at pH 4.6, 4.2, 3.8, 3.4, 3.0; same results for 2 yrs.
			Williams	0	0	33	BNL	k	no effect at pH 4.1, 3.3, 2.7
			Williams	0	0	46	BNL	k	no effect at pH 4.1, 3.3, 2.7
			Williams	0	0	31	BNL	l	no effect at pH 4.1, 3.3, 2.7
			Amsoy	0	0, –	33	BNL	k	reduced yield at pH 2.7; no effect at 3.3, 4.1
			Amsoy	0	–,0	46	BNL	k	no effect at 2.7, 4.1; apparent reduced yield at pH 3.3
			Amsoy	–	–	31	BNL	m	reduced yield at pH 4.4, 4.1, 3.3
			Amsoy 71	–	–	32	IL	h	reduced yield at pH 4.2, 3.8, 3.4, 3.0; no effect at 4.6
			Amsoy 71	0	0	32	IL	j	no effect at pH 4.6, 4.2, 3.8, 3.4, 3.0
			Amsoy 71	0	0	31	IL	i	no effect at pH 4.6, 4.2, 3.8, 3.4, 3.0
			Amsoy 71	0	0	21	ANL	n,o	no effect at pH 4.2, 3.7, 3.4; same results for 2 yrs.
			Amsoy	0	–	31	BNL	l	reduced yield at pH 3.3, 2.7; no effect at 4.1
			Amsoy	–	–	38	BNL	p	toxic metals in simulated rain; reduced yield at pH 4.1, 3.3, 2.7
			Amsoy	–,0	0	31	BNL	q	no effect at 4.1, 3.3; apparent reduced yield at pH 4.4
			Davis	0	0	23	NC	r	ambient rain (pH 4.3, 4.1) not excluded; same results for 2 yrs.
			Davis	0	0	26	ORNL	s	no effect at pH 4.2 and 3.2; ambient O_3 did not affect response
			Forrest	0	0	57	NC	t	ambient rain (pH 4.1) not excluded; no effect pH 4.5, 4.2, 3.8, 3.4, 2.9
			Forrest	0	0	57	NC	t	no effect at pH 4.2, 3.7, 3.2, 2.7
			Asgrow	0	0	31	BNL	q	no effect at pH 4.4, 4.1, 3.3
			Asgrow	–	–	31	BNL	m	reduced yield at pH 4.4, 4.1, 3.3
			Corsoy	0	0	31	BNL	q	no effect at pH 4.4, 4.1, 3.3
			Corsoy	–	0	31	BNL	m	no effect at pH 3.3; apparent reduced yield at pH 4.4, 4.1
			Hobbit	0	0	31	BNL	q	no effect at pH 4.4, 4.1, 3.3
			Hobbit	–	–	31	BNL	m	reduced yield at pH 4.4, 4.1, 3.3
			Hodgson	0	0, +	41	OME	zz	no difference at pH 4.3, 3.6, 3.3, 3.2, 3.0
hay	10.7	24.9	FS954/Arlington (timothy/clover)	0	+	52	ANL	u	sandy loam soil; perennial crop, snow included; no effect pH 4.5, 4.2, 3.5; yield stimulation pH 3.2
			FS954/Arlington	0	0	52	ANL	u	loam soil; perennial crop, snow included; no effect pH 4.5, 4.2, 3.5, 3.2
wheat	8.7	27.1	Arthur	0	0	17	ORNL	v	yield as foliage biomass not grain; no effect at pH 4.2 and 3.2, no interaction with ambient O_3
tobacco[g]	3.1	0.32	Kupchunos	0	0	–	CAES	w	no yield effect at pH 4.0, 3.5, 3.0, 2.5, 2.0; quality reduced at pH 2.0
potato	2.0	0.53	Norchip	0	0	16	PEN	x	no effect at pH 4.6, 3.8, 2.8; same results for 2 yrs.
oat	0.8	3.3	Ogle	0	0	16	PEN	y	no effect at pH 4.8, 4.2, 3.6, 3.0
beans, snap	0.1	0.09	Provider	0	+,0	9,14	BTI	z	no effect at pH 4.2, 3.4 and 2.6 in 1st yr., increase in yield at 3.4 and 2.6 in 2nd yr.

[a]Value based on 1984 season average price received by farmers in billions of dollars. (U.S. Bur. Census. 1985)
[b]Harvested in 1984. (U.S. Bur. Census, 1985)
[c]Effects are reported for differences (PR<0.05) in marketable yield from plants exposed to simulated acidic rain compared to rain with no strong acids.
[d]Results using simulated rain with pH values within the range of current ambient acidic rain (pH 3.8 to 4.5).
[e]Results using simulated rain with pH values less than 3.8.
[f]IL = Univ. of IL, BNL = Brookhaven Nat'l Lab, ANL = Argonne Nat'l Lab, NC = N. Carolina State Univ., ORNL = Oak Ridge Nat'l Lab, OME = Ontario Ministry of Environ., PEN = Penn. State Univ., BTI = Boyce Thompson Institute, CAES = CT Agr. Exp. Station.
[g]Applied in 10 applications to runoff (exact amount not determined)
[h]Banwart, 1987
[i]Banwart, 1986
[j]Banwart, 1984, 1985
[k]Evans et al. 1985
[l]Evans, Lewin and Patti, 1984
[m]Evans et al. 1986
[n]Irving et al. 1986
[o]Miller and Irving, 1984
[p]Evans and Curry, 1979
[q]Evans, Lewin, and Hendrey, 1986
[r]Heagle et al. 1983
[s]Johnston and Shriner, 1986
[t]DuBay et al. 1984
[u]Irving et al. 1984
[v]Shriner and Johnston, 1984
[w]Rathier and Frink, 1984
[x]Pell et al. 1986
[y]Pell and Puente, 1987
[z]Troiano et al. 1984
[zz]Irving, 1986b

Source: U.S. NAPAP (1987).

The preceding discussion serves to demonstrate the limitations in the knowledge base. In much of the experimentation, one, two, or three pollutants have been evaluated singly or jointly as appropriate. An ideal field assessment should consist of a number of field sites with known, but variable, exposure dynamics of multiple pollutants. Intensive site specific characterization of the physical and chemical variables should then be coupled to the growth dynamics of the crop in establishing numerical relationships between cause and effects. Mathematical techniques are presently available to segregate the contributions of individually measured variables to the overall growth dynamics and productivity of crops (Krupa 1987).

2.5.1.2 Forests and Trees

While the present research effort does not include evaluation of tree or forest responses to sulphur and nitrogen oxides and their products in Alberta, for completeness the following brief discussion of the subject matter is provided.

A forest stand is a dynamic unit which changes over time as individual trees die from a variety of competing biotic and abiotic causes thus allowing space for the continuing addition of new trees. Species composition may change completely, or species composition and age distribution of the stand may be essentially maintained by plant succession. A successional sequence is made up of a series of stands which replace one another until a final stand develops which is capable of replacing itself. This final stand or type is referred to as the climax (Daubenmire 1968). In pure, even-aged stands, planted and intensively managed, the concept of succession has little utility.

A wide variety in forest stands implies a similar variety of responses following injury by air pollutants. The destructive influence of an air pollutant on the stability of a stand comprised of a complex variety of species compared with a single species plantation may have widely differing ecologic and economic importance. The relative susceptibility of each species and its importance in a successional sequence, plus intended uses of the stand, are a few of the variables which must be considered when determining the significance of injury (Miller and McBride 1975).

Much of the information on the effects of air pollutants on forests comes from investigations during the first half of this century near large point sources (Hepting 1968). Mortality and decline of forests were found in the immediate vicinity of these unregulated sources. While these studies provide evidence that relatively high concentrations of pollutants such as SO_2 can cause severe effects, they do not really reflect the present-day situation. In general, focus has shifted from site specific studies of the effects of gaseous pollutants from individual point sources to the evaluation of chronic exposures and effects of multiple pollutants from multiple sources within regional airsheds (Shugart and McLaughlin 1985).

One of the best known studies on regional scale impacts of air pollutants on forests concerns the effects of photochemical oxidants on the Ponderosa-Jeffrey pine ecosystem in southern California (Miller 1973). More recently, widespread tree decline has been observed at a number of locations in Europe and North America (Table 2.34). A number of hypotheses have been proposed for the cause of this decline, including gaseous air pollutants and acidic precipitation.

Table 2.34 Summary of case studies of forest decline in North America.

Authors	Species	Location	Findings/Evaluation
1. Johnson & Siccama (1983)	Red Spruce	Appalachians	▪ decreased growth post-1965 ▪ authors concluded that decline is "stress-related" ▪ several hypotheses presented; none proven
2. Raynal et al. (1980)	Red Spruce	New York	▪ seedling mortality observed ▪ reduced tree ring growth post-1965 ▪ limited measurements, so hypotheses for decline and seedling mortality cannot be tested
3. Siccama et al. (1982)	Red Spruce	Vermont	▪ resampled red spruce plots on Camels Hump exhibit decline in basal area from 1965 to 1979 ▪ variability not reported, so statistical significance cannot be assessed
4. Scott et al. (1985)	Red Spruce, Fir, Birch	New York	▪ basal area decline in red spruce and balsam fir above 900 m from 1964 to 1982

Continued...

Table 2.34 (Continued).

Authors	Species	Location	Findings/Evaluation
			▪ stand size not reported ▪ no hypothesis for cause of decline presented
5. Cook (1985)	Red Spruce	New York	▪ compared annual tree ring growth to climate factors for period 1750 to 1976 ▪ concluded that climate may be a contributing, but not sole, cause of observed decline in ring width post-1967 ▪ much of ring width variation pre-1968 explained by climate factors
6. Friedland et al. (1984)	Red Spruce	Vermont, New York	▪ red spruce decline evident in 1980s ▪ present hypothesis that winter damage, compounded by "over-stimulation" with nitrogen deposition may be a causative factor ▪ hypothesis requires further testing
7. Bruck (1984)	Red Spruce, Fir	Southern Appalachians	▪ mortality and defoliation of both species increase with elevation ▪ no drought observed ▪ metal toxicity to soil micro-organisms suggested ▪ basal area not reported, declining cannot be quantified ▪ reported metals appear too low for toxicity, based on laboratory studies
8. Stephenson & Adams (1984)	Red Spruce	Virginia	▪ no decline on Mt. Rodgers reported from 1954 to 1983 ▪ basal area of larger trees increased
9. Adams et al. (1985)	Red Spruce, Fir	Virginia, West Virginia	▪ annual ring width declined post-1965 ▪ growth rates 1930-65 show correlation with drought events ▪ no conclusions on cause for decline post-1965 presented

Continued...

Table 2.34 (Concluded).

Authors	Species	Location	Findings/Evaluation
10. Johnson et al. (1981)	Pine	New Jersey	▪ 2/3 of trees show some growth decline post-1965 ▪ no causative factors confirmed
11. Puckett (1982)	Pine, others	New York	▪ relationship between tree growth and climate had changed from 1900 to 1973 ▪ author suggests acid rain and/or air pollution as contributing factor, but no data on pollution used in the analysis
12. McClenahen & Dochinger (1985)	White Oak	Ohio	▪ tree growth during period 1900-78 correlated with climate and air pollution factors ▪ climate explains less of the growth variance post-1930 ▪ authors suggest that air pollution may be contributing; but insufficient data presented for hypothesis testing
13. McLaughlin, et al. (1985)	Sugar Maple	Ontario	▪ decline in annual growth from 1956-1976 ▪ drought, pathogens, and caterpillar investation suggested as contributor
14. Vogelmann et al. (1985)	Maple, Fir Beech	Vermont	▪ declines in basal area, biomass and density reported for maple and beech ▪ insects and fungal diseases eliminated as possible contributing factors ▪ no defintive conclusions drawn on role of air pollution or drought

Source: Hakkarinen and Allan (1986). Refer to that document for complete references.

Among the numerous hypotheses proposed for the tree decline in Europe are: acidic precipitation; O_3; organic pollutants in the atmosphere; heavy metal deposition; aluminum toxicity through the soil; mineral deficiencies; biotic pathogens, including viruses and mycoplasma; sudden changes in climatology; over-management; natural growth cycle; and even microwaves used in the communication by the NATO (North Atlantic Treaty Organization) forces.

Given the non-specificity of tree species involved in the decline in Europe (e.g. silver fir, spruce, and beech), non-specificity of the sites involved (calcareous and granitic soils) and distance between locations of observations (e.g., Bavaria and southern Sweden), one wonders about the validity of a single cause-effect relationship. As with crops, multiple causes with corollary relationships may be more valid. Recently, Prinz (1987) has summarized the knowledge of the possible causes of tree decline in West Germany. Similarly, Table 2.35 provides a summary of what we know and do not know about the relationship between acidic deposition and tree decline. In addition, Mayo (1987) has reviewed the world literature on the effects of acidic deposition on forests.

As with crops, a number of investigators have examined the responses of tree species to O_3 and SO_2 exposures under field, greenhouse, and chamber conditions (Davis and Wilhour 1976). More recently, Mayo (1987) provided a summary of the results of specific studies on pollutant exposure or deposition and adverse tree responses (Table 2.36).

As opposed to the previous discussion, many of the studies conducted on the effects of air pollutants on tree species in Alberta pertain to point source oriented problems (Addison et al. 1984; Legge et al. 1981; Nosal 1984). At the present time, there are no published articles on regional scale assessments of the effects of sulphur and nitrogen oxides and their products on Alberta's forests.

2.5.2 Soils

Tabatabai (1985), Reuss and Johnson (1986), and Turchenek et al. (1987) have reviewed the knowledge on the effects of acidic deposition on soils.

The acidification of soils is one aspect of the effects of acidic deposition on the terrestrial ecosystem that has caused considerable concern in various parts of the world in recent years. The consequent studies have mainly focussed on the acid forest soils of Scandinavia, northern Europe, northeastern United States and southeastern Canada. Reasons for the concern have related mainly to the possibility of reduced forest productivity and to the role of soils in stream and lake acidification. Interest concerning the impact of acidic deposition has also been expressed in relation to the possibility of reduced productivity in agricultural soils.

Table 2.35 Current state of knowledge about acidic deposition and forests.

What Is Known	What Is Not Known
1. Some tree species in North American forests are experiencing dieback, mortality and apparent growth decline.	1. The primary cause(s) of forest decline, the geographic extent of damage or the percentage of tree population affected.
2. In Germany, Norway spruce and fir, as well as broad leaved species have been experiencing dieback and increased mortality in rapidly increasing severity since 1982, with symptoms affecting an estimated 50% of forests in 1984.	2. The cause(s) of forest decline and associated symptoms in Germany, the extent of damage to forests, or why the addition of calcium and magnesium has initially alleviated some symptoms.
3. Forests in which the decline has been studied receive elevated airborne pollutant concentrations and deposition rates.	3. The historic rates and trends of airborne pollutant concentrations deposition; whether some areas which have high pollutant loading do not have tree damage; whether some areas with low loading do have tree damage.
4. Although sulphur and nitrogen are soil nutrients, excess concentrations in their oxidized states can leach other nutrients from the soil.	4. The extent to which historic and current rates of deposition of these materials have depleted or are depleting nutrients.
5. Many tree species undergo periodic cycles in mortality.	5. The extent to which natural (e.g., climate, pathogens) and man-made (e.g., land-use history, forest management practices) variables are observed symptoms.
6. The movement of nutrients and other chemicals through the forest ecosystem, and the relationship between that movement and chemical input through precipitation, is highly complex and variable in time and among forest types.	6. Mechanisms within these complex systems in order to ascribe symptoms to specific causes.

Continued...

Table 2.35 (Concluded).

What Is Known	What Is Not Known
7. Trees react to water availability, but annual average precipitation (the only commonly available measure of moisture) is insufficient for evaluating the relationship of seasonality and timing of soil moisture on tree growth and health.	7. Detailed historical rainfall patterns for most forest areas exhibiting dieback, decline, and mortality.
8. Many soils in forest study areas are acidic and nutrient poor for reasons unrelated to pollutants.	8. Whether pollutants are further exacerbating poor soil conditions.
9. Experiments using simulated acid rain have demonstrated that: - rain more acidic than typical acid rain can produce measureable effects on plant leaves; - rain chemically similar to typical acid rain can leach nutrients from soils which have characteristics of those found in areas where trees are declining; - tree leaves and needles change substances with incoming precipitation, altering the chemistry of the rain and leaves.	9. How to extrapolate data collected in controlled experiments to different species under field conditions
10. Some trees exhibiting yellowing needles (chlorosis) and subsequent dieback contain lower than expected levels of essential nutrients such as calcium and magnesium.	10. Critical levels of essential nutrients in foliage and whether the observed lower than expected levels are the result of pollutants.

Source: Hakkarinen and Allan (1986).

Table 2.36 References to negative tree response due to dry acidic deposition related pollutants or due to wet deposition.

Species/ Community	Location	Nutrient or Pollutant	Response	Treatment or Condition	Conclusion	Reference
Picea abies, *Larix*	Germany Bavaria	Mg, Ca, K, Zn	Low Mg & relatively low Ca, K, Zn in needles	S concentration in toxic range, up to 238 ppm Al in needles	Mg deficiency from acid deposition affects trees	Zech et al. (1985)
Forests	Europe	NH_4	Soft growth	Ammonia from agriculture, industry, etc.	Decreased frost hardiness. Increased insect & disease attack. Nitrogen stress	Nihlgard (1985)
Lodgepole pine	Canada	S-gas	Reduced growth	Downwind from a sour gas plant approx. 15 y	S-gas has reduced photosynthesis and growth	Legge et al.(1976) Legge et al.(1978) Legge (1980)
Red spruce	U.S.A.	S, Al	Poor foliage condition	S content of poor foliage .12%, while good condition foliage is .10%	Higher aluminum at Camels Hump site	Johnson et al.(1982)
Forest ecosystem	Solling Germany		Dieback		Acidification of the soil and aluminum toxicity	Ulrich - numerous articles - see Mayo (1987)

Continued...

Table 2.36 (Continued).

Species/ Community	Location	Nutrient or Pollutant	Response	Treatment or Condition	Conclusion	Reference
Green ash, Paper birch, Red pine	Wisconsin Greenhouse	SO_2	Relative growth rate (RGR) of Green ash not affected. RGR of pine roots lowered.	Greenhouse fumigation at different temperatures	Resistance varied with temperature	Norby & Kozlowski (1981a)
Green ash Eucalyptus *Pinus resinosa*	Wisconsin Greenhouse	SO_2	Growth reduced except for Green ash	SO_2 at 0.0, 0.2, 0.35 or 0.5 ppm for 30 h at 22°C	Post fumigation temperature had no effect	Norby & Kozlowski (1981b)
Scots pine	UK	O_3 and acid mist	O_3 reduced fine root biomass & increased senescence	56 day exposure	O_3 caused nutrient leaching from leaves, not likely a cause of forest decline	Skeffington & Roberts (1985)
Scots pine, Norway spruce	Sweden	S	Reduced growth	Field soil susceptible to acid deposition	Conclude no reason to attribute reduced growth to anything other than acid deposition	Jonsson (1977)

Continued...

Table 2.36 (Concluded).

Species/ Community	Location	Nutrient or Pollutant	Response	Treatment or Condition	Conclusion	Reference
Seedlings of Jeffrey pine, Monterey pine, Shore pine, Lodgepole pine, Sugar pine, Western white pine, Ponderosa pine, Douglas fir	Oregon	O_3	Only Ponderosa & western White pine growth, reduced	O_3 exposure chambers outdoors for 22 weeks to seedlings	Increased seedling emergence, necrosis in all species	Wilhour & Neely (1977)
Mixed conifer Ecosystem	California	Oxidants	Decline of Ponderosa & Jeffrey pine	Ambient oxidant exposures	Changes in ecosystem components, structure & processes are occurring	U.S. EPA (1986)
Yellow poplar seedlings	Virginia Greenhouse	SO_2, O_3 SAR*	Changes in relative growth rate, leaf area ratio, specific leaf area, leaf weight ratio, root-shoot ratio, & unit leaf rate	Exposure chambers	O_3 was more active at lower pH, SO_2 more active at higher pH	Chevone et al. (1986)

* SAR = simulated acidic rain.
Source: Mayo (1987). For complete references, refer to that publication.

The question of the nature and extent of acidic deposition effects on soils is controversial. Convincing evidence for deleterious effects of acid inputs on soils and forest or agricultural productivity is scant. It has been argued that the acidification of soils and lakes in the eastern United States has not been the consequence of atmospheric acidic deposition as much as of changes in land use over the past several decades (Krug and Frink 1983a). With regard to agricultural soils, a commonly expressed opinion is that effects will be insignificant because cultivated soils are predominantly well buffered. Moreover, the acidifying influence of fertilizers is generally much greater than that of atmospheric acidic inputs. The consequent necessity for liming farmed soils renders any concern about atmospheric deposition unnecessary. The issue is further complicated by the possibility that the major acidifying components of atmospheric deposition, namely SO_2 and NO_x, are plant nutrients and may therefore be beneficial to crop or forest growth. The impact of deposition on soils near emitters or point sources may be greater than on soils under regional conditions, but there also is little knowledge of these. In Alberta there is insufficient information concerning the proportions of acidic deposition due to dryfall and to wetfall. There is likewise insufficient understanding of the interactions of these different forms of inputs with soil.

The main precursors of acid forming atmospheric pollutants in Alberta are anthropogenic emissions of SO_x, reduced sulphur compounds, and NO_x (Hunt et al. 1982). Within wet and dry deposition, there are sub-categories of mechanisms by which materials reach plant and soil surfaces (Fowler 1980; Galloway and Parker 1980):

1. Wet deposition
   a. Incident wet deposition - gravitational transfer of water to earth surfaces.
   b. Throughfall - water that has passed through a leaf canopy.
   c. Net throughfall - difference between throughfall and incident deposition for a particular area.
2. Dry deposition
   a. Dry fallout - soil and seasalt particles of $>10\text{-}30\,\mu m$ diameter which settle by gravity.

b. Aerosol impaction - particles of <10 μm diameter which impact or deposit on to surfaces. These are commonly particles of $(NH_4)_2SO_4$, NH_4NO_3, and others.
c. Gaseous adsorption - gases sorbed by foliage or soils (SO_2, NO_x, and CO_2).

Recent reviews of deposition mechanisms and interactions of gaseous and particulate matter in relation to plant and soil surfaces include those of Fowler (1980), Chamberlain (1986), and Weidensaul and McClenahen (1986). Dry deposition is probably the most important atmospheric removal process for SO_2. Dry deposition measurement is difficult compared with wet deposition, and few reliable data for the Province of Alberta have been obtained (Sandhu et al. 1980).

The concept of resistance has been used in mathematical descriptions of flux of airborne gaseous substances to surfaces (Fowler 1980). Different substances have different resistance factors to deposition which can be empirically determined. Properties of the deposited substance itself and meteorological conditions also have resistance factors. Surface resistances of soils have not been measured due to their heterogeneity, but calcareous soils apparently have negligible surface resistance. However, resistance increases with decreasing pH and increasing dryness. More gases will be sorbed by soils of high moisture content because the resistance of water is negligible, at least for SO_2, NO_2, and HNO_3. Other gases which can be adsorbed include N_2O, NO, H_2S, CH_2SH, CH_3SSCH_3, COS, and CS_2 (U.S. Committee on the Atmosphere and the Biosphere, 1981).

Particulate dry materials are deposited by impaction and gravitational forces unless the diameter is less than 1 μm, in which case transport occurs by Brownian motion. Deposition rates thus increase with increasing particle size and also with an increase in roughness of the surface. This latter factor is important because particles, upon deposition, may be returned to the free atmosphere by bounce-off or blow-off unless obstacles are encountered (Fowler 1980).

As indicated previously, few quantitative data on dry deposition are available for Alberta. Total deposition has been measured in some studies, and gaseous sorption measurement by plant and soil surfaces has been the object of others. Caiazza et al. (1978) found dry-wet deposition ratios of 4.8 for sulphate and 5.6 for unfiltered total N in the Edmonton area.

A greater role of dry deposition in atmospheric inputs than had previously been considered was recently reported for a mixed hardwood forest in the eastern United States (Lindberg et al. 1986). Precipitation was found to be the most important deposition process for SO_4^{2-} and NH_4^+, dry deposition of vapours was most important for NO_3^- and H^+, and dry deposition of coarse particles was most important for K^+ and Ca^{2+}. The amounts of both total and dry deposition were found to be underestimated significantly by standard bulk deposition collectors.

Chaudry et al. (1982) reviewed some aspects of the reactions of atmospheric deposition of sulphur with soils. Extensive literature on the subject has shown that soils generally have a high sulphur sorption capacity. The moisture content of soil, among other factors, was found to influence the amount and rate of S sorbed. Nyborg et al. (1977, 1980) and Hsu and Hodgson (1977) have demonstrated in a number of experiments that Alberta soils have high capacities for sorbing S from the air and that soil pH can be depressed as a result. Total S gains of 12-53 kg ha^{-1} were reported for sites downwind from SO_2 emitters. The distances from SO_2 emitters ranged from 2-37 km; the highest S gains were obtained for soils nearest to and farthest from the emitters, while intermediate distances were somewhat lower (Nyborg et al. 1977). Generally, little of the sorbed S was found in SO_4^{2-}. In addition, it was found that snowpacks had low S content (<1 kg ha^{-1}), rainfall contributed up to 4 kg S ha^{-1} y^{-1} near emitters, water surfaces sorbed 4-15 kg S ha^{-1} y^{-1}, and forest throughfall had S content three to four times greater than incident wet deposition (Nyborg et al. 1977).

Some limitations were indicated in the studies of Nyborg et al. (1977, 1980). The method of total S determination was considered to be imprecise. The pH depressions of 0.1 or 0.2 units were recognized as being within normal spatial and temporal variability of this soil property. In addition, the errors involved in calculating S deposition on a kg h^{-1} basis could be largely due to large multiplication factors involved in converting from measurements of small quantities of soil. Thus, while the results of these SO_2 deposition and acidification studies should be considered as only approximate, they demonstrate a potential for acidification of soils. Further research not only into levels and types of deposition but also of mechanisms of sorption and transformations in soils appears to be desirable. To date, knowledge and investigation of these aspects has been hampered by inadequate data on quantities and types of pollutants in the air.

The molecular forms of S were not identified in the aforementioned experiments. In chamber studies of SO_2 and NO_2 sorption, Ghiorse and Alexander (1976) suggested that some of the non-sulphate S was in an organic form. These authors found experimental evidence that the initial products of SO_2 sorption were sulphite, bisulphite, and hydrogen ions. These species were rapidly converted to SO_4^{2-} in O_2^- containing soil water. The investigators speculated that organic S may be formed due to the sorption sites on organic matter or due to the presence of carbonyl groups with which sorption products can form sulphite-addition compounds. They suggested that the role of microflora is a passive one where SO_2 reactive substances are produced.

In contrast, in NO_2 sorption studies, microflora were found to have an active role (Ghiorse and Alexander 1976). In sterile soil, NO_2 was initially converted non-biologically to both NO_2^- and NO_3^-. Nitrification was incomplete in sterile soil, but in non-sterile soils, nitrifying microorganisms converted almost all NO_2 to NO_3^-, and nitrification of native soil N was apparently stimulated.

In most natural and agricultural ecosystems, air pollutants will first encounter vegetation surfaces. These, like soil, have a tremendous SO_2 sorption capacity which diminishes with the time of exposure. Conifers apparently can accumulate much more sulphur than deciduous species (Horntvedt et al. 1980). Interactions of acid forming substances with vegetation are not the subject of this section, but it is clear that studies of soil effects must take the vegetation factor into account. Indeed, in natural and agricultural ecosystems, it would be desirable to conduct integrated rather than independent studies of these components. A general conclusion of this brief review is that sorption mechanisms and reactions of acid forming substances in soils are not well understood and further research specific to individual situations is required.

Several factors affect the ultimate soil acidifying influence of atmospheric deposition. The role of water movement in soil, for example, has been reviewed by Bache (1980). The effect of a reacting soil solution will depend on the pattern of infiltration and flow through the soil. This is generally determined by the structure of the soil in each layer and by the rate of application of water. Bache (1980) concluded that more influent solution will react with soil surfaces at low than at high infiltration rates. In field situations, amounts of surface runoff, lateral flow, and channel flow are determining factors.

Components of the water balance in an ecosystem will also influence the transport and transformation of acidic deposition. In areas where evapotranspiration exceeds precipitation, materials from rainfall are not moved to the subsoil but are accumulated (Reuss 1980). If the excess of evapotranspiration over precipitation was consistent and continuous, only accumulation would occur. However, in most areas, precipitation varies such that at times there may be sufficient soil input to leach materials to subsoils. In a study of Brunisolic sandy soils under a jack pine forest in northeastern Alberta, McGill et al. (1980) reported that rainwater percolated below the root zone only when precipitation exceeded 30 mm over several days. On an annual basis, drainage accounted for 20-40% of input water at different locations. It was considered that the water content of the snowpack plus precipitation occurring during and immediately following the thaw was the prime input to deep soil drainage. In addition, during the growing season, it was found that about 40% of rainfall was intercepted by the canopy and the soil litter layer from which it re-evaporated. Thus, little moisture entered the mineral soil during most (over half) rainfall events. In this type of water balance situation, atmospheric deposition would be expected to accumulate in upper litter and mineral soil horizons over most of the year. During spring thaw and episodes of high rainfall, a substantial proportion of accumulated materials may be flushed down to lower soil layers and possibly to the water table.

Snow composition and snowmelt have been found to be important factors in acidification of natural waters. Studies in the SNSF (The Norwegian Interdisciplinary Research Programme) project demonstrated that lakes and rivers showed particularly low pH values during snowmelt (Overrein et al. 1980). Moreover, the first fractions of meltwater contained higher concentrations of ions than the bulk snow. The composition of runoff was considerably changed by contact with soil and vegetation. Estimates of effects of snowmelt on water bodies, as well as on soils and vegetation, require data on surface runoff, subsurface flow, and deep drainage. Water and wind erosion are other factors which influence the fate of atmospheric deposition.

In conclusion, not only soil chemical processes, but soil physical factors, water dynamics, and landscape processes influence the transformation and movement of acidifying substances. Studies which include all these factors will be necessary in any thorough investigation of acidic deposition impacts on soils.

Table 2.37 provides a summary of the potential impacts of acidic deposition on soils.

Table 2.37 Summary of the potential impacts of acidic deposition on soils.

Process or Property	Hypothetical Impact of Acidic Deposition
I. Soil Exchange Complex	
Exchange Capacity	Decrease in CEC resulting from the influence of clay aluminum
	Increase in CEC of soils with oxy-hydroxides due to sulphate adsorption
Exchangeable Acidity	Increase
Base Saturation	Decrease
Clay Mineral Morphology	Increased formation of hydroxy-Al interlayers and acid weathering
Aluminum	Increased mobilization and leaching
	Increased availability and toxicity
II. Organic Matter	
Organic Matter Turnover	Decreased rate of C mineralization due to acidification and/or associated trace metal toxicity
	Decreased CO_2 flux from land to atmosphere
	Increased retention of organic matter
Microbial Community Dynamics	Shift from bacteria to more acid - tolerant fungi
Organo-Mineral Associations	Reduced organo-clay interaction due to disruption of cation bridge linkages
Root Uptake	Trace metal toxicity due to acidification
III. Plant Nutrients	
Nitrogen	Decreased ammonification
	Decreased nitrification
	Changes in products of denitrification
	Increase in leaching
	Enhanced cation leaching due to NO_3^- inputs
	Reduced plant availability
Sulphur	Increased SO_4^{2-} reduction in low S, anoxic systems
	Increased reduced-S flux and reduced CH_4 flux to atmosphere

Continued...

Table 2.37 (Concluded).

Process or Property	Hypothetical Impact of Acidic Deposition
Sulphur (Continued).	Decreased leaching of S
	Decreased leaching of cations in sesquioxidic soils; increased leaching in others
	Reduced plant availability
Phosphorus	Decreased leaching and $AlPO_4$ precipitation in soil with high Al
	Increased PO_4^{3-} solubilization, plant availability and leaching in calcareous soils
	Reduced availability with pH reduction
Fe, Mn, Zn, Cu, Co	Increased availability
	Increased leaching
Mo, B	Reduced availability
Ca, Mg, K	Reduced availability
	Increased leaching
Toxic Elements	Some micronutrients may reach toxic levels due to increased solubility
	Increased concentrations, toxicity, and leaching of heavy metals
	Increased Al toxicity
IV. Weathering	
Carbonates	Increased dissolution
Primary Minerals	Increased dissolution
Clay Minerals	Increased influence of aluminum (formation of Al interlayers)
	Reduced surface charge

Source: Turchenek et al. (1987).

2.5.3 Surface Waters

Telang (1987) provided an analysis of the literature on surface water acidification. There is general agreement among investigators concerning two aspects of surface water acidification. Near a point source of acid-forming emissions, surface water acidification is caused by anthropogenic input of acidifying substances (e.g., Sudbury, Ontario, Beamish 1976). Factors of quantity and quality of atmospheric deposition affect the process of surface water acidification. However, disagreement exists concerning the mechanism of acidification of remotely located lakes and streams by long range transport of atmospheric acid precursors. Inputs of acidic substances to surface waters can occur by three pathways: (1) directly from the atmosphere; (2) indirectly from the atmosphere via runoff over or through the watershed; and (3) from the watershed itself, from internal generation of hydrogen ions, for example, acidification of the soil which in turn can lead to water acidification (Marcus et al. 1983; Mason and Seip 1985). The disagreement is due to a lack of quantitative data on the inputs of acidic substances to surface waters by each of these pathways. Gorham and McFee (1980) recognized the lack of such quantitative data and suggested the need for mass-balance studies to determine origins of acids, metals, and organic molecules entering aquatic ecosystems. At the present time, attempts are being made to quantify these inputs either by direct measurements of surface waters or by modelling (Dillon et al. 1982; Wright 1983; Gherini et al. 1984). Dillon et al. (1982) conducted mass-balance studies on 11 lakes and their watersheds in the Sudbury and Muskoka-Haliburton areas of Ontario. Wright (1983) studied input-output budgets for a small acidified lake and its catchment area at Langtjern, southern Norway. While these authors reported that inputs of acidic substances to lakes by direct atmospheric deposition (Pathway 1) were quantified, it appears that they determined the input of substances to lakes from the entire watershed via streamflow. Indirect input of acidic substances from the atmosphere via runoff over and through the watershed (Pathway 2) and input of substances from the watershed due to internal generation of hydrogen ions (Pathway 3) were not quantified. To predict the relative contribution of acidifying substances from the various hydrologic pathways in a watershed, Gherini et al. (1984), as a part of the Integrated Lake-Watershed Acidification Study (ILWAS), developed a model for two lakes (Woods and Panther) in the Adirondack Mountains. The simulation model showed that the routing of water through soils (shallow versus deep flow) largely determined the extent of lake acidification. Analysis of the two lake basins using the model and field data indicated that the internal production of acidity was approximately two-thirds the amount of the atmospheric loading. The ILWAS was the first effort to quantify inputs of acidifying substances to surface waters by various hydrologic pathways, using a simulation model.

Precipitation volume or depth influences surface water chemistry depending upon the relative inputs of acidifying substances by the three pathways mentioned previously. Elzerman (1983) reported that changes in the chemical composition of streamwater during precipitation events are important for the following three reasons: (1) significant portions of total annual fluxes of dissolved and particulate materials can occur during storm events; (2) some products of neutralization and other reactions of rainwater with watershed components will appear in stormflows; and (3) excursions in chemical composition beyond important threshold values may occur, for example, transient events in which aluminum reaches toxic concentrations in the streamwater.

It has been observed that inputs of hydrogen and other related ions to aquatic environments increase as precipitation volume increases (Powers 1929; Likens et al. 1977; Rosenqvist 1978a,b; Elzerman 1983; Harte et al. 1985). During 1925-26, Powers (1929) observed that rapid increases in runoff volume following heavy precipitation to the Great Smoky Mountains was accompanied by a rapid drop in stream pH from 4.89-4.29. During the heavy precipitation events the streamwater developed a yellow tinge which disappeared after a few days. Powers attributed this to leaching of chemical constituents from naturally acidic soils.

Likens et al. (1977) observed that during the period 1964-74, hydrogen ion concentrations in Hubbard Brook were directly correlated with discharge volumes ($R^2=0.73$). Likens et al. suggested that increase in H^+ can be partially explained by the amount of precipitation and that factors other than precipitation may also be important for an increased load of H^+.

Rosenqvist (1978b) reported that in the summer of 1972, the pH in the Birkenes Creek, southern Norway, dropped from 5.6-4.4 (i.e., 5 times the acidity of rainwater, from $10^{-5.1}$ moles H^+ to $10^{-4.4}$ moles H^+) after a heavy deposition of mainly unpolluted rain. While Rosenqvist did not report the chemical composition of the rainwater, he attributed the changes in the pH of the creek after the precipitation event to cation exchange processes in the soil and increased rates of leaching of naturally occurring hydrogen ions.

These observations suggest that the changes in chemical composition of stream water during precipitation events are greatly influenced by the existing soil chemistry rather than strictly by hydrogen ion concentration in the rain. For example, large amounts of H^+ and Al^{3+} occur in soluble form in naturally highly acidic coniferous forest watersheds and all that may be required to flush these ions out is rain, whether it be pH 7.0 or 4.3. During storm events, large amounts of these ions will be washed out, particularly on steep slopes because of subsurface flow through the O and A horizons where most of the internally produced H^+ and Al^{3+} ions are generated. Moreover, if soil has a pH of 4.1 and contains large amounts of H^+ and Al^{3+} and is highly buffered by weak acids (e.g., 300-2000 keq ha^{-1}), the influence of rain of pH 4.3 on changes in chemical composition of streamwater will be difficult to assess, particularly when the rain event contains far less than one keq ha^{-1} of hydrogen ions. During the first phase of a rain event, all soluble chemical components of the soil (e.g., SO_4^{2-}, NO_3^-, Cl^-) will be at an elevated concentration and it may have no relationship to the chemistry of the rain.

Elzerman (1983) studied storm-induced chemical changes in streamwater of the Lake Issaqueema watershed in South Carolina for the period 1981-82. He observed that during the stormflow, pH dropped slightly (maximum 0.6 pH units), and so did the concentration of bicarbonate ion (base concentration, 198 μeq L^{-1}; stormflow concentration 147 μeq L^{-1}). The concentration of major cations (Mg^{2+}, Ca^{2+}, and Na^+), showed a dilution effect due to increased streamflow, causing temporary acidification of stream water. The concentrations of SO_4^{2-} and aluminum showed increases (baseflow concentrations of 85 and 10 μeq L^{-1}, respectively, and stormflow concentrations of 143 and 100 μeq L^{-1}, respectively). Harte et al. (1985) made similar observations for surface waters in a high elevation watershed in the Colorado Rockies. They observed that the volume of precipitation caused changes in alkalinity by varying the hydrologic flushing rates in the watershed; however, they did not specify the degree of change in alkalinity. Changes in alkalinity can occur due to increased physical and chemical weathering of soils, subsoils, and bedrock as a consequence of increased hydrogen ion loadings and runoff rates accompanying increased rainfall volumes. This can result in increased concentrations of weathered chemicals in surface waters.

Like precipitation quantity, precipitation quality can also potentially affect surface water chemistry. Numerous chemical species in addition to acids, are transported in the atmosphere and deposited with precipitation. For example, in Norway, Lunde et al. (1977) detected more than 450 organic compounds in precipitation. Organic compounds detected include alkanes, polycyclic aromatic hydrocarbons, phthlates, fatty acid esters, aldehydes, amines, pesticides, and polychlorinated biphenyls (Lunde et al. 1977;

Strachan and Huneault 1979; Alfheim et al. 1980). Haines (1981) suggested that these compounds, which are of anthropogenic origin, are probably produced by combustion processes and thus are likely to be related to acidic precipitation. However, it is important to note that these compounds, unless present as organic acids or converted organic acids, do not contribute to surface water acidification.

The most noticeable examples of the impacts of atmospheric precipitation on watersheds and surface waters are associated with localized point source emissions. Examples include the area surrounding smelters near Sudbury, Ontario, and Halifax County, Nova Scotia. Beamish (1976) cited Harvey's unpublished work on lakes near Sudbury. Harvey measured the sulphate concentration of over 100 lakes and found concentrations were highest close to the smelters and decreased with distance from the smelters. Beamish (1976) made a similar observation for the concentration of Ni and Cu in lakes in the La Cloche Mountains, southwest of Sudbury. The mountain lakes contained high concentrations of Ni (5-15 $\mu g\ L^{-1}$) and Cu (2-4 $\mu g\ L^{-1}$) when compared with concentrations measured in lakes remote from major areas of industrialization (Ni: $<3\ \mu g\ L^{-1}$; Cu: $<2\ \mu g\ L^{-1}$).

Watt et al. (1979) compared concentrations of hydrogen and sulphate ions in 16 lakes from Halifax County. Concentrations of both ions decreased significantly with distance from the emission sources. They also observed that lake pH values had decreased significantly since the study of these lakes 21 years earlier. The decrease in pH was approximately 0.34 units for lakes and ponds located on granitic rocks (average pH in 1955=4.66; in 1977=4.32), and 0.65 units for those located on metamorphic rocks (average pH in 1955=5.62; in 1977=4.97). The increase in SO_4^{2-} concentrations was about 27.91 $\mu eq\ L^{-1}$, for lakes and ponds located on the granitic rocks (average SO_4^{2-} in 1955=128.52 $\mu eq\ L^{-1}$; in 1977=156.43 $\mu eq\ L^{-1}$), and 47.5 $\mu eq\ L^{-1}$ for those on metamorphic rocks (average SO_4^{2-} in 1955=116.65 $\mu eq\ L^{-1}$; in 1977=164.14 $\mu eq\ L^{-1}$). Watt et al. (1979) attributed the elevated hydrogen ion concentrations in lakes to sulphur emissions from industries around Halifax. However, since these lakes and ponds contained high concentrations of dissolved organic carbon (average DOC 10,000 $\mu g\ L^{-1}$), natural humic substances may also have contributed to surface water acidification. It is difficult to speculate on this aspect since Watt et al. (1979) did not provide colour values for these waters.

In addition to hydrogen, sulphate, copper, and nickel, other ions have been measured above the background concentrations in surface waters as a result of local emissions. These include lead, zinc, manganese, iron, nickel, mercury, vanadium, and cadmium (Gorham and McFee 1980). Haines (1981) observed that metal concentrations are higher in acidic precipitation than in nonacidic precipitation.

Although highest metal concentrations are known to occur in precipitation near smelters, highways, and urban areas, elevated levels have also been detected in remote areas. For example, mercury has been detected in precipitation at remote areas of Quebec and Ontario at concentrations, an order of magnitude higher than in surface waters of those areas. Although mercury in the atmosphere may originate from natural sources, increased acidity of precipitation appears to result in increased scavenging of mercury from the air (Tomlinson et al. 1980).

Snowmelts have been shown to affect surface water (Galloway et al. 1980; Hendrey et al. 1980b; Johannes and Altwicker 1980; Johannessen et al. 1980; Bjarnborg 1983; Cadle et al. 1984; and Schofield 1984). pH values of snow can range from 4.5-5.0, and may be as low as 3.3 (Seip 1980). The average sulphate and nitrate values can vary from 528-1056 $\mu g\ L^{-1}$ and 686-1584 $\mu g\ L^{-1}$, respectively (Cadle et al. 1984). The rapid release of acids from snowpack during the spring thaw can cause a temporary and sudden drop in the pH of poorly buffered lakes and streams. These decreases in pH produce drastic effects on fish populations and are also suggested to be the main cause of lake invasion by sphagnum moss (Schofield 1976a,b; Hultberg 1977). While it is generally recognized that most of the acidity in snow is from HNO_3 and H_2SO_4, there is uncertainty regarding their relative importance to the springtime acidification of lakes and streams. The stability of these chemical species in the snowpack is questionable. Johannes et al. (1980) showed that sulphate is preferentially leached from the snowpack during the winter, thus increasing the importance of HNO_3. However, it is not clear as to how much of the HNO_3 is biologically utilized during the melt period. Added to these uncertainties is the importance of the leaching of acidic substances from the soil by meltwater (Tetra Tech Inc. 1984). Cadle et al. (1984) inspected the snowpack concentrations of SO_4^{2-}, NO_3^-, Cl^-, H^+, NH_4^+, Ca^{2+}, Mg^{2+}, Na^+, and K^+ on a weekly basis and observed no loss in these species from the snowpack before the thaw period.

Aging snow has been suggested to become progressively less acidic, perhaps due to the exchange of hydrogen ions with cations originating from organic materials (such as cone scales, seeds, leaves, twigs, bark fragments) in the snowpack (Hornbeck et al. 1977).

During the major spring thaw, the first meltwaters contain higher concentrations of ions than does bulk snow (Hendrey et al. 1980; Johannes and Altwicker 1980; Cadle et al. 1984). It has been reported that 66-83% of SO_4^{2-}, 50-61% of NO_3^-, and 40-52% of H^+ in snow are released during the first 21-35% of the snowmelt. This input of acid (pH "shock") results in a sudden drop of pH in streams or lakes (Galloway et al. 1980). The pH "shock" is commonly followed by peak flow and dilution of all major ions. Johannessen et al. (1980) observed that during and after snowmelt the streamwater concentrations of ionic species varied relative to the baseflow value, and they suggested three stages of snowmelt. During the first stage, as the water flow increases with the initial snowmelt, the concentrations of Ca^{2+}, Mg^{2+}, and HCO_3^- ions supplied by weathering are often, for a short time, at concentrations above the corresponding values for baseflow. This is due to the release of older water in the catchment area (piston flow). This process is followed by a dilution of the ions due to mixing with snowmelt water, which contains a very low concentration of ions produced by weathering but a peak concentration of the snowmelt ions H^+, NH_4^+, Na^+, NO_3^-, SO_4^{2-}, and Cl^-. Later, at the third stage, because the remaining snow contains most of the water but low concentration of the ions, the snowmelt becomes more dilute with respect to these ions, and consequently a dilution in concentration of the snowmelt ions in the streamwater is observed. After spring snowmelt is complete and discharge decreases again to just above baseflow levels, the concentrations of Ca^{2+}, Mg^{2+}, Na^+, SO_4^{2-}, Cl^-, and HCO_3^- are usually lower than in the baseflow prior to snowmelt. This suggests that the "baseflow reservoir" has been renewed and the catchment needs time to recover.

Jeffries et al. (1979, cited in Marcus et al. 1983) showed that the hydrogen ion discharge from three Canadian watersheds varied proportionally with the discharge volume during the two month spring snowmelt. Precipitation was very light during this period and thus had little effect on stream discharge volume and composition. It was found that runoff acidity was not relatively high in the early periods of snowmelt as one would expect. This suggests that sources in addition to snowpack accumulation (for example, leaching of hydrogen ions from soil) contributed to the discharge hydrogen ion loadings. Seip (1980) suggested that increased hydrogen ion concentrations in meltwater reflect not only snow accumulations of hydrogen ions, but also the influences of the accumulation of ions in soils over the winter. He pointed out that the contact between soil and meltwater is very important, and this contact determines the meltwater chemistry. A number of factors which contribute to meltwater chemistry through this contact consist of the degree of soil freezing, air temperature, thickness of snow cover, texture of soil, and type of vegetation. Seip (1980) cited the work of Rueslatten and Jorgensen in Norway, who explained the increase of hydrogen ion concentrations in meltwater as resulting from leaching of organic acids and from cation exchange processes in the humus layer of the soil. However, the investigators did not analyze their samples for dissolved organic carbon, sulphate, or chloride thus making their discussion of the mechanism speculative. Schofield (1984) reported temporal acidification of three lakes in the Adirondack Mountains during snowmelt. He attributed the process to base cation dilution and increased strong acid anion concentrations, particularly NO_3^-. These changes in surface water chemistry were related to an upward shift in flow paths from groundwater-dominated baseflow (mineral horizon) to shallow interflow (humus layer) during increased snowmelt discharge. During the snowmelt period, elevated levels of aluminum were also observed. Most of the mobilized aluminum was in the inorganic form and very little was in the complexed organic form.

One of the major controversies on the subject of surface water acidification concerns the source of protons or hydrogen ions. Some researchers believe that atmospheric inputs of anthropogenic acidic substances cause acidification of surface waters leading to serious effects on fish (Beamish 1976; Overrein et al. 1980; Rahel and Magnuson 1983; Somers and Harvey 1984); others believe that internal proton production in soils, intensified by changes in land use practices, is the primary contributor to surface water acidification (Rosenqvist 1978a,b; Krug and Frink 1983a,b).

Acidic rain falling directly on lake surfaces can contribute to acidification, even though such an influence is considered to be only minor. This process depends on the alkalinity of the watershed and the ratio of drainage area to lake area (Dillon et al. 1978; Panel on Processes of Lake Acidification 1984). In watersheds with granitic soils, the rate of weathering is very low and so is the alkalinity of their drainage water. For lakes in such areas, with small ratios of drainage area to lake area, direct inputs of acidic rain may exceed inputs of alkalinity from the watershed, resulting in acidification of those lakes. One such case has been identified in south-central Ontario, Canada, where the lake occupies 42% of the watershed area and the ratio of drainage area to lake area is 1:4. The alkalinity of the lake in 1967 was 33 μeq L^{-1}; in 1977 it was approximately 10 μeq L^{-1}. From its watershed characteristics, it seems reasonable to assume, that in this case acidification is occurring through direct inputs of acidic rain to the lake (Dillon et al. 1978). Krug and Frink (1983b) suggested that for low alkalinity lakes with drainage area to lake area ratios of 3:7, inputs of acid by rain may be equivalent to a measurable fraction of the alkalinity supplied by the watershed. Hence, the relative contribution of external hydrogen ion inputs versus internal hydrogen ion generation in the acidification of lakes of this type must be assessed on an individual basis.

Rosenqvist (1978b) pointed out that surface water acidification is a commonly occurring natural process and is not necessarily dependent upon external atmospheric sources of hydrogen ions. He proposed that acidic precipitation accounts for only a minor portion of the acidity of brook, river, and lake waters, and that the major contributor to the acidity of lakes is acidic runoff from acid soils. Rosenqvist further pointed out that runoff from such soil will be acidic regardless of the pH of rain, and experimentally demonstrated that it is the content of electrolytes and not the pH of precipitation which is important in the acidification of the runoff water. Wilkander (1946, cited by Krug and Frink 1983b) has shown that additions of neutral salts generally acidify the leachate from acid soils more effectively than do strong acids, at the concentration present in acidic rain. Rosenqvist (1978b) described acidification of runoff in terms of hydrology and soil acidity. Runoff from rapid snowmelts and heavy rains moves primarily through the organic-rich, acidic humus horizons and, therefore, is more acidic than runoff from gentle rains or slower snowmelts which percolate through the near-neutral subsoil. He pointed out that areas of Scandinavia believed to be affected by acidic rain are areas where there are also naturally acidic waters. This also appears to be true for some parts of North America (Patrick et al. 1981). Rosenqvist et al. (1980) reported that for high altitude regions of Norway where extensive areas of exposed bedrock exist, natural hydrogen ion production can range between 2000 and 5000 eq ha^{-1}, and airborne acidic inputs can range between 200 to 500 eq ha^{-1}. Based on these values, natural processes account for 85 to 95% of the total hydrogen ion flux in the region. Marcus et al. (1983) cited observations of Groterud (1981) which support the findings of Rosenqvist et al. (1980) for Norway. Groterud conducted lysimeter experiments in topsoils during heavy rains and snowmelt in the lake area near Oslo, Norway. He found that topsoils contributed about four times higher hydrogen ion concentration to drainage in comparison to the contribution by precipitation at an average of 70 cm y^{-1} and pH 4.4.

Rosenqvist (1978b) also attributed acidification of surface waters in Norway to changes in land use practices; agricultural land reverted to forest or heathland. The vegetation removes bases from the soil and at the same time contributes considerable amounts of humus which becomes thicker and more acidic with time. The acidification potential of the humus is very high. Rosenqvist estimated that the exchange acidity of the surface humus accumulated under a 90-year-old spruce forest is equivalent to approximately 1000 years of acidic rain at 1 m y^{-1} with pH 4.3. Rosenqvist (1980) documented an increase in forested land and an increase of over 70% in the volume of standing wood in southern Norway from 1927 to 1973. According to Rosenqvist, particularly in the southern part of Norway, the natural production of acid is many times higher than that derived from precipitation, and thus there is no reason to believe that acidic deposition is the major reason for the acidification of surface waters. Based on this observation, it appears that humic soils acidify runoff in watersheds where there is incomplete neutralization of acids by mineral weathering.

Based on the reports of land use practices in the watersheds of acidified lakes, Krug and Frink (1983b), like Rosenqvist, have pointed out that regions experiencing acidic rain are also undergoing dramatic changes in land use and, therefore, acidic rain is not the only source of increasing acidity. One such example they cited concerns the acidification of Nova Scotia lakes. Watt et al. (1979) reported that a series of lakes and ponds in Nova Scotia were more acidic in 1977 compared to the observations in 1955. Because the appropriate watersheds had been largely undisturbed between the two studies, acidification was attributed to acidic rain. However, Krug and Frink (1983b) pointed out that the important factor in this acidification relates to what happened in the watersheds before 1955. They noted that this area of Nova Scotia was treeless and barren due to destructive fires by miners in the early 1900's (Woodward 1906, cited by Krug and Frink 1983b). Therefore, they suggested that forest regrowth, accumulation of humus, and increased acidity of runoff water may have been the cause of the acidification.

Natural soil acidification has been reported by many investigators. Gorham et al. (1979) reported that augmented organic matter accumulation and soil microbial biomass can increase carbonic acid and organic acid accumulations in the soil. They also noted that nitrification of organic matter and conversion of ammonia to nitrate in the soil generates nitric acid. This process approximately adds 475 eq ha^{-1} of hydrogen ions (Oden 1976). According to Gorham et al. (1979), in terms of biomass, organic sulphur is about 10% of the accumulated nitrogen. Oxidation of these sulphur compounds can yield enough sulphate ion and sulphuric acid to acidify soils. These acids are washed out during runoff.

Because of natural acidification processes, forest soils have become acidic over time. Soils in some conifer forests have natural pH levels around 3.0 (Brady 1974). Sollins et al. (1980, cited by Marcus et al. 1983) developed a hydrogen ion budget for a coniferous watershed in Oregon and established the predominance of internal hydrogen ion sources. They found that biological processes of cation exchange by vegetation and microbial decomposition results in a net hydrogen ion flux of 3000 eq ha^{-1}, in the range found by Rosenqvist (1980). During the study, the hydrogen ion concentration in the precipitation was 6.3 $\mu g\ L^{-1}$ (pH 5.2). They suggested that an increase in hydrogen ion concentration in precipitation to 100 $\mu g\ L^{-1}$ (pH 4.0) would add only 300 eq ha^{-1} to the total natural hydrogen ion flux in the soil solution or about 10%.

Studies by Likens et al. (1977) at Hubbard Brook, New Hampshire, have shown that the contribution of acid from external atmospheric sources and internal watershed sources can be approximately equal. For the Hubbard Brook Experimental Station they calculated that about 2000 eq ha^{-1} of hydrogen ions are consumed by geological weathering. During 1963-74, the supply of hydrogen ions by bulk precipitation averaged 865 eq ha^{-1}. Therefore, the remaining 1135 eq ha^{-1} appeared to be contributed by internal watershed sources such as chemical and biological processes occurring within the soil.

Organic acids (humic acids) in humic soils can contribute to the lowering of surface water pH through runoff. An example of this effect was reported by Ramberg (1981) for a small watershed in Central Sweden. A runoff brook fed by the water flowing through the upper organic layer of the forest soil showed a two year mean pH of 4.9. When one metre deep drainage ditches were constructed to prevent water flow through the organic-rich soil, the pH of the stream increased to a mean value of 5.8. The report, however, did not indicate the proportion of hydrogen ions contributed by the peat soils in the study versus atmospheric deposition. A high proportion of acidity in this Swedish brook may have been generated internally. Gorham et al. (1984b) have suggested that peat soils can provide a substantial contribution to the hydrogen ions.

Johnson (1981) made similar observations for waters draining from bog soils. He reported that natural, unpolluted waters draining from surface and bog soils near Petersburg, Alaska, show pH levels below 4.7. He attributed the low pH values to leaching of organic acids.

Gorham et al. (1984b) reported that where acidic deposition is severe, the acidity of bog waters can be dominated by mineral acids. They cited, as an example, two bogs in the southern Pennine Mountains of England, where the waters were light coloured but high in hydrogen (160 μeq L^{-1}) and sulphate ions (273 μeq L^{-1}). Vangenechten et al. (1984) made similar observations for 25 poorly buffered campine-bog pools in the province of Antwerp, Belgium. A strong correlation between hydrogen and the excess sulphate, high aluminum concentration (8000 μg L^{-1}), and unusually high calcium content (6400 μg L^{-1}) in these bog pools suggested that acidification was due to atmospheric input. However, lack of information on precipitation pH, geological and soil characteristics, and hydrological conditions makes interpretation of their results difficult.

Patrick et al. (1981) suggested that naturally-occurring acidic lakes fall into two groups. The first group occurs in areas of igneous rock or sand substrata. Lakes of this type are generally small with poorly buffered waters and low conductivity. Such lakes generally have a pH of 5.6-5.7 and are susceptible to acidification by carbonic acid. Acid runoff can depress the pH of these lakes even further. These lakes are even susceptible to plankton metabolic activity and, because of this, pH can change by as much as four units within a period of two hours (Allen 1972, cited by Marcus et al. 1983).

The second group of lakes occurs in geographic locations similar to the first group, but is associated with peat and bog vegetation. Such lakes can have pH levels as low as 3.7 (Gorham et al. 1984b). In some bogs, pH as low as 3.0 has also been reported (Wetzel 1975).

Bog acidification can occur in three ways involving: (a) cation exchange by sphagnum (*Sphagnum* sp.); (b) organic acids; and (c) oxidation of reduced sulphur compounds (Patrick et al. 1981; Gorham et al. 1984b). *Sphagnum* sp. contain polygalacturonic acids. These plant species can remove metal cations from precipitation, replacing them with hydrogen ions from the polygalacturonic acids, making bogs more acidic. Organic acids that contribute acidity to bog waters are mainly "humic" substances resulting from organic decay. Patrick et al. (1981) reported that although organic acids can contribute to bog acidity, it is the sulphate, derived from the oxidation of reduced sulphur through the biological activity in the sediments, that causes low pH in bog waters.

Paleolimnological records suggest that natural processes have been acidifying lakes and ponds for at least 10,000 years. Whitehead et al. (1981) estimated historical pH levels of 3.8 in Adirondack lakes by studying sediment diatoms. Based on their analyses, they suggested that all lakes were basic in the late glacial period (pH 7.8), and became more acidic during the early Holocene (pH $\leq$6). Diatom analyses suggested that in some lakes pH levels below 5 have existed for the past 2000 to 10,000 years. Whitehead et al. hypothesized that historical depressions in pH were related to changes in terrestrial vegetation, which, in turn, depressed pH levels in soils. Sediment core pollen analyses indicated that lake pH levels decreased concurrently with the movement of hemlock and fir into watersheds.

However, Battarbee and Charles (1986) pointed out that of 11 Adirondack lakes studied, six of the seven lakes with a current pH of 5.0 or less showed clear evidence of recent pH decreases. The remaining four lakes (current pH $>$5.2) indicated no substantial pH decrease. Data on ^{210}Pb indicate that the most rapid pH decreases (0.5 to 1.0 units) occurred between 1930 and 1970. Battarbee and Charles (1986) suggested that although the primary cause of recent acidification appears to be atmospheric deposition of acidic substances, at some sites the role played by watershed disturbance in acidification needs to be taken into account.

Ford (1981) investigated sediment cores of two ponds; one was located in New Hampshire and the other in Vermont. Diatom analyses indicated that the New Hampshire pond had been acidic for the last 10,000 years. Analyses of the Vermont pond indicated that the alkalinity in that pond declined to zero about 5000 years ago. The diatom analyses indicated that pH levels of both ponds decreased around 5000 years ago.

A diatom stratigraphy and reconstruction study of 13 New England lakes suggested that for some seven lakes, pH declines started between 1920 and 1950 (Battarbee and Charles 1986). These lakes are situated in areas receiving the greatest acid loading, suggesting deposition of acidic substances as the cause of the pH decline. For two additional lakes, the decline appears to be caused by heavy timber harvesting in the catchment areas. The four remaining lakes showed no change in diatom-inferred pH since the 1770's.

Most of the paleolimnological data suggest that acidification of many surface waters is caused by internal generation of hydrogen ions within watersheds and that atmospheric deposition may be the secondary source of acidification of surface waters. Table 2.38 provides a summary of watershed characteristics that influence surface water susceptibility to acidification.

Table 2.38 Watershed characteristics that influence surface water susceptibility to acidification.

Category	Increased Susceptibility	Decreased Susceptibility
Bedrock geology	Resistant to weathering (metamorphic, igneous)	Easily weathered (sedimentary, calcite-containing)
Soils		
Buffering capacity	Low	High
Depth	Shallow	Deep
SO_4 adsorption capacity	Low	High

Continued...

Table 2.38 (Concluded).

Category	Increased Susceptibility	Decreased Susceptibility
Topography	Steep-sloped	Shallow-sloped
Watershed to surface water area ratio	Low	High
Lake flushing rate	High	Low
Watershed vegetation and land use		
Dominant vegetation	Coniferous	Deciduous
Forest management	Reforestation	Clearcutting
Water quality		
Alkalinity	Low (<200 μeq L^{-1})	High (>200 μeq L^{-1})
Trophic status	Highly oligotrophic	Less oligotrophic, mesotrophic, eutrophic
Cultural eutrophication	Forestry	Agriculture, municipal
Humic substances	Absent	Present
Sphagnum moss	Present	Absent
Sulfate reduction potential	Low	High
Climate/meteorology		
Precipitation	High	Low
Snow accumulation	High	Low
Growing season	Short	Long
Alkaline dusts	Low	High

Source: Marcus et al. (1983).

A number of investigators have studied the effects of increasing acidity on aquatic ecosystems. Results of some of these are summarized in Table 2.39.

2.6 CONCLUDING REMARKS

Acidic (acidic precipitation) and acidifying (SO_x and NO_x) pollutants in the atmosphere and their observed and/or predicted impacts on the sensitive environmental components locally and on a regional scale are of great concern at this time nationally and internationally. Much knowledge has been gained on the atmospheric chemical transformation of SO_2 and the photo-oxidation cycle. However, a similar knowledge of the conversion of NO_2 to NO_3^- is somewhat limited. As with the SO_2 chemistry, air pollutant

Table 2.39 Effects of increasing acidity on aquatic ecosystems.

Taxon or Process	Type of Evidence: Field Observation	Field Experiment	Lab Experiment	Observed Effects
BENTHOS				
Molluscs (most species except fingernail clams, family *Sphaeriidae*)	-	-		The calcareous shell of these animals is soluble under acidic conditions, making this group highly sensitive to low pH. Few species present below pH 6.0 except for several species of fingernail clams which may persist down to pH 4.5-5.0.
Crayfish	Almer et al. (1978)	Mills (1982)	Malley (1980) *	In soft water lakes, calcium uptake and exo-skeleton formation inhibited in pH range 5.0-5.8. Reproduction impaired at pH 5.4
Amphipods (*Gammarus*)	Okland (1980c) Sutcliffe & Carrick (1973)	-	Costa (1967) Borgstrom & Hendrey (1976)	Absent below pH 6.0. In the laboratory avoids pH 6.2 and lower.
Mayfly larvae (*Ephemeroptera*)		Hall et al. (1980)	Bell & Nebeker (1969) Bell (1971)	Most species decline or are absent in pH range 4.5 to 5.5.

Continued...

Table 2.39 (Continued).

Taxon or Process	Type of Evidence			Observed Effects
	Field Observation	Field Experiment	Lab Experiment	
Water striders (*Gerridae*), back-swimmers (*Notonectidae*), waterboatmen (*Corixidae*), beetles (*Dytiscidae*, *Gyrinidae*), dragonflies (*Odanata*).	*	-	-	Tolerant of acidity. Increase in abundance in acidified lakes (below pH 5.0) after other invertebrate groups and fish have been eliminated.
Benthos community structure	*	Hall et al. (1980)	Bell & Nebeker (1969) Bell (1971)	With increasing acidity, species richness declines. Entire groups of aquatic organisms are absent or poorly represented below pH 5.0 (e.g., molluscs, amphipods, crayfish, may-flies). Other taxa become dominant, particularly after loss of fishes (e.g., predacious beetles and true bugs).
Benthic algae (periphyton)	*	Hall et al. (1980)	Hendrey (1976) Schindler (1980)	Algal mats overgrow rooted plants and cover bottom substrates in acidified lakes below pH 5.0

Continued...

Table 2.39 (Continued).

Taxon or Process	Type of Evidence			Observed Effects
	Field Observation	Field Experiment	Lab Experiment	
MACROPHYTES				
Eriocaulon sp. *Lobelia* sp.	Grahn (1977) Best & Peverly (1981) Miller et al. (1982)	-	Laake (1976)	Rosette plant communities may become overgrown by algal mats. Tissue aluminum concentrations increase as pH decreases. Photosynthesis of rosette species decreases by 75% as pH declines from 5.5 to 4.0.
Daphnia	*	-	Davis & Ozburn (1969) Parent & Cheetham (1980)	Most species are acid-sensitive and absent below pH 7.0 to 5.5.
Zooplankton community structure	*	-	-	The number of species declines as acidity increases. Taxa characteristic of acid conditions include certain genera of rotifers (*Keratella*, *Kellicottia*, *Polyarthra*); cladocerans (*Bosmina*); and copepods (*Biaptonus*).

Continued...

Table 2.39 (Continued).

Taxon or Process	Type of Evidence: Field Observation	Field Experiment	Lab Experiment	Observed Effects
Phytoplankton community structure	*	Yan & Stokes (1978)	-	The number of species declines as acidity increases. Dinoflagellates (Phylum *Pyrrophyta*) frequently dominate acidified lakes (pH 4.0-5.0). Dinoflagellates are a less palatable food source for zooplankton compared with the phytoplankton they frequently replace.
FISHES				
Fathead Minnow (*Pimephales promelas*)	Rahel & Magnuson (1983)	Mills (1982)	Mount (1973)	One of the most acid-sensitive fish species. Reproductive failure occurs near pH 6.0. Generally absent in waters below pH 6.5.
Darters (*Etheostoma exile*, *E. nigrum*, *Percina capnodes*) and Minnows (several *Notropis* spp. *Pimephales notatus*).	Harvey (1980)	-	Rahel & Magnuson (1983)	Very acid-sensitive. Generally absent below pH 6.0 in both naturally acidic and anthropogenically acidified waters.

Continued...

Table 2.39 (Continued).

Taxon or Process	Type of Evidence: Field Observation	Field Experiment	Lab Experiment	Observed Effects
Smallmouth Bass (*Micropterus dolomieui*)	Beamish (1976) Harvey (1980), Rahel & Magnuson (1983)	-	-	Reproduction ceases and populations become extinct below pH 5.2-5.5.
Lake Trout (*Salvelinus namayeusch*)	Beamish (1976) Beamish et al. (1975)	Mills (1982)	Beamish (1972) Trojnar (1977a)	Experiences reproductive failure near pH 5.0. Generally absent below pH 5.0 in both naturally acidic and anthropogenically acidified waters.
White Sucker (*Catostomus commersoni*)	Harvey (1980) Rahel & Magnuson (1983)	Mills (1982)	-	
Rainbow Trout (*Salmo gairdneri*)			*	Adversely affected by pH below 5.0-5.5.
Atlantic Salmon (*Salmo salar*)	*	-		Adversely affected by pH below 5.0.
Brown Trout (*Salmo trutta*)	*	*	*	Lower pH limit between 4.5 to 5.0.
Brook Trout (*Salvelinus fontinalis*)		Hall et al. (1980)		Lower pH limit between 4.2 to 5.0.

Continued...

Table 2.39 (Concluded).

Taxon or Process	Type of Evidence: Field Observation	Type of Evidence: Field Experiment	Type of Evidence: Lab Experiment	Observed Effects
Sunfishes (*Ambloplites rupestris*, *Micropterus salmoides*, *Lepomis* spp.)	Harvey (1980) Rahel & Magnuson (1983)	Smith (1957)	-	Lower pH limit near 4.5.
Yellow Perch (*Perca flavescens*)	Svardson (1976) Keller et al. (1980) Harvey (1980) Rahel & Magnuson (1983)	-	Rahel (1983)	Lower pH limit 4.2 to 4.5. May become very abundant after other species have become extinct.
DECOMPOSITION	Hendrey (1976) Leivestad et al. (1976)	Scheider et al. (1976) Gahnstrom et al. (1980) Hall et al. (1980)	Leivestad et al. (1976)	Bacterial decomposition is significantly reduced in the pH range 4.0 to 5.0. In many cases, fungi replace bacteria as the primary decomposers.

* numerous research
\- not studied
Source: Magnuson (1983).

transport and deposition across international boundaries has also been the subject of intensive research.

In the past several years, numerous papers have been published on wet deposition of air pollutants, but corresponding publications on dry deposition have not attained a similar momentum. A great deal of preoccupation appears to exist at this time concerning the characterization of dry deposition velocities and pollutant fluxes, but there exists a lack of effects or response models sensitive enough to respond to such precise measurements.

Many response models are driven by long-term pollutant averaging techniques. While soils may require many, many years to respond to the atmospheric deposition of air pollutants, other receptors and processes, such as vegetation and stream chemistry, appear to depend upon pollutant episodicity and much shorter response time. Such relationships, however, are stochastic in nature.

Long term pollutant averaging techniques, and the subsequent computation of deposition values, are statistically invalid since many pollutants under ambient conditions exhibit "non-normal" frequency distributions. However, numerous response models have not considered this, due to a lack of the required communication between disciplines.

While ambient wet deposition by itself does not appear to cause adverse effects on vegetation (crops and forests, with the possible exception of those at high elevations) and with some exceptions, soils, there appears to be a debate regarding its effects on surface waters. This controversy is mainly centered on the mechanisms of surface water acidification.

At least among the vegetation effects scientists, there has been an immense interest to perform experiments with simulated acidic rain (SAR). However, these studies are of little value since they have no resemblance to ambient phenomena. Unfortunately, many of the studies on the joint effects of multiple pollutants have failed to consider the ambient patterns (real time series) of the occurrences of such pollutants on a site specific basis.

As opposed to those scientists interested in health and visibility effects, others interested in responses of various ecosystem components appear to have almost totally ignored the significance of fine particle aerosols and vapours. Where these pollutants are considered, the methodology used is inadequate.

While field chambers, greenhouse, controlled environment, and experimental simulation studies are valuable in gaining fundamental knowledge of response processes, in the end, under ambient conditions, cause and effect relationships must be viewed holistically. Such an analysis can only be performed by the intensive qualitative and quantitative characterization of all important atmospheric variables (both physical and chemical) and response variables in common space and time. Equally important is a need to develop assessive and predictive models which can be driven by these valuable data in establishing cause and effects relationships.

2.7 LITERATURE CITED

Addison, P.A., S.S. Malhotra, and A.A. Khan. 1984. Effect of sulfur dioxide on woody boreal forest species grown on native soils and tailings. *Journal of Environmental Quality* 13(3): 333-336.

Adedipe, N.O. and D.P. Ormrod. 1974. Ozone-induced growth suppression in radish plants in relation to pre- and post-fumigation temperatures. *Zeitschrift Pflanzenphysiologie* 71: 281-287.

Adedipe, N.O., R.A. Fletcher, and D.P. Ormrod. 1973. Ozone lesions in relation to senescence of attached and detached leaves of tobacco. *Atmospheric Environment* 7: 357-361.

Alfheim, I., M. Stobet, N. Gjos, A. Bjorseth, and S. Wilhelmsen. 1980. An analysis of organic micropollutants in aerosols. In: **Ecological Impact of Acid Precipitation,** Proceedings of an International Conference, eds. D. Drabløs and A. Tollan. 1980 March 11-14; Sandefjord, Norway; SNSF Project. Oslo, Norway; pp. 100-101.

Allen, H.L. 1972. Phytoplankton photosynthesis, micronutrient interactions, and inorganic carbon availability in a soft-water Vermont lake. In: **Nutrients and Eutrophication, the Limiting-Nutrient Controversy**, ed. G.E. Likens. Special Symposium, Vol. 1. American Society of Limnology and Oceanography. Lawrence, Kansas. pp. 68-83. (Original not seen; information taken from Marcus et. al. 1983.)

Almer, B., W. Dickson, C. Ekstrom, and E. Hornstrom. 1978. Sulfur pollution and the aquatic ecosystem. In: **Sulfur in the Environment.** Part II, ed. J. Nriagu. New York: John Wiley and Sons, pp. 273-311.

Ashenden, T.W. and I.A.D. Williams. 1980. Growth reductions in *Lolium multiflorum* Lam. and *Phleum pratense* L. as a result of sulfur dioxide and nitrogen dioxide pollution. *Environmental Pollution*, Series A 21: 131-139.

Bache, B.W. 1980. The acidification of soils. In: **Effects of Acid Precipitation on Terrestrial Ecosystems.** NATO Conference Series. Volume 4, eds. T.C. Hutchinson and M. Havas. New York: Plenum Press, pp. 183-202.

Barrie, L.A. 1981. The prediction of rain acidity and SO_2 scavenging in eastern North America. *Atmospheric Environment* 15: 31-41.

Battarbee, R.W. and D.F. Charles. 1986. Diatom-based pH reconstruction studies of acid lakes in Europe and North America: A synthesis. *Water, Air, and Soil Pollution* 30: 347-354.

Beamish, R.J. 1972. Lethal pH for the white sucker *Catostomus commersoni* (Lacepede). *Transactions of the American Fisheries Society* 101: 355-358.

Beamish, R.J. 1976. Acidification of lakes in Canada by acid precipitation and the resulting effects on fish. *Water, Air, and Soil Pollution* 6: 501-514.

Beamish, R.J., W.L. Lockhart, J.C. Van Loon, and H.H. Harvey. 1975. Long-term acidification of a lake and resulting effects on fishes. *Ambio* 4: 98-102.

Bell, G.R. 1971. Effect of low pH on the survival and emergence of aquatic insects. *Water Resources* 5: 313-319.

Bell, H.L. and A.V. Nebeker. 1969. Primary studies on the tolerance of aquatic insects to low pH. *Journal of the Kansas Entomological Society* 42(2): 230-237.

Bennett, J.P. and V.C. Runeckles. 1977. Effects of low levels of ozone on plant competition. *Journal of Applied Ecology* 14: 877-880.

Berger, A., J.L. Melice, and Cl. Demuth. 1982. Statistical distributions of daily and high atmospheric SO_2 concentrations. *Atmospheric Environment* 16: 2863-2877.

Best, M.D. and J.H. Peverly. 1981. Water and sediment chemistry and elemental composition of macrophytes in thirteen Adirondack lakes (abstract). In: **Proceedings of International Symposium on Acidic Precipitation and Fishery Impacts in North Eastern North America**, North Eastern Division American Fisheries Society. 1981 August 2-5; Cornell University, Ithaca, New York; p. 348.

Bielski, B.H.J, D.E. Cabelli, R.L. Arudi, and A.B. Ross. 1985. Reactivity of $H_2O_2/O_2^{\cdot}$ radicals in aqueous solutions. *Journal of Physical Chemistry Reference Data* 14: 1041-1100.

Bjarnborg, B. 1983. Dilution and acidification effects during the spring flood of four Swedish mountain brooks. *Hydrobiologia* 101: 19-26.

Blum, U. and W.W. Heck. 1980. Effects of acute ozone exposure on snap bean at various stages of its life cycle. *Environmental and Experimental Botany* 20: 73-85.

Borgstrom, R. and G.R. Hendrey. 1976. pH tolerance of the first larval stages of *Lepidurus arcticus* (Pallas) and adult *Gammarus lacustris* G.O. Sars. SNSF Project, Internal Report 22/76. Oslo-As, Norway. 37 pp.

Bowersox, V.C. and R.G. de Pena. 1980. Analysis of precipitation chemistry at a central Pennsylvania site. *Journal of Geophysical Research* 85: 5614-5620.

Brady, N.C. 1974. *The Nature and Properties of Soils*. New York: Macmillan Publishing. 639 pp.

Buttazzoni, C., I. Lavagnini, A. Marani, F.Z. Grandi, and A. Del Turco. 1986. Probability model for atmospheric sulphur dioxide concentrations in the area of Venice. *Journal of the Air Pollution Control Association* 36: 1028-1030.

Cadle, S.H., J.M. Dasch, and N.E. Grossnickle. 1984. Retention and release of chemical species by a northern Michigan snowpack. *Water, Air, and Soil Pollution* 22: 303-319.

Caiazza, R., K.D. Hage, and D. Gallup. 1978. Wet and dry deposition of nutrients in Central Alberta. *Water, Air, and Soil Pollution* 9: 309-314.

Cass, G.R. and R.H. Shair. 1984. Sulfate accumulation in a sea breeze/land breeze circulation system. *Journal of Geophysical Research* 89(D1): 1429-1438.

Chamberlain, A.C. 1986. Deposition of gases and particles on vegetation and soils. In: **Air Pollutants and Their Effects on the Terrestrial Ecosystem,** eds. A.H. Legge and S.V. Krupa. New York: John Wiley and Sons, pp. 189-209.

Chamberlain, J., H. Foley, D. Hammer, G. MacDonald, D. Rothaus, and M. Ruderman. 1981. The physics and chemistry of acid precipitation. Technical Report JSR-81-25. SRI International, Palo Alto, California. 195 pp.

Chang, T.Y., J.M. Norbeck, and B. Weinstock. 1979. An estimate of the NO_x removal rate in an urban atmosphere. *Environmental Science and Technology* 13: 1534-1537.

Chapman, E.G., D.S. Sklaren, and J.S. Flickinger. 1986. Organic acids in springtime Wisconsin precipitation samples. *Atmospheric Environment* 20: 1717-1725.

Charlson, R.J. and H. Rodhe. 1982. Factors controlling the acidity of natural rainwater. *Nature* 295: 683-685.

Chaudry, M., M. Nyborg, M. Molina-Ayala, and R.W. Parker. 1982. The reaction of SO_2 emissions with soils. In: **Acid Forming Emissions in Alberta and Their Ecological Effects.** Symposium-Workshop Proceedings. Alberta Environment, Canadian Petroleum Association, and the Oil Sands Environmental Study Group. 1982 March 9-12; Edmonton, Alberta; pp. 415-433.

Cheng, L., A. Davis, E. Peake, and D. Rogers. 1987. The Use of Aircraft Measurements to Determine Transport, Dispersion and Transformation Rates of Pollutants Emitted from Oil Sands Extraction Plants in Alberta. Final Report to Research Management Division, Alberta Environment, Edmonton, Alberta. 117 pp.

Chevone, B.I., D.E. Herzfeld, S.V. Krupa, and A.H. Chappelka. 1986. Direct effects of atmospheric sulfate deposition on vegetation. *Journal of the Air Pollution Control Association* 36: 813-816.

Cogbill, C. and G. Likens. 1974. Acid precipitation in northeastern United States. *Water Resources Research* 190: 1133-1137.

Committee on the Atmosphere and the Biosphere. 1981. *Atmosphere-Biosphere Interactions: Toward a Better Understanding of the Ecological Consequences of Fossil Fuel Combustion.* Washington, D.C.: National Academy Press. 263 pp.

Costa, H.H. 1967. Responses of *Gammarus pulex* (L.) to modified environment II. Reactions to abnormal hydrogen ion concentrations. *Crustaceana* 13: 1-10.

Crutzen, P.J. 1974. Photochemical reactions by and influencing ozone in the troposphere. *Tellus* 26: 47-57.

Cullis, C.F. and M.M. Hirschler. 1980. Atmospheric sulphur: Natural and man-made sources. *Atmospheric Environment* 14: 1263-1278.

Dana, M.T. and R.C. Easter. 1987. Statistical summary and analyses of event precipitation chemistry from the MAP3S network, 1976-1983. *Atmospheric Environment* 21: 113-127.

Dana, M.T., J.M. Hales, and M.A. Wolf. 1975. Rain scavenging of SO_2 and sulphate from power plant plumes. *Journal of Geophysical Research* 80: 4119-4129.

Daubenmire, R. 1968. *Plant Communities.* New York: Harper. 300 pp.

Davidson, C.I. and Y. Wu. 1988. Dry Deposition of Particles and Vapors. In: **Acid Deposition, Volume 2**. Sources, Emissions, and Mitigation, ed. D.C. Adriana. Advances in Environmental Sciences Series. New York: Springer-Verlag (in press).

Davis, D.D. and R.G. Wilhour. 1976. Susceptibility of woody plants to sulfur dioxide and photochemical oxidants. U.S. Environmental Protection Agency Publication No. EPA-600/3-76-102, Corvallis, Oregon, 71 pp.

Davis, P. and G. Ozburn. 1969. The pH tolerance of *Daphnia pulex* (Leydig, emend., Richard). *Canadian Journal of Zoology* 47: 1173-1175.

Dillon, P.J., D.S. Jeffries, and W.A. Scheider. 1982. The use of calibrated lakes and watersheds for estimating atmospheric deposition near a large point source. *Water, Air, and Soil Pollution* 18: 241-258.

Dillon, P.J., D.S. Jeffries, W. Snyder, R. Reid, N.D. Yan, D. Evans, J. Moss, and W.A. Scheider. 1978. Acidic precipitation in south-central Ontario: recent observations. *Journal of the Fisheries Research Board of Canada* 35: 809-815.

Dreisinger, B.R. and P.C. McGovern. 1970. Monitoring SO_2 and correlating its effects on crops and forestry in the Sudbury area. In: **Proceedings of the Conference on Impact of Air Pollution on Vegetation,** ed. S.N. Linzon. 1970 April 7-9; Toronto, Ontario; pp. 12-28.

EPA, U.S. Environmental Protection Agency. 1978. Effects of photochemical oxidants on vegetation and certain microorganisms. In: **Air Quality Criteria for Ozone and Other Photochemical Oxidants.** pp. 253-259.

EPA, U.S. Environmental Protection Agency. 1983. The acidic deposition phenomenon and its effects. EPA 600/8-83-016B. pp. 1-1 to 7-49.

EPA, U.S. Environmental Protection Agency. 1986. Air quality criteria for ozone and other photochemical oxidants, Volume 3. EPA/600/8-84-020 cF. 6-1 to 8-56 pp.

Eatough, D.J., R.J. Arthur, N.L. Eatough, M.W. Hill, N.F. Mangelson, B.E. Richter, L.D. Hansen, and J.A. Cooper. 1984. Rapid conversion of $SO_{2(g)}$ to sulphate in a fog bank. *Environmental Science and Technology* 18: 855-859.

Elzerman, A.W. 1983. Effects of acid deposition (rain) on a piedmont aquatic ecosystem: acid inputs, neutralization, and pH changes. Final Technical Completion Report B-141-SC to Bureau of Reclamation, U.S. Department of the Interior, Washington, D.C. 79 pp.

Engle R.L. and W.H. Gabelman. 1966. Inheritance and mechanism of resistance to ozone damage to onion, *Allium cepa* L. *Proceedings of the American Society of Horticultural Science* 89: 423-430.

Engle, R.L. and W.H. Gabelman. 1967. The effects of low levels of ozone on pinto beans, *Phaseolus vulgaris* L. *Proceedings of American Society of Horticultural Science* 91: 304-309.

Evans, L.S. 1973. Bean leaf growth response to moderate ozone levels. *Environmental Pollution* 4: 17-26.

Finlayson-Pitts, B.J. and J.N. Pitts. 1986. *Atmospheric Chemistry: Fundamentals and Experimental Techniques*. New York: John Wiley and Sons. 1098 pp.

Ford, J. 1981. Interglacial perspectives on ecosystem acidification. *Bulletin of Ecological Society of America* 62(2): 154.

Forrest, J., R.W. Garber, and L. Newman. 1981. Conversion rates in power plant plumes based on filter pack data: The coal-fired Cumberland plume. *Atmospheric Environment* 15: 2273-2282.

Fowler, D. 1980. Wet and dry deposition of sulphur and nitrogen compounds from the atmosphere. In: **Effects of Acid Precipitation on Terrestrial Ecosystems.** NATO Conference Series. Volume 4, eds. T.C. Hutchinson and M. Havas. New York: Plenum Press, pp. 9-27.

Fowler, D. and J.N. Cape. 1982. Air pollutants in agriculture and horticulture. In: **Effects of Gaseous Air Pollution in Agriculture and Horticulture,** eds. M.H. Unsworth and D.P. Ormrod. London: Butterworth Scientific, pp. 3-26.

Fowler, D. and J.N. Cape. 1984. On the episodic nature of wet deposited sulphate and acidity. *Atmospheric Environment* 18: 1859-1866.

Fujiwara, T. 1975. Studies on the development of injury symptoms caused by sulphur dioxide at low level concentrations in plants and the diagnosis of injury. *Bulletin of the Agricultural Laboratory CRIEPI,* No. 74001.

Gahnstrom, G., G. Andersson, and S. Fleischer. 1980. Decomposition and exchange processes in acidified lake sediment. In: **Ecological Impact of Acid Precipitation, Proceedings of an International Conference,** eds. D. Drabløs and A. Tollan. 1980 March 11-14; Sandefjord, Norway; SNSF Project, Oslo, Norway; pp. 306-307.

Galloway, J.N. and G.E. Likens. 1981. Acid precipitation: The importance of nitric acid. *Atmospheric Environment* 15: 1081-1085.

Galloway, J.N. and G.G. Parker. 1980. Difficulties in measuring wet and dry deposition on forest canopies and soil surfaces. In: **Effects of Acid Precipitation on Terrestrial Ecosystems.** NATO Conference Series Volume 4, eds. T.C. Hutchinson and M. Havas. New York: Plenum Press, pp. 57-68.

Galloway, J.N. and D.M. Whelpdale. 1980. An atmospheric sulfur budget for eastern North America. *Atmospheric Environment* 14: 409-417.

Galloway, J.N., G.E. Likens, and E.S. Edgerton. 1976. Acid precipitation in northeastern United States: pH and acidity. *Science* 194: 722-724.

Galloway, J.N., C.L. Schofield, G.R. Hendrey, N.E. Peters, and A.H. Johannes. 1980. Sources of acidity in three lakes acidified during snowmelt. In: **Ecological Impact of Acid Precipitation, Proceedings of International Conference,** eds. D. Drabløs and A. Tollan. 1980 March 11-14; Sandefjord, Norway; SNSF Project, Oslo, Norway; pp. 264-265.

Garland, J.A. 1978. Dry and wet removal of sulfur from the atmosphere. *Atmospheric Environment* 12: 349-362.

Garrels, R.M. and F.T. MacKenzie. 1971. *Evolution of Sedimentary Rocks*. New York: Norton. 397 pp.

Gherini, S.A., C.W. Chen, L. Mok, R.A. Goldstein, R.J.M. Hudson, and G.F. Davis. 1984. The ILWAS Model: Formulation and Application in the Integrated Lake-Watershed Acidification Study, Volume 4. Summary of Major Results. Electric Power Research Institute. EA-3221. pp. 7-1 to 7-45.

Ghiorse, W.C. and M. Alexander. 1976. Effect of microorganisms on the sorption and fate of sulfur dioxide and nitrogen dioxide in soil. *Journal of Environmental Quality* 50: 227-230.

Gillani, N.V. 1978. Project MISTT: Meso scale plume modelling of the dispersion, transformation, and ground removal of SO_2. *Atmospheric Environment* 12: 569-588.

Gmur, N.F., L.S. Evans, and E.A. Cunningham. 1983. Effects of ammonium sulfate aerosols on vegetation. II. Mode of entry and responses of vegetation. *Atmospheric Environment* 17: 715-721.

Gorham, E. and W.W. McFee. 1980. Effects of acid deposition upon outputs from terrestrial to aquatic ecosystems. In: **Effects of Acid Precipitation on Terrestrial Ecosystems,** eds. T.C. Hutchinson and M. Havas. NATO Conference Series 1: Ecology, Volume 4. New York: Plenum Press, pp. 465-480.

Gorham, E., F.B. Martin, and J.T. Litzau. 1984a. Acid rain: Ionic correlations in the eastern United States, 1980-1981. *Science* 225: 407-409.

Gorham, E., S.E. Bayley, and D.W. Schindler. 1984b. Ecological effects of acid deposition upon peatlands: A neglected field in "acid rain" research. *Canadian Journal of Fisheries and Aquatic Sciences* 41: 1256-1268.

Gorham, E., P.M. Vitousek, and W.A. Reiners. 1979. The regulation of chemical budgets over the course of terrestrial ecosystem succession. *Annual Review of Ecological Systems* 10: 53-84.

Graedel, T.E. and K.I. Goldberg. 1983. Kinetic studies of raindrop chemistry. I. Inorganic and organic processes. *Journal of Geophysical Research* 88: 865-882.

Graedel, T.E., M.L. Mandich, and C.J. Weschler. 1986. Kinetic model studies of atmospheric droplet chemistry. II. Homogeneous transition metal chemistry in raindrops. *Journal of Geophysical Research* 91(D4): 5021-5205.

Grahn, O. 1977. Macrophyte succession in Swedish lakes caused by deposition of airborne acid substances. *Water, Air, and Soil Pollution* 7: 295-305.

Groterud, O. 1981. The dynamics of acidification from a new view elucidated by studies in a lake area of Norway. *Verhandlungen Internationale Vereinignung für Theoretische und Angewandte Limnologie* 21(1): 406-411.

Guderian, R. 1977. *Air Pollution - Phytotoxicity of Acidic Gases and its Significance in Air Pollution Control.* New York: Springer-Verlag. 127 pp.

Guderian, R. 1985. *Air Pollution by Photochemical Oxidants.* New York: Springer-Verlag. 346 pp.

Guderian, R. and H. Strattmann. 1968. Freilandversuche zur Ermittlung von Schwefeldioxidwirkungen auf die Vegetation. III. Teil Grenzwerte schadlicher SO_2 Immissionen fur Obstund Forstkulturen sowie fur landwirtschaftliche und gartnerische Pflanzenarten. Forschungsberichte des Landes Nordrhein Westfalen. 114 pp.

Guiang, S.F., S.V. Krupa, and G.C. Pratt. 1984. Measurement of S(IV) and organic anions in Minnesota rain. *Atmospheric Environment* 18: 1677-1682.

Haines, T.A. 1981. Acidic precipitation and its consequences for aquatic ecosystems: A review. *Transactions of the American Fisheries Society* 110: 669-707.

Hakkarinen, C. and M.A. Allan. 1986. Forest Health and Acidic Deposition. EPRI EA-4813-SR. Palo Alto, California. pp. 1-1 to 6-4.

Hall, R.J., G.E. Likens, S.B. Fiance, and G.R. Hendrey. 1980. Experimental acidification of a stream in the Hubbard Brook Experimental Forest, New Hampshire. *Ecology* 61(4): 976-989.

Hansen, D.A., G.M. Hidy, and G.J. Stensland. 1981. Examination of the basis for trend interpretation of historical rain chemistry in the eastern United States. Environmental Research and Technology Incorporation. Document P-A097R. pp. 1-1 to A-1.

Harte, J., G.P. Lockett, R.A. Schneider, H. Michaels, and C. Blanchard. 1985. Acid precipitation and surface-water vulnerability on the western slope of the high Colorado Rockies. *Water, Air, and Soil Pollution* 25: 313-320.

Harvey, H.H. 1980. Widespread and diverse changes in the biota of North American lakes and rivers coincident with acidification. In: **Ecological Impact of Acid Precipitation, Proceedings of an International Conference,** eds. D. Drabløs and A. Tollan. 1980 March 11-14; Sandefjord, Norway; SNSF Project, Oslo, Norway; pp. 93-98.

Heagle, A.S. and J.W. Johnston. 1979. Variable responses of soybeans to mixtures of ozone and sulfur dioxide. *Journal of the Air Pollution Control Association* 29: 729-732.

Heagle, A.S., D.E. Body, and G.E. Neely. 1974. Injury and yield responses of soybeans to chronic doses of ozone and sulfur dioxide in the field. *Phytopathology* 64: 132-136.

Heagle, A.S., D.E. Body, and E.K. Punds. 1972. Effect of ozone on yield of sweet corn. *Phytopathology* 62: 583-687.

Heagle, A.S., R.B. Philbeck, H.H. Rogers, and M.B. Letchworth. 1979a. Dispensing and monitoring ozone in open-top field chambers for plant-effects studies. *Phytopathology* 69: 15-20.

Heagle, A.S., S. Spencer, and M.B. Letchworth. 1979b. Yield response of winter wheat to chronic doses of ozone. *Canadian Journal of Botany* 57: 1999-2005.

Heagle, A.S., V.M. Lesser, J.O. Rawlings, W.W. Heck, and R.B. Philbeck. 1986. Responses of soybeans to chronic doses of ozone applied as constant or proportional additions to ambient air. *Phytopathology* 76: 51-56.

Hegg, D.A. and P.V. Hobbs. 1979. Some observations of particulate nitrate concentrations in coal-fired power plant plumes. *Atmospheric Environment* 13: 1715-1716.

Heggestad, H.E. 1973. Photochemical air pollution injury to potatoes in the Atlantic coastal states. *American Potato Journal* 50: 315-328.

Hendrey, G. 1976. Effects of low pH on the growth of periphytic algae in artificial stream channels. SNSF Project Report IR 2576, Oslo, Norway.

Hendrey, G.R., J.N. Galloway, and C.L. Schofield. 1980. Temporal and spatial trends in the chemistry of acidified lakes under ice cover. In: **Ecological Impact of Acid Precipitation, Proceedings of an International Conference,** eds. D. Drabløs and A. Tollan. 1980 March 11-14; Sandefjord, Norway; SNSF Project, Oslo, Norway; pp. 266-267.

Hepting, G.H. 1968. Diseases of forest and tree crops caused by air pollutants. *Phytopathology* 58: 1098-1101.

Herzfeld, D.E. 1982. Interactive effects of sub-micron sulfuric acid aerosols and ozone on soybean and pinto bean. St. Paul, Minnesota: University of Minnesota. 105 pp. M.Sc. Thesis.

Hicks, B.B. 1984. The acidic deposition phenomenon and its effect. A-7 Dry deposition processes. Critical Assessment Review Papers, Vol. 1, Atmospheric Sciences, EPA-600/8-83-016AF, pp. 7-1 to 7-70.

Hidy, G.M. and P.K. Mueller. 1986. The sulfur oxide - particulate matter complex. In: **Air Pollutants and Their Effects on the Terrestrial Ecosystem,** eds. A.H. Legge and S.V. Krupa. New York: John Wiley and Sons. pp. 51-104.

Hill, A.C., H.E. Heggestad, and S.N. Linzon. 1970. Ozone. In: **Recognition of Air Pollution Injury to Vegetation: A Pictorial Atlas,** eds. J.S. Jacobson and A.C. Hill. Pittsburgh, Pennsylvania: Air Pollution Control Association. pp. B1-B22.

Hoffman, G.J., E.V. Mass, and S.L. Rawlins. 1973. Salinity-ozone interactive effects on yield and water relations of pinto bean. *Journal of Environmental Quality* 2: 148-152.

Hornbeck, J.W., G.E. Likens, and J.S. Eaton. 1977. Seasonal patterns in acidity of precipitation and their implications for forest stream ecosystems. *Water, Air, and Soil Pollution* 7: 355-365.

Horntvedt, R., G.J. Dollard, and E. Joranger. 1980. Effects of acid precipitation on soil and forest, SNSF-project, Norway 2. Atmosphere-vegetation interactions. In: **Ecological Impact of Acid Precipitation, Proceedings of an International Conference,** eds. D. Drabløs and A. Tollan. 1980 March 11-14; Sandefjord, Norway; SNSF Project, Oslo, Norway; pp. 192-193.

Horsman, D.C., A.O. Nicholls, and D.M. Calder. 1980. Growth responses of *Dactylis glomerata, Lolium perenne* and *Phalaris aquatica* to chronic ozone exposure. *Australian Journal of Plant Physiology* 7: 511-517.

Hosker, R.P. and S.E. Lindberg. 1982. Review: atmospheric deposition and plant assimilation of gases and particles. *Atmospheric Environment* 16: 889-910.

Hsu, H. and G.W. Hodgson. 1977. Organic compounds of sulphur: Initial data for soils and streams in Alberta. In: **Proceedings of Alberta Sulphur Gas Research Workshop III,** eds. H.S. Sandhu and M. Nyborg. 1977 November 17-18; University of Alberta, Edmonton. Edmonton, Alberta: Research Secretariat, Alberta Environment; pp. 246-263.

Hultberg, H. 1977. Thermally stratified acid water in late winter--a key factor inducing self-accelerating processes which increase acidification. *Water, Air, and Soil Pollution* 7: 279-294.

Hunt, J.E., R.G. Wright, and R.L. Desjardins. 1982. A measurement system for SO_2 dry deposition rates. In: **Proceedings. Acid Forming Emissions in Alberta and Their Ecological Effects.** eds. H.S. Sandhu, J.R. Clements, and B.L. Magill. 1982 March 9-12; Edmonton, Alberta; pp. 121-152.

Husar, R.B., J.P. Lodge, Jr., and D.J. Moore (eds.). 1978. *Sulphur in the Atmosphere*. New York: Pergamon Press. 816 pp.

International Electric Research Exchange. 1981. Effects of SO_2 and its derivatives on health and ecology. Volume 2 - Natural ecosystems, agriculture and fisheries. 291 pp.

Irving, P.M. 1983. Acidic precipitation effects on crops: a review and analysis of research. *Journal of Environmental Quality* 12: 442-453.

Irving, P.M. 1985. Modeling the response of greenhouse-grown radish plants to acidic rain. *Environmental and Experimental Botany* 25: 327-328.

Jacobson, J.S. 1984. Effects of acidic aerosol, fog, mist and rain on crops and trees. *Philosophical Transactions of the Royal Society of London* 305: 327-338.

Jeffries, D.S., C.M. Cox, and P.J. Dillon. 1979. The depression of pH in lakes and streams in central Ontario during snowmelt. *Journal of the Fisheries Research Board of Canada* 36: 640-646. (Original not seen; information taken from Marcus et al. 1983.)

Johannes, A.H. and E.R. Altwicker. 1980. Atmospheric Inputs to Three Adirondack Lake Watersheds. In: **Ecological Impact of Acid Precipitation, Proceedings of an International Conference,** eds. D. Drabløs and A. Tollan. 1980 March 11-14; Sandefjord, Norway; SNSF Project, Oslo, Norway; pp. 256-257.

Johannes, A.H., E.R. Altwicker, and N.L. Clesceri. 1984. Atmospheric inputs to the ILWAS lake watershed in the Integrated Lake-Watershed Acidification Study. Vol. 4. Summary of major results. Prep. by Tetra Tech Inc., Lafayette, California, pp. 2-13 to 2-14.

Johannes, A.H., J.N. Galloway, and D.E. Troutman. 1980. Snowpack storage and ion release. In: **Ecological Impact of Acid Precipitation, Proceedings of an International Conference**, eds. D. Drabløs and A. Tollan. 1980 March 11-14; Sandefjord, Norway; SNSF Project, Oslo, Norway; pp. 260-261.

Johannessen, M., A. Skartveit, and R.F. Wright. 1980. Streamwater chemistry before, during, and after snowmelt. In: **Ecological Impact of Acid Precipitation, Proceedings of an International Conference,** eds. D. Drabløs and A. Tollan. 1980 March 11-14; Sandefjord, Norway; SNSF Project, Oslo, Norway; pp. 224-225.

Johnson, D.W. 1981. The natural acidity of some unpolluted waters in southeastern Alaska and potential impact of acid rain. *Water, Air, and Soil Pollution* 16: 243-252.

Johnson, D.W., G.S. Henderson, D.D. Huff, S.E. Lindberg, D.D. Richter, D.S. Shriner, D.E. Todd, and J. Turner. 1982. Cycling of organic and inorganic sulphur in a chestnut oak forest. *Oecologia* 54: 141-148.

Jones, H.C., F.P. Weatherford, and J.C. Noggle. 1979. Power plant siting: assessing risks of SO_2 effects on agriculture. *Proceedings of the 72nd Annual Conference of the Air Pollution Control Association.* 1979 June 24-29, Cincinnati, Ohio.

Jonsson, B. 1977. Soil acidification by atmospheric pollution and forest growth. *Water, Air, and Soil Pollution* 7: 497-501.

Junge, C.E. 1963. *Air Chemistry and Radioactivity.* New York, London: Academic Press. 382 pp.

Kasina, S. 1980. On precipitation acidity in southern Poland. *Atmospheric Environment* 14: 1217-1221.

Katz, M. and G.A. Ledingham. 1939. Effects of SO_2 on yield of barley and alfalfa. In: **Effects of SO_2 on Vegetation.** National Research Council Canada, Publication No. 815. Ottawa. 457 pp.

Kawamura, K. and I.R. Kaplan. 1983. Organic compounds in the rainwater of Los Angeles. *Environmental Science and Technology* 17: 497-501.

Keene, W.C. and J.N. Galloway. 1984. Organic acidity in precipitation of North America. *Atmospheric Environment* 18(11): 2491-2497.

Keller, W., J. Gunn, and N. Conroy. 1980. Acidification impacts on lakes in the Sudbury, Ontario, Canada area. In: **Ecological Impact of Acid Precipitation, Proceedings of an International Conference,** eds. D. Drabløs and A. Tollan. 1980 March 11-14; Sandefjord, Norway; SNSF Project, Oslo, Norway; pp. 228-229.

Khatamian, H., N.O. Adedipe, and D.P. Ormrod. 1973. Soil-plant-water aspects of ozone phytotoxicity in tomato plants. *Plant and Soil* 38: 531-541.

Klemm, R.F. and J.M.L. Gray. 1982. Acidity and chemical composition of precipitation in Central Alberta, 1977-1978. In: **Proceedings Acid Forming Emissions in Alberta and Their Ecological Effects**, eds. H.S. Sandhu, J.R. Clements, and B.L. Magill. 1982 March 9-12; Edmonton, Alberta; pp. 153-179.

Knapp, W.W., V.C. Bowersox, B.I. Chevone, S.V. Krupa, J.A. Lynch, and W.W. McFee. 1988. Precipitation Chemistry in the United States: Summary of Ion Concentration Variability 1978-1984. Technical Bulletin, Water Resources Research Institute, Cornell University, Ithaca, New York. 165 pp.

Kochhar, M., U. Blum, and R.A. Reinert. 1980. Effects of O_3 on fescue and (or) ladino clover: Interactions. *Canadian Journal of Botany* 58: 241-249.

Kociuba, P.J. 1984. Estimate of sulphate deposition in precipitation for Alberta. Atmospheric Environment Service, Environment Canada, Edmonton, Alberta. 6 pp.

Kress, L.W. and J.E. Miller. 1981. Impact of ozone on soybean yield. Argonne National Laboratory. Radiological and Environmental Research Division Annual Report, Part III Ecological. 1980. January-December. pp. 11-14.

Krug, E.C. and C.R. Frink. 1983a. Acid rain on acid soil: A new perspective. *Science* 221: 520-525.

Krug, E.C. and C.R. Frink. 1983b. Effects of acid rain on soil and water. Bulletin 811. The Connecticut Agricultural Experiment Station, New Haven. November 1983. 45 pp.

Krupa, S.V. 1987. Responses of alfalfa to sulfur dioxide exposures from the emissions of the NSP-SHERCO Coal-Fired Power Plant Units 1 and 2. Final report to the Northern States Power Company, Minneapolis, Minnesota, 255 pp.

Krupa, S.V. and A.H. Legge. 1986. Single or Joint Effects of Coarse and Fine Particle Sulfur Aerosols and Ozone on Vegetation. In: **Aerosols: Research, Risk Assessment and Control Strategies.** Proceedings of the Second U.S.-Dutch International Symposium, Williamsburg, Virginia, U.S.A., eds. S.D. Lee, T. Schneider, L.D. Grant and P.J. Verkerk. 1985. Chelsea, Michigan: Lewis Publishers Inc., 1221 pp.

Krupa, S.V. and W.J. Manning. 1988. Atmospheric ozone: Formation and effects on vegetation. *Environmental Pollution* 50(1): 101-137.

Krupa, S.V., M.R. Coscio, Jr., and F.A. Wood. 1976. Evidence for a multiple hydrogen-ion donor system in rain. *Water, Air, and Soil Pollution 6*:415-422.

Krupa, S.V., J.P. Lodge, Jr., M. Nosal, and G.E. McVehil. 1987. Characteristics of aerosols and rain chemistry in north central U.S.A. In: **Proceedings of International Conference on Acidic Rain: Scientific and Technical Advances,** Lisbon, Portugal, eds. R. Perry, R.M. Harrison, J.N.B. Bell, and J.N. Lester. London: Selper Ltd., pp. 121-128.

Laake, M. 1976. Effekter av lav pH på produksjon, nedbrytning og stoffkretslop i littoralsonen. SNSF Project. Internal Report 29/76. Oslo-As, Norway.

Larson, T.V., R.J. Charlson, E.J. Knudson, G.D. Christian, and H. Harrison. 1975. The influence of a single sulphur dioxide point source on the rain chemistry of a single storm in the Puget Sound region. *Water, Air, and Soil Pollution* 4: 319-328.

Lau, Y.K. 1985. A 5-year (1978-1982) summary of precipitation chemistry measurements in Alberta. Pollution Control Division, Alberta Environment, Edmonton, Alberta.

Lau, Y.K. and N.C. Das. 1985. Precipitation quality monitoring in Alberta. In: **Impact of Air Toxics on the Quality of Life.** Proc. 1985 Annual Meeting, CPANS/PNWIS Sections, Air Pollution Control Association. Pittsburgh, Pennsylvania. pp. 213-233.

Law, R.M. and T.A. Mansfield. 1982. Oxides of nitrogen and the greenhouse atmosphere. In: **Effects of Gaseous Air Pollution in Agriculture and Horticulture,** eds. M.H. Unsworth and D.P. Ormrod. London: Butterworth Scientific, pp. 93-112.

Lefohn, A.S. and H.M. Benedict. 1982. Development of mathematical index that describes ozone concentration, frequency and duration. *Atmospheric Environment* 16: 2529-2532.

Lefohn, A.S. and S.V. Krupa. 1984. Sulfate as a surrogate for hydrogen in rainfall - An analysis of rainfall chemistry data. Draft Report U.S. EPA, Washington, D.C. 60 pp.

Lefohn, A.S. and S.V. Krupa. 1988. The relationship between hydrogen and sulfate ions in precipitation - a numerical analysis of rain and snowfall chemistry. *Environmental Pollution* 49(4): 289-311.

Legge, A.H. 1980. Primary productivity, sulfur dioxide, and the forest ecosystem: an overview of a case study. In: **Proceedings of the Symposium on Effects of Air Pollutants on Mediterranean and Temperate Forest Ecosystems.** 1980 June 22-27; Riverside, California; General Technical Report PSW-43, Pacific Southwest Forest and Range Experiment Station, Forest Service, U.S. Department of Agriculture, Berkeley, California; pp. 51-62.

Legge, A.H. and S.V. Krupa, eds. 1986. *Air Pollutants and Their Effects on the Terrestrial Ecosystem.* New York: John Wiley and Sons. 662 pp.

Legge, A.H., G.W. Harvey, P.F. Lester, D.R. Jaques, H.R. Krouse, J. Mayo, A.P. Hartgerink, R.G. Amundson, and R.B. Walker. 1976. Quantitative assessment of the impact of sulfur gas emissions on a forest ecosystem. A final report submitted to the Whitecourt Environmental Study Group, March 1976. 159 pp.

Legge, A.H., D.R. Jaques, H.R. Krouse, E.C. Rhodes, H.U. Schellhase, J. Mayo, A.P. Hartgerink, P.F. Lester, R.G. Amundson, and R.B. Walker. 1978. Sulphur gas emissions in the boreal forest: The West Whitecourt case study. A final report submitted to the Whitecourt Environmental Study Group. October 1978. 615 pp.

Legge, A.H., D.R. Jaques, G.W. Harvey, H.R. Krouse, H.M. Brown, E.C. Rhodes, M. Nosal, H.U. Schellhase, J. Mayo, A.P. Hartgerink, P.F. Lester, R.G. Amundson, and R.B. Walker. 1981. Sulphur gas emissions in the boreal forest: The West Whitecourt case study I. Executive Summary. *Water, Air, and Soil Pollution* 15:77-85.

Leivestad, H., G. Hendrey, I.P. Muniz, and E. Snekvik. 1976. Effects of acidic precipitation on freshwater organisms. In: **Impact of Acid Precipitation on Forest and Freshwater Ecosystems in Norway**. SNSF Research Report No. 6, ed. F.H. Braekke. Oslo, Norway: SNSF Project. pp. 87-111.

Likens, G.E. and F.H. Bormann. 1974. Acid rain: A serious environmental problem. *Science* 184: 1176-1179.

Likens, G.E. and T.J. Butler. 1981. Recent acidification of precipitation in North America. *Atmospheric Environment* 15: 1103-1109.

Likens, G.E, F.H. Bormann, and J.S. Eaton. 1980. Variations in precipitation and stream water chemistry at the Hubbard Brook Experimental Forest during 1964-1977. In: **Effects of Acid Precipitation on Terrestrial Ecosystems, Proceedings of the NATO Conference on Effects of Acid Precipitation on Vegetation and Soils,** eds. T.C. Hutchinson, and M. Havas. 1978 May 21-27; Toronto, Canada; New York: Plenum Press. pp. 443-464.

Likens, G.E., F.H. Bormann, R.S. Pierce, J.S. Eaton, and N.M. Johnson. 1977. *Biogeochemistry of a Forested Ecosystem*. New York: Springer-Verlag. 146 pp.

Lindberg, S.E., G.M. Lovett, D.D. Richter, and D.W. Johnson. 1986. Atmospheric deposition and canopy interactions of major ions in a forest. *Science* 231: 141-145.

Logan, J.A. 1983. Nitrogen oxides in the troposphere: global and regional budgets. *Journal of Geophysical Research* 88: 10,785-10,807.

Lovett, G.M. and S.E. Lindberg. 1984. Dry deposition and canopy exchange in a mixed oak forest as determined by analysis of throughfall. *Journal of Applied Ecology* 21: 1013-1027.

Lunde, G.J., N.G. Gether, and M. Lande. 1977. Organic micropollutants in precipitation in Norway. *Atmospheric Environment* 11: 1007-1014.

MAP3S/RAINE Research Community. 1982. The MAP3S/RAINE precipitation chemistry network: Statistical overview for the period 1976-1980. *Atmospheric Environment* 16: 1603-1631.

Maas, E.V., G.J. Hoffman, and G. Ogata. 1973. Salinity-ozone interactions on pinto bean: Integrated response to ozone concentration and duration. *Journal of Environmental Quality* 2: 400-404.

Madsen, B.C. 1981. Acid rain at Kennedy Space Center, Florida: Recent observations. *Atmospheric Environment* 15: 853-862.

Magnuson, J. J. 1983. Effects on aquatic biology. In: **The Acidic Deposition Phenomenon and its Effects. Critical Assessment Review Papers. Volume II. Effects Sciences.** U.S. Environmental Protection Agency, Washington, D. C. EPA-600/8-83-016B. pp. 5-1 to 5-203.

Male, L.M. 1982. An experimental method for predicting a plant yield response to pollution time series. *Atmospheric Environment* 16: 2247-2252.

Malley, D.F. 1980. Decreased survival and calcium uptake by the crayfish *Oronectes virilis* in low pH. *Canadian Journal of Fisheries and Aquatic Sciences* 37: 364-372.

Manning, W.J. 1978. Chronic foliar ozone injury: Effects on plant root development and possible consequences. *California Air Environment* 7: 3-4.

Manning, W.J., W.A. Feder, P.M. Papia, and I. Perkins. 1971. Influence of foliar ozone injury on root development and root surface fungi of pinto bean plants. *Environmental Pollution* 1: 305-312.

Marcus, M.D., B.R. Parkhurst, and F.E. Payne. 1983. An assessment of the relationship among acidifying depositions, surface water acidification, and fish populations in North America. Volume 1. Prepared by Western Aquatics, Inc. for Electric Power Research Institute, as EA-3127, Research Project 1910-2. 105 pp.

Martin, A. and F.R. Barber. 1977. Some observations on acidity and sulphur in rainwater from rural sites in central England and Wales. *Atmospheric Environment* 12: 1481-1487.

Mason, J. and H.M. Seip. 1985. The current state of knowledge on acidification of surface waters and guidelines for further research. *Ambio* 14: 45-51.

Mayo, J.M. 1987. The Effects of Acid Deposition on Forests. Prep. for the Alberta Government-Industry Acid Deposition Research Program by the Department of Biology, Emporia State University, Emporia, Kansas. ADRP-B-09-87. 80 pp.

McGill, W.B., A.H. Maclean, L.W. Turchenek, and C.A. Gale. 1980. Interim report of soil research related to revegetation of the oil sands area. Prep. for Alberta Oil Sands Environmental Research Program by the Department of Soil Science, University of Alberta, Edmonton. AOSERP Report O.F. 7. 181 pp.

McNaughton, D.J. 1981. Relationships between sulfate and nitrate ion concentrations and rainfall pH for use in modelling applications. *Atmospheric Environment* 15: 1075-1079.

Mészáros, E. 1981. *Atmospheric Chemistry. Fundamental Aspects*. Amsterdam: Elsevier Scientific Publishing Company. 201 pp.

Meurrens, A. 1974. Sulfites et sulfates dan la pluie. Paper presented at the COST Technical Symposium, Ispra, Yugoslavia. (As cited by J.A. Garland, 1978).

Meyers, P.A. and R.A. Hites. 1982. Extractable organic compounds in Midwest rain and snow. *Atmospheric Environment* 16: 2169-2175.

Miles, L.J. and K.H. Yost. 1982. Quality analysis of USGS precipitation chemistry data for New York. *Atmospheric Environment* 16: 2889-2898.

Miller, G.E., I. Wile, and G.G. Hitchin. 1982. Patterns of accumulation of selected metals in members of the soft-water macrophyte flora of central Ontario lakes. *Aquatic Botany* 15: 53-64.

Miller, H.G. 1983. Studies of proton flux in forests and heaths in Scotland. In: **Effects of Accumulation of Air Pollutants in Forest Ecosystems,** eds. B. Ulrich and J. Pankrath. Boston, Massachusetts: D. Reidel, pp. 183-193.

Miller, P.R. 1973. Oxidant-induced community change in a mixed conifer forest. *Advances in Chemistry Series* 122: 101-117.

Miller, P.R. and J.R. McBride. 1975. Effects of Air Pollutants on Forests. In: **Responses of Plants to Air Pollution,** eds. J.B. Mudd and T.T. Kozlowski. New York: Academic Press. pp. 195-235.

Mills, K.H. 1982. Fish population responses during the experimental acidification of a small lake. Presented at American Chemical Society Meeting, Las Vegas, Nevada. 1982 March.

Mount, D.I. 1973. Chronic effect of low pH on fathead minnow survival, growth and reproduction. *Water Resources* 7: 987-993.

National Academy of Sciences. 1977. *Ozone and other photochemical oxidants.* Committee on Medical and Biological Effects of Environmental Pollutants. Washington, D.C. 719 pp.

Neely, G.E., D.T. Tingey, and R.G. Wilhour. 1977. Effects of ozone and sulfur dioxide singly and in combination on yield, quality, and N-fixation of alfalfa. In: **International Conference on Photochemical Oxidant Pollution and Its Control,** ed. B. Dimitriades. EPA-600/3-77-001b, U.S. Environmental Protection Agency, Research Triangle Park, North Carolina. pp. 663-673.

Nihlgard, B. 1985. The ammonium hypothesis - an additional explanation to the forest dieback in Europe. *Ambio* XIV(1): 2-8.

Norby, R.J. and T.T. Kozlowski. 1981a. Interactions of SO_2 concentrations and post fumigation temperature on growth of five species of woody plants. *Environmental Pollution* 25: 27-39.

Norby, R.J. and T.T. Kozlowski. 1981b. Relative sensitivity of three species of woody plants to SO_2 at high or low exposure temperature. *Oecologia* 51(1): 33-36.

Nosal, M. 1984. Atmosphere-biosphere interface: analytical design and a computerized regression model for lodgepole pine response to chronic, atmospheric SO_2 exposure. RMD Report 83/26 and 83/27. Research Management Division, Alberta Environment, Edmonton, Alberta. 97 pp.

Nosal, M. and S.V. Krupa. 1986. Numerical methodology in the risk assessment of air pollutant-induced ecological effects. Proceedings of the Annual Meeting of the Air Pollution Control Association, Pittsburgh, Pennsylvania 86-32.1:1-16.

Nyborg, M., J. Crepin, D. Hocking, and J. Baker. 1977. Effect of sulphur dioxide on precipitation and on the sulphur content and acidity of soils in Alberta, Canada. *Water, Air, and Soil Pollution* 7: 439-448.

Nyborg, M., R.W. Parker, L.W. Hodgins, D.H. Laverty, and S. Tayki. 1980. Soil acidification by SO_2 emissions in Alberta, Canada. In: **Ecological Impact of Acid Precipitation, Proceedings of an International Conference,** eds. D. Drabløs and A. Tollan. 1980. March 11-14; Sandefjord, Norway; SNSF Project, Oslo, Norway; pp. 180-181.

Oden, S. 1976. The acidity problem. An outline of concepts. *Water, Air, and Soil Pollution* 6: 137-166.

Ogata, G. and E.V. Maas. 1973. Interactive effects of salinity and ozone on growth and yield of garden beet. *Journal of Environmental Quality* 2: 518-520.

Okland, K.A. 1980. Okologi og utbredelse til *Gammarus lacustris* G. O. Sars i Norge, med vekt pa forsuringsproblemer. SNSF Project, Internal Report 67/80. Oslo-As, Norway.

Olson, M.P. and E.C. Voldner. 1982. AES-LRD Model Profile. U.S./Canada Memorandum of Intent on Transboundary Air Pollution, Work Group 1, Atmospheric Science and Analysis, Report 2-5.

Ormrod, D.P., N.O. Adedipe, and G. Hofstra. 1971. Responses of cucumber, onion, and potato cultivars to ozone. *Canadian Journal of Plant Science* 51: 283-288.

Oshima, R.J. 1973. Effect of ozone on a commercial sweet corn variety. *Plant Disease Reporter* 57: 719-723.

Oshima, R.J., M.P. Poe, P.K. Braegelmann, D.W. Balding, and V. van Way. 1976. Ozone dosage-crop loss function for alfalfa: a standardized method for assessing crop losses from air pollutants. *Journal of the Air Pollution Control Association* 26: 861-865.

Overrein, L.N., H.M. Seip, and A. Tollan. 1980. Acid precipitation effects on forests and fish. Final Report of the SNSF-Project 1972-1980. Research Report 19. As, Norway. 175 pp.

Panel on Processes of Lake Acidification: Environmental Studies Board; Commission on Physical Sciences, Mathematics, and Resources. 1984. Acid deposition: Processes of lake acidification. Available from the National Technical Information Service as PB84-216175. 11 pp.

Parent, S. and R. Cheetham. 1980. Effects of acid precipitation on *Daphnia magna. Bulletin of Environmental Contamination and Toxicology* 25: 298-304.

Patrick, R., V.P. Binetti, and S.G. Halterman. 1981. Acid lakes from natural and anthropogenic causes. *Science* 211: 446-448.

Pell, E.J., W.C. Weissberger, and J.J. Speroni. 1980. Impact of ozone on quantity and quality of greenhouse-grown potato plants. *Environmental Science and Technology* 14: 568-571.

Peters, N.E., R. Schroeder, and D. Troutman. 1982. Temporal trends in the acidity of precipitation and surface waters of New York. U.S. Geological Survey, Water Supply Paper 2188.

Powers, E.B. 1929. Fresh water studies. 1. The relative temperature, oxygen content, alkali reserve, the carbon dioxide tension and pH of the waters of certain mountain streams at different altitudes in the Smoky Mountain National Park. *Ecology* 10(1): 97-111.

Pratt, G.C. 1982. Effects of Ozone and Sulfur Dioxide and Soybeans. St. Paul, Minnesota: University of Minnesota. 152 pp. Ph.D. Thesis.

Pratt, G.C. and S.V. Krupa. 1983. Seasonal trends in precipitation chemistry. *Atmospheric Environment* 17: 1845-1847.

Pratt, G.C., and S.V. Krupa. 1985. Aerosol chemistry in Minnesota and Wisconsin and its relation to rainfall chemistry. *Atmospheric Environment* 19: 962-971.

Pratt, G.C., M.R. Coscio, and S.V. Krupa. 1984. Regional rainfall chemistry in Minnesota and West Central Wisconsin. *Atmospheric Environment* 18: 173-182.

Pratt, G.C., M.R. Coscio, D.W. Gardner, B.I. Chevone, and S.V. Krupa. 1983. An analysis of the chemical properties of rain in Minnesota. *Atmospheric Environment* 17: 347-355.

Pratt, G.C., R. C. Hendrickson, B.I. Chevone, D.A. Christopherson, M.V. O'Brien, and S.V. Krupa. 1983. Ozone and oxides of nitrogen in the rural upper-midwestern U.S.A. *Atmospheric Environment* 17: 2013-2023.

Prinz, P. 1987. Causes of forest damage in Europe. *Environment* 29(9): 10-15; 32-37.

Pruppacher, H.R., R.G. Semonin, and W.G Slinn. 1983. *Precipitation Scavenging, Dry Deposition, and Resuspension*. New York: Elsevier Science Publishing Co. Vol. 1, 729 pp. Vol. 2, 1462 pp.

Rahel, F.J. 1983. Population differences in acid tolerance between yellow perch, *Perca flavescens,* from naturally acidic and alkaline lakes. *Canadian Journal of Zoology* 61: 147-152.

Rahel, F.J. and J.J. Magnuson. 1983. Low pH and the absence of fish species in naturally acidic Wisconsin lakes: Inferences for cultural acidification. *Canadian Journal of Fisheries and Aquatic Sciences* 40: 3-9.

Ramberg, L. 1981. Increase in stream pH after a forest drainage. *Ambio* 10(1): 34-35.

Reuss, J.O. 1980. Simulation of soil nutrient losses resulting from rainfall acidity. *Ecological Modelling* 11: 15-38.

Reuss, J.O. and D.W. Johnson. 1986. *Acid Deposition and the Acidification of Soils and Waters.* New York: Springer-Verlag. 119 pp.

Richards, L.W., J.A. Anderson, D.L. Blumenthal, A.A. Brandt, J.A. MacDonald, N. Watus, E.S. Macias, and P.S. Bhardwaja. 1981. The chemistry, aerosol physics, and optical properties of a western coal-fired power plant plume. *Atmospheric Environment* 15: 2111-2134.

Roedel, W. 1979. Measurements of sulfuric acid saturation vapor pressure: Implications for aerosol formation by heteromolecular nucleation. *Journal of Aerosol Science* 10: 375-386.

Rosenqvist, I.Th. 1978a. Acid precipitation and other possible sources for acidification of rivers and lakes. *The Science of the Total Environment* 10: 271-272.

Rosenqvist, I.Th. 1978b. Alternative sources for acidification of river water in Norway. *The Science of the Total Environment* 10: 39-49.

Rosenqvist, I.Th. 1980. Influence of forest vegetation and agriculture on the acidity of fresh water. *Advances in Environmental Science and Engineering* 3: 56-79.

Rosenqvist, I.Th., P. Jorgensen, and H. Reuslatten. 1980. The importance of natural H^+ production for acidity in soil and water. In: **Ecological Impact of Acid Precipitation, Proceedings of an International Conference,** eds. D. Drabløs and A. Tollan. 1980 March 11-14; Sandefjord, Norway; SNSF Project, Oslo, Norway; pp. 240-241.

Sandhu, H.S., H.P. Sims, R.A. Hursey, W.R. MacDonald, and B.R. Hammond. 1980. Environmental Sulphur Research in Alberta: A Review. Research Secretariat. Edmonton: Alberta Department of Environment. 90 pp.

Scheider, W.A., J. Jones, and B. Cave. 1976. A preliminary report on the neutralization of Nelson lake near Sudbury, Ontario. Ontario Ministry of the Environment. Rexdale, Ontario.

Schertz, T.L. and R.M. Hirsch. 1985. Trend Analysis of Weekly Acid Rain Data - 1978-1983. Water Resources Investigations Report 85-4211. U.S. Geological Survey. 64 pp.

Schindler, D.W. 1980. Experimental acidification of a whole lake: A test of the oligotrophication hypothesis. In: **Ecological Impact of Acid Precipitation, Proceedings of an International Conference,** eds. D. Drabløs and A. Tollan. 1980 March 11-14; Sandefjord, Norway; SNSF Project, Oslo, Norway; pp. 370-374.

Schofield, C.L. 1976a. Acid precipitation: Effects on fish. *Ambio* 5(5-6): 228-230.

Schofield, C.L. 1976b. Lake acidification in the Adirondack mountains of New York: Causes and consequences. In: **Proceedings of the First International Symposium on Acid Precipitation and the Forest Ecosystem.** U.S. Department of Agriculture, USDA Forest Service General Technical Report NE-23, eds. L.S. Dochinger and T.A. Seliga. 1975 May 12-15; Columbus, Ohio; 477 pp.

Schofield, C.L. 1984. Surface water chemistry in the ILWAS basins. In: **The Integrated Lake-Watershed Acidification Study.** Volume 4. Summary of Major Results. Electric Power Research Institute, Report EPRI EA-3221, pp. 6-1 to 6-31.

Sehmel, G.A. 1980. Particle and gas dry deposition: A review. *Atmospheric Environment* 14: 983-1011.

Seip, H.M. 1980. Acid snow-snowpack chemistry and snowmelt. In: **Effects of Acid Precipitation on Terrestrial Ecosystems,** eds. T.C. Hutchinson and M. Havas. NATO Conference Series 1: Ecology, Volume 4. New York: Plenum Press. pp. 77-94.

Sequeira, R.A. 1982. Acid rain: An assessment based on acid-base considerations. *Journal of the Air Pollution Control Association* 32: 241-245.

Shannon, J.G. and C.L. Mulchi. 1974. Ozone damage to wheat varieties at anthesis. *Crop Science* 14: 335-337.

Shertz, R.D., W.J. Kender, and R.C. Musselman. 1980. Effects of ozone and sulfur dioxide on grapevines. *Science and Horticulture* 13: 37-45.

Shinohara, T., Y. Yamamoto, H. Kitano, and M. Fukuda. 1974. Interactions of light and ozone injury in tobacco. *Proceedings of Crop Science Society of Japan* 43: 433-438.

Shugart, H.H. and S.B. McLaughlin. 1985. Modeling SO_2 Effects on Forest Growth and Community Dynamics. In: **Sulfur Dioxide and Vegetation,** eds. W.E. Winner, H.A. Mooney, and R.A. Goldstein. California: Stanford University Press, pp. 478-491.

Skeffington, R.A. 1983. Soil properties under three species of trees in southern England in relation to acid deposition in through-fall. In: **Effects of Accumulation of Air Pollutants in Forest Ecosystems,** eds. B. Ulrich and J. Pankrath. Boston, Massachusetts: D. Reidel. pp. 219-231.

Smith, R.F. 1957. Lakes and ponds. Fishery survey report number 3. New Jersey Department of Conservation and Economic Development, Division of Fish and Game. Trenton, New Jersey. 198 pp.

Snedecor, G.W. and W.G. Cochran. 1978. *Statistical Methods.* Ames, Iowa: Iowa State University Press, 593 pp.

Sollins, P., C.C. Grier, F.M. McCorison, K. Cromack, and R. Fogel. 1980. The internal element cycles of an old-growth Douglas-fir ecosystem in western Oregon. *Ecological Monographs* 50: 261-285.

Somers, K.M. and H.H. Harvey. 1984. Alteration of fish communities in lakes stressed by acid deposition and heavy metals near Wawa, Ontario. *Canadian Journal of Fisheries and Aquatic Sciences* 41: 20-29.

Spicer, C.W. 1982a. The distribution of oxidized nitrogen in urban air. *Science of the Total Environment* 24: 183-192.

Spicer, C.W. 1982b. Nitrogen oxide reactions in the urban plume of Boston. *Science* 215: 1095-1097.

Stensland, G.J., D.M. Whelpdale, and G. Oehlert. 1986. Precipitation chemistry. In: **Acid Deposition. Long Term Trends.** Washington, D.C.: National Academy Press. pp. 128-199.

Stevens, R.K., T.G. Dzubay, G. Russwarm, and D. Rickel. 1978. Sampling and analysis of atmospheric sulfates and related species. *Atmospheric Environment* 12: 55-68.

Strachan, W. and H. Huneault. 1979. Polychlorinated biphenyls and organochlorine pesticides in Great Lakes precipitation. *Journal of Great Lakes Research* 5: 61-68.

Sutcliffe, D.W. and T.R. Carrick. 1973. Studies on mountain streams in the English Lake District. *Freshwater Biology* 3: 437-462.

Svardson, G. 1976. Interspecific population dominance in fish communities of Scandinavian lakes. Report of the Institute of Freshwater Research at Drottningholm 55: 144-171.

Tabatabai, M.A. 1985. Effect of acid rain on soils. *Critical Reviews in Environmental Control* 15: 65-110.

Teasley, J.I., ed. 1984. *Acid Precipitation Series.* Vol. 1-9. Boston, Massachusetts: Butterworth Scientific Publishers.

Telang, S.A. 1987. Surface Water Acidification Literature Review. Prep. for the Alberta Government-Industry Acid Deposition Research Program by the Kananaskis Centre for Environmental Research. The University of Calgary, Calgary, Alberta. ADRP-B-01-87. 123 pp.

Tetra Tech, Inc. 1984. The Integrated Lake-Watershed Acidification Study. Volume 3. Lake chemistry program. Electric Power Research Institute Report, Lafayette, California. EPRI EA-3221, pp. 7-1 to 7-7.

Thomas, M.D. 1935. Absorption of sulfur dioxide by alfalfa and its relation to leaf injury. *Plant Physiology* 10: 291-307.

Tingey, D.T., J.A. Dunning, and G.M. Jividen. 1973. Radish root growth reduced by acute ozone exposures. In: **Proceedings of the Third International Clean Air Congress.** 1973 October 8-12; Dusseldorf, Germany; pp. A154-A156.

Tomlinson, G., R. Brouzes, R. McLean, and J. Kadlecek. 1980. The role of clouds in atmospheric transport of mercury and other pollutants. In: **Ecological Impact of Acid Precipitation, Proceedings of an International Conference,** eds. D. Drabløs and A. Tollan. 1980 March 11-14; Sandefjord, Norway; SNSF Project, Oslo, Norway; pp. 134-137.

Torn, M.S., J.E. Degrange, and J.H. Shinn. 1987. The Effects of Acidic Deposition on Alberta Agriculture: A Review. Prep. for the Alberta Government-Industry Acid Deposition Research Program by the Environmental Sciences Division, Lawrence Livermore National Laboratory, Livermore, California. ADRP-B-08-87. 160 pp.

Trojnar, J.R. 1977. Egg and larval survival of white suckers *(Catostomus commersoni)* at low pH. *Journal of Fisheries Research Board of Canada* 34: 262-266.

Turchenek, L.W., S.A. Abboud, C.J. Tomas, R.J. Fessenden, and N. Holowaychuk. 1987. Effects of Acid Deposition on Soils in Alberta. Prep. for the Alberta Government-Industry Acid Deposition Research Program by the Alberta Research Council, Edmonton. ADRP-B-05-87. 202 pp.

U.S. NAPAP. The National Acid Precipitation Assessment Program. 1987. Refer to Volumes 1 through 4, Interim Assessment Report.

U.S. National Academy Press. 1986. *Acid Deposition. Long Term Trends*. Washington, D.C. 506 pp.

U.S. National Academy of Sciences. 1977. *Ozone and Other Photochemical Oxidants*. Committee on Medical and Biological Effects of Environmental Pollutants. Washington, D.C. 719 pp.

U.S. National Research Council. 1983. *Acidic Deposition. Atmospheric processes in eastern North America*. National Academy Press, Washington, D.C. 375. pp.

Unsworth, M.H. and D.P. Ormrod (eds.). 1982. *Effects of Gaseous Air Pollution in Agriculture and Horticulture*. London, United Kingdom: Butterworth Scientific. 532 pp.

Van Haut, H. and H. Stratmann. 1967. Experimentelle Untersuchungen über die Wirkung von Stickstoffdioxid auf Pflanzen. Immissions Bodennutzungssch des Landes Nordrhein-Westfalen 7: 50-70.

Vangenechten, J.H.D., S. Van Puymbroeck, and O.L.J. Vanderborght. 1984. Acidification in Campine bog lakes. In: **Belgian Research on Acid Deposition and the Sulphur Cycle,** ed. O.L.J. Vanderborght. Belgium: SCOPE Publication, pp. 251-262.

Voldner, E.C. and J.D. Shannon. 1983. Evaluation of predicted wet deposition fields of sulfur in eastern Canada. Transactions: The Meteorology of Acid Deposition, an APCA Specialty Conference. pp. 387-399.

Voldner, E.C., L.A. Barrie, and A. Sirois. 1986. A literature review of dry deposition of oxides of sulphur and nitrogen with emphasis on long-range transport modelling in North America. *Atmospheric Environment* 20: 2101-2123.

Watt, W.D., C.D. Scott, and S. Ray. 1979. Acidification and other chemical changes in Halifax county lakes after 21 years. *Limnology and Oceanography* 24(6): 1154-1161.

Weidensaul, T.C. and J.R. McClenahen. 1986. Soil-air pollutant interactions. In: **Air Pollutants and Their Effects on the Terrestrial Ecosystem,** eds. A.H. Legge and S.V. Krupa. New York: John Wiley and Sons, pp. 397-414.

Wetzel, R.G. 1975. *Limnology*. Philadelphia: W.B. Saunders Company. 743 pp.

Whitby, K.T. 1978. The physical characteristics of sulfur aerosols. *Atmospheric Environment* 12: 135-159.

Whitehead, D.R., S.E. Reed, and D.F. Charles. 1981. Late glacial and post glacial pH changes in Adirondack lakes. *Bulletin of Ecological Society of America* 62(2): 154.

Whitfield, M. 1979. Activity coefficients in natural waters. In: **Activity Coefficients in Electrolyte Solutions,** ed. R.M. Pytkowicz. Boca Raton, Florida: CRC Press, II: 153-299.

Wiklander, L. 1946. Studies on ionic exchange with special reference to the conditions in soils. *Kungliga Lantbrukshog Skolans Annaler* 14: 1-171. (Original not seen; information taken from Krug and Frink 1983b.)

Wilhour, R.G. and G.E. Neely. 1977. Growth response of conifer seedlings to low ozone concentrations. In: **International Conference on Photochemical Oxidant Pollution.** Proceedings: Volume II, 1976 September 12-17; Raleigh, North Carolina; U.S. Environmental Protection Agency Report EPA-600/3-77-0016; pp. 635-645.

Winner, W.E., H.A. Mooney, and R.A. Goldstein (eds.). 1985. *Sulfur Dioxide and Vegetation: Physiology, Ecology, and Policy Issues*. California: Stanford University Press. 593 pp.

Woodward, K.W. 1906. Forestry in Nova Scotia. *Journal of Forestry* 4: 10-13.

Wright, R.F. 1983. Input-output budgets at Langtjern, a small acidified lake in southern Norway. *Hydrobiologia* 101: 1-12.

Wright, R.F. and A. Henriksen. 1983. Restoration of Norwegian lakes by reduction in sulphur deposition. *Nature* 305: 422-424.

Yan, N. and P. Stokes. 1978. Phytoplankton of an acidic lake, and its responses to experimental alterations of pH. *Environmental Conservation* 5: 93-100.

Zahn, R. 1961. Wirkung von Schwefel dioxid auf die vegetation, Ergebnisse aus Begasungsversuchen. *Staub* 21: 56-60.

Zech, W., T. Suttner, and E. Popp. 1985. Elemental analysis and physiological responses of forest trees in SO_2 - polluted areas of NE-Bavaria. *Water, Air, and Soil Pollution* 25: 175-183.

CHAPTER 3

CHARACTERISTICS OF THE BACKGROUND AIR QUALITY

A. H. Legge, E. Peake, M. Strosher, M. Nosal, G. E. McVehil, and M. Hansen

3.1 INTRODUCTION

Certain chemical constituents (e.g., photochemical oxidants) occur naturally in the troposphere, the background air quality. From a mechanistic sense, background air quality is defined as that monitored at remote sites. While there is some question as to whether any location on the earth is truly untouched by human activities in some way, we refer to the atmosphere in such remote areas as natural troposphere (Finlayson-Pitts and Pitts 1986). The background concentrations of trace gases in this "clean" troposphere are essentially determined by the competitive physical, chemical, and biological processes occurring within atmospheric and geologic systems (Demerjian 1986). By contrast, in the urban polluted atmospheres, pollutant concentrations are determined predominantly by anthropogenic sources acted upon by physical and chemical processes over limited scales of time and space. Tables 3.1 and 3.2 provide a summary of the comparative concentrations of some trace gases in the clean troposphere and in the polluted atmosphere.

The chemistry of the clean troposphere and the mathematical simulation of the relevant physical and chemical processes have been reviewed extensively by Chameides and Walker (1973, 1976), Crutzen (1974), Fishman and Crutzen (1977), Levy (1971), Logan (1983), Logan et al. (1981), Stedman and Shetter (1983), Stewart et al. (1977), and Wofsy et al. (1972).

3.2 THE PHOTOCHEMISTRY OF THE CLEAN TROPOSPHERE

According to Demerjian (1986), the photochemistry of the unpolluted troposphere develops around a chain reaction sequence

Table 3.1 Concentrations of some trace gases in the clean troposphere and in typical urban polluted atmospheres.

Species	Clean Troposphere (ppb)	Polluted Atmosphere (ppb)	References
O_3	30	100-(200)	National Research Council (1976); Singh et al. (1978)
HO	1.0×10^{-5}-2.0×10^{-4}	4.1×10^{-4}-2.4×10^{-3}	Wang et al. (1975); Davis et al. (1976)
HO_2	10^{-4}-10^{-2}	0.1-0.2	Calvert and McQuigg (1975); Cox et al. (1976)
N_2O	330	--	Cicerone et al. (1978)
NO	0.01-0.05	60-740	Drummond (1977); Ritter et al. (1978); Air Quality Criteria for Oxides of Nitrogen (1980)
NO_2	0.1-0.5	40-220	Ritter et al. (1978); Noxon (1978); Air Quality Criteria for Oxides of Nitrogen (1980)
HONO	10^{-3}-10^{-1}	4-21	Nash (1974)
$HONO_2$	0.02-0.3	6-20	Huebert and Lazrus (1978); Doyle et al. (1979)
PAN[a]	1	10-65	Lonneman et al. (1976)
NH_3	0.1-1	20-80	Dawson (1977); Doyle et al. (1979)
NH_4NO_3	0.03-0.5	8-30	Doyle et al. (1979)
H_2	500	--	Schmidt (1974)
H_2O_2	0.1-1	5-40	Bufalini et al. (1972); Kok et al. (1978)
CH_4	1.6-1.7×10^3	2-3×10^3	Fink et al. (1964); Altshuller et al. (1973)
NMHC[b]	5-10	10^2-10^3	Robinson et al. (1973); Leonard et al. (1976)
H_2CO	0.1-1	10-40	Altshuller and McPherson (1963)
CO	50-200	10^2-10^4	Seiler (1974); EPA National Air Quality and Emissions Trends Report (1977)
CO_2	3.3×10^5	--	Lowe et al. (1979)

[a] PAN = peroxyacetyl nitrate
[b] NMHC = non-methane hydrocarbons

Source: Demerjian (1986), refer to that paper for full references cited in the table. (Reprinted with permission, John Wiley & Sons, Inc. Copyright 1986).

Table 3.2 Typical peak concentrations of gas phase criteria pollutants observed in the troposphere over the continents.[a]

	Type of Atmosphere			
Pollutant	Remote	Rural	Moderately Polluted	Heavily Polluted
CO	$\lesssim$0.2 ppm[f]	0.2-1 ppm[h]	~1-10 ppm[g]	10-50 ppm
NO_2	$\lesssim$ <1 ppb[b,f,j]	1-20 ppb[c,d]	0.02-0.2 ppm[g]	0.2-0.5 ppm
O_3	$\lesssim$0.05 ppm[b,f]	0.02-0.08 ppm	0.1-0.2 ppm	0.2-0.5 ppm
SO_2	$\lesssim$1 ppb[e]	~1-30 ppb[d]	0.03-0.2 ppm	0.2-2 ppm
NMHC	$\lesssim$65 ppbC[f]	100-500 ppbC[h]	300-1500 ppbC[i]	$\gtrsim$ 1.5 ppmC

[a] As 1 h averages
[b] Kelly et al. 1980
[c] Spicer et al. 1982; Pratt et al. 1983
[d] Martin and Barber 1981
[e] Ludwick et al. 1980; Maroulis et al. 1980
[f] Kelly et al. 1982; Hoell et al. 1984
[g] Ferman et al. 1981
[h] Seila 1979
[i] Sexton et al. 1982
[j] Johnston and McKenzie 1984

Source: Finlayson-Pitts and Pitts (1986). Refer to that book for full references cited in the table.

involving NO, CH_4, CO, and O_3. A brief discussion of the more important reaction steps involved in the mechanistic sequence is presented below. The photochemical reaction chain sequence in the troposphere is initiated by hydroxyl radicals (HO or OH) formed from the interaction of $O(^1D)$, the product of photolysis of ozone in the short-end portion of the solar spectrum, with water.

$$O_3 + h\nu(\lambda \leq 310 \text{ nm}) \rightarrow O(^1D) + O_2 \tag{1}$$

$$O(^1D) + H_2O \rightarrow 2HO \tag{2}$$

The HO produced reacts with CH_4 and CO present in the clean troposphere, resulting in the generation of peroxy radical species.

$$HO + CH_4 \rightarrow CH_3 + H_2O \tag{3}$$

$$HO + CO \rightarrow H + CO_2 \tag{4}$$

$$CH_3 + O_2 + M \rightarrow CH_3O_2 + M \tag{5}$$

$$H + O_2 + M \rightarrow HO_2 + M \tag{6}$$

The peroxy radicals in turn participate in a chain-propagating sequence, which converts nitric oxide (NO) to nitrogen dioxide (NO_2) and in the process produces additional hydroxyl and peroxy radical species.

$$CH_3O_2 + NO \rightarrow CH_3O + NO_2 \tag{7}$$

$$HO_2 + NO \rightarrow HO + NO_2 \tag{8}$$

$$CH_3O + O_2 \rightarrow HO_2 + H_2CO \tag{9}$$

$$H_2CO + h\nu(\lambda \leq 370 \text{ nm}) \rightarrow H + HCO \tag{10}$$

$$HCO + O_2 \rightarrow HO_2 + CO \tag{11}$$

The major chain terminating steps include:

$$HO + NO_2 + M \rightarrow HONO_2 + M \tag{12}$$

$$HO_2 + HO_2 \rightarrow H_2O_2 + O_2 \quad (13)$$

$$H_2O_2 + HO \rightarrow H_2O + HO_2 \quad (14)$$

The role that the chemistry of the clean troposphere plays in controlling the abundance of tropospheric ozone has been a subject of considerable debate. The model calculations indicate that photochemical processes can produce and destroy tropospheric O_3 at rates equivalent to those estimated for stratospheric injection and depositional losses at the earth's surface. The reaction sequence for O_3 production involves converting NO to NO_2 at a rate sufficiently high to maintain a NO_2/NO ratio to sustain the observed background levels of O_3.

$$HO_2 + NO \rightarrow NO + NO_2 \quad (8)$$

$$NO_2 + h\nu(\lambda \leq 430 \text{ nm}) \rightarrow NO + O \quad (15)$$

$$O + O_2 + M \rightarrow O_3 + M \quad (16)$$

$$NO + O_3 \rightarrow NO_2 + O_2 \quad (17)$$

$$HO + CO \rightarrow H + CO_2 \quad (4)$$

In general, reactions (15) through (17) govern the ozone concentrations present in the sunlight-irradiated atmosphere at any instant and to a first approximation the steady state relationship

$$(NO_2)K_{15}/(NO)K_{17} = (O_3)$$

provides a good estimate of ozone given the ratio $(NO_2)/(NO)$ and K_{15}/K_{17} (Leighton 1961). The photolytic rate constant K_{15} is directly related to the integrated actinic solar flux over the wavelength range 290-430 nm.

The paths for ozone destruction in the troposphere include the reaction sequence

$$HO_2 + O_3 \rightarrow HO + 2O_2 \quad (18)$$

$$HO + O_3 \rightarrow HO_2 + O_2 \quad (19)$$

Background concentrations of O_3 observed in a number of locations around the world typically show average daily 1 hour maxima of ~20-60 ppb (Singh et al. 1978). To explain the behaviour of O_3 in remote locations, Singh et al. (1977, 1978) developed a schematic representation of the variations in the O_3 concentration at the surface and in the free troposphere (Figure 3.1) and variations in O_3 concentrations by season (Figure 3.2). Figure 3.1 depicts a large O_3 reservoir with no average diurnal variation, except near the earth's surface where O_3 concentrations are controlled by surface destruction and mixed layer dynamics. Idealized seasonal variations in O_3 concentrations are depicted in Figure 3.2, where natural O_3 effects (curve A) are expected to be at a maximum in the early spring. Perturbations in O_3 levels due to localized O_3 production and transport from urban centres, both resulting from photochemical processes, are depicted by curves B and C.

Two sources have been suggested for the O_3 in the natural troposphere: (1) injection from the stratosphere, and (2) photochemical production via NO_x-NMOC (non-methane organic compounds) reactions in sunlight involving naturally occurring NO_x and hydrocarbons or CO. There has been much controversy regarding the relative importance of these two sources (Vukovich et al. 1985; Levy et al. 1985; Logan 1985).

Finlayson-Pitts and Pitts (1986) have provided, as follows, an excellent analysis of the subject matter. The occasional occurrence of stratospheric injection of O_3 at particular times and in certain locations is now reasonably well accepted. The stratosphere contains relatively high concentrations of O_3 compared to the troposphere; indeed, the troposphere contains only ~10% of the total O_3 in the atmosphere (Crutzen and Gidel 1983). Although mixing of stratospheric air into the troposphere is inhibited by the temperature increase at the tropopause, meteorological phenomena can lead to periodic short-term "break downs" of this temperature discontinuity in a particular location, leading to a temporary mixing of stratospheric air containing O_3 into the troposphere (Singh et al. 1980). Such episodes are often characterized by relatively rapid changes in the O_3 concentration with peaks occurring at times not expected from photochemical processes. For example, the data for the Zugspitze, West Germany, showed a maximum 1 h average O_3 concentration of 196 ppb which occurred at midnight on January 8, 1975. Such high O_3 concentrations were not observed in the region

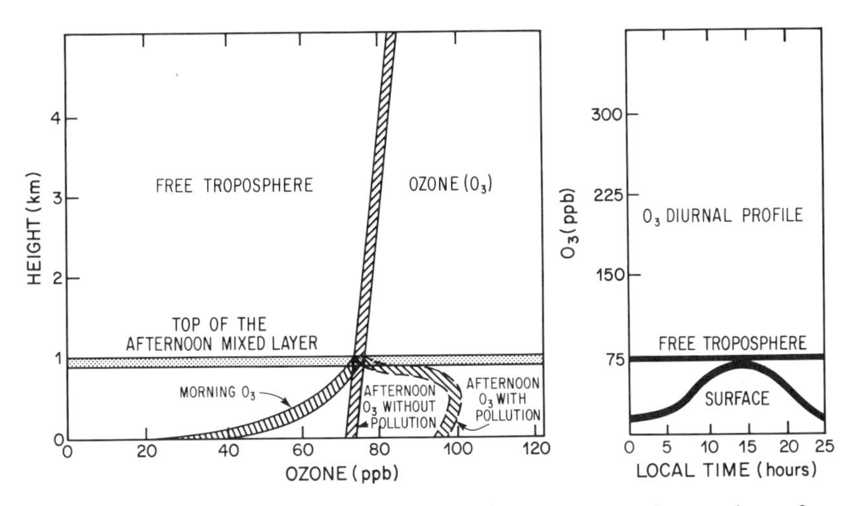

Figure 3.1 A schematic diagram of variations in O_3 concentrations at the surface and in the free troposphere.
Source: Singh et al. (1978). (Reprinted with permission, Copyright 1978 Pergamon Press PLC).

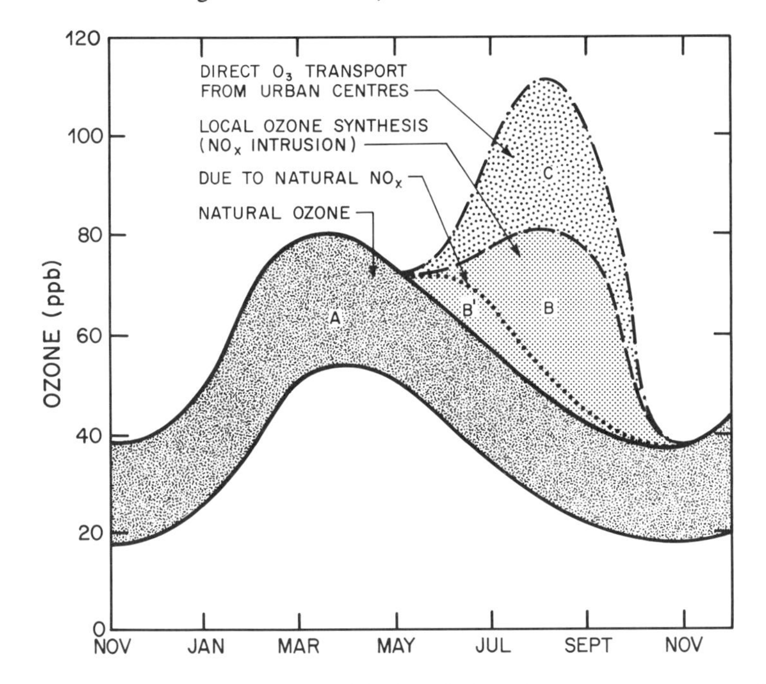

Figure 3.2 A schematic diagram of the idealized variations in O_3 concentrations at remote locations.
Source: Singh et al. (1978). (Reprinted with permission, Copyright 1978 Pergamon Press PLC).

surrounding the Zugspitze on the previous day, suggesting that photochemical processes combined with transport could not account for this peak. At the same time as the O_3 increased, the concentration of the isotope ^7Be increased; ^7Be has been suggested as a tracer of stratospheric air, although its applicability as a tracer is controversial at present (Singh et al. 1980; Dutkiewicz and Husain 1985). In any event, such rapid changes producing high O_3 peaks at times inconsistent with a photochemical source, have been observed at a number of locations (Chung and Dann 1985), and have been taken as evidence for the injection of air from the stratosphere into the troposphere.

The seasonal variation in O_3, with a peak in the late winter and early spring, and the lack of a diurnal variation in the O_3 at remote sites are also cited as evidence for a non-photochemical source of O_3. Thus, exchange of stratospheric and tropospheric air is most effective during late winter and early spring (Danielsen and Mohnen 1977); in addition, a photochemical source would be expected to peak with the increase in solar intensity in the summer and in the afternoon.

Tropospheric O_3 concentrations tend to be higher in the northern hemisphere than in the southern hemisphere (Fishman and Crutzen 1978), which is consistent with model predictions that the downward flux of O_3 from the stratosphere should be larger in the northern hemisphere (Mahlman et al. 1980; Gidel and Shapiro 1980).

Finally, the O_3 mixing ratio (defined as the ratio of the concentration of O_3 in molecules cm^{-3} to that of air) increases with height in the troposphere at all latitudes and seasons in the northern hemisphere (Chatfield and Harrison 1977). Since atmospheric mixing tends to carry O_3 down toward the earth's surface, the increase with altitude implies a stratospheric source.

However, while these arguments have been made for the contribution of stratospheric injection of O_3 to the earth's surface, other data strongly suggest that ground-level impacts are infrequent and do not generally lead to O_3 concentrations exceeding 100 ppb. Thus, long-range transport of photochemical oxidant precursors from urban areas is well known; for example, long-range transport from western Europe has been postulated to be partially responsible for some of the elevated ozone concentrations observed in Norway (Schjoldager et al. 1978; Schjoldager 1981).

In addition, aircraft measurements of four stratospheric O_3 intrusion events suggest that more than half of the O_3 mass injected into the upper troposphere by these events is likely mixed and diluted into the troposphere above the 3 km elevation level (Viezee et al. 1983). Viezee et al. suggest that elevated ground-level concentrations of O_3 due to stratospheric intrusions occur less than 1% of the time.

There are a number of indications that O_3 produced in the troposphere by photochemical processes is the most significant factor for some locations and times. First, as discussed previously, NO_x-NMOC reactions are well known to produce O_3 in rural and urban areas; there is no *a priori* reason to think the same types of reactions might not occur with naturally emitted NO_x and hydrocarbons. Secondly, the higher O_3 concentrations in the northern hemisphere, which are often cited as evidence for stratospheric injection, may in fact be due to photochemical reactions which involve partial contributions from anthropogenically emitted pollutants. Thirdly, the broad maxima occurring in the summer at many locations, especially within a few hundred kilometres of urban areas in Europe and the United States, has been suggested to support the production of ozone via photochemical processes involving anthropogenic emissions (Logan 1985).

Anthropogenic emissions in the highly industrialized northern hemisphere exceed those in the southern hemisphere, thus, higher O_3 concentrations might be expected in the north if there is a contribution from an anthropogenic source. The lack of a diurnal variation in O_3 at remote sites, however, does not rule out photochemical processes; Liu et al. (1980) point out that the photochemical lifetime expected for O_3 is sufficiently long (~10 days) that a large diurnal variation in its concentration would not be expected.

Finally, the major loss processes assumed for O_3 in the natural troposphere are loss at the surface (i.e., dry deposition) and, to a lesser extent, photochemical reactions (1) and (2a):

$$O_3 + h\nu(\lambda \leq 320\ \text{nm}) \rightarrow O(^1D) + O_2 \qquad (1)$$

$$O(^1D) + H_2O \xrightarrow{(a)} 2\ OH \qquad (2)$$

$$\xrightarrow{(b)} O(^3P) + H_2O$$

At 50% relative humidity (in air at 1 atmospheric pressure and 298°K) approximately 10% of the $O(^1D)$ produced in Reaction (1) reacts with water vapour; the rest is deactivated to the ground state, $O(^3P)$, by air. Of the 10% which reacts with H_2O, $\geq$ 95% proceeds via Reaction (2a) to produce 2 OH. The remaining 5%, Reaction (2b), does not represent a net loss since the $O(^3P)$ recombines with O_2 to give O_3:

$$O(^3P) + O_2 \xrightarrow{M} O_3 \tag{3}$$

The calculated loss of O_3 at the surface and by photolysis is approximately a factor of 4 larger than the estimated stratospheric flux, implying that another major O_3 source must be present to maintain relatively constant O_3 levels in the troposphere (Liu et al. 1980).

It has been shown that O_3 and CO concentration profiles, as a function of altitude above the earth's surface, frequently show either strong positive or negative correlations (Fishman and Seiler 1983). Fishman, Seiler, and co-workers attribute the positive CO-O_3 correlations to photochemical production of O_3 in the troposphere and the negative correlations to stratospheric injection. Thus, the stratosphere has very low CO mixing ratios (40-50 ppb), so that mixing of stratospheric air into the troposphere should be characterized by lower CO and higher O_3 concentrations. On the other hand, since the oxidation of CO can lead to the formation of O_3 if sufficient NO is present, photochemical production of O_3 should be associated with higher CO levels. While the correlations between O_3 and CO can thus be suggestive of the source of O_3, Fishman and Seiler (1983) point out that these correlations do not provide conclusive proof.

Negative correlations between CO and O_3 were found for high altitudes at high latitudes in both hemispheres, suggesting that recent stratospheric injection may have occurred. The region of negative correlation close to the earth's surface is attributed to production of CO at the surface where O_3, however, is destroyed. The strongest positive correlations between CO and O_3 were found in the middle troposphere between latitudes ~20°N and 45°N and these results are consistent with the larger sources of precursors of O_3, that is, NO_x, NMOC, and CO, in the northern hemisphere compared with the southern. Model calculations by Fishman and Seiler

(1983) suggested that ~15-25 ppb O_3 is generated throughout the troposphere north of ~30°N by photochemical processes. This modelling also showed that the increasing O_3 concentrations with altitude were consistent with a photochemical source if the surface loss of O_3 was sufficiently large compared with its rate of production.

In summary, it appears likely that O_3 in the natural troposphere has two sources: stratospheric injection and NO_x-NMOC-CO chemistry. The contribution of each varies from location to location, as well as from time to time. However, the larger anthropogenic emissions in the northern hemisphere are expected to lead to a relatively greater importance of the photochemical source in this portion of the globe.

Figure 3.3 provides a graphic summary of the concentrations of tropospheric O_3 as a function of latitude (Pruchniewicz 1973).

As with O_3, the chemistry of NO_x in the remote troposphere is similar to that in more polluted atmospheres, except that the concentrations are obviously smaller (Finlayson-Pitts and Pitts 1986). Figure 3.4 provides a summary of the gas phase chemistry

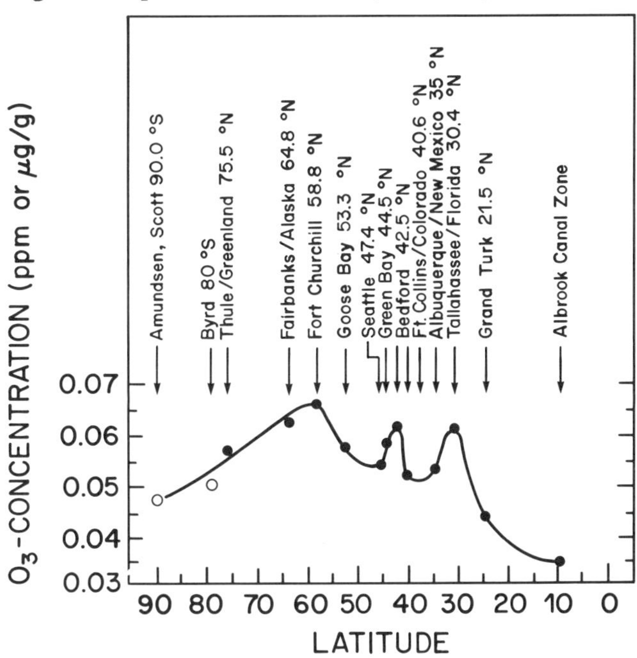

Figure 3.3 The concentration of tropospheric ozone as a function of latitude. Source: Pruchniewicz (1973). (Reprinted with permission, Copyright 1973 Birkhäuser Verlag AG).

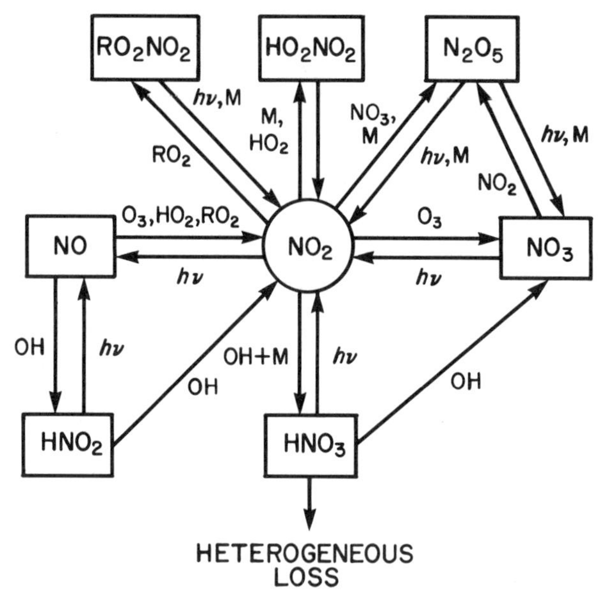

Figure 3.4 Summary of the gas phase chemistry of the oxides of nitrogen in the clean troposphere.
Source: Logan (1983). (Reprinted with permission, Copyright 1983 American Geophysical Union).

of the oxides of nitrogen in the clean troposphere. The relative concentrations of various oxides of nitrogen at remote locations are summarized in Table 3.3.

3.3 SULPHUR COMPOUNDS IN THE CLEAN TROPOSPHERE

There are substantial natural emissions of sulphur compounds into the troposphere from biological activity in vegetation, soils, and aquatic ecosystems. However, in contrast to anthropogenic emissions which are almost entirely in the form of SO_2 (except for example, gas processing plants), natural emissions are predominantly in the form of reduced sulphur compounds.

Tables 3.4 and 3.5 provide a summary of the approximate tropospheric concentration range of selected sulphur compounds in unpolluted air. Interestingly, while compilations of global sources and sinks for COS (OCS) and CS_2 (Figures 3.5 and 3.6) were available, similar listings were not found for H_2S and DMS, with the exception of what is provided in Table 3.5.

Table 3.3 Typical concentrations of some gas phase trace species and non-criteria pollutants reported in the troposphere over the continents.

Pollutant	Type of Atmosphere: Remote	Rural	Moderately Polluted	Heavily Polluted
NO	$\lesssim$50 ppt[a,v]	~0.05-20 ppb[g,i,v]	0.02-1 ppm[j]	~1-2 ppm
PAN	$\lesssim$50 ppt[q]	2 ppb[g]	2-20g ppb	20-70 ppb[e]
NH_3	15 ppt[m]	1-10 ppb[c,o,r]	1-10 ppb[c,o,r]	10-100 ppb[f]
HNO_3	≤0.03-0.1 ppb[a]	~0.1-4 ppb[b,r]	1-10 ppb[p,r]	10-50 ppb[e]
NO_3	$\lesssim$5 ppt[l]	5-10 ppt	10-100 ppt[k,l]	100-430 ppt[m]
HONO	<30 ppt[o]	0.03-0.8 ppb[o]	0.8-2 ppb	2-8 ppb[h,n]
HCHO	≤0.5-2 ppb[d]	2-10 ppb[o]	10-20 ppb[d,u]	20-75 ppb[e]
OH	(4-40) x 10^{-3} ppt[t]	0.01-0.10 ppt[t]	0.05-0.4 ppt[s,t]	≥0.4 ppt
(Midday average peak for sunny conditions)	(1-10) x 10^5 mole cm^{-3}	(2.5-25) x 10^5 mole cm^{-3}	(1-10) x 10^6 mole cm^{-3}	1 x 10^7 mole cm^{-3}

[a] Kelly et al. 1980; Huebert and Lazrus 1980; Bollinger et al. 1984
[b] Shaw et al. 1982; Huebert and Lazrus 1980
[c] Levine et al. 1980
[d] National Research Council 1981
[e] Tuazon et al. 1981
[f] Doyle et al. 1979
[g] Temple and Taylor 1983; Spicer 1982
[h] Pitts et al. 1983
[i] Martin and Barber 1981; Pratt et al 1983
[j] Ferman et al. 1981
[k] Platt et al. 1981
[l] Platt et al. 1982
[m] Platt et al. 1980b; Atkinson et al. 1985
[n] Gras 1983
[o] Harward et al. 1982
[p] Spicer 1977
[q] Singh and Salas 1983
[r] Cadle et al. 1982
[s] Hard et al. 1984
[t] Hübler et al. 1984, and references therein
[u] Schjoldager 1984
[v] Logan 1983

Source: Finlayson-Pitts and Pitts (1986). Refer to that book for full references cited in the table.

Table 3.4 Approximate tropospheric concentration range of selected sulphur compounds in unpolluted air.

Location	Atmospheric Concentrations[a] ($ng\ m^{-3}$)				
	H_2S	DMS	CS_2	COS	SO_2
Ocean boundary layer	<5-150	<2-200	1200-1550	50-70	<15-300
Temperate continental boundary layer	20-200	PD	1200-1550	PD	<15-300
Tropical coastal boundary layer	100-9000	PD	1200-1550	PD	PD
Free tropsophere	PD	<2(PD)	1200-1550	<10	30-300

[a] PD indicates poorly determined at this review.

To convert $ng\ m^{-3}$ to ppb, multiply by appropriately: H_2S-717 x 10^{-6}; DMS-394 x 10^{-6}; CS_2-332 x 10^{-6}; COS-408 x 10^{-6}; and SO_2-382 x 10^{-6}.

Source: Harriss and Niki (1984)

Table 3.5 Global atmospheric sulphur.

Sulphur Compounds	Concentration*	Pool Size Tg(S)**	Residence Time (Days)
SO_4^{2-}	0.1 - 0.56	0.26	7
SO_2	45 - 340	0.25	4
COS	100 - 560	2.20	160
CS_2	70 - 370	0.60	45
$(CH_3)_2S$	58	0.021	0.75
H_2S	0 - 320	0.041	1.5

* SO_4^{2-} concentration in $\mu g\ m^{-3}$; others in parts per trillion (ppt) by volume. Values reported for remote locations.
** 1 Tg = 10^{12}g.

Source: Sze and Ko (1980). (Reprinted with permission, Copyright 1980 Pergamon Press PLC).

Dimethylsulphide (DMS) is the most abundant volatile sulphur compound in seawater with an average concentration of ~100 x 10^{-9} g L^{-1}. This compound is produced by both algae and bacteria. The evidence for a biogenic origin for DMS has come from laboratory measurements of emissions produced in pure, axenic cultures of marine planktonic algae and field measurements of emissions from soils, benthic macroalgae, decaying algae, and corals. Extensive oceanographic studies have shown direct correlations between DMS concentrations in seawater and indications of phytoplankton activity. The vertical distribution, local patchiness, and distribution of DMS in oceanic ecozones exhibit a pattern very similar to primary productivity. Selected groups of marine organisms such as *Coccolithophorids* (i.e., a type of marine planktonic algae) and stressed corals are particularly prolific producers of DMS. The calculated global sea-to-air flux of sulphur as DMS is ~0.1 g S m^2 y^{-1}, which totals to approximately 39 x 10^{12} g S y^{-1}. A more limited set of measurements has been made in coastal salt marshes with DMS emissions commonly in the range of 0.006 to 0.66 g S m^2 y^{-1}.

Knowledge of natural sources of H_2S to the troposphere is still rudimentary. Preliminary studies have shown that anaerobic, sulphur-rich soils (e.g., coastal soils and sediments) emit H_2S to the atmosphere, albeit with strong temporal and spatial variations.

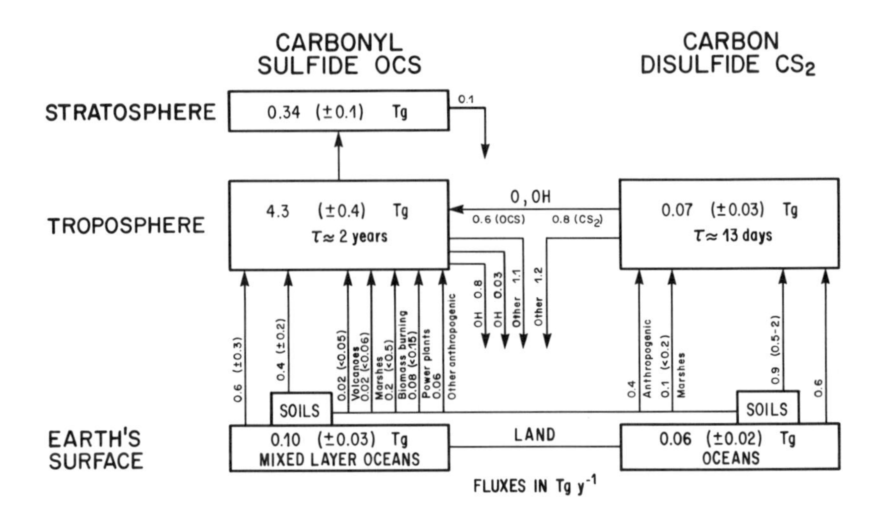

Figure 3.5 Global sources, sinks, and environmental reservoirs of carbonyl sulphide (OCS) and carbon disulphide (CS_2). Oceanic and atmospheric burdens are in teragrams (1 Tg = 10^{12} g) and fluxes are in Tg y^{-1}.
Source: Khalil and Rasmussen (1984). (Reprinted with permission, Air Pollution Control Association Copyright 1984).

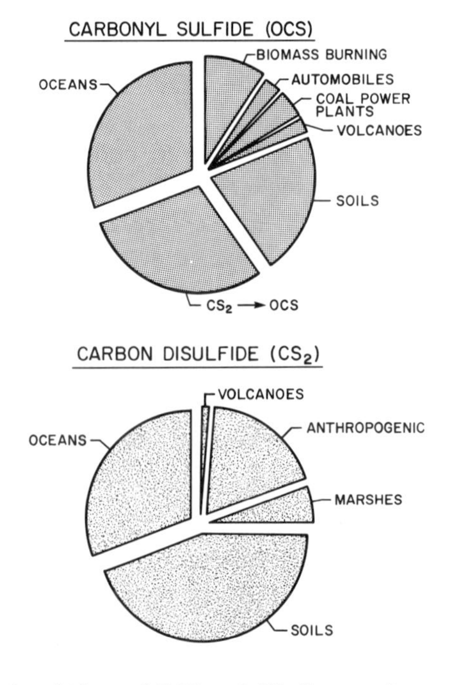

Figure 3.6 The global emissions of OCS and CS_2 from various sources.
Source: Khalil and Rasmussen (1984). (Reprinted with permission, Air Pollution Control Association Copyright 1984).

Hydrogen sulphide fluxes at a single location can vary by a factor of up to 10^4 depending on variables such as light, temperature, Eh, pH, O_2, and rate of microbial sulphate reduction in the sediment. The presence of active photosynthetic organisms or a layer of oxygenated water at the sediment surface can reduce or stop emissions due to rapid oxidation of H_2S.

Agricultural and forest soils can also be a source of H_2S to the atmosphere. Measurements by several investigators suggest that maximum emissions from nonmarine soils are associated with wet tropical forest soils. It is likely that many soils that appear to be aerated contain anaerobic microhabitats suitable for microbial sulphate reduction; the magnitude of H_2S emissions will depend on the net effects of many processes that influence production, transport in the soil, oxidation rates, and exchange at the soil-air interface.

Photochemical sources of atmospheric H_2S have been proposed to occur through a combination of the following reactions:

$$OH + COS \rightarrow SH + CO_2,$$

$$OH + CS_2 \rightarrow COS + SH,$$

$$SH + HO_2 \rightarrow H_2S + O_2,$$

$$SH + CH_2O \rightarrow H_2S + HCO,$$

$$SH + H_2O_2 \rightarrow H_2S + HO_2,$$

$$SH + CH_3OOH \rightarrow H_2S + CH_3O_2,$$

and

$$SH + SH \rightarrow H_2S + S.$$

Removal of H_2S is thought to be accomplished by:

$$OH + H_2S \rightarrow H_2O + SH,$$

resulting in a lifetime of approximately 1 day. *In situ* photochemical production from COS and CS_2 precursors is the most likely source of H_2S measured in remote ocean air. Atmospheric concentrations of H_2S in continental air are highly variable, resulting from a complex interaction of factors determining ground emissions, *in situ* photochemical production, and atmospheric lifetime.

Figure 3.7 provides a schematic representation of the oxidation processes of volatile sulphur compounds in the atmosphere.

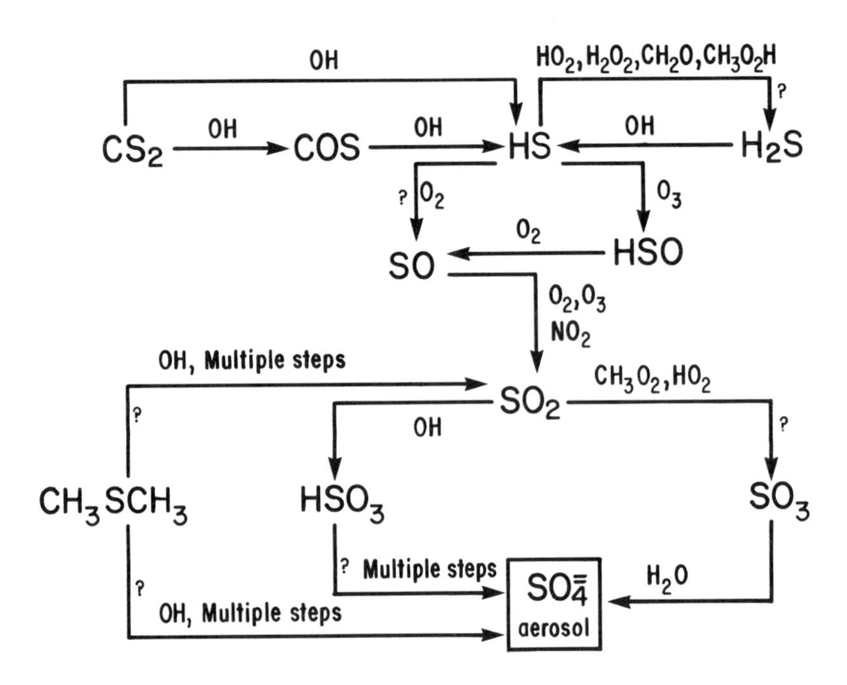

Figure 3.7 Oxidation of volatile S compounds in the atmosphere.
Source: Sze and Ko (1980). (Reprinted with permission, Copyright 1980 Pergamon Press PLC).

3.4 THE COMPOSITION OF AEROSOLS AT REMOTE LOCATIONS

An aerosol can be defined as a solid or a liquid particle in a gas phase. In this section, however, the discussion is restricted to solid particles. With regard to their formation process and size, these solid particles can be divided into two distinct groups: fine and coarse particles (Whitby 1978). Fine particles with a radius smaller than 0.5-1.0 μm are formed by condensation and coagulation, while coarse particles arise mostly from surface disintegration.

The effects of atmospheric aerosols depend not only on their concentration and size but also on their chemical composition. The chemical composition of the tropospheric aerosols results from the interactions of many formative and dynamic processes. For this reason, particles are often composed of several materials, and the composition varies as a function of time and location. This is further complicated under polluted urban conditions. Table 3.6 provides for example, a comparison of mean concentrations of some major chemical constituents in fine and coarse particles at four rural sites in Minnesota. The chemical constituents which are governed by coagulative processes, SO_4^{2-}, NO_3^-, and NH_4^+, are generally found in higher concentrations in the fine particles (<2.4 μm), while others produced by mechanical or disintegrative processes, Ca^{2+}, Si^{4+}, Al^{3+} are found in higher concentrations in the coarse fraction (>2.5 μm).

Barrie and Hoff (1985) examined the chemical composition of particulate matter at three sites in the Canadian Arctic over a five year period (Table 3.7). According to the authors, anthropogenic pollution, typified by SO_4^{2-} and V, exhibited a persistent seasonal cycle. The seasonal variation of a given aerosol constituent depended on its source. There were four distinctive seasonal variations for: (1) anthropogenic constituents Cr, Cu, Mn, Ni, Pb, Sr, V, H, NH_4, SO_4, NO_3; (2) halogens Br, I, F; (3) sea salt elements Na, Mg, Cl; and (4) soil constituents Al, Ba, Ca, Fe, and Ti.

Throughout the Norwegian and North American Arctic during the winter half of the year, there were elevated levels of SO_4^{2-} (Table 3.8). In the High Arctic, airborne concentrations of total nitrate were at least 7 times lower on a molar basis than SO_4^{2-} concentrations (Barrie and Hoff 1985).

Observations of the elemental content of aerosols were made by Barrie (1986) at Birch Mountain, north of Fort McMurray, Alberta under background northerly wind conditions. The proximity of Birch Mountain to the periphery of the arctic air mass meant that both pacific and arctic air masses do influence that site. The results from Birch Mountain are compared with summertime measurements made by Rahn (1971) at Fort Smith, Alberta (Table 3.9).

Air-parcel back-trajectories were examined for the 3 sampling periods on Birch Mountain. It was found that the first period was distinct from the other two in that the air came predominantly from the North Pacific, south of Alaska rather than from the Alaskan Arctic via the Mackenzie Valley. Chemically, this manifested itself as lower aerosol V concentrations and higher Na and Cl concentrations indicative of sea salt. Higher V concentrations were expected in the polluted arctic air rather than in the relatively unpolluted

Table 3.6 Comparison of the mean concentrations of chemical constituents in fine (<2.5 μm) and coarse (>2.5 μm) particles at four rural locations in Minnesota. F = fine; C = coarse particles.

Sampling Location		Concentration of Chemical Constituents ($\mu g\ m^{-3}$)							
		SO_4^{2-}	NO_3^-	NH_4^+	Ca^{2+}	Si^{4+}	Al^{3+}	Mn^{2+}	Zn^{2+}
Sherburne	F	2.1	1.4	1.6	0.04	0.1	0.09	0.001	0.008
	C	0.3	0.4	0.1	0.3	1.4	0.4	0.02	0.005
Wright	F	2.3	1.6	1.6	0.1	0.1	0.03	0.001	0.03
	C	0.3	0.5	0.2	0.3	1.1	0.3	0.01	0.004
Sandstone	F	1.8	0.06	0.7	0.05	0.1	0.04	0.008	0.007
	C	0.15	0.1	0.5	0.2	1.0	0.3	0.01	0.002
Ely	F	2.3	0.08	0.8	0.05	0.1	0.02	0.0004	0.006
	C	0.5	0.3	0.06	0.2	0.7	0.2	0.007	0.004

Source: Krupa and Nosal (1984)

Table 3.7 Annual arithmetic mean concentrations (ng m^{-3}) of aerosol constituents at three Canadian arctic sites (a year includes July-June) for 1980-81 and 1981-82.

	Alert		Igloolik		Mould Bay	
	80-81	81-82	80-81	81-82	80-81	81-82
Anthropogenic						
*Cr	0.16	0.30	0.062	0.15	0.31	0.32
*Cu	0.78	1.33	0.69	1.13	1.61	1.02
Mn	1.60	1.49	0.45	0.71	0.79	0.87
Ni	0.32	0.38	0.14	0.27	0.40	0.45
*Pb	1.72	1.70	3.61	3.84	2.00	3.22
*Sr	0.30	0.57	0.35	0.67	0.35	0.41
V	0.36	0.62	0.21	0.47	0.28	0.56
*Zn	2.80	3.48	3.80	3.45	2.82	3.79
H	6.3	7.4	4.0	5.9	6.5	9.2
NH_4	82	102	77	95	72	125
SO_4	890	1052	667	966	738	1036
NO_3	55	74	62	71	59	67
Halogens						
Br	3.8	8.9	5.7	14	7.0	16
I	0.43	0.49	0.96	1.03	0.45	0.46
F	2.0	7.0	7.2	5.8	5.4	6.7
Seasalt						
Na	121	158	229	313	179	243
Cl	114	182	368	505	251	354
K	11	16	17	31	15	11
*Mg	92	84	99	129	69	63
Soil						
Al	--	146	13	--	45	--
*Ba	0.50	0.37	0.16	0.21	0.42	0.47
*Ca	201	139	83	94	35	29
*Fe	86	86	16	21	36	29
*Ti	0.49	1.31	0.68	--	0.35	0.24
*P	2.79	1.67	1.37	0.93	3.11	1.12

For convenience, valence signs have been omitted from the ions.
* The fraction soluble in concentrated nitric acid.

Source: Barrie and Hoff (1985). (Reprinted with permission, Copyright 1985 Pergamon Press PLC).

Table 3.8 A comparison of the arithmetic mean atmospheric concentration of anthropogenic aerosol constituents (ng m^{-3}) during the winter months (December-April) of 1980-81 and 1981-82 in the Canadian arctic with measurements elsewhere in the arctic and in southern Sweden near major sources.

	Canadian Arctic			Norwegian Arctic		S. Sweden ++
	Winter Months			Winter* (3-weeks)	March+ 1983	Annual
Constituent	Alert	Igloolik	Mould Bay	Spitsbergen*		Mean
Cr	0.37	0.27	0.41	2.6	1.2	2.8
Cu	1.88	1.53	2.01	1.9	2.5	2.4
Mn	1.63	0.95	1.22	1.5	2.7	6.7
Ni	0.62	0.42	0.59	0.7	1.3	1.5
Pb	3.49	6.38	5.42	4.9	12.2	21.0
Sr	0.61	0.88	0.61	--	--	--
V	0.96	0.76	0.87	--	2.8	2.5
Zn	6.30	6.56	7.15	3.2	13.7	24.0
S	599	520	544	690	--	800

* Heintzenberg et al. (1981), 2-5 day samples in late winter.
\+ Pacyna et al. (1985), 1 month study.
++ Lannefors et al. (1983).
Source: Barrie and Hoff (1985). (Reprinted with permission, copyright 1985 Pergamon Press PLC).

North Pacific air. In comparison to the summer values measured by Rahn (1971), V concentrations in the winter arctic air were about 20 times higher. This is consistent with the seasonal variation in V observed in the arctic air mass. Another feature of the winter measurements at Birch Mountain was that, relative to Al (a crustal reference element), the following elements were enriched in the aerosol by more than 10-fold compared to soil: V, Sb, Zn, As, Cl, I, and Br. The first four elements are likely of anthropogenic origin. Cl originates from the sea. Iodine and Br are either natural or anthropogenic in origin (Barrie and Hoff 1985).

Table 3.9 A comparison of background aerosol composition (ng m^{-3}) measured at Birch Mountain 80 km northwest of Fort McMurray under northerly flow conditions and at Fort Smith, by Rahn (1971). Rahn's data represent the average of four two-week samples.

	Birch Mountain 1976			
Element	Sample 1 North Pacific	Sample 2 Alaskan Arctic	Sample 3 Alaskan Arctic	Fort Smith Summer 1970
Al	32.0	43.0	54.0	54.0
As	0.29	0.85	0.34	0.25
Br	3.6	1.6	1.4	0.44
Ca	15.0	32.0	31.0	33.0
Cl	142.0	42.0	47.0	7.0
Cu	<0.78	<1.0	<1.2	0.7
I	0.40	0.36	0.42	0.16
K	25.0	19.0	27.0	44.0
Mg	24.0	18.0	20.0	--
Mn	0.51	0.72	0.78	1.2
Na	101.0	59.0	52.0	15.0
Sb	<0.085	0.13	<0.6	0.11
Sc	<0.010	0.0085-0.013	0.0093-0.0160	0.36
Ti	<3.8	5.2	6.1	4.1
V	0.40	3.5	3.6	0.17
Zn	--	3.2-6.0	<4.4	3.1

Source: Barrie (1986). (Reprinted with permission, Copyright 1986 Kluwer Academic Publishers).

3.5 THE COMPOSITION OF PRECIPITATION AT REMOTE LOCATIONS

It is often assumed (Garrels and MacKenzie 1971; Galloway et al. 1976; Likens and Bormann 1974; Likens et al. 1981) that the pH of natural rainwater is controlled by the dissociation of dissolved CO_2, has a value of 5.6, and that decreases below this are due to the addition of acidic components by human activity (Charlson and Rodhe 1982). "However, decreases could be due to the removal by rainwater of naturally occurring acids from the air (notably H_2SO_4 in the natural portion of the sulphur cycle). Consideration of the cycling of water and sulphate through the atmosphere and the amount and composition of sulphate aerosol expected to be scavenged by a given amount of cloud water in remote locations indicates that, in the absence of basic materials (such as NH_3 and $CaCO_3$), average pH values of ~5 are expected to occur in pristine locations. This value must vary considerably due to variability in scavenging efficiencies as well as geographic patchiness of the sulphur, nitrogen, and water cycles. Thus, pH values might range from 4.5 to 5.6 due to variability of sulphur cycle alone." (Charlson and Rodhe 1982).

Sequeira (1981, 1982) also concluded that a pH of 5.6 may not be a reasonable reference value for unpolluted precipitation. According to Sequeira, the pH of atmospheric precipitation at a given location depends on the chemical nature and relative proportions of acids and bases in solution (Tables 3.10 and 3.11). For additional details on the nature and chemistry of precipitation, the reader is referred to Chapter 2.3.

Table 3.10 Chemistry of precipitation at high altitudes. Acidity parameters and correlations.

	Sampling Station		
	Monte Cimone Italy	Mauna Loa Hawaii	Alamosa Colorado
Data period	1975-76	1973-78	1973-75
Total no. of monthly samples	20	35	25
Parameters			
1) pH			
Max	6.90	6.90	7.91
Min	4.20	3.33	4.64
Weighted mean*	4.97	4.47	6.51
Samples with pH <5.00 (%)	25	51	4
2) H^+: corr. (r) with:			
SO_4^{2-}	0.50(19)[b++]	0.82(31)[a]	0.012(25)
SO_4^{2-} (pH <5)	0.76(5)	0.88(18)[a]	--

Continued

Table 3.10 (Concluded).

	Sampling Station		
	Monte Cimone Italy	Mauna Loa Hawaii	Alamosa Colorado
Data period	1975-76	1973-78	1973-75
Total no. of monthly samples	20	35	25
Parameters			
2) H^+: corr. (r) with: (cont.)			
SO_4^{2-} (xs)+	0.51(19)[b]	0.66(25)[a]	c
SO_4^{2-} (xs, pH <5)	0.76(5)	0.81(15)[a]	--
NO_3^-	-0.14(17)	0.65(18)[a]	0.009
NO_3^- (pH <5)	-0.12(5)	0.64(12)[b]	--
Cl^- (pH <5)	0.91(5)[b]	0.67(16)[a]	--
3a) NH_4^+: corr. (r) with:			
SO_4^{2-}	-0.019(19)	0.36(19)	0.018
SO_4^{2-} (xs)	-0.0016(19)	0.28(14)	--
Acidity	--	0.87(11)[a]	0.86(12)a
NO_3^-	0.70(17)[a]	0.77(13)[a]	0.088(16)
3b) Ratio NH_4^+/SO_4^{2-} (xs):			
Max	2.47	0.32	1.83
Min	0.054	0.007	0.00077
A*	0.573	0.117	0.560
M*	0.21	0.12	0.27
G*	0.295	0.0775	0.150
σ_g*	3.4	3.0	8.8
4) Alkalinity: corr. (r) with:			
Ca^{2+}	0.74(15)[a]	--	--

* Weighted mean pH = $-\log_{10} (\Sigma[H^+]_i/\Sigma P_i)$, where $[H^+]_i$ is the hydrogen ion concentration for the ith month and P_i is the corresponding precipitation.

\+ SO_4^{2-}(xs) ≡ SO_4^{2-} (non-marine) = SO_4^{2-} (reported) -0.25Na^+ (reported), where 0.25 is SO_4^{2-}/Na^+ ratio in seawater.

++ Figures in parentheses are the number of data pairs/groups used. Significance determined using statistical tables (Snedecor and Cochran 1967).

[a] Significant at 0.01 level; [b] significant at 0.05 level, [c] not calculated since (SO_4^{2-} ≈ SO_4^{2-}(xs)). The above apply for data in the following table.

* A = arithmetic mean; M = median; G = geometric mean, σ_g = geometric standard deviation.

Source: Sequeira (1982). (Reprinted with permission, Copyright 1982 Pergamon Press PLC).

Table 3.11 Correlations of SO_4^{2-} and Ca^{2+} with other salt/mineral constituents for precipitation at high altitude sites.

Correlation	Sampling Station		
	Monte Cimone Italy	Mauna Loa Hawaii	Alamosa Colorado
Simple (r):			
$SO_4^{2-}.Na^+$	0.10(19)	0.64(25)[a]	0.80(24)[a]
$SO_4^{2-}.K^+$	0.10(15)	0.78(17)[a]	0.63(24)[a]
$SO_4^{2-}.Ca^{2+}$	-0.12(19)	0.72(28)[a]	0.65(24)[a]
$Ca^{2+}.Na^+$	0.83(20)[a]	0.25(27)	0.66(25)[a]
$Ca^{2+}.K^+$	0.71(15)[a]	0.59(21)[a]	0.69(25)[a]
$Na^+.K^+$	0.97(15)[a]	0.63(18)[a]	0.55(25)[a]
Multiple (r):			
$SO_4^{2-}.Na^+K^+Ca^{2+}$	--	0.92(17)[a]	0.85(24)[a]
$Ca^{2+}.Na^+K^+$	0.92(15)[a]	0.77(18)[a]	0.77(25)[a]

[a] Significant at 0.01 level.

Source: Sequeira (1982). (Reprinted with permission, Copyright 1982 Pergamon Press PLC).

Galloway et al. (1982) examined the composition of precipitation in remote areas of the world (Table 3.12). Precipitation composition (excluding seasalt) at St. Georges, Bermuda, was primarily controlled by anthropogenic processes; composition and acidity at San Carlos, Venezuela, Katherine, Australia, Poker Flat, Alaska, and Amsterdam Island were controlled by unknown mixtures of natural and anthropogenic processes. Precipitation was acidic; average volume-weighted pH values were 4.8 for Bermuda; 5.0, Alaska; 4.9, Amsterdam Island; 4.8, Australia; and 4.8, Venezuela. Precipitation acidity at Bermuda and Alaska appeared to be from long-range transport of sulphate aerosol; at Venezuela, Australia, and Amsterdam Island, from mixtures of weak organic and strong mineral acids, primarily H_2SO_4. Relative proportions of weak to strong acids were large at Venezuela and small at Amsterdam Island. Weak and strong acids were from mixtures of natural and anthropogenic processes. Once contributions from human activities were removed, the lower limit of natural contributions was probably $\geq$ pH 5.

Table 3.12 Minima, maxima, arithmetic means, standard deviations, and volume-weighted means of the composition of precipitation events collected at the GPCP sites* (μeq L^{-1}).

	**SO_4^{2-}	SO_4^{2-}	NO_3^-	Cl^-	Mg^{2+}	Na^+	K^+	Ca^{2+}	NH_4^+	H^+
Amsterdam Island, Indian Ocean (N = 26; vwm pH = 4.92)										
Minimum	0.0	7.4	0.3	12.6	2.6	10.9	0.2	0.0	0.3	4.4
Maximum	38.4	150	8.7	1480	255	1190	25.8	53.0	27.8	55.0
Arithmetic mean	11.5	52.6	2.7	406	72.8	334	7.2	15.5	5.1	18.2
Standard deviation	9.5	44.8	2.1	467	80.0	371	7.8	17.3	7.0	12.7
Volume-weighted mean	8.8	30.6	1.7	208	38.7	177	3.7	7.4	2.1	12.0
Poker Flat, Alaska (N = 16; vwm pH = 4.96)										
Minimum	0.0	0.0	1.2	0.5	0.0	0.0	0.0	0.0	0.3	6.5
Maximum	28.0	29.9	5.6	22.9	3.0	15.4	7.3	6.6	10.4	18.6
Arithmetic mean	10.2	10.5	2.4	4.8	0.5	2.1	1.2	0.5	2.0	11.8
Standard deviation	7.9	9.2	1.1	5.6	0.9	3.9	1.9	1.7	2.5	4.0
Volume-weighted mean	7.1	7.2	1.9	2.6	0.2	1.0	0.6	0.1	1.1	11.0
Katherine, Australia (N = 40; vwm pH = 4.78)										
Minimum	0.0	0.0	0.4	1.7	0.0	0.0	0.0	0.0	0.3	4.1
Maximum	18.7	24.6	20.5	84.8	15.1	64.9	7.7	50.3	18.6	61.7
Arithmetic mean	6.9	8.3	5.9	20.6	3.6	11.3	1.2	5.7	2.8	19.2
Standard deviation	4.3	5.3	4.8	22.2	4.0	14.2	1.5	10.7	3.3	12.3
Volume-weighted mean	5.5	6.3	4.3	11.8	2.0	7.0	0.9	2.5	2.0	16.6
San Carlos, Venezuela (N = 14; vwm pH = 4.81)										
Minimum	0.0	0.0	0.3	0.0	0.0	0.1	0.0	0.0	0.1	5.8
Maximum	6.7	7.3	14.5	11.6	2.6	6.8	6.2	2.6	12.9	39.8
Arithmetic mean	3.0	3.3	3.5	4.3	0.8	2.7	1.1	0.5	2.3	17.0
Standard deviation	2.4	2.5	3.6	3.4	0.9	2.0	1.7	0.7	3.4	10.0
Volume-weighted mean	2.7	2.9	2.6	2.5	0.5	1.8	0.8	0.3	2.3	15.5
St. Georges, Bermuda (N = 67; vwm pH = 4.79)										
Minimum	0.0	5.0	0.0	19.7	4.0	18.7	0.4	0.0	0.0	0.6
Maximum	159	337	50.5	1780	321	1490	47.3	67.5	38.9	162
Arithmetic mean	21.6	48.8	7.9	264	49.3	221	6.5	14.4	4.8	21.1
Standard deviation	25.3	47.7	9.1	337	60.8	282	8.0	14.2	7.1	26.9
Volume-weighted mean	18.3	36.3	5.5	175	34.5	147	4.3	9.7	3.8	16.2

Continued. . .

Table 3.12 (Concluded).

$H_2PO_4^-$ values for San Carlos only were minimum, 0.0; maximum, 4.64; arithmetic mean, 0.60; standard deviation, 1.29; volume-weighted mean (vwm), 0.6.
Only those samples with enough volume were analyzed for all elements. All values are for chloroform added aliquots, except the Bermuda values; samples in Bermuda were refrigerated in the dark at 4°C until analyzed.

* GPCP: Global Precipitation Chemistry Project.
**SO_4^{2-} = Concentration of SO_4^{2-} in excess of that supplied by seasalt.

Source: Galloway et al. (1982). (Reprinted with permission, Copyright 1982 American Geophysical Union).

Similarly, Sequeira (1981) from an analysis of global precipitation chemistry data from WMO (World Meteorological Organization) concluded that even remote maritime baseline stations, far removed from major continents, could become predisposed to acidic rain if there is a deficiency of non-marine calcium relative to non-marine sulphate. Sequeira also concluded that not all non-marine sulphate and nitrate could be present as acid. Most recently Lefohn and Krupa (1988) found that even in polluted sites, 30-98% of the total anion (SO_4^{2-} + NO_3^- + Cl^-) concentrations were most likely associated with cations other than H^+.

3.6 BACKGROUND AIR QUALITY AT FORTRESS MOUNTAIN

3.6.1 Monitoring Site Description and Methodology

To relate observed air quality conditions (and modification to "best case" situations), it is necessary to monitor the composition of the air and precipitation entering the province of Alberta from the outside. This "background air quality" is by no means constant; air from parts of the United States will have a different composition than air from British Columbia, and both will differ from air from the Northwest Territories. In addition, individual sources will give rise to variations in composition of air from each of these three major areas. Accordingly, a fixed station should be located near the British Columbia border, representing the most frequent source area of influent air (the reader is referred to *Researching Acid Deposition: Workshop Proposals, Acid Deposition Research Program*, ADRP, Calgary, Alberta, Canada, June 1984).

Thus, a background air quality monitoring station was established at Fortress Mountain, Alberta (Figures 3.8, 3.9, 3.10 and Table 3.13). Air pollutants (both dry and wet--rain only) and meteorological parameters were measured (Table 3.14) at this site for two full years (1985-87).

A quality control and quality assurance program was implemented during data collection and archiving:

1. Routine multipoint calibration of automated ambient air quality monitors was performed approximately once every month. Meteorological sensors were calibrated every six months.
2. Automatic two-point zero and span calibration of continuous ambient air quality monitors occurred daily. All ambient air quality data were corrected for errors, based on valid daily zero and span readings.
3. The status of all instrumentation was checked on a weekly basis. A number of parameters indicative of the instrument status were recorded and used in both real-time and during the quality assurance process to ascertain proper operation of the equipment.
4. Hourly air quality data were retrieved daily, including weekends and holidays, from each monitoring site by telephone and inspected to ascertain instrument status and performance.
5. The air quality monitoring sites were visited by a technician at least once per week and usually more frequently as required by maintenance schedules, and after daily inspection of the data.
6. Independent audits of the ambient air quality monitors were performed by the personnel from Alberta Department of Environment four times over the two-year monitoring period.
7. After collection, data were processed and summarized according to hour of the day and month of the year to verify their physical meaning. Values recorded during periods of instrument malfunction were flagged. Periods of malfunction were determined from documentation maintained by the field technician. Inconsistencies between data for related parameters were searched for and flagged, after a physical basis for the inconsistency was found. These activities were part of a rigorous verification procedure that was performed before the data were archived and analyzed.

Figure 3.8 Map of western Canada and colour satellite image showing the locations of the ADRP Biophysics Research air quality and meteorological monitoring stations (A, B, C) along with major towns and cities (1 to 9). The air quality and meteorological monitoring stations are identified as (A) **Fortress Mountain**, (B) **Crossfield West**, and (C) **Crossfield East**. The towns and cities are identified as (1) Banff, (2) Canmore, (3) Exshaw, (4) Cochrane, (5) Crossfield, (6) Airdrie, (7) Calgary, (8) Okotoks, and (9) High River. The Rocky Mountains are seen to the left, while the agricultural land is seen to the right as the patch-work pattern. The satellite image is a combination of images acquired on September 28, 1985, and on September 6, 1986, by LANDSAT 5. The scale of the image is approximately 1:1,200,000. Satellite image courtesy of Advanced Satellite Productions Inc. and Canada Centre for Remote Sensing Company.

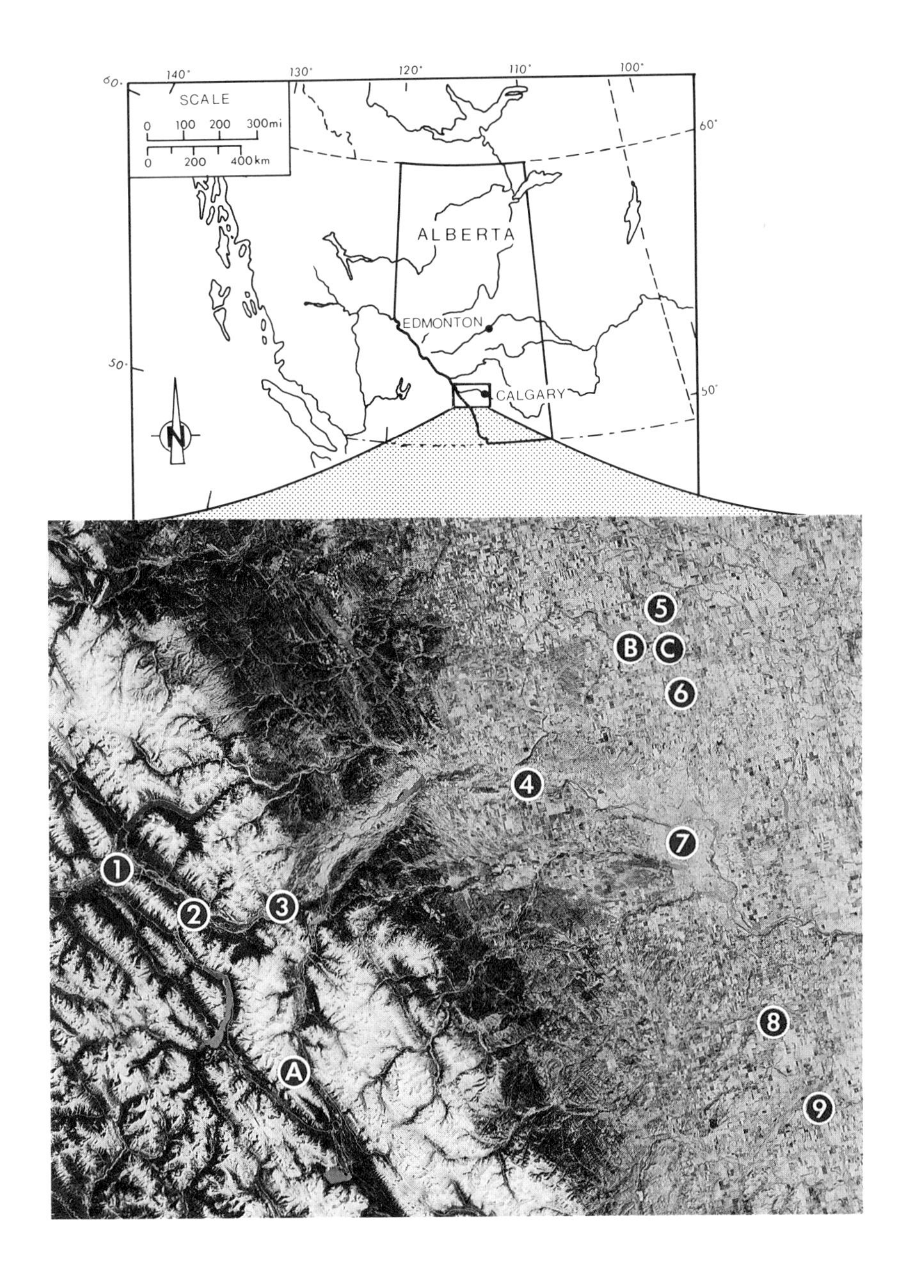

140°
130°
120°
110°
100°
60°
SCALE
0 100 200 300mi
0 200 400km
ALBERTA
EDMONTON
CALGARY
50°
N
1
2
3
4
5
6
7
8
9
A
B
C

Figure 3.9 View from a helicopter, looking west, of the Fortress Mountain, Alberta, air quality monitoring station.

Figure 3.10 (A) View from a helicopter, looking north, of the Fortress Mountain air quality monitoring station. (B) Wide-angle view of some of the equipment located on top of the air quality monitoring station. (C) Wide-angle view inside the air quality monitoring station showing the station computer operator and the rack-mounted air quality and meteorological instrumentation.

Table 3.13 Description of the sampling site at Fortress Mountain, Alberta.

Site Name:	Fortress Mountain
Coordinates:	50°49'40" 115°12'20"
Elevation:	2103 m MSL
Site Description:	
a) Location:	The monitoring site is located on the northernmost edge of a ridge, at the treeline, paralleling the Kananaskis River valley, which has a southwest to northeast orientation.
b) Access:	Access to the Fortress site from Calgary is via the TransCanada Highway west, and south on Highway 40 to the Fortress Junction. A 10 km gravel road proceeds to the Fortress ski resort. The monitoring site is located approximately 1 km beyond the resort facilties.
c) Topography:	Terrain in the vincity is mountainous. Elevation of the valley floor, ridge, and mountain peaks of the surrounding area are approximately 1200, 2100, and 3000 m, respectively.
d) Vegetation:	The monitoring site lies within the subalpine and alpine ecoregions. The vegetation in this area is sparse. The more common vegetation components of the two ecoregions consist of lodgepole pine (secondary succession by Engelman spruce) within the sub-alpine region and low growing shrubs, fescue grass, and *Phyllodoce* communities, within the alpine region (vegetation is dependent on substrate, snow cover, and protection from the environment within the alpine ecoregion).
e) Sources in Vicinity:	Summer: human activity is low, mainly hiking and camping. Winter: human activity is much higher due to the large number of skiers.
f) Snowfall:	$\bar{x}$ 263.7 mm $\pm$ 97.2 mm y^{-1}
g) Rainfall:	$\bar{x}$ 319.3 mm $\pm$ 85.3 mm y^{-1}
h) Soil:	Brunisolic soils are most abundant in this region. In some areas, soils are poorly developed due to harsh alpine environment.
i) Emissions:	▪ The total combined NO_x and SO_2 emissions within the subalpine and alpine ecoregions as a percentage of total provincial NO_x and SO_2 emissions is 0.3%. ▪ Total NO_x and SO_2 emissions within these 2 ecoregions are 2.6 and 4.2 t d^{-1}, respectively. ▪ Percentage of the total area of Alberta occupied by these 2 ecoregions is 6.3%.

Table 3.14 Summary of air quality and meteorology measurement methodology.

Measurement Category and Parameter	Measurement Method	Time Period of Data Resolution
1. Gases		
a) Near Continuous		
H_2S	Flame photometry	2 min.
SO_2	Flame photometry	2 min.
NO_x (NO, NO_2)	Chemiluminescence	2 min.
O_3	UV photometry	2 min.
CO_2	IR absorption	2 min.
b) Instantaneous		
COS	Gas chromatography*	
CS_2	Gas chromatography	
H_2S	Gas chromatography	
SO_2	Gas chromatography	
c) Integrated		
SO_2	Annular denuder + IC**	24 h or 48 h
HNO_3 (vapour)	Annular denuder + IC	24 h or 48 h
HNO_2 (vapour)	Annular denuder + IC	24 h or 48 h
NH_3	Annular denuder + colorimetry	24 h or 48 h
NMHC[1]	Gas chromatography*	1 h
2. Particulate Matter (Integrated)		
SO_4^{2-}	Annular denuder/filter + IC	24 h or 48 h
NO_3^-	Annular denuder/filter + IC	24 h or 48 h
NH_4^+	Annular denuder/filter + colorimetry	24 h or 48 h
Cations (24 ions)	Dichot filter + XRF**	48 h or 96 h
3. Precipitation (rain only) [SO_4^{2-}, NO_3^-, Cl^-, Ca^{2+}, Mg^{2+}, Na^+, K^+, NH_4^+, H^+]	Automated, refrigerated sequential sampler + IC + AA + colorimetry	32 mm sequential, samples of individual rain events

Continued...

Table 3.14 (Concluded).

Measurement Category and Parameter	Measurement Method	Time Period of Data Resolution
4. Meteorology (10 m) (Surface)		
Wind direction	Wind vane	2 min.
Wind speed	Anemometer	2 min.
Air temperature	Thermistors	2 min.
Dew point	Dew cell	2 min.
Barometric pressure	Analog output barometer	2 min.
Radiation	Pyrradiometer	2 min.
Rain	Tipping bucket	--

* Identity of individual substances was confirmed by mass spectrometry.
**IC = Ion chromatography; XRF = x-ray fluorescence spectroscopy; AA = Atomic absorption spectroscopy.
1 Non-methane hydrocarbons.

8. Before being archived, data were subjected to a second rigorous verification procedure in which they were scrutinized on a statistical basis. Results for each data subset were compared with each other and interpreted with respect to air quality data collected from other monitoring programs. Only after passing these two verification steps were data entered into the archive for use in subsequent statistical analyses.

3.6.2 Automated Measurements of Gaseous Air Pollutants and Meteorological Parameters

Initially the pollutant concentrations were evaluated for their frequency distributions. Such distributions did not fit a "normal" or "lognormal" frequency distribution. They were best described by the distribution functions of the Weibull family (Table 3.15).

Historically, most air pollutant concentrations were reported as hourly averages (e.g., U.S. EPA-SAROAD [Storage and Retrieval of Aerometric Data]). In certain cases, such as with SO_2 and H_2S downwind from point sources, pollutant episodes can occur for periods much shorter than one hour. Under these conditions, computations of hourly averages can provide distorted expressions of the actual episodes or exposure regimes. To prevent this possible problem in the present study, gaseous pollutant concentrations and meteorological parameters were evaluated as both 10 minute and hourly means. Such means were computed from 2 minute data points or observations.

Table 3.15 Summary statistics of the tests of the Weibull Distribution of the concentrations of gaseous air pollutants.

Monitoring Location	Weibull Parameters Shape	Scale	Loc.	KS Distance -Raw[1]	% Below DL[2]	KS Distance-Synthetic[3]	% Below DL[2]
Fortress Mountain							
NO_2	1.2746	1.2885	0	0.0054439	98.12	0.0072190	98.57
NO	0.8387	0.4731	0	0.0009713	99.79	0.0072200	99.69
O_3	5.2635	44.028	0	0.0363230	00.02	0.0072200	00.03
SO_2	0.3092	0.7925	0	0.0006531	99.45	0.0082004	99.56
H_2S	1.2318	0.5866	0	0.0000118	100.00	0.0062600	100.00
Crossfield West							
NO_2	0.6266	1.94670	0	0.0220210	74.35	0.0072190	79.29
NO	0.1981	0.35390	0	0.0022240	97.13	0.0083000	97.68
O_3	2.8848	27.3688	0	0.0303918	02.89	0.0072200	00.74
SO_2	0.2316	0.01797	0	0.0027986	96.36	0.0111400	96.50
H_2S	0.4716	0.81067	0	0.0015200	99.32	0.0113500	99.54
Crossfield East							
NO_2	0.6664	5.0746	0	0.0281463	52.19	0.0079200	57.50
NO	0.2425	0.2067	0	0.0069340	85.96	0.0063800	87.29
O_3	1.6544	26.642	0	0.0666501	11.48	0.0062600	5.76
SO_2	0.3619	0.5543	0	0.0073998	91.72	0.0098419	93.18
H_2S	0.2138	0.4996	0	0.0021576	97.62	0.0083070	98.42

[1] Least Kolmogorov-Smirnov distance between raw data and Weibull Distribution family.
[2] DL = Detection Limit.
[3] Kolmogorov-Smirnov distance between synthetically generated data and Weibull Distribution with parameters from (1).

Cumulative probability distribution function for the ADRP data:

$F(x) = 1 - \exp[-\{(x-\xi)/\alpha\}^c]$

where c is the shape, α is the scale, ξ is the location,

mean $= \Gamma(c^{-1} + 1)$,

variance $= \Gamma(2c^{-1} + 1) - [\Gamma(c^{-1} + 1)]^2$,

median $= \alpha[\log_e 2]^{1/c} + \xi$,

mode $= \alpha[(c-1)/c]^{1/c} + \xi$, and Γ is the Gamma function.

Continued...

Table 3.15 (Concluded).

Accuracy of the Weibull Generator Method

Percent of Lost Data	Relative Accuracy % Scale	Shape
0	97	97
5	97	97
10	97	97
20	99	99
30	100	98
50	100	99
60	99	100
70	98	99
80	97	98
90	89	89
95	91	86

At Fortress Mountain (with the exception of O_3 and CO_2) >99% of the 10 minute mean values of all pollutant concentrations measured by automated methods were below the minimum detection limit (DL) for the appropriate methodology used in each case. Frequently, when air pollutant concentrations are below the detection limit, they are reported in the literature as a zero value. In reality, the pollutant concentrations most likely lie between the zero and the minimum detection limit. Because of this, the computation of a mean statistic using a number of zero values leads to artifacts or underestimates of the true mean. In some cases, individual investigators have used a value equal to one-half the minimum detection limit (1/2 DL) as a substitute for the zero. The 1/2 DL was substituted uniformly for all values below the minimum detection limit.

As previously stated, in the present case the pollutant frequency distributions were best defined by the Weibull function. Therefore, in the case of the primary pollutants with the exception of CO_2, which was never observed below the minimum detection limit, a stochastic Weibull probability generator was used to substitute all values below the minimum detection limit with values predicted by the probability generator. This effort was performed after identifying the Kolmogorov-Smirnov (KS) least distance between the theoretical and empirical Weibull distributions for each primary pollutant (Table 3.15). The unbias and robustness of the Weibull generator method was also tested prior to its application. Totally synthetic data of the Weibull distribution, with known shape and scale parameters, were generated. These data closely resembled the actual ambient pollutant distribution. A certain percentage of the lowest values were then discarded from the synthetic data set, thus simulating the effect of the minimum detection limit of the ambient pollutant characterization. These truncated data were very similar to the actual observations in the present case. It was then assumed that the values for the discarded data were not known and were estimated from the truncated distributions through procedures analogous to those used with the actual ambient data. Table 3.15 provides summary data on the high level of accuracy in the statistical parameters of measurement when observations below the detection limit were replaced by values obtained through the Weibull generator method. Tables 3.16 through 3.23 provide a comparison of the results of the computation of mean values using three different approaches: (1) substitution through the probability generator; (2) uniform substitution with the value for 1/2 DL; and (3) uniform substitution with zero. As one would expect, the computations with 1/2 DL substitution provided the highest mean values, while the use of "zero" for all values below the minimum detection limit resulted in the lowest mean values. The application of the probability generator resulted in intermediate values which may be considered as more realistic. The summary statistics on the reconstructed Weibull frequency distributions are presented in Tables 3.24 through 3.29.

Table 3.16 Automated air quality measurement data: comparison of mean values of SO_2 (ppb) using different computational methods.

	Type of Computation*	Fortress Mountain		Crossfield West		Crossfield East	
		10 min	hour	10 min	hour	10 min	hour
Year 1	Weibull	0.125	0.134	0.595	0.592	1.389	1.373
	1/2 DL	2.020	2.021	2.365	2.359	2.839	2.823
Year 2	Weibull	0.170	0.179	0.520	0.518	1.309	1.299
	1/2 DL	2.055	2.055	2.287	2.281	2.831	2.822
Both Years	Weibull	0.147	0.156	0.558	0.555	1.349	1.335
	1/2 DL	2.038	2.038	2.327	2.321	2.835	2.822
	0-Subst.	0.054	0.056	0.416	0.408	0.955	0.942

Continued...

Table 3.16 (Concluded).

* Weibull: Values below the detection limit (DL) (4 ppb) substituted through the use of a stochastic probability generator for a Weibull distribution; 1/2 DL: one-half of the detection limit (2 ppb) uniformly substituted for all values 4 ppb; 0-Subst: values below the detection limit (4 ppb) uniformly substituted with zero.

Table 3.17 Automated air quality measurement data: comparison of median values of SO_2 (ppb) using different computational methods.

	Type of Computation*	Fortress Mountain		Crossfield West		Crossfield East	
		10 min	hour	10 min	hour	10 min	hour
Year 1	Weibull	0.100	0.100	0.100	0.200	0.400	0.400
	1/2 DL	2.000	2.000	2.000	2.000	2.000	2.000
Year 2	Weibull	0.000	0.100	0.100	0.200	0.400	0.400
	1/2 DL	2.000	2.000	2.000	2.000	2.000	2.000
Both Years	Weibull	0.000	0.100	0.100	0.200	0.400	0.400
	1/2 DL	2.000	2.000	2.000	2.000	2.000	2.000
	0-Subst.	0.000	0.000	0.000	0.000	0.000	0.000

* Weibull: Values below the detection limit (DL) (4 ppb) substituted through the use of a stochastic probability generator for a Weibull distribution; 1/2 DL: one-half of the detection limit (2 ppb) uniformly substituted for all values 4 ppb; 0-Subst: values below the detection limit (4 ppb) uniformly substituted with zero.

Table 3.18 Automated air quality measurement data: comparison of mean values of H_2S (ppb) using different computational methods.

	Type of Computation*	Fortress Mountain		Crossfield West		Crossfield East	
		10 min	hour	10 min	hour	10 min	hour
Year 1	Weibull	0.551	0.551	0.307	0.304	0.297	0.298
	1/2 DL	2.000	2.000	2.052	2.050	2.150	2.147
Year 2	Weibull	0.553	0.554	0.247	0.246	0.363	0.366
	1/2 DL	2.002	2.002	2.003	2.003	2.213	2.211
Both Years	Weibull	0.552	0.552	0.278	0.276	0.330	0.333
	1/2 DL	2.001	2.001	2.028	2.027	2.182	2.180
	0-Subst.	0.001	0.001	0.036	0.035	0.232	0.230

Continued...

Table 3.18 (Concluded).

* Weibull: Values below the detection limit (4 ppb) substituted through the use of a stochastic probability generator for a Weibull distribution; 1/2 DL: one-half of the detection limit (2 ppb) uniformly substituted for all values 4 ppb; 0-Subst: values below the detection limit (4 ppb) uniformly substituted with zero.

Table 3.19 Automated air quality measurement data: comparison of median values of H_2S (ppb) using different computational methods.

	Type of Computation*	Fortress Mountain		Crossfield West		Crossfield East	
		10 min	hour	10 min	hour	10 min	hour
Year 1	Weibull	0.500	0.500	0.200	0.200	0.000	0.100
	1/2 DL	2.000	2.000	2.000	2.000	2.000	2.000
Year 2	Weibull	0.500	0.500	0.200	0.200	0.000	0.100
	1/2 DL	2.000	2.000	2.000	2.000	2.000	2.000
Both Years	Weibull	0.500	0.500	0.200	0.200	0.000	0.100
	1/2 DL	2.000	2.000	2.000	2.000	2.000	2.000
	0-Subst.	0.000	0.000	0.000	0.000	0.000	0.000

* Weibull: Values below the detection limit (4 ppb) substituted through the use of a stochastic probability generator for a Weibull distribution; 1/2 DL: one-half of the detection limit (2 ppb) uniformly substituted for all values 4 ppb; 0-Subst: values below the detection limit (4 ppb) uniformly substituted with zero.

Table 3.20 Automated air quality measurement data: comparison of mean values of NO (ppb) using different computational methods.

	Type of Computation*	Fortress Mountain		Crossfield West		Crossfield East	
		10 min	hour	10 min	hour	10 min	hour
Year 1	Weibull	0.522	0.522	0.466	0.469	4.240	4.212
	1/2 DL	2.015	2.015	2.298	2.297	5.656	5.628
Year 2	Weibull	0.526	0.525	0.283	0.288	3.011	2.992
	1/2 DL	2.018	2.017	2.135	2.135	4.463	4.445
Both Years	Weibull	0.524	0.523	0.374	0.377	3.601	3.578
	1/2 DL	2.016	2.016	2.215	2.215	5.036	5.013
	0-Subst.	0.021	0.020	0.261	0.261	3.293	3.268

Continued...

Table 3.20 (Concluded).

* Weibull: Values below the detection limit (4 ppb) substituted through the use of a stochastic probability generator for a Weibull distribution; 1/2 DL: one-half of the detection limit (2 ppb) uniformly substituted for all values 4 ppb; 0-Subst: values below the detection limit (4 ppb) uniformly substituted with zero.

Table 3.21 Automated air quality measurement data: comparison of median values of NO (ppb) using different computational methods.

	Type of Computation*	Fortress Mountain		Crossfield West		Crossfield East	
		10 min	hour	10 min	hour	10 min	hour
Year 1	Weibull	0.500	0.500	0.100	0.100	0.400	0.400
	1/2 DL	2.000	2.000	2.000	2.000	2.000	2.000
Year 2	Weibull	0.500	0.500	0.000	0.100	0.300	0.400
	1/2 DL	2.000	2.000	2.000	2.000	2.000	2.000
Both Years	Weibull	0.500	0.500	0.000	0.100	0.300	0.400
	1/2 DL	2.000	2.000	2.000	2.000	2.000	2.000
	0-Subst.	0.000	0.000	0.000	0.000	0.000	0.000

* Weibull: Values below the detection limit (4 ppb) substituted through the use of a stochastic probability generator for a Weibull distribution; 1/2 DL: one-half of the detection limit (2 ppb) uniformly substituted for all values 4 ppb; 0-Subst: values below the detection limit (4 ppb) uniformly substituted with zero.

Table 3.22 Automated air quality measurement data: comparison of mean values of NO_2 (ppb) using different computational methods.

	Type of Computation*	Fortress Mountain		Crossfield West		Crossfield East	
		10 min	hour	10 min	hour	10 min	hour
Year 1	Weibull	1.213	1.212	2.932	2.945	6.567	6.572
	1/2 DL	2.058	2.059	3.650	3.661	6.911	6.917
Year 2	Weibull	1.184	1.183	2.648	2.658	6.409	6.433
	1/2 DL	2.036	2.035	3.390	3.398	6.771	6.795
Both Years	Weibull	1.198	1.197	2.789	2.800	6.486	6.501
	1/2 DL	2.047	2.047	3.518	3.528	6.839	6.854
	0-Subst.	0.077	0.076	2.005	2.017	5.766	5.782

Continued...

Table 3.22 (Concluded).

* Weibull: Values below the detection limit (4 ppb) substituted through the use of a stochastic probability generator for a Weibull distribution; 1/2 DL: one-half of the detection limit (2 ppb) uniformly substituted for all values 4 ppb; 0-Subst: values below the detection limit (4 ppb) uniformly substituted with zero.

Table 3.23 Automated air quality measurement data: comparison of median values of NO_2 (ppb) using different computational methods.

	Type of Computation*	Fortress Mountain		Crossfield West		Crossfield East	
		10 min	hour	10 min	hour	10 min	hour
Year 1	Weibull	1.100	1.100	1.300	1.200	2.600	2.700
	1/2 DL	2.000	2.000	2.000	2.000	2.800	3.100
Year 2	Weibull	1.100	1.100	1.200	1.200	2.300	2.500
	1/2 DL	2.000	2.000	2.000	2.000	2.400	2.900
Both Years	Weibull	1.000	1.100	1.300	1.200	2.400	2.600
	1/2 DL	2.000	2.000	2.000	2.000	2.600	3.000
	0-Subst.	0.000	0.000	0.000	0.000	1.000	1.700

* Weibull: Values below the detection limit (4 ppb) substituted through the use of a stochastic probability generator for a Weibull distribution; 1/2 DL: one-half of the detection limit (2 ppb) uniformly substituted for all values 4 ppb; 0-Subst: values below the detection limit (4 ppb) uniformly substituted with zero.

Table 3.24 Best-fit Weibull distribution parameters and empirical frequency distribution of 10-minute SO_2 concentrations (ppb) 1985-87.

	Fortress Mountain	Crossfield West	Crossfield East
Shape	0.396	0.447	1.141
Scale	0.042	0.145	0.509
Range			
0 - 4	99.271	96.786	93.883
4 - 10	0.629	1.997	3.327
10 - 20	0.099	0.870	1.514
20 - 30	0.001	0.185	0.562
30 - 40		0.070	0.299
40 - 50		0.032	0.158
50 - 60		0.017	0.104
60 - 70		0.013	0.040

Continued...

Table 3.24 (Concluded).

	Fortress Mountain	Crossfield West	Crossfield East
70 - 80		0.010	0.027
80 - 90		0.010	0.034
90 - 100		0.006	0.014
100 - 110		0.001	0.011
110 - 120		0.001	0.008
120 - 130		0.001	0.005
130 - 140		0.001	0.004
140 - 150			0.005
150 - 160			0.003
160 - 170			0.002

Minimum detection limit for SO_2: 4.0 ppb. Data include substitution of Weibull probability generated values for all values <4.0 ppb.

Table 3.25 Best-fit Weibull distribution parameters and empirical frequency distribution of 10-minute H_2S concentrations (ppb) 1985-87.

	Fortress Mountain	Crossfield West	Crossfield East
Shape	3.222	0.793	0.324
Scale	0.594	0.269	0.049
Range			
0 - 4	99.985	99.657	97.826
4 - 8	0.011	0.241	1.337
8 - 12	0.002	0.027	0.417
12 - 16	0.001	0.021	0.198
16 - 20	0.001	0.014	0.115
20 - 24		0.009	0.042
24 - 28		0.007	0.024
28 - 32		0.009	0.013
32 - 36		0.006	0.010
36 - 40		0.000	0.005
40 - 44		0.001	0.003
44 - 48		0.004	0.003
48 - 52		0.000	0.002
52 - 56		0.001	0.002
56 - 60		0.000	0.001
60 - 64		0.002	0.001
64 - 68		0.000	0.000
68 - 72		0.001	0.001

Minimum detection limit for H_2S: 4.0 ppb. Data include substitution of Weibull probability generated values for all values <4.0 ppb.

Table 3.26 Best-fit Weibull distribution parameters and empirical frequency distribution of 10-minute NO concentrations (ppb) 1985-87.

	Fortress Mountain	Crossfield West	Crossfield East
Shape	2.216	0.365	0.611
Scale	0.552	0.061	0.616
Range			
0 - 4	99.842	98.009	88.333
4 - 20	0.155	1.656	6.917
20 - 40	0.003	0.286	2.407
40 - 60	0.026	1.007	
60 - 80	0.012	0.512	
80 - 100	0.009	0.283	
100 - 120	0.002	0.192	
120 - 140	0.129		
140 - 160	0.072		
160 - 180	0.063		
180 - 200	0.033		
200 - 220	0.029		
220 - 240	0.015		
240 - 260	0.006		
260 - 280	0.002		

Minimum detection limit for NO: 4.0 ppb. Data include substitution of Weibull probability generated values for all values <4.0 ppb.

Table 3.27 Best-fit Weibull distribution parameters and empirical frequency distribution of 10-minute NO_2 concentrations (ppb) 1985-87.

	Fortress Mountain	Crossfield West	Crossfield East
Shape	3.358	1.188	1.017
Scale	1.262	1.992	4.569
Range			
0 - 4	99.191	79.835	58.433
4 - 10	0.783	14.792	21.045
10 - 15	0.025	2.776	7.011
15 - 20	0.001	1.098	4.429
20 - 25		0.646	3.419
25 - 30		0.401	2.664
30 - 35		0.300	1.458
35 - 40		0.094	0.810
40 - 45		0.026	0.413
45 - 50		0.011	0.210
50 - 55		0.017	0.055
55 - 60		0.004	0.035
60 - 65			0.013

Continued...

Table 3.27 (Concluded).

	Fortress Mountain	Crossfield West	Crossfield East
65 - 70			0.003
70 - 75			0.001
75 - 80			0.000
80 - 85			0.001

Minimum detection limit for NO_2: 4.0 ppb. Data include substitution of Weibull probability generated values for all values <4.0 ppb.

Table 3.28 Best-fit Weibull distribution parameters and empirical frequency distribution of 10-minute O_3 concentrations (ppb) 1985-87.

	Fortress Mountain	Crossfield West	Crossfield East
Shape	4.375	2.755	1.692
Scale	45.723	31.310	28.939
Range			
0 - 5	0.010	2.024	10.651
5 - 10	0.006	2.865	5.775
10 - 20	1.126	18.027	18.986
20 - 30	8.339	37.575	27.775
30 - 40	34.242	25.569	24.747
40 - 50	33.565	8.989	8.698
50 - 60	13.114	3.879	2.864
60 - 70	5.727	0.934	0.452
70 - 80	2.341	0.122	0.049
80 - 90	0.915	0.016	0.001
90 - 100	0.422		0.000
100 - 110	0.156		0.001
110 - 120	0.032		0.000
120 - 130	0.005		0.000
130 - 140			0.001

Minimum detection limit for O_3: 5 ppb. Data include substitution of Weibull probability generated values for all values <5.0 ppb.

Table 3.29 Best-fit Weibull distribution parameters and empirical frequency distribution of 10-minute CO_2 concentrations (ppm) 1985-87.

	Fortress Mountain	Crossfield West	Crossfield East
Shape	50.743	37.527	32.791
Scale	348.872	351.049	352.623
Range			
285 - 300	0.000	0.020	0.000
300 - 315	0.188	0.193	0.093
315 - 330	2.318	5.053	5.929
330 - 345	37.871	36.160	33.653
345 - 360	57.859	46.910	46.335
360 - 375	1.608	10.146	9.902
375 - 390	0.156	1.193	2.298
390 - 405		0.220	0.815
405 - 420		0.063	0.429
420 - 435		0.025	0.243
435 - 450		0.010	0.138
450 - 465		0.004	0.071
465 - 480		0.002	0.057
480 - 495		0.001	0.036
495 - 500			0.001

Minimum detection limit for CO_2: none at known ambient concentrations. Weibull probability substitution was not used.

Tables 3.30 through 3.35 provide summary statistics of Weibull substituted 10 minute and 1 hour data at Fortress Mountain during 1985-86, 1986-87, and the combined years. The 10 minute and hourly mean SO_2, H_2S, NO, NO_2, and O_3 concentrations were identical to each other during a given year (compare the results in Table 3.30 with those in Table 3.31; results in Table 3.32 with those in Table 3.33). Similarly, the median 10 minute or 1 hour pollutant concentrations during a given year were identical to the corresponding value for the other year (Table 3.30 vs 3.32, and Table 3.31 vs 3.33). These results suggested that the overall air quality at Fortress Mountain was fairly homogeneous when smoothed annual characteristics were evaluated. Table 3.36 provides a comparison of the mean hourly pollutant concentrations observed over two years (1985-87) at Fortress Mountain with the corresponding values reported in the literature for unpolluted or least polluted sites (remote or rural). The mean hourly concentrations at Fortress Mountain for SO_2, NO_2, and O_3 were below or nearly equal to the

Table 3.30 Summary statistics of automated gaseous pollutant measurements and meteorological variables*. Sampling location: Fortress Mountain. Sampling period: 1985-86. Time resolution of data: 10 minutes.

Variable	(n)	Mean	S.D.	Range		Median	Percentile						Skewness	Kurtosis
				Min.	Max.		10	30	70	90	95	99		
1. SO_2	44637	.13	.48	.00	23.6	.00	.00	.00	.10	.30	.40	.80	17.2	443.5
2. H_2S	48060	.55	.20	.00	2.9	.50	.30	.40	.60	.80	.90	1.1	.65	.99
3. NO	48530	.52	.40	.00	20.8	.50	.20	.30	.60	.90	1.00	1.4	14.5	489.3
4. NO_2	46128	1.2	.64	.10	14.0	1.1	.70	.90	1.3	1.7	1.9	4.0	5.37	56.0
5. CO_2	49067	345.7	6.7	320.0	387.0	346.6	336.2	341.6	350.0	354.0	355.0	358.2	- .23	- .54
6. O_3	47883	40.9	9.2	.00	73.0	40.8	28.8	36.6	45.6	52.8	56.0	60.4	- .15	.05
7. Wind Speed	50498	14.0	9.5	.00	88.2	12.8	3.0	8.4	17.6	25.6	30.8	45.2	1.1	2.48
8. Wind Direction	50718	203.2	66.1	1.00	358.2	201.6	145.8	179.2	231.5	279.4	318.2	349.8	- .62	1.59
9. Temperature	50735	1.1	9.4	-33.10	22.8	.90	- 9.80	- 2.50	6.5	12.8	15.4	18.8	- .66	.85
10. Barometric pressure	50735	782.5	7.6	748.0	815.2	783.6	773.0	779.0	787.0	791.0	793.0	796.0	- .72	.96
11. Precipitation	21336	.001	.01	.00	.90	.00	.00	.00	.00	.00	.00	.00	26.7	1254.4
12. Humidity	49982	37.6	8.7	20.0	66.6	37.0	26.0	32.6	42.2	50.2	52.4	55.2	.12	- .72
13. Radiation	41176	347.2	496.5	- 2.00	2218.2	69.8	.00	.00	426.2	1137.5	1484.6	1900.4	1.53	1.46

* All gas concentrations in ppb, except CO_2 (ppm); wind speed: km h^{-1}; wind direction: degrees; temperature: °C; pressure: mbars; precipitation: mm; humidity: %; radiation: Langleys min^{-1}.

Table 3.31 Summary statistics of automated gaseous pollutant measurements and meteorological variables*. Sampling location: Fortress Mountain. Sampling period: 1985-86. Time resolution of data: 1 hour.

Variable	(n)	Mean	S.D.	Range		Median	Percentile						Skewness	Kurtosis
				Min.	Max.		10	30	70	90	95	99		
1. SO_2	7692	.13	.44	.00	14.8	.10	.00	.10	.10	.20	.20	.40	17.065	362.008
2. H_2S	8271	.55	.09	.10	1.3	.50	.40	.50	.60	.70	.70	.80	.150	.817
3. NO	8354	.52	.20	.00	5.9	.50	.40	.40	.60	.70	.7	01.1	10.281	208.550
4. NO_2	7942	1.2	.50	.10	13.0	1.1	.90	1.1	1.2	1.4	1.5	3.5	8.893	116.208
5. CO_2	8435	345.8	6.7	320.7	378.9	346.6	336.3	341.6	350.2	353.8	355.0	358.3	- .235	- .584
6. O_3	8378	40.9	9.1	2.8	70.9	40.9	28.7	36.7	45.6	52.6	55.9	60.3	- .159	.039
7. Wind Speed	8437	14.0	8.9	.00	78.4	13.0	3.5	8.8	17.5	25.0	29.8	43.5	1.106	2.502
8. Wind Direction	8473	203.2	55.1	2.0	352.7	200.6	159.0	181.4	227.0	264.2	293.8	332.3	- .697	2.285
9. Temperature	8474	1.1	9.4	-33.0	22.4	.90	- 9.70	- 2.40	6.4	12.8	15.4	18.8	- .663	.853
10. Barometric pressure	8474	782.6	7.6	748.0	813.8	783.6	773.0	779.0	787.0	791.3	793.0	796.0	- .719	.961
11. Precipitation	3569	.001	.01	.00	.40	.00	.00	.00	.00	.00	.00	.00	18.370	461.498
12. Humidity	8378	37.6	8.6	20.0	64.1	37.0	26.0	32.6	42.3	50.1	52.2	55.2	.105	- .715
13. Radiation	6920	345.6	483.9	- 2.00	2154.3	73.9	.00	.00	441.2	1128.7	1430.9	1817.5	1.444	1.149

* All gas concentrations in ppb, except CO_2 (ppm); wind speed: km h^{-1}; wind direction: degrees; temperature: °C; pressure: mbars; precipitation: mm; humidity: %; radiation: Langleys min^{-1}.

Table 3.32 Summary statistics of automated gaseous pollutant measurements and meteorological variables*. Sampling location: Fortress Mountain. Sampling period: 1986-87. Time resolution of data: 10 minutes.

Variable	(n)	Mean	S.D.	Range Min.	Range Max.	Median	Percentile 10	Percentile 30	Percentile 70	Percentile 90	Percentile 95	Percentile 99	Skewness	Kurtosis
1. SO_2	43494	.17	.79	.00	17.8	.00	.00	.00	.10	.30	.40	3.4	11.072	139.028
2. H_2S	40392	.55	.25	.00	16.6	.50	.30	.40	.60	.80	.90	1.1	14.888	766.182
3. NO	47265	.53	.42	.00	23.0	.50	.20	.30	.60	.90	1.00	1.5	14.050	430.722
4. NO_2	45695	1.2	.54	.00	17.4	1.1	.70	.90	1.3	1.7	1.9	2.8	5.377	79.686
5. CO_2	49364	345.1	9.1	312.0	384.4	347.0	333.8	340.0	350.0	354.0	357.0	369.3	- .346	1.194
6. O_3	48152	45.8	14.6	10.2	129.4	41.4	31.8	37.6	49.0	67.2	74.6	91.8	1.184	1.531
7. Wind Speed	51110	15.1	9.4	.00	80.4	14.0	4.2	9.4	18.6	27.0	31.6	45.4	1.089	2.250
8. Wind Direction	51029	198.1	65.5	1.00	357.8	198.6	136.41	78.0	229.0	269.2	294.4	345.8	- .785	1.555
9. Temperature	51005	3.3	8.2	-22.8	24.6	3.3	- 6.70	- 2.20	8.4	14.3	16.5	20.0	- .047	- .558
10. Barometric pressure	51020	783.8	6.9	749.2	799.0	784.0	774.2	781.0	788.0	792.8	794.2	797.0	- .431	.194
11. Precipitation	51115	.001	.01	.00	.80	.00	.00	.00	.00	.00	.00	.00	25.823	1033.686
12. Humidity	48365	34.6	8.9	20.0	79.6	33.4	24.0	29.6	38.2	45.0	48.0	67.0	1.040	2.328
13. Radiation	51106	297.0	465.5	- 1.00	2221.6	13.2	.00	.00	318.0	1049.1	1374.9	1827.4	1.702	2.081

* All gas concentrations in ppb, except CO_2 (ppm); wind speed: km h^{-1}; wind direction: degrees; temperature: °C; pressure: mbars; precipitation: mm; humidity: %; radiation: Langleys min^{-1}.

Table 3.33 Summary statistics of automated gaseous pollutant measurements and meteorological variables*. Sampling location: Fortress Mountain. Sampling period: 1986-87. Time resolution of data: 1 hour.

				Range			Percentile							
Variable	(n)	Mean	S.D.	Min.	Max.	Median	10	30	70	90	95	99	Skewness	Kurtosis
1. SO_2	7515	.18	.75	.00	13.0	.10	.00	.10	.10	.20	.2	03.1	11.4	142.0
2. H_2S	6956	.55	.13	.30	5.7	.50	.40	.50	.60	.70	.70	.80	16.3	569.1
3. NO	8140	.53	.21	.00	7.2	.50	.40	.40	.60	.70	.70	1.1	10.4	212.2
4. NO_2	7874	1.2	.35	.20	8.6	1.1	.90	1.1	1.2	1.4	1.5	2.4	8.4	111.7
5. CO_2	8456	345.1	9.1	312.2	381.3	347.0	333.8	340.1	350.0	354.2	357.0	369.0	- .36	1.2
6. O_3	8472	45.8	14.6	13.1	122.4	41.4	31.9	37.5	49.0	67.3	74.4	92.1	1.2	1.5
7. Wind Speed	8535	15.1	8.9	.00	70.1	14.0	4.7	9.9	18.4	26.2	30.7	43.7	1.1	2.3
8. Wind Direction	8522	198.1	55.9	1.00	352.3	199.2	150.6	180.3	224.3	257.9	277.6	323.0	- .98	2.2
9. Temperature	8519	3.3	8.2	-22.60	24.4	3.4	- 6.60	- 2.10	8.4	14.3	16.5	19.9	- .05	- .56
10. Barometric pressure	8520	783.8	6.9	751.9	799.0	784.0	774.4	781.0	788.0	792.8	794.4	797.0	- .43	.18
11. Precipitation	8534	.001	.01	.00	.30	.00	.00	.00	.00	.00	.00	.00	15.9	312.0
12. Humidity	8160	34.4	8.9	20.0	78.8	33.5	23.6	29.5	38.1	45.0	47.9	66.1	1.00	2.2
13. Radiation	8534	297.6	454.0	- 1.00	2148.1	18.4	.00	.00	336.2	1035.9	1341.2	1752.0	1.6	1.7

* All gas concentrations in ppb, except CO_2 (ppm); wind speed: km h^{-1}; wind direction: degrees; temperature: °C; pressure: mbars; precipitation: mm; humidity: %; radiation: Langleys min^{-1}.

Table 3.34 Summary statistics of automated gaseous pollutant measurements and meteorological variables*. Sampling location: Fortress Mountain. Sampling period: 1985-87. Time resolution of data: 10 minutes.

Variable	(n)	Mean	S.D.	Range		Median	Percentile						Skewness	Kurtosis
				Min.	Max.		10	30	70	90	95	99		
1. SO_2	88131	.15	.65	.00	23.6	.00	.00	.00	.10	.30	.40	1.00	13.236	215.160
2. H_2S	88452	.55	.22	.00	16.6	.50	.30	.40	.60	.80	.90	1.19	.262	508.588
3. NO	95795	.52	.41	.00	23.0	.50	.20	.30	.60	.90	1.00	1.4	14.295	458.538
4. NO_2	91823	1.2	.59	.00	17.4	1.1	.70	.90	1.3	1.7	1.9	3.4	5.442	65.860
5. CO_2	98431	345.4	8.0	312.0	387.0	347.0	335.0	341.0	350.0	354.0	356.0	364.2	- .356	1.146
6. O_3	96035	43.4	12.5	.00	129.4	41.2	30.2	37.2	46.8	59.4	67.4	84.4	1.176	2.762
7. Wind Speed	101608	14.6	9.4	.00	88.2	13.4	3.6	9.0	18.2	26.4	31.2	45.2	1.101	2.346
8. Wind Direction	101747	200.7	65.8	1.00	358.2	200.0	143.2	178.6	230.2	274.0	306.6	348.4	- .697	1.579
9. Temperature	101740	2.2	8.9	-33.1	24.6	1.9	- 8.0	- 2.30	7.5	13.7	15.9	19.4	- .453	.550
10. Barometric pressure	101755	783.2	7.3	748.0	815.2	784.0	773.0	780.0	787.4	792.0	794.0	797.0	- .620	.767
11. Precipitation	72451	.001	.01	.00	.90	.00	.00	.00	.00	.00	.00	.00	26.192	1121.652
12. Humidity	98347	36.1	8.9	20.0	79.6	35.2	25.0	31.0	40.4	48.0	51.6	58.8	.551	.502
13. Radiation	92282	319.4	480.2	- 2.00	2221.6	35.0	.00	.00	369.2	1089.41	426.6	1862.4	1.621	1.787

* All gas concentrations in ppb, except CO_2 (ppm); wind speed: km h^{-1}; wind direction: degrees; temperature: ^{o}C; pressure: mbars; precipitation: mm; humidity: %; radiation: Langleys min^{-1}.

Table 3.35 Summary statistics of automated gaseous pollutant measurements and meteorological variables*. Sampling location: Fortress Mountain. Sampling period: 1985-87. Time resolution of data: 1 hour.

Variable	(n)	Mean	S.D.	Range		Median	Percentile						Skewness	Kurtosis
				Min.	Max.		10	30	70	90	95	99		
1. SO_2	15207	.16	.61	.00	14.8	.10	.00	.10	.10	.20	.20	.90	13.5	207.1
2. H_2S	15227	.55	.11	.10	5.7	.50	.40	.50	.60	.70	.70	.80	12.2	499.3
3. NO	16494	.52	.21	.00	7.2	.50	.40	.40	.60	.70	.70	1.1	10.4	211.7
4. NO_2	15816	1.2	.43	.10	13.0	1.1	.90	1.1	1.2	1.4	1.5	3.0	9.21	29.6
5. CO_2	16891	345.4	8.0	312.2	381.3	346.9	335.1	341.1	350.0	354.0	355.8	364.0	-.37	1.1
6. O_3	16850	43.4	12.4	2.8	122.4	41.2	30.3	37.1	46.7	59.4	67.4	84.2	1.2	2.8
7. Wind Speed	16971	14.5	8.9	.00	78.4	13.5	4.1	9.4	18.0	25.7	30.2	43.5	1.1	2.4
8. Wind Direction	16995	200.6	55.5	1.00	352.7	199.9	156.0	180.9	225.7	261.0	285.5	328.6	-.84	2.3
9. Temperature	16993	2.2	8.9	-33.00	24.4	1.9	- 8.00	- 2.30	7.4	13.7	15.9	19.3	-.46	.55
10. Barometric pressure	16994	783.2	7.3	748.0	813.8	784.0	773.3	780.0	787.5	792.0	794.0	797.0	-.62	.76
11. Precipitation	12103	.00	.01	.00	.40	.00	.00	.00	.00	.00	.00	.00	17.6	428.8
12. Humidity	16538	36.0	8.9	20.0	78.8	35.2	24.7	31.0	40.4	47.8	51.5	58.2	.53	.46
13. Radiation	15454	319.1	468.2	-2.00	2154.3	37.8	.00	.00	384.8	1077.2	1380.6	1789.3	1.5	1.4

* All gas concentrations in ppb, except CO_2 (ppm); wind speed: km h^{-1}; wind direction: degrees; temperature: ^oC; pressure: mbars; precipitation: mm; humidity: %; radiation: Langleys min^{-1}.

Table 3.36 A comparison of mean hourly pollutant concentrations at Fortress Mountain (1985-87) with values published in the literature for remote and rural locations. All concentrations in ppb.

Pollutant	Remote Locations	Fortress Mountain	Rural Locations
SO_2	<1.0	0.16	1.0 - 30
H_2S	0.05 - 0.10	0.55	-
NO	<0.05	0.52	0.05 - 20
NO_2	<1.0	1.2	1.0 - 20
O_3	<50	43	20 - 80

For details and literature regarding the pollutant concentrations in remote and rural locations, the reader is referred to Section 3.2 of this Chapter.

concentrations reported for remote sites. The concentrations of other pollutants were within the range reported for rural sites. The relatively high values for H_2S and NO are most likely due to the mathematical artifacts caused by the very high number of missing observations. Nevertheless, the Fortress Mountain site may be considered as representative of an acceptable background air quality site for Alberta and possibly for northwestern North America.

As opposed to the smoothed (annual mean) pollutant statistics, the range and percentiles of pollutant concentrations, both 10 minute and one hour, varied between the two monitoring years (Tables 3.30 through 3.33). This is to be expected due to the stochasticity and variability in the climatology at any given site between years. The computations of long-term means tend to mask the underlying pollutant episodicity.

At the Fortress Mountain site, while the mean hourly O_3 concentrations ranged from 40.9 ppb in 1985-86 to 45.8 ppb in 1986-87, corresponding maximum concentrations ranged from 70.9 to 129.4 ppb (Tables 3.31 and 3.33 and Figure 3.11).

These high O_3 concentrations were associated with episodes on two or more consecutive days (Table 3.37 and Figure 3.21) each time during April 15-June 30, 1987. Such O_3 episodes were observed on approximately 14 of the 45 days examined. During these episodes the highest daily O_3 maxima were observed during both daylight and nighttime hours. Unusually large, site specific O_3 episodes have been reported for other high elevation monitoring sites (U.S. EPA 1984). The high O_3 concentrations at these sites

FORTRESS MOUNTAIN

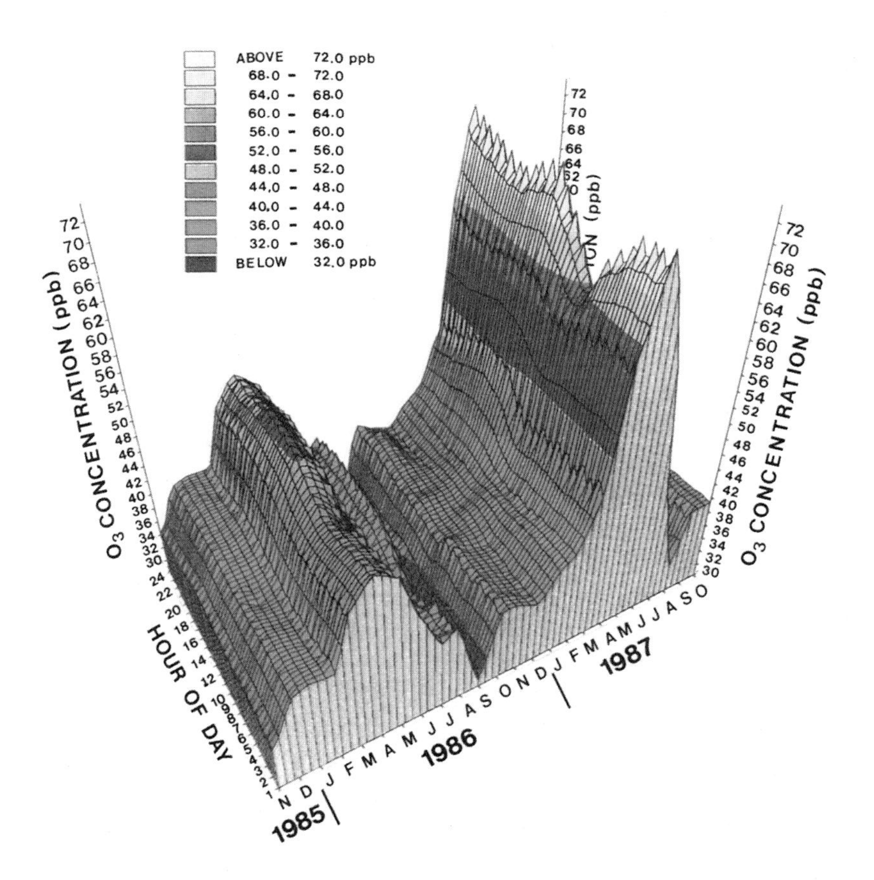

Figure 3.11 Fortress Mountain: Ozone concentration as a function of hour of the day and the month of the year (November 1, 1985 - October 31, 1987). The seasonal episodic nature of the O_3 concentration is clearly seen in this three dimensional plot.

have been attributed to stratospheric intrusion: NO_x-NMOC (non-methane organic carbon)-CO chemistry, long-range transport, or their combination (Finlayson-Pitts and Pitts 1986).

Table 3.37 Characteristics of representative O_3 episode versus non-episode days at Fortress Mountain, 1987.

Episode			Non-Episode		
Date	Highest $\bar{x}$ Hourly O_3 (ppb)	Hour	Date	Highest $\bar{x}$ Hourly O_3 (ppb)	Hour
Apr 24	93.5	1800	Apr 23	66.5	0300
Apr 25	89.3	0100	Apr 28	68.1	1300
Apr 26	86.8	1200	Apr 29	63.5	2400
May 12	105.8	2400	May 05	81.1	2400
May 13	122.4	0100	May 10	72.6	0100
May 25	113.4	2200	May 21	63.2	0400
May 26	101.1	0100	Jun 04	67.6	1500
May 27	108.3	1200	Jun 08	69.7	0200
May 28	107.6	2000	Jun 09	49.2	2400
May 29	107.3	1000	Jun 11	56.4	2400
May 30	110.3	0400	Jun 13	66.0	1700
May 31	100.3	1100	Jun 16	68.8	1400
Jun 06	102.7	0800	Jun 22	52.8	0100
Jun 07	91.4	0100	Jun 30	53.3	0700

Table 3.38 provides a comparative summary of cumulative daily average, one hour pollutant concentrations, and surface meteorological parameters for 14 O_3-episode versus 14 non-episode days at the Fortress Mountain monitoring site. With the exception of the O_3 concentrations, values for all other parameters were very comparable during the two types of processes. These results suggest that the observed high O_3 concentrations were most likely not due to local, photochemically generated O_3. The correlation coefficients (r) between NO_2:NO versus O_3 were <0.02 independent of the class of O_3 concentrations (<50.0 ppb; 50.1-90.0 ppb or >90.1 ppb) used in the statistical analysis. However, in many cases, hourly concentrations of NO_2-NO during the early morning hours were much higher on episode versus non-episode days.

Examples of boundary layer air trajectories in 3-hour time steps are shown for specific periods of high O_3 concentrations measured at Fortress Mountain (Figures 3.12 through 3.20). All figures are labelled with the ending time of the trajectory at Fortress Mountain

Table 3.38 A comparison of average daily one hour pollutant concentrations and meteorological parameters for O_3 episode and non-episode days at Fortress Mountain during April 15-June 30, 1987.

Pollutant or Meteorological Parameter*	Episode	Non-Episode
O_3 (max)**	103	64
O_3	82	55
NO_2	1.2	1.1
NO	0.50	0.51
SO_2	0.11	0.10
H_2S	0.60	0.57
Wind Speed	13.7	12.4
Temperature	7.0	9.1
Humidity	29.7	32.4
Radiation	444	423
Barometric Pressure	783	787
Number of Days	14	14

* All gaseous pollutants in ppb; wind speed: km h^{-1}; temperature: °C; humidity: %; radiation: Langleys min^{-1}; barometric pressure: mbar.
** Mean of highest daily O_3 concentration.

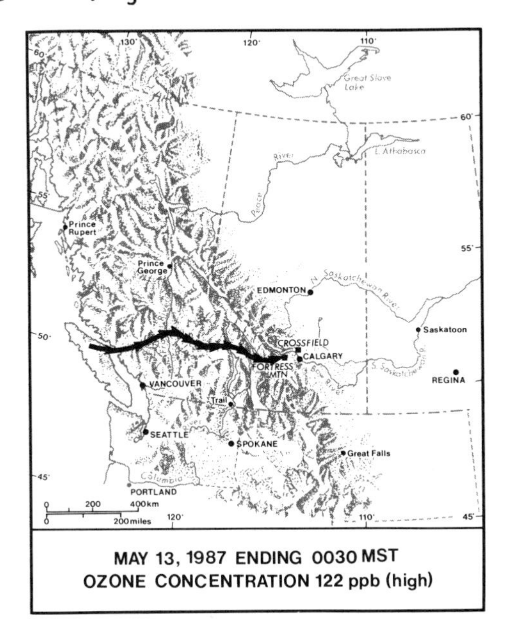

Figure 3.12 Trajectories for the air masses arriving at Fortress Mountain on May 13, 1987, at 0030 MST. The individual arrows represent 3-hour steps in travel time of the air masses.

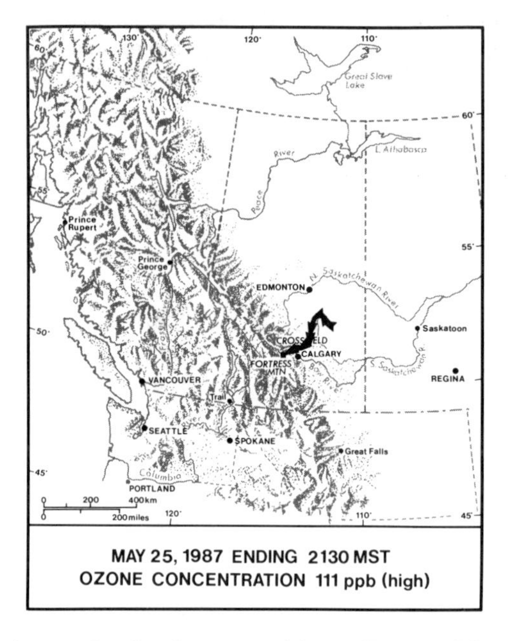

Figure 3.13 Trajectory for the air mass arriving at Fortress Mountain on May 25, 1987, at 2130 MST hours. The individual arrows represent 3-hour steps in travel time of the air mass.

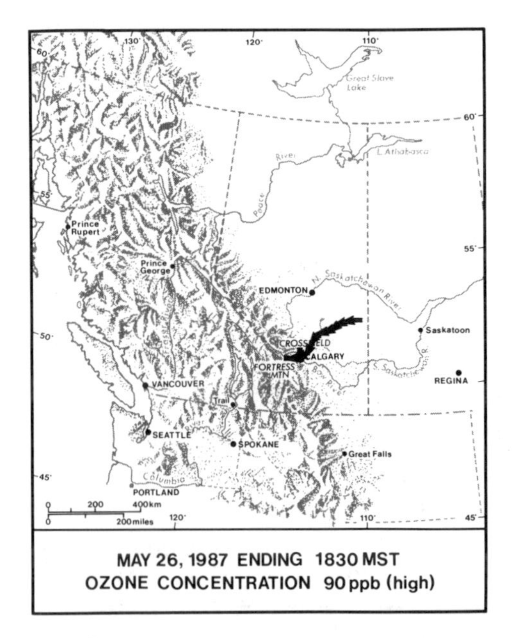

Figure 3.14 Trajectory for the air mass arriving at Fortress Mountain on May 26, 1987, at 1830 MST hours. The individual arrows represent 3-hour steps in travel time of the air mass.

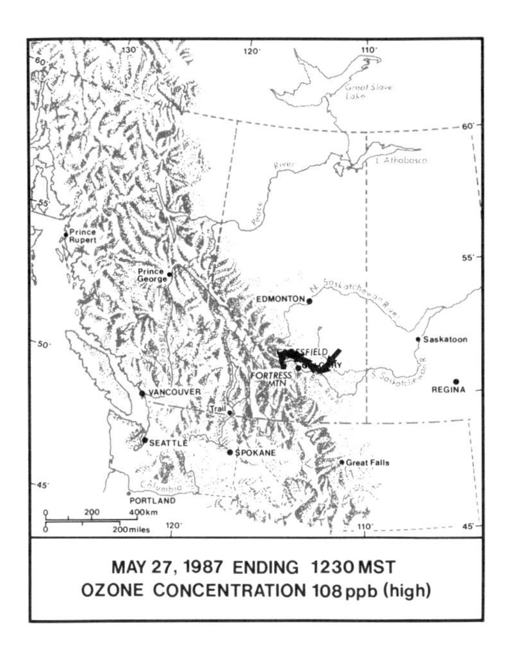

Figure 3.15 Trajectory for the air mass arriving at Fortress Mountain on May 27, 1987, at 1230 MST hours. The individual arrows represent 3-hour steps in travel time of the air mass.

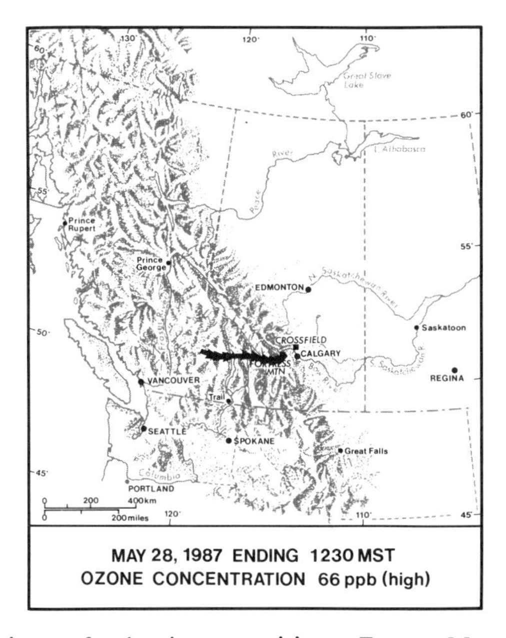

Figure 3.16 Trajectory for the air mass arriving at Fortress Mountain on May 28, 1987, at 1230 MST hours. The individual arrows represent 3-hour steps in travel time of the air mass.

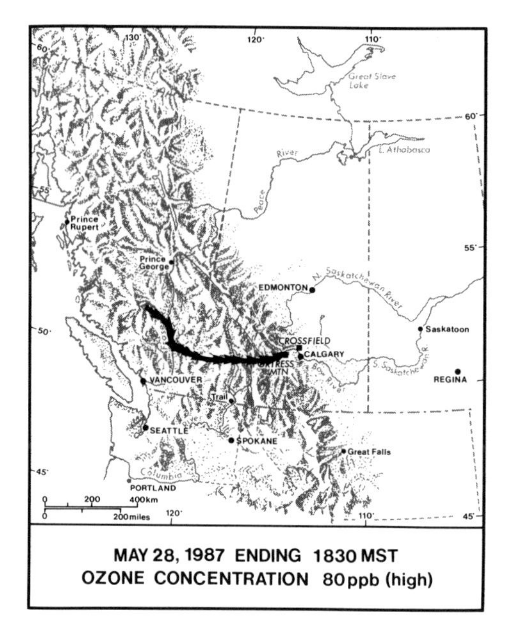

Figure 3.17 Trajectory for the air mass arriving at Fortress Mountain on May 28, 1987, at 1830 MST hours. The individual arrows represent 3-hour steps in travel time of the air mass.

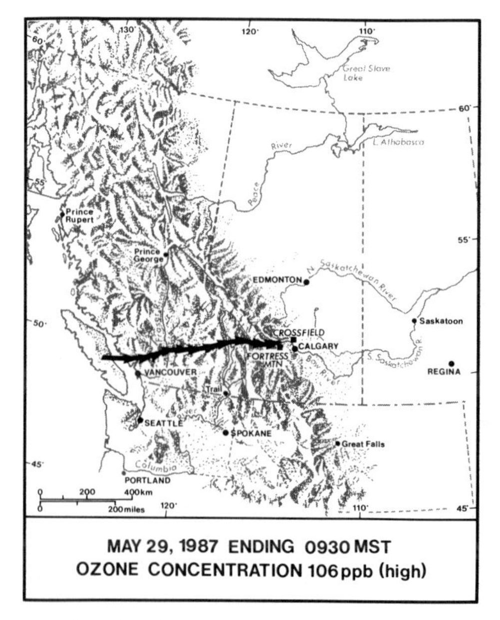

Figure 3.18 Trajectory for the air mass arriving at Fortress Mountain on May 29, 1987, at 0930 MST hours. The individual arrows represent 3-hour steps in travel time of the air mass.

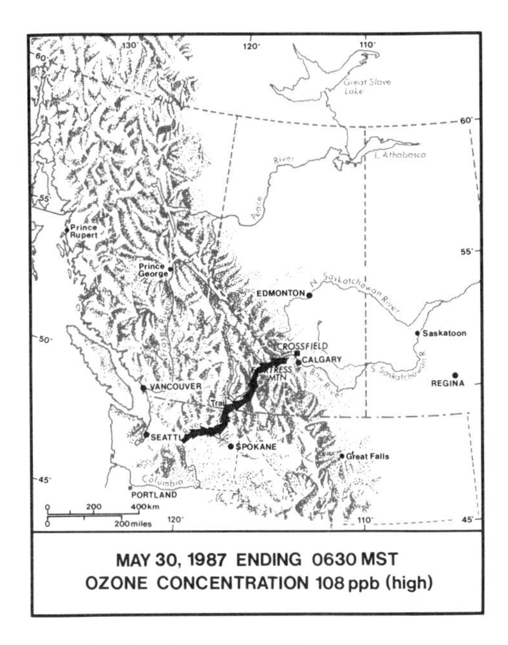

Figure 3.19 Trajectory for the air mass arriving at Fortress Mountain on May 30, 1987, at 0630 MST hours. The individual arrows represent 3-hour steps in travel time of the air mass.

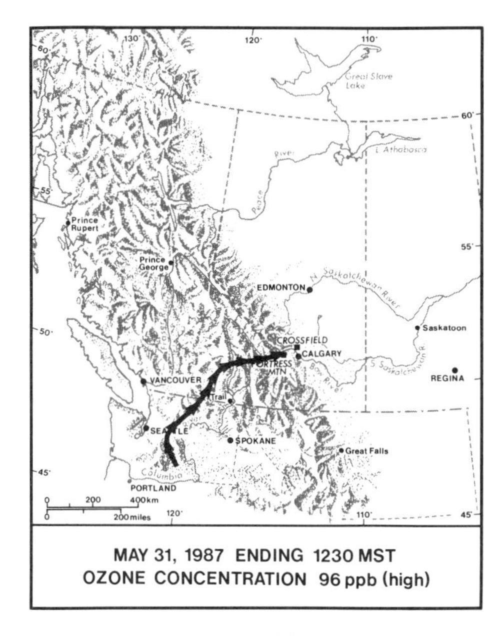

Figure 3.20 Trajectory for the air mass arriving at Fortress Mountain on May 31, 1987, at 1230 MST hours. The individual arrows represent 3-hour steps in travel time of the air mass.

and the average one hour O_3 concentration for that time. Of the 14 O_3-episode days examined, the trajectories showed a variety of directions and speeds of air movement. While the trajectories did not indicate that a single unique condition resulted in high O_3 concentrations, it was possible to suggest source areas for the air mass.

It should be recognized that the trajectories are estimates based upon observed and forecasted wind fields, and may contain significant errors, especially during times of rapidly changing flow conditions. Furthermore they represent the air flow only in the lower layers (lowest 500 m) of the atmosphere. In mountainous regions, as at Fortress, there are likely to be local variations in air flow not accounted for by the large-scale wind analyses. Thus, the trajectories should be interpreted as an estimate of the general direction and speed of air masses arriving at Fortress.

O_3 episode of May 25-31, 1987 (Figure 3.21): The trajectories for the periods of peak O_3 concentrations on May 25, 26, and 27 all showed an easterly flow from the Calgary area, upslope to Fortress Mountain (Figures 3.13 through 3.15). This air flow was probably the result of a weak synoptic scale pressure gradient in combination with daytime heating, resulting in upslope winds. The combination of trajectory analysis and correlation of the O_3 concentrations the wind direction at Fortress Mountain (Figure 3.22) leaves little doubt that the high O_3 concentrations on these days occurred due to rising air movements from the eastern plains near Calgary to Fortress Mountain.

The wind conditions were totally different during the remainder of the week, with generally west to southwest winds at Fortress (Figure 3.22). The trajectories for May 28 through May 31 (Figures 3.16 through 3.20) suggest that the air masses arriving at the Fortress Mountain monitoring site had their origin in the Puget Sound area. While it was not possible to identify specific source areas for these days, it appeared plausible that the elevated O_3 concentrations at Fortress Mountain were related to urban air pollutants from the populated areas of the Pacific Northwest in the U.S. and southwestern British Columbia.

In other cases of O_3 episodes at Fortress Mountain, the air masses also appeared to come from the west and southwesterly directions. For example, during April 24-26 (Table 3.38) the trajectories showed that the air flow was from southwest to west. Similarly on May 13 (Figure 3.12), the initial trajectory crossed

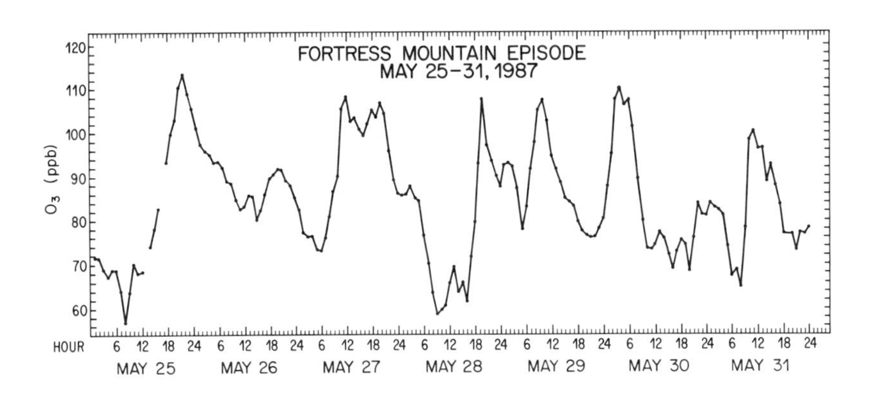

Figure 3.21 Profile of a multi-day O_3 episode (May 25-31, 1987) at Fortress Mountain.

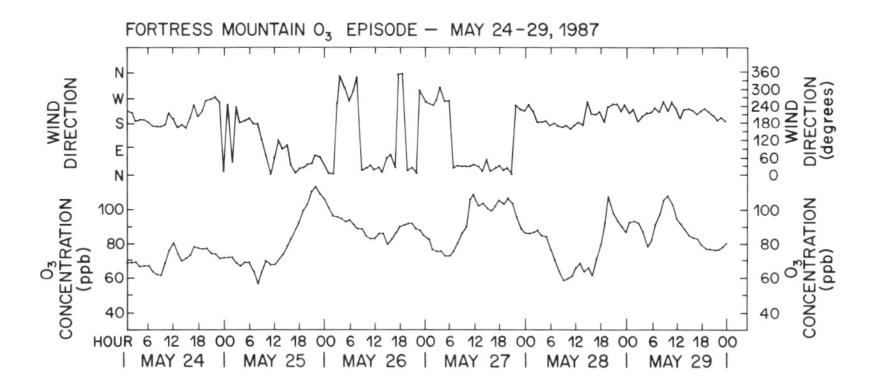

Figure 3.22 Profile of a multi-day O_3 episode (May 24-29, 1987) and the corresponding patterns in wind directions at Fortress Mountain.

Vancouver Island, while a later trajectory seemed to originate from further north (Fraser Valley) in British Columbia. On June 6-7 (Table 3.38) the trajectory for the first day originated in northwestern Washington, while a later trajectory showed air flow from areas northwest of Washington. Taken as a group, these results indicated that the high O_3 concentrations at Fortress Mountain were generally associated with air masses from the west and southwest.

Because of the elevation of the Fortress Mountain monitoring site (2103 m MSL), air movement at altitudes above the boundary layer was also considered in the analysis of air mass trajectories. For two cases of high ozone concentrations at Fortress Mountain (May 25-26, Figures 3.13 and 3.14); and May 29-30, 1987, (Figures 3.18 and 3.19), trajectories were calculated from LFM wind data for the 700 millibar level (approximately 3000 m above sea level). These higher altitude trajectories confirmed that general air movement during these periods was from northwestern Washington state to Fortress Mountain.

The timing of the high ozone concentrations at Fortress Mountain in these cases was consistent with the transport of daytime urban emissions from the Puget Sound region. Air arriving at Fortress Mountain near midnight on May 12 would have originated near Seattle or Vancouver on the morning of the 12th, according to the 700 mb wind speeds.

An air mass leaving the Seattle area during the afternoon on May 28 would have, if transported with the 700 mb wind flow, arrived at Fortress Mountain near 0800 on May 29th. Similarly, air leaving western Washington during the morning of May 29 would have arrived at Fortress Mountain in the early morning hours of May 30. Surface weather conditions in the Pacific Northwest during these times included periods of precipitation and frontal passages during the night, but relatively clear weather with weak high pressure behind the fronts during the daytime.

A set of trajectories was also developed for non-episode O_3 days at Fortress Mountain (Table 3.38). On April 23 (Figure 3.23) and April 28, air masses at Fortress Mountain arrived from the south and southeast. This was also the case for May 22 (Figure 3.24), June 4, and June 9 (Figure 3.25).

On a number of other non-episode O_3 days, air masses originated from the west, southwest, or northwest. In general, trajectories for these non-episode days tended to have lower wind speeds and less well defined flow directions compared to the trajectories for the

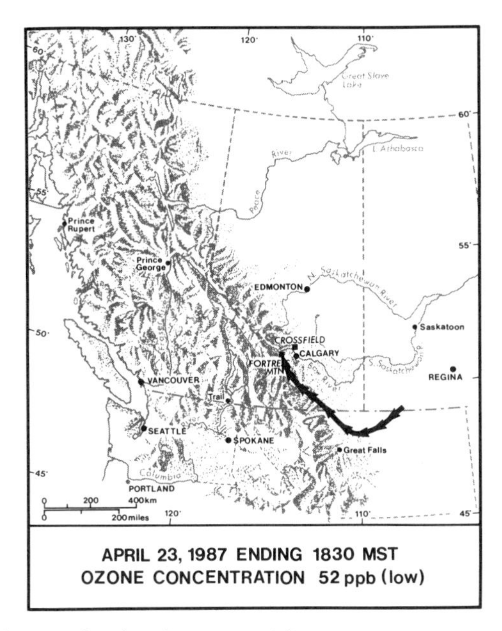

Figure 3.23 Trajectory for the air mass arriving at Fortress Mountain on April 23, 1987, at 1830 MST hours. The individual arrows represent 3-hour steps in travel time of the air mass.

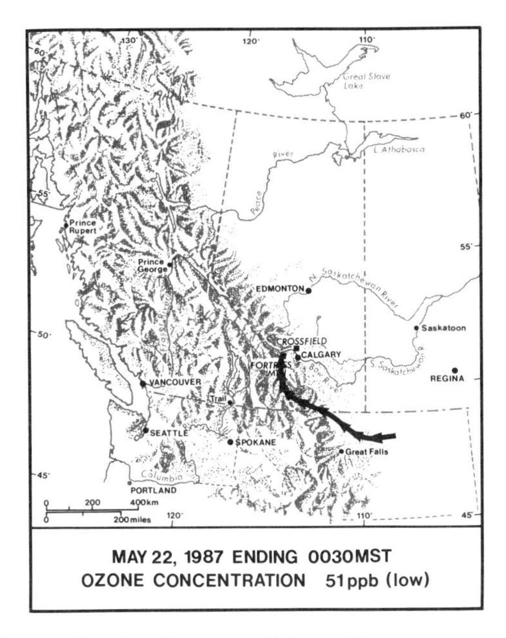

Figure 3.24 Trajectory for the air mass arriving at Fortress Mountain on May 22, 1987, at 0030 MST hours. The individual arrows represent 3-hour steps in travel time of the air mass.

Figure 3.25 Trajectory for the air mass arriving at Fortress Mountain on June 9, 1987, at 0030 MST hours. The individual arrows represent 3-hour steps in travel time of the air mass.

O_3 episode days.

From the information presented in the previous sections, it was possible to evaluate whether the O_3 episodes at Fortress Mountain during 1987 were the result of long-range transport, stratospheric intrusion, or their combination.

1. Generally at Fortress Mountain O_3 episodes occurred over two or more consecutive days. During these episodes, highest daily maxima occurred both during daylight and nighttime hours. These patterns were not consistent with known characteristics of stratospheric intrusion. Stratospheric intrusions are often characterized by relatively rapid changes in the O_3 concentration with peaks occurring at times not expected from photochemical processes. For example, the data for Zugspitze, West Germany, showed a maximum 1 hour average O_3 concentration of 196 ppb at midnight on January 8, 1975 (Singh et al. 1978). Such high O_3 concentrations were not observed in the region surrounding Zugspitze on the previous day.

   While O_3 data were not available for the region surrounding Fortress Mountain, the type of phenomenon observed at Zugspitze should result in the modification of the daily O_3 profile. At Fortress Mountain there were no distinct differences in the daily O_3 profile of episode versus non-episode days (the reader is referred to the section that follows).

2. If there was stratospheric intrusion, it should result in the dilution of tropospheric NO and NO_2. This was not the case at Fortress Mountain.
3. Stratospheric air is known to be virtually free of moisture. Thus, its intrusion into the troposphere should result in a reduction in the relative humidity. This was not the case at Fortress Mountain when episode and non-episode days were compared.
4. At Fortress Mountain while, in general, episodes were governed by air masses originating from the west-southwest direction, non-episode situations were governed by air masses originating from the south-southeast.

   Data are available to strongly suggest that ground-level impacts of stratospheric intrusion are infrequent and do not generally lead to O_3 concentrations exceeding 100 ppb (Finlayson-Pitts and Pitts 1986). Thus, long-range transport of photochemical oxidant precursors from urban areas is well known: for example, long-range transport from western Europe has been postulated to be primarily responsible for some of the elevated ozone concentrations observed in Norway (Schjoldager et al. 1978 and Schjoldager, 1981). In this context, Liu et al. (1980) point out that the photochemical lifetime expected for O_3 is fairly long (~10 days).
5. According to Viezee et al. (1983) aircraft measurements of four stratospheric O_3 intrusion events suggest that half of the O_3 mass injected into the upper troposphere by these events is likely mixed and diluted into the troposphere above the 3 km elevation. The Fortress Mountain monitoring site was located at a 2.1 km elevation. Viezee et al. suggested that ground-level elevated concentrations of O_3 due to stratospheric intrusions occur less than 1% of the time. At Fortress Mountain O_3 episodes were observed on 30% of the 45 days examined.
6. At Fortress Mountain the highest O_3 concentrations were observed in 1987 during months with the highest net solar

radiation. The decay of water molecules and photochemical scavenging of O_3 were considered to be unimportant in the region due to the lack of any major sources (Julius London, University of Colorado, personal communication).

Diurnal O_3 profiles: Both on episode and non-episode days, at the Fortress Mountain Site the diurnal O_3 profiles (Figure 3.26) were not typical of the known patterns for low elevation monitoring sites (refer to U.S. National Academy of Sciences 1977 and Chapter 4). While during the daytime high O_3 concentrations occurred between roughly 1100 and 1500 hours, after a decrease during the subsequent hours, concentrations increased again sharply into the nighttime on days with O_3 episodes. Such a pattern was also observed on non-episode days, but with a lag time in the hour-to-hour variation in comparison to the pattern on the days of the episode (Figure 3.26). The periods of rise in O_3 concentrations during the day (0800-1000 to 1300-1400 hours), the subsequent decrease, and the occurrence of much higher concentration mostly at night (Table 3.38 and Figure 3.26) appeared to be consistent with the patterns of diurnal changes in synoptic versus katabatic air flows in the mountains and the adjacent Kananaskis Valley (E.C. Rhodes, University of Calgary, personal communication). Results similar to these diurnal variations have been previously obtained by Angle and Sandhu (1986) at Birch Mountain, Alberta.

Because of the nature of the computation, these diurnal patterns were not observed when the O_3 concentration was averaged for each hour of the day cumulatively for each year (Figure 3.27). The flat diurnal profile was similar to that which has been previously reported at Whiteface Mountain, NY and Whitetop Mountain, VA (U.S. NAPAP 1987).

According to Singh et al. (1978) variation in O_3 concentrations at remote sites consists of a yearly cycle with a maximum occurring in late winter or early spring. In Alberta, Angle and Sandhu (1986) observed O_3 maxima in May at Bitumount (350 m MSL, 57° 41' 30"N) and Birch Mountain (850 m MSL, 57° 51' 30"N) and in April at Ellerslie (687 m MSL, 53° 19'N). At Fortress Mountain (2103 m MSL, 50° 49' 40"N) peak monthly mean O_3 concentrations were observed during April in 1986 and in May during 1987 (Figure 3.27). Singh et al. (1978) concluded that while a cyclic pattern of O_3 concentration at remote locations was quite consistent, it was not identical from year to year. For example, at a monitoring site near Colestrip, Montana, the maximum one-hour O_3 concentration

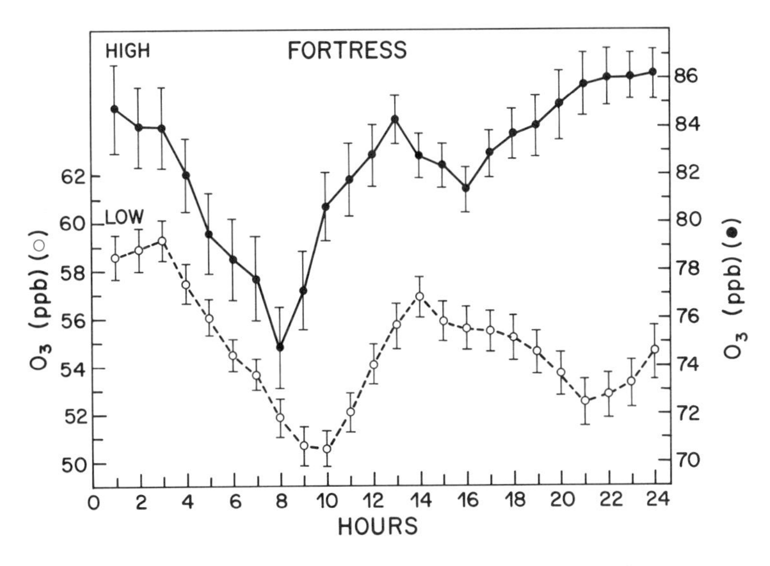

Figure 3.26 Diurnal O_3 profiles on episode versus non-episode days at Fortress Mountain. The vertical bars represent S.D. of the mean for each hour.

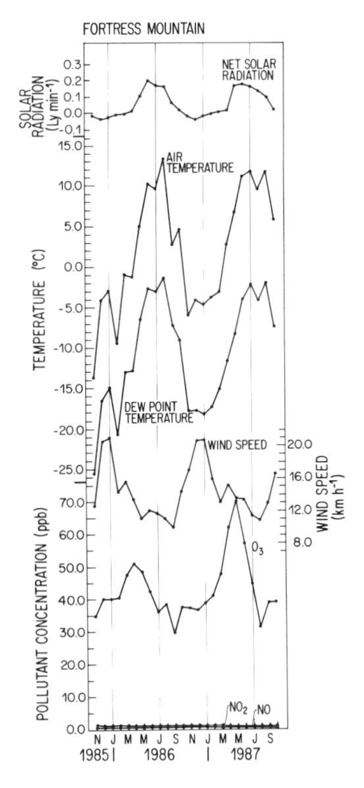

Figure 3.27 Mean monthly patterns of O_3, NO, and NO_2, and selected meteorological parameters at Fortress Mountain.

for the month was 93 ppb during 1976 while it was ~80 ppb in 1975. Similarly at Mauna Loa, Hawaii, both the hourly-average O_3 concentration for the month and the maximum one-hour concentration for the month in 1975 were above their corresponding values for 1974. Similarly at Fortress Mountain the highest monthly mean (70.8 ppb) and the 99.5th percentile (112 ppb) of O_3 concentrations were observed during May 1987--in comparison to 1986 when the values were, respectively, 48.9 ppb and 61.0 ppb (Table 3.43).

Tables 3.39 through 3.44 provide summary statistics on the occurrences of major pollutants by month. Note that the data for SO_2 and H_2S are based on 10-minute averages, while all other pollutants (NO, NO_2, O_3 and CO_2) are based on hourly averages. During February 1986 there was a sharp peak in the monthly mean SO_2 and values $\geq$ 95th percentile. Such a pattern was repeated again during March 1987, even though all the values were only in the sub-ppb range (Table 3.39). During the period December to May, there was a strong relationship (R^2=0.68 for 1986 and 0.82 for 1987) between the monthly SO_2 values and the monthly mean temperature (Table 3.45). A similar pattern was exhibited by NO_2, (R^2 values between NO_2 and temperature were 0.81 and 0.88 for 1986 and 1987, respectively) (Table 3.46). There was also a strong relationship between the monthly 10-minute average SO_2 and hourly NO_2 values during December to May. The R^2 values were 0.97 and 0.99 for 1986 and 1987, respectively (Table 3.47), suggesting a common source of origin for these two pollutant during the cold months.

The Fortress Mountain Ski Resort is located approximately 1 km from the monitoring site. During peak demand for residential heating a combination of propane and wood combustion were used at the resort. It appears that this small source might have contributed to the sub-ppb monthly peak SO_2 and NO_2 concentrations observed during the two coldest months (February and March) at the Fortress Mountain air quality monitoring site.

Table 3.39 Summary statistics of 10-minute SO_2 concentrations (ppb) by month. Sampling location: Fortress Mountain.

			Percentile				
Year	Month	Mean	75th	90th	95th	99th	99.5th
1985	Nov	.17	.10	.30	.50	4.00	5.0
	Dec	.11	.10	.30	.40	.80	.88

Continued...

Table 3.39 (Concluded).

Year	Month	Mean	Percentile 75th	90th	95th	99th	99.5th
1986	Jan	.09	.10	.30	.40	.70	.80
	Feb	.37	.10	.40	2.4	7.0	8.0
	Mar	.09	.10	.30	.40	.70	.80
	Apr	.10	.10	.20	.40	.73	.80
	May	.09	.10	.30	.40	.70	.80
	Jun	.09	.10	.30	.40	.70	.80
	Jul	.09	.10	.20	.40	.70	.80
	Aug	.09	.10	.30	.40	.70	.80
	Sep	.09	.10	.30	.40	.80	.90
	Oct	.09	.10	.30	.40	.74	.90
	Nov	.10	.10	.30	.40	.70	.80
	Dec	.12	.10	.30	.40	.80	4.0
1987	Jan	.10	.10	.30	.40	.77	.83
	Feb	.15	.10	.30	.40	.90	5.1
	Mar	.76	.20	.70	7.0	11.0	12.2
	Apr	.09	.10	.30	.40	.70	.80
	May	.09	.10	.30	.40	.70	.80
	Jun	.10	.10	.30	.40	.80	.90
	Jul	.11	.10	.30	.40	.80	2.1
	Aug	.09	.10	.30	.40	.70	.80
	Sep	.09	.10	.30	.40	.70	.80
	Oct	.12	.10	.30	.40	.84	1.6

Table 3.40 Summary statistics of 10-minute H_2S concentrations (ppb) by month. Sampling location: Fortress Mountain.

Year	Month	Mean	Percentile 75th	90th	95th	99th	99.5th
1985	Nov	.55	.70	.80	.90	1.1	1.2
	Dec	.55	.70	.80	.90	1.1	1.2
1986	Jan	.55	.70	.80	.90	1.1	1.2
	Feb	.55	.70	.80	.90	1.1	1.2
	Mar	.55	.70	.80	.90	1.1	1.2
	Apr	.55	.70	.80	.90	1.1	1.2
	May	.55	.70	.80	.90	1.1	1.2
	Jun	.55	.70	.80	.90	1.1	1.2
	Jul	.55	.70	.80	.90	1.1	1.2
	Aug	.55	.70	.80	.90	1.1	1.2
	Sep	.55	.70	.80	.90	1.1	1.2
	Oct	.55	.70	.80	.90	1.1	1.2
	Nov	.55	.70	.80	.90	1.1	1.2
	Dec	.55	.70	.80	.90	1.1	1.2

Continued...

Table 3.40 (Concluded).

Year	Month	Mean	Percentile 75th	90th	95th	99th	99.5th
1987	Jan	.55	.70	.80	.90	1.1	1.2
	Feb	.55	.70	.80	.90	1.1	1.2
	Mar	.56	.70	.80	.90	1.1	1.2
	Apr	.55	.70	.80	.90	1.1	1.2
	May	.63	.70	.80	1.0	3.2	9.3
	Jun	.56	.70	.80	.90	1.1	1.3
	Jul	.55	.70	.80	.90	1.1	1.2
	Aug	.55	.70	.80	.90	1.1	1.2
	Sep	.55	.70	.80	.90	1.1	1.2
	Oct	.55	.70	.80	.90	1.1	1.2

Table 3.41 Summary statistics of hourly NO concentrations (ppb) by month. Sampling location: Fortress Mountain.

Year	Month	Mean	Percentile 75th	90th	95th	99th	99.5th
1985	Nov	.53	.70	.90	1.0	1.6	2.8
	Dec	.55	.70	.90	1.0	1.9	3.2
1986	Jan	.53	.70	.90	1.0	1.6	2.3
	Feb	.53	.70	.90	1.0	1.5	2.0
	Mar	.53	.70	.90	1.0	1.6	2.6
	Apr	.54	.70	.90	1.0	1.5	2.4
	May	.50	.60	.90	1.0	1.3	1.4
	Jun	.51	.70	.90	1.0	1.3	1.4
	Jul	.50	.60	.80	1.0	1.3	1.4
	Aug	.52	.70	.90	1.0	1.3	1.5
	Sep	.50	.60	.80	1.0	1.3	1.4
	Oct	.51	.60	.90	1.0	1.3	1.4
	Nov	.58	.70	.90	1.1	2.6	4.1
	Dec	.55	.70	.90	1.1	2.0	3.5
1987	Jan	.53	.70	.90	1.0	1.6	2.2
	Feb	.55	.70	.90	1.0	1.8	3.3
	Mar	.54	.70	.90	1.0	2.0	3.2
	Apr	.51	.70	.80	1.0	1.3	1.4
	May	.50	.70	.80	1.0	1.3	1.4
	Jun	.51	.70	.90	1.0	1.3	1.4
	Jul	.51	.70	.90	1.0	1.3	1.5
	Aug	.51	.70	.90	1.0	1.3	1.5
	Sep	.51	.60	.90	1.0	1.3	1.5
	Oct	.51	.70	.80	1.0	1.3	1.5

Table 3.42 Summary statistics of hourly NO_2 concentrations (ppb) by month. Sampling location: Fortress Mountain.

			Percentile				
Year	Month	Mean	75th	90th	95th	99th	99.5th
1985	Nov	1.5	1.5	2.3	4.4	6.0	6.7
	Dec	1.2	1.4	1.7	1.9	2.4	3.3
1986	Jan	1.2	1.4	1.7	1.9	3.0	3.6
	Feb	1.4	1.5	1.9	2.9	7.4	9.0
	Mar	1.2	1.4	1.7	1.9	2.6	3.7
	Apr	1.2	1.4	1.7	1.9	3.7	4.4
	May	1.1	1.4	1.6	1.8	2.2	2.4
	Jun	1.1	1.4	1.6	1.8	2.3	2.6
	Jul	1.1	1.4	1.6	1.8	2.2	2.3
	Aug	1.1	1.4	1.6	1.8	2.1	2.3
	Sep	1.2	1.4	1.7	1.9	2.7	4.2
	Oct	1.1	1.4	1.6	1.8	2.1	2.2
	Nov	1.3	1.4	1.8	2.2	4.7	6.1
	Dec	1.2	1.4	1.7	1.9	2.8	3.8
1987	Jan	1.2	1.4	1.7	1.9	2.4	2.8
	Feb	1.2	1.4	1.7	1.9	2.4	2.9
	Mar	1.4	1.5	2.0	3.3	5.6	6.2
	Apr	1.2	1.4	1.7	1.8	2.2	2.4
	May	1.1	1.4	1.6	1.8	2.1	2.3
	Jun	1.2	1.4	1.7	1.8	2.3	2.8
	Jul	1.1	1.4	1.6	1.8	2.2	2.3
	Aug	1.1	1.4	1.7	1.8	2.2	2.4
	Sep	1.2	1.4	1.6	1.8	2.2	2.4
	Oct	1.1	1.4	1.6	1.8	2.1	2.3

Table 3.43 Summary statistics of hourly O_3 concentrations (ppb) by month. Sampling location: Fortress Mountain.

			Percentile				
Year	Month	Mean	75th	90th	95th	99th	99.5th
1985	Nov	35.0	40.0	44.0	47.2	52.0	53.8
	Dec	40.4	42.8	44.0	44.8	46.6	47.0
1986	Jan	40.4	43.2	45.2	46.4	47.6	47.8
	Feb	40.7	45.2	47.0	47.6	48.6	50.0
	Mar	47.8	50.4	52.8	53.6	57.6	58.6
	Apr	51.3	56.2	59.2	60.2	62.2	63.0
	May	48.9	54.2	57.0	58.4	61.0	64.4
	Jun	42.4	50.2	55.2	59.0	66.2	68.2
	Jul	36.3	40.8	46.4	50.0	57.0	59.6

Continued...

Table 3.43 (Concluded).

Year	Month	Mean	Percentile 75th	90th	95th	99th	99.5th
	Aug	38.7	44.0	54.4	57.4	64.6	65.4
	Sep	29.6	35.2	38.4	40.0	42.6	43.6
	Oct	37.8	43.6	46.0	48.0	51.2	52.0
	Nov	37.5	39.8	42.8	44.6	48.0	48.6
	Dec	36.9	39.2	41.6	42.8	46.4	47.5
1987	Jan	39.4	41.6	43.2	44.2	45.8	46.2
	Feb	41.3	43.2	45.3	47.2	50.5	51.8
	Mar	48.5	61.4	69.6	73.4	81.8	82.8
	Apr	62.6	67.8	74.4	82.6	90.7	91.5
	May	70.8	78.4	90.5	98.2	109.8	112.0
	Jun	58.8	66.0	78.0	84.2	97.2	100.2
	Jul	45.9	53.2	61.0	63.6	67.8	69.2
	Aug	31.5	35.8	40.2	45.0	50.4	52.8
	Sep	39.1	44.0	47.2	49.6	56.2	58.8
	Oct	39.3	43.2	45.4	47.4	50.6	51.4

Table 3.44 Summary statistics of hourly CO_2 concentrations (ppm) by month. Sampling location: Fortress Mountain.

Year	Month	Mean	Percentile 75th	90th	95th	99th	99.5th
1985	Nov	342	344	353	355	360	362
	Dec	349	351	352	353	357	359
1986	Jan	347	349	351	353	355	357
	Feb	342	347	350	353	356	357
	Mar	352	353	355	356	359	361
	Apr	351	354	355	356	358	359
	May	352	354	356	357	361	364
	Jun	345	347	349	350	353	354
	Jul	339	342	344	346	350	352
	Aug	338	341	343	345	352	355
	Sep	339	341	344	346	355	362
	Oct	351	354	355	356	361	363
	Nov	355	358	370	372	378	378
	Dec	348	350	352	354	368	372
1987	Jan	348	349	352	356	367	372
	Feb	348	350	353	355	359	360
	Mar	351	354	358	359	362	365

Continued...

Table 3.44 (Concluded).

Year	Month	Mean	Percentile 75th	90th	95th	99th	99.5th
	Apr	349	350	352	353	355	355
	May	352	354	356	357	362	366
	Jun	350	353	356	359	364	365
	Jul	339	343	350	353	358	360
	Aug	329	335	337	339	341	343
	Sep	337	339	341	342	348	349
	Oct	337	339	341	342	345	346

Table 3.45 Relationship between mean of 10-minute SO_2 concentrations by the month and monthly mean temperature at Fortress Mountain.

Year	Month	$\bar{x}$ SO_2 (ppb)		Temperature (°C)
1985	Dec	0.11		- 4.17
1986	Jan	0.09		-14.80
	Feb	0.37		-20.60
	Mar	0.09		-12.90
	Apr	0.10		-12.80
	May	0.10		- 6.40
			R^2 = 0.68	
	Dec	0.12		- 4.15
1987	Jan	0.10		- 4.60
	Feb	0.15		- 3.70
	Mar	0.76		-14.90
	Apr	0.09		+ 2.80
	May	0.10		+ 6.43
			R^2 = 0.82	

Table 3.46 Relationship between mean hourly NO_2 concentrations by the month and monthly mean temperature at Fortress Mountain.

Year	Month	$\bar{x}$ SO_2 (ppb)		Temperature (°C)
1985	Dec	1.17		- 4.17
1986	Jan	1.19		-14.80
	Feb	1.41		-20.60
	Mar	1.19		-12.90
	Apr	1.20		-12.80
	May	1.14		- 6.40
			R^2 = 0.81	

Continued...

Table 3.46 (Concluded).

Year	Month	$\bar{x}$ SO_2 (ppb)	Temperature (°C)
	Dec	1.19	- 4.15
1987	Jan	1.17	- 4.60
	Feb	1.17	- 3.70
	Mar	1.39	-14.90
	Apr	1.15	+ 2.80
	May	1.14	+ 6.43
		R^2 = 0.88	

Table 3.47 Relationship between mean of 10-minute SO_2 and hourly NO_2 concentrations by the month at Fortress Mountain.

Year	Month	$\bar{x}$ SO_2 (ppb)	$\bar{x}$ NO_2 (ppb)
1985	Dec	0.11	1.17
1986	Jan	0.09	1.19
	Feb	0.37	1.41
	Mar	0.09	1.19
	Apr	0.10	1.20
	May	0.10	1.14
		R^2 = 0.97	
	Dec	0.12	1.19
1987	Jan	0.10	1.17
	Feb	0.15	1.17
	Mar	0.76	1.39
	Apr	0.10	1.15
	May	0.10	1.14
		R^2 = 0.99	

In addition to the seasonal statistics, summary data on the daily patterns of the major air pollutants are presented in Tables 3.48 through 3.53. No major patterns could be discerned from these data. This was also the case with correlation analyses between pollutant concentrations for the entire two year study period and the meteorological parameters for the corresponding time period.

Table 3.48 Summary statistics of 10-minute SO_2 concentrations (ppb) by daily hour. Sampling location: Fortress Mountain.

Hour	Mean	Percentile				
		75th	90th	95th	99th	99.5th
1	.17	.10	.30	.50	4.0	6.0
2	.20	.10	.30	.50	5.4	9.5
3	.15	.10	.30	.40	1.2	5.9
4	.16	.10	.30	.45	3.4	5.0
5	.16	.10	.30	.50	1.5	5.0
6	.15	.10	.30	.40	1.0	4.0
7	.15	.10	.30	.50	.9	4.0
8	.16	.10	.30	.40	1.1	7.2
9	.14	.10	.30	.40	.9	5.8
10	.14	.10	.30	.40	.9	5.0
11	.14	.10	.20	.40	1.0	6.1
12	.15	.10	.30	.40	1.0	6.0
13	.14	.10	.30	.40	1.0	4.2
14	.11	.10	.30	.40	.8	1.2
15	.11	.10	.30	.40	.8	.9
16	.12	.10	.30	.40	.8	.9
17	.11	.10	.30	.40	.8	1.0
18	.12	.10	.30	.40	.8	1.0
19	.12	.10	.30	.40	.8	1.6
20	.14	.10	.30	.40	.9	5.3
21	.15	.10	.30	.40	1.0	6.6
22	.18	.10	.30	.50	4.0	7.0
23	.17	.10	.30	.50	4.0	6.0
24	.16	.10	.30	.50	3.2	6.0

Table 3.49 Summary statistics of 10-minute H_2S concentrations (ppb) by the daily hour. Sampling location: Fortress Mountain.

Hour	Mean	Percentile				
		75th	90th	95th	99th	99.5th
1	.55	.70	.80	.90	1.1	1.2
2	.55	.70	.80	.90	1.1	1.2
3	.55	.70	.80	.90	1.1	1.2
4	.56	.70	.80	.90	1.2	1.3
5	.56	.70	.80	.90	1.1	1.2
6	.55	.70	.80	.90	1.1	1.2
7	.55	.70	.80	.90	1.1	1.2
8	.55	.70	.80	.90	1.1	1.2
9	.55	.70	.80	.90	1.1	1.2

Continued...

Table 3.49 (Concluded).

Hour	Mean	Percentile 75th	90th	95th	99th	99.5th
10	.55	.70	.80	.90	1.1	1.2
11	.55	.70	.80	.90	1.1	1.2
12	.55	.70	.80	.90	1.1	1.2
13	.56	.70	.80	.90	1.1	1.2
14	.55	.70	.80	.90	1.1	1.2
15	.55	.70	.80	.90	1.1	1.2
16	.55	.70	.80	.90	1.1	1.2
17	.55	.70	.80	.90	1.1	1.2
18	.57	.70	.80	.90	1.2	1.2
19	.55	.70	.80	.90	1.1	1.2
20	.55	.70	.80	.90	1.1	1.2
21	.55	.70	.80	.90	1.1	1.2
22	.55	.70	.80	.90	1.1	1.2
23	.55	.70	.80	.90	1.1	1.2
24	.55	.70	.80	.90	1.1	1.2

Table 3.50 Summary statistics of hourly NO concentrations (ppb) by the daily hour. Sampling location: Fortress Mountain.

Hour	Mean	Percentile 75th	90th	95th	99th	99.5th
1	.51	.60	.90	1.0	1.4	1.6
2	.51	.70	.90	1.0	1.4	1.5
3	.50	.60	.80	1.0	1.3	1.5
4	.50	.70	.90	1.0	1.2	1.4
5	.50	.60	.80	1.0	1.4	1.5
6	.52	.70	.90	1.0	1.3	1.5
7	.53	.60	.90	1.0	1.7	2.8
8	.57	.70	.90	1.1	2.9	4.5
9	.56	.70	.90	1.1	2.2	3.7
10	.53	.70	.90	1.0	1.6	2.1
11	.53	.70	.90	1.0	1.5	2.6
12	.53	.70	.90	1.0	1.4	1.9
13	.51	.60	.80	1.0	1.4	1.8
14	.53	.70	.90	1.0	1.4	1.7
15	.53	.70	.90	1.0	1.3	1.5
16	.51	.60	.85	1.0	1.4	1.7
17	.54	.70	.90	1.0	1.6	2.4
18	.52	.70	.80	1.0	1.4	1.6
19	.51	.60	.80	1.0	1.3	1.5

Continued...

Table 3.50 (Concluded).

Hour	Mean	Percentile				
		75th	90th	95th	99th	99.5th
20	.52	.70	.90	1.0	1.4	1.6
21	.52	.70	.90	1.0	1.5	1.8
22	.52	.70	.90	1.0	1.4	1.7
23	.52	.70	.90	1.0	1.3	1.6
24	.51	.70	.80	1.0	1.3	1.5

Table 3.51 Summary statistics of hourly NO_2 concentrations (ppb) by the daily hour. Sampling location: Fortress Mountain.

Hour	Mean	Percentile				
		75th	90th	95th	99th	99.5th
1	1.2	1.4	1.7	1.9	3.4	5.4
2	1.2	1.4	1.7	1.9	3.1	4.2
3	1.2	1.4	1.7	1.9	3.6	.3
4	1.2	1.4	1.7	1.9	3.8	.4
5	1.2	1.4	1.7	1.9	3.4	.6
6	1.2	1.4	1.7	1.9	4.0	5.0
7	1.2	1.4	1.7	2.0	3.7	4.6
8	1.3	1.4	1.8	2.1	4.4	5.9
9	1.2	1.4	1.7	2.0	3.7	4.9
10	1.2	1.4	1.7	1.9	2.7	3.2
11	1.2	1.4	1.7	1.9	2.9	3.6
12	1.2	1.4	1.7	1.8	2.4	2.7
13	1.2	1.4	1.7	1.8	2.3	2.6
14	1.2	1.4	1.7	1.9	2.2	2.5
15	1.1	1.4	1.6	1.8	2.2	2.5
16	1.2	1.4	1.7	1.8	2.3	2.9
17	1.2	1.4	1.7	1.9	3.0	4.0
18	1.2	1.4	1.7	1.9	3.9	5.0
19	1.2	1.4	1.7	1.9	4.6	5.6
20	1.2	1.4	1.7	1.9	4.8	6.0
21	1.2	1.4	1.7	2.0	4.4	6.0
22	1.2	1.4	1.7	1.9	4.4	6.3
23	1.2	1.4	1.7	2.0	4.4	5.4
24	1.2	1.4	1.7	1.9	4.2	6.2

Table 3.52 Summary statistics of hourly O_3 concentrations (ppb) by the daily hour. Sampling location: Fortress Mountain.

		Percentile				
Hour	Mean	75th	90th	95th	99th	99.5th
1	43.8	49.0	59.8	68.6	88.7	92.8
2	43.6	49.4	59.5	68.5	88.3	96.0
3	42.0	47.4	57.0	64.8	84.1	93.2
4	42.4	47.2	59.0	66.6	83.3	93.5
5	42.9	48.2	59.4	67.8	83.4	95.6
6	42.5	48.0	58.2	66.6	82.9	93.0
7	42.1	47.4	57.8	64.8	81.7	91.9
8	41.6	47.0	57.0	63.5	80.3	88.1
9	41.6	47.2	57.2	63.4	80.8	87.2
10	42.1	47.6	57.8	64.6	83.2	91.0
11	42.8	48.4	58.6	66.0	82.4	94.0
12	43.6	49.0	60.0	68.3	82.3	92.7
13	44.0	49.2	60.4	69.4	85.4	89.9
14	44.3	49.6	60.6	69.0	84.0	88.0
15	44.5	50.0	60.6	69.2	84.2	89.1
16	44.4	50.2	60.6	69.2	84.8	88.4
17	44.4	50.2	60.6	68.8	85.2	89.0
18	44.4	50.0	60.2	68.0	82.4	92.8
19	44.2	49.8	60.0	67.6	86.6	93.9
20	44.0	49.4	59.6	67.4	85.3	94.0
21	44.0	49.4	59.8	67.9	86.6	93.5
22	43.9	49.2	60.0	68.6	88.3	92.6
23	43.7	48.8	60.0	68.4	88.2	91.2
24	43.4	8.8	59.2	68.4	86.5	92.1

Table 3.53 Summary statistics of hourly CO_2 concentrations (ppm) by the daily hour. Sampling location: Fortress Mountain.

		Percentile				
Hour	Mean	75th	90th	95th	99th	99.5th
1	346	351	354	355	360	369
2	346	351	354	356	361	370
3	347	351	354	356	362	369
4	346	351	354	356	368	370
5	346	351	354	355	363	369
6	346	352	354	356	362	370
7	346	352	354	356	362	370
8	346	352	354	356	363	368
9	346	352	354	356	363	368

Continued...

Table 3.53 (Concluded).

Hour	Mean	Percentile				
		75th	90th	95th	99th	99.5th
10	346	352	354	356	362	368
11	346	352	355	357	364	369
12	346	352	355	357	366	371
13	346	351	355	359	370	372
14	345	351	354	359	371	375
15	345	351	354	358	370	374
16	344	350	353	356	366	370
17	344	350	353	355	366	368
18	344	350	353	354	363	369
19	344	350	353	354	360	368
20	344	350	353	354	360	368
21	345	350	353	354	359	360
22	345	350	353	355	359	360
23	345	351	353	355	359	368
24	346	351	354	355	360	369

3.6.3 Measurement of Sulphur Gases by Instantaneous Or By Integrated Sampling and Gas Chromatography

In addition to the near-continuous measurements previously discussed, sulphur gases were also quantified at discrete points in time (every 8 minutes) by gas chromatography using sulphur specific flame photometric detection. A Tracor Model 270 HA Atmospheric Sulphur gas analyzer was automated to sample 9 mL of air at every 8th minute. Automated instrument calibration was also performed 4-5 times per day. The instrument detection limit for COS, CS_2 and H_2S was considered to be 1-2 ppb, while the detection limit for SO_2 was 2-3 ppb. No interferences were expected in these analyses.

To determine the absolute concentrations of each sulphur gas, a further study using a cryofocusing technique (Thornton et al. 1986; Driedger et al. 1987; Pleil et al. 1987) with lower limits of detection was carried out. In this experiment 100-500 mL of air was drawn through a two stage Nafion porous polymer dryer to remove moisture which is known to interfere with the detection of very low concentrations of sulphur gases. The dried air samples were then cryogenically trapped in a Teflon sample loop and subsequently introduced into the gas chromatograph separation column through revolatilization of the sample. The limit of detection of the sulphur gases by this method was then entirely dependent upon the sample volume used in the analysis. In the present case, the sample volume used was considered to provide values in the ppt range.

In addition to the samples collected for sulphur gas analyses, air samples were also trapped in a Tenax column (U.S. EPA Method TO-1, 1984) for the identification of organic compounds. The chemical constituents in these samples, after thermal desorption, cryogenic trapping and flash volatilization in the laboratory were collected on a gas chromatograph column (Brown and Purnell 1979; Pankow 1983; Pellizzari et al. 1984; Riggin and Markle 1985). Subsequently the samples were analyzed by combined gas chromatography and mass spectrometry (Finnigan Model 4023). The detection limits for these organic compounds on an average were 0.1 $\mu g\ m^{-3}$ of sampled air.

Table 3.54 provides summary statistics on the sulphur gas concentrations measured by conventional sampling and gas chromatography.

Table 3.54 Summary statistics on the ambient concentrations of sulphur gases at Fortress Mountain measured by conventional sampling and gas chromatography (1985-87).

		Concentration (ppb)					
Pollutant	(n)	Mean	S.D.	Median	75%	99%	Range
COS	(4748)	1.0	0.14	1.0	1.0	2.0	1.0 - 3.0
CS_2	(3845)	1.0	0.23	1.0	1.0	2.0	1.0 - 4.0
H_2S	(7809)	1.1	0.36	1.0	1.0	3.0	1.0 - 4.0
SO_2	(13,466)	1.2	0.52	1.0	1.0	3.0	1.0 - 9.0

Table 3.55 provides summary statistics on the sulphur gas concentrations in cryogenic samples.

Table 3.55 Summary statistics on the ambient concentrations of sulphur gases at Fortress Mountain measured by cryogenic sampling and gas chromatography.

		Concentration (ppt)					
Pollutant	(n)	Mean	S.D.	Median	75%	99%	Range
COS	32	315	37	317	331	360	230 - 423
CS_2	32	18	4	18	21	24	12 - 26
H_2S	32	73	15	73	85	96	44 - 99
SO_2	32	530	280	347	853	913	251 - 955

The mean pollutant concentrations presented in Table 3.54 were based strictly on observations above the minimum detection limit for the sampling/analytical method used. In this case, the minimum detection limits for all pollutants were >1.0 ppb. No attempt was made to adjust the data for those observations below the minimum detection limit (>85-90% of the total observations). Thus, the mean values presented in Table 3.54 reflect inflated representations of the true ambient concentrations. In contrast, even though "n" was small for the cryogenic sampling method, there were no values below the detection limits for that approach. Thus, the means presented in Table 3.55 should be considered as the more realistic representation of Fortress Mountain air quality. These values compare favourably with the global background concentrations of sulphur species reported by Sze and Ko (1980): SO_2 45-340 ppt; H_2S 0-320 ppt; COS 100-560 ppt; and CS_2 70-370 ppt (Table 3.5).

Table 3.56 provides a summary of the correlation analysis between the measured sulphur species by cryogenic sampling at Fortress Mountain.

Table 3.56 Correlation coefficients between sulphur gases collected by cryogenic sampling at Fortress Mountain.

Variables	(r)
COS vs H_2S	0.53*
COS vs CS_2	0.39
COS vs SO_2	0.25
H_2S vs SO_2	0.28
H_2S vs CS_2	0.27
CS_2 vs SO_2	-0.14

* Significant at 0.01 level.

The results presented in Table 3.57 on principal component analysis substantiate the relationship between COS and H_2S. Both COS and H_2S loaded high in Factor 1 which, however, accounted for only 47% of the total variability. CS_2 also loaded relatively high in Factor 1. To the contrary SO_2 loaded very high in Factor 2. These results are discussed in detail in Chapter 4 in the context of air quality at Crossfield, the urban monitoring site.

Table 3.57 Results of the principal component analysis of sulphur gases characterized by cryogenic sampling at Fortress Mountain.

Variable	Factor Loading	
	1	2
COS	0.85	0.10
H_2S	0.78	0.26
CS_2	0.67	-0.58
SO_2	0.26	0.87
Pct. of variability	47	29
Eigenvalue	1.88	1.15

The organic compounds identified at Fortress Mountain consisted of hexane, cyclohexane, benzene, methyl benzene, ethyl benzene, and 1,4-dimethyl benzene. Hexane, cyclohexane, and benzene accounted for a predominant portion of the mixture. It was not possible to make a sufficient number of measurements of these compounds to provide acceptable quantitative data. However, Table 3.58 provides a summary from the published literature of the ambient concentrations of some of the organic compounds identified at Fortress Mountain.

Table 3.58 Typical ambient concentrations of selected hydrocarbons (ppb) as reported in published literature.

Hydrocarbon	Urban	Rural
Hexane	9 - 117	0.2 - 2.0
Cyclohexane	<0 - 16	0.1 - 1.0
Benzene	8 - 212	0.4 - 5.0
Ethyl benzene	4 - 48	1.0 - 2.0

Extracted from Demerjian 1986; Finlayson-Pitts and Pitts 1986; Arnts and Meeks 1981.

3.6.4 Integrated Measurements of Gaseous Air Pollutants and Aerosols (Annular Denuder and Dichotomous Sampler)

A better understanding of atmospheric processes and the present and potential impacts of certain air pollutants (e.g. HNO_3 and NH_3) on the environment has been limited by the lack of satisfactory measurement methods for the dry deposition of these pollutants. However, over the past several years a number of investigators have developed impregnated filter and denuder technology for the measurement of pollutants such as HNO_3 and NH_3 (Ferm 1979; Forrest et al. 1982; Lewin and Klockow 1982; Possanzini et al. 1983; Shaw et al. 1978, 1982; Stevens et al. 1978; U.S. EPA 1986). Stevens (1986) provided a critical review of the methods for measuring chemical species that contribute to acidic dry deposition. He concluded that one of the most promising systems is based on the use of annular denuders and filter packs operated in series (Table 3.59).

3.6.4.1 KAPS Annular Denuder-Filter Pack System

Using a serial annular denuder-filter pack system (Kananaskis Air Pollutant Measurement System, KAPS) 48-hour integrated samples of SO_2, HNO_2, HNO_3, NH_3 and fine particle SO_4^{2-} and NO_3^- were collected at Fortress Mountain. These samples were analyzed by the appropriate method described in Table 3.14. A summary statistic of the observed concentrations of the various chemical constituents are presented in Table 3.61. The abbreviations used for the variables are presented in Table 3.60. The mean value for the SO_2 concentration measured by the denuder method (2.0 $\mu g\ m^{-3}$ SO_4^{2-} value converted to SO_2, 0.5 ppb) was much higher than the mean of the 10-minute values (0.16 ppb) reported with the continuous monitoring method (Table 3.35).

The mean value with the continuous measurement was low, most probably because >99% of the total observations were below the limit of minimum detection (4 ppb) for the method used (Table 3.24). For the computation of 10-minute averages, values predicted by the Weibull probability generator, were used for substitution of the values below the limits of detection. In contrast, with KAPS the SO_2 values used for computation of the mean were obtained from the analysis of air samples collected over a 48-hour period,

Table 3.59 Description of sampling systems and instruments that measure chemical species that contribute to acidic dry deposition.

Method	Sampling Rate (LPM)	Sampling Period	Denuder Type and Coating	Filter Type	Species Measured
DDM cyclone Denuder Filter Pack System	Denuder, 1 Filter Pack, 2	7 day	300 mm x 4 mm tube coated with $MgCO_3$	Nylon (25 mm) Teflon (47 mm) TEA treated Whatman	Fine particle nitrate, HNO_3, SO_4^{2-}, SO_2, NO_2
AIHL, Filter Pack	20	4-24 hrs	N/A	Teflon (47 mm Followed by NaCl treated W41)	Fine particle nitrate, HNO_3, SO_4^{2-}, SO_2, NO_2
Modified Canadian Filter Pack	14 1.5 1.0	1 day 7 day 28 day	N/A	1) Teflon 2) Nylon, 3) W41, Na_2CO_3 treated	SO_4^{2-}, Total Nitrate SO_2
Cyclone-Denuder Pack Concentration Monitor	15	>12 hr	320 mm x 10 mm Nylon lined tube	Teflon, Nylon TEA treated W41	HNO_3, NH_3, H^+, SO_4^{2-}Filter SO_2, NO_2, Nitrate

Continued...

Table 3.59 (Concluded).

Method	Sampling Rate (LPM)	Sampling Period	Denuder Type and Coating	Filter Type	Species Measured
Cyclone-Annular Denuder-Filter Pack in Series	1 4 15	7 day 24 hr 12-24 hr	Annular design[1] Na_2CO_3 (SO_2, HNO_3, HNO_2) Citric acid (NH_3)	Teflon, Nylon, TEA treated W41	HNO_3, HNO_2, SO_2, HCl, Nitrate, NO_2, NH_3, H^+, SO_4^{2-} and fine particle mass
NOAA Filter Pack	3	7 day	N/A	Teflon Nylon W41, Na_2CO_3	SO_4^{2-} Total nitrate, SO_2
Tungstic Acid Diffusion Tube	1	30 min	200 mm x 3 mm WO_2		HNO_3 and NH_3
Chemiluminescent Methods	1	1 min	N/A	N/A	O_3, SO_2 NO_x

[1] Annular design is 20 times more efficient in removing gases than tubular designs of equivalent length and Reynolds number operation.
[2] Sensitivity of the methods listed in this table which directly derive the analysis value from a filter or denuder is approximately 0.1 mg m^{-3} assuming at least 2 m^3 of air is sampled.

Source: Stevens (1986).

Table 3.60 KAPS annular denuder system. Explanation of the abbreviations used in Tables 3.61 through 3.65.

NY	= Nylon filter
TF	= Teflon filter
T1	= Denuder Tube #1
T2	= Denuder Tube #2
	(T1 and T2 were coated with sodium carbonate)
T3	= Denuder Tube #3
	(T3 was coated with citric acid)
DNO_2	= Corrected denuder NO_2^- concentration (HNO_2)
DNO_3	= Corrected denuder NO_3^- concentration (HNO_3)
DSO_4	= Corrected denuder SO_4^{2-} concentration (2/3 SO_4^{2-} = SO_2)
FNO_3	= Total fine particulate NO_3^- concentration
FSO_4	= Total fine particulate SO_4^{2-} concentration

two (winter) or three (summer) times per week, and there were very few values below the measurement limit. Independent of these differences, the SO_2 concentrations quantified by the denuder method were still within the range reported for remote locations (Table 3.36).

The mean SO_2 value (0.51 ppb) reported from the denuder method (Table 3.61) was close to the mean of 0.53 ppb reported from the cryogenic sampling method (Table 3.55). The observed difference may be because the "n" for KAPS (154) was relatively high compared to the cryogenic samples (32).

The minimum SO_2 concentrations reported by the denuder method and the cryogenic sampling were 191 ppt and 167 ppt, respectively (Tables 3.61 and 3.55). However, the maximum SO_2 concentration obtained by the denuder method (6.5 ppb) was approximately 7 times greater than the corresponding value for the cryogenic sampling (955 ppt). The high values ($>$6.0 ppb) reported by the denuder method, however, occurred on only two occasions in the entire two year study period.

Mean concentrations of HNO_2 (DNO_2), HNO_3 (DNO_3), and NH_3 at Fortress Mountain over the two year period were 0.05 $\pm$ 0.06; 0.31 $\pm$ 0.32, and 0.26 $\pm$ 0.15 $\mu g\ m^{-3}$. Similarly the mean concentrations of fine particulate NO_3^- and SO_4^{2-} concentrations were 0.13 $\pm$ 0.09 and 0.51 $\pm$ 0.37 $\mu g\ m^{-3}$ (Table 3.61). All these values were in the low end of the range for typical ambient concentrations (Table 3.62).

Table 3.61 Summary statistics of integrated (annular denuder system, KAPS) measurements of gaseous and particulate pollutants. Sampling location: Fortress Mountain. Sampling period: 1985-87. Sampling duration: 48 h. All units, except (n), skewness and kurtosis, in $\mu g\ m^{-3}$. For convenience, valence signs have been omitted.

Variable	(n)	Mean	S.D.	Range		Median	Percentile						Skewness	Kurtosis
				Min.	Max.		10	30	70	90	95	99		
NO_2 NY	148	.02	.03	.001	.16	.01	.003	.005	.02	.06	.09	.152	.2	5.4
NO_2 TF	81	.01	.02	.001	.14	.004	.002	.003	.01	.01	.02	--	7.0	54.9
NO_3 NY	152	.09	.06	.01	.38	.08	.03	.06	.11	.17	.22	.37	2.1	6.3
NO_3 TF	155	.03	.04	.002	.44	.02	.01	.01	.03	.07	.09	.30	6.3	54.8
SO_4 NY	148	.08	.15	.005	1.2	.04	.01	.03	.04	.13	.40	1.0	4.9	28.9
SO_4 TF	157	.43	.35	.02	1.7	.34	.06	.19	.59	.93	1.2	1.7	1.2	1.5
NO_2 T1	149	.09	.08	.003	.56	.06	.03	.05	.10	.19	.25	.53	2.6	9.6
NO_2 T2	141	.04	.04	.002	.19	.03	.01	.02	.05	.10	.13	.19	1.6	2.4
NO_3 T1	155	.26	.31	.02	3.4	.20	.10	.15	.27	.44	.54	2.3	7.8	77.4
NO_3 T2	155	.07	.05	.01	.36	.05	.03	.04	.06	.13	.15	.33	3.0	11.0
SO_4 T1	155	1.6	3.5	.04	26.3	.56	.19	.36	1.3	3.4	5.2	26.1	5.3	31.7
SO_4 T2	155	.22	.40	.02	4.1	.11	.04	.07	.19	.49	.71	2.8	6.9	61.7
NH_3 T3	154	.26	.15	.04	1.3	.26	.08	.18	.32	.41	.47	.91	2.0	11.4
DNO_2[a]	137	.05	.06	.001	.46	.03	.01	.02	.05	.12	.16	.44	3.7	18.8
DNO_3[b]	154	.31	.32	.001	3.5	.25	.12	.19	.33	.52	.67	2.3	7.0	65.3
DSO_4[c]	154	2.0	3.9	.001	27.6	.74	.23	.46	1.63	4.1	8.3	27.4	6.9	36.5
FNO_3[d]	149	.13	.09	.02	.66	.11	.04	.08	.14	.21	.26	.59	2.7	11.9
FSO_4[e]	146	.51	.37	.05	2.0	.45	.10	.25	.66	1.1	1.2	1.9	1.1	1.3

[a] DNO_2 = NO_2T1 - NO_2T2
[b] DNO_3 = NO_3T1 + NO_3T2
[c] DSO_4 = SO_4T1 + SO_4T2
[d] FNO_3 = NO_3TF + NO_3NY
[e] FSO_4 = SO_4TF + SO_4NY

Table 3.62 Typical ambient concentrations of chemical species related to acidic dry deposition.

Pollutant	Range of Concentrations ($\mu g\ m^{-3}$)
SO_2	0.20 - 30
HNO_2	0.20 - 2.0
HNO_3	0.10 - 15
NH_3	0.10 - 20
NO_3^- (fine particle)	0.10 - 10
SO_4^{2-} (fine particle)	0.20 - 40

Source: Stevens (1986).

Diurnal variations in the concentrations of pollutants are expected to occur as a result of differences in atmospheric chemistry and in meteorology between day and night. These variations are shown in Table 3.63. Atmospheric concentrations of nitric acid (DNO_3) at the Fortress Mountain, Crossfield West, and Crossfield East sampling sites were greater during the day. Conversely, concentrations of nitrous acid (DNO_2) were greater at night at Crossfield West and Crossfield East, and concentrations at Fortress Mountain were too low to show a difference. These results are consistent with the known chemistry of nitrogen compounds in the atmosphere (Sjödin and Ferm 1985).

$$NO + NO_2 + H_2O \rightarrow 2HNO_2 \quad (1)$$

$$HNO_2 + h\nu \rightarrow OH\bullet + NO \quad (2)$$

Table 3.63 Summary of integrated (annular denuder system, KAPS) measurements of gaseous and particulate pollutants with daytime (0200-1900 MST) and nighttime (1900-0700 MST) sampling. Sampling period: 1985-87. Units: $\mu g\ m^{-3}$.

	Fortress		Crossfield West		Crossfield East	
	Day	Night	Day	Night	Day	Night
NO_2 T1	0.12	0.10	0.23	0.34	0.38	0.61
NO_2 T2	0.05	0.05	0.08	0.09	0.13	0.16
NO_3 T1	0.30	0.26	0.43	0.34	0.42	0.37

Continued...

Table 3.63 (Concluded).

	Fortress		Crossfield West		Crossfield East	
	Day	Night	Day	Night	Day	Night
NO_3 T2	0.08	0.08	0.11	0.11	0.11	0.13
SO_4 T1	1.33	1.65	7.27	5.14	9.83	6.03
SO_4 T2	0.32	0.25	0.54	0.50	0.81	0.56
NH_3 T3	0.31	0.25	1.49	1.61	1.93	1.97
NO_3 NY	0.10	0.12	0.33	0.35	0.32	0.23
NO_3 TF	0.03	0.05	0.26	0.37	0.30	0.48
SO_4 NY	0.05	0.08	0.19	0.18	0.14	0.13
SO_4 TF	0.41	0.45	0.96	0.84	0.90	0.90
D NO_2 [a]	0.07	0.05	0.15	0.25	0.25	0.48
D NO_3 [b]	0.38	0.34	0.54	0.45	0.53	0.50
D SO_4 [c]	1.63	1.90	7.81	5.60	10.64	6.59
F NO_3 [d]	0.13	0.17	0.59	0.72	0.62	0.71
F SO_4 [e]	0.46	0.53	1.15	1.02	1.04	1.03

[a] D NO_2 = NO_2 T1 - NO_2 T2
[b] D NO_3 = NO_3 T1 + NO_3 T2
[c] D SO_4 = SO_4 T1 + SO_4 T2
[d] FNO_3 = NO_3 TF+ NO_3 NY
[e] FSO_4 = SO_4 TF + SO_4 NY

Sulphur dioxide concentrations were highest during the day at the two Crossfield sampling sites. This may be the result of meteorological conditions or perhaps higher rates of SO_2 emissions by the natural gas processing plants in the vicinity during the day.

The results of the correlation analysis between the variables sampled with the KAPS system are presented in Table 3.64. The key features of this table are summarized in Table 3.65.

Table 3.64 Pearson's correlation coefficients (r) between variables measured through the annular denuder system (KAPS). Sampling location: Fortress Mountain. Sampling period: 1985-87. Sampling duration: 48 hours. For convenience, valence signs have been omitted.*

Variables	(r)	Variables	(r)
NO_3 NY vs NO_3 TF	.35	NO_2 T1 vs NO_2 T2	.71
NO_3 NY vs SO_4 NY	.16	NO_2 T1 vs NO_3 T1	.52
NO_3 NY vs SO_4 TF	.54	NO_2 T1 vs NO_3 T2	.26

Continued...

Table 3.64 (Continued).

Variables	(r)	Variables	(r)
NO_3 NY vs NO_2 T1	.40	NO_2 T1 vs SO_4 T1	.52
NO_3 NY vs NO_2 T2	.18	NO_2 T1 vs SO_4 T2	.76
NO_3 NY vs NO_3 T1	.41	NO_2 T2 vs NO_3 T1	.11
NO_3 NY vs NO_3 T2	.20	NO_2 T2 vs NO_3 T2	.18
NO_3 NY vs SO_4 T1	.26	NO_2 T2 vs SO_4 T1	.24
NO_3 NY vs SO_4 T2	.46	NO_2 T2 vs SO_4 T2	.39
NO_3 TF vs SO_4 NY	.06	NO_3 T1 vs NO_3 T2	.27
NO_3 TF vs SO_4 TF	.40	NO_3 T1 vs SO_4 T1	.50
NO_3 TF vs NO_2 T1	.62	NO_3 T1 vs SO_4 T2	.58
NO_3 TF vs NO_2 T2	.29	NO_3 T2 vs SO_4 T1	.09
NO_3 TF vs NO_3 T1	.84	NO_3 T2 vs SO_4 T2	.41
NO_3 TF vs NO_3 T2	.17	SO_4 T1 vs SO_4 T2	.47
NO_3 TF vs SO_4 T1	.59	NH_3 T3 vs NO_3 NY	.47
NO_3 TF vs SO_4 T2	.59	NH_3 T3 vs NO_3 TF	.36
SO_4 NY vs SO_4 TF	- .06	NH_3 T3 vs SO_4 NY	.16
SO_4 NY vs NO_2 T1	- .04	NH_3 T3 vs SO_4 TF	.31
SO_4 NY vs NO_2 T2	- .08	NH_3 T3 vs NO_2 T1	.43
SO_4 NY vs NO_3 T1	.08	NH_3 T3 vs NO_2 T2	.37
SO_4 NY vs NO_3 T2	- .005	NH_3 T3 vs NO_3 T1	.31
SO_4 NY vs SO_4 T1	.04	NH_3 T3 vs NO_3 T2	.29
SO_4 NY vs SO_4 T2	.02	NH_3 T3 vs SO_4 T1	.005
SO_4 TF vs NO_2 T1	.15	NH_3 T3 vs SO_4 T2	.57
SO_4 TF vs NO_2 T2	- .02	DNO_2 vs NO_3 NY	.43
SO_4 TF vs NO_3 T1	.49	DNO_2 vs NO_3 TF	.68
SO_4 TF vs NO_3 T2	.07	DNO_2 vs SO_4 NY	- .02
SO_4 TF vs SO_4 T1	.28	DNO_2 vs SO_4 TF	.23
SO_4 TF vs SO_4 T2	.26	DNO_2 vs NO_2 T1	.91
DNO_2 vs NO_2 T2	.36	FNO_3 vs NO_2 T2	.27
DNO_2 vs NO_3 T1	.64	FNO_3 vs NO_3 T1	.71
DNO_2 vs NO_3 T2	.30	FNO_3 vs NO_3 T2	.23
DNO_2 vs SO_4 T1	.56	FNO_3 vs SO_4 T1	.48
DNO_2 vs SO_4 T2	.78	FNO_3 vs SO_4 T2	.62
DNO_3 vs NO_3 NY	.43	FSO_4 vs NO_3 NY	.60
DNO_3 vs NO_3 TF	.52	FSO_4 vs NO_3 TF	.34
DNO_3 vs SO_4 NY	.08	FSO_4 vs SO_4 NY	.33
DNO_3 vs SO_4 TF	.48	FSO_4 vs SO_4 TF	.93
DNO_3 vs NO_2 T1	.52	FSO_4 vs NO_2 T1	.13

Continued...

Table 3.64 (Concluded).

Variables			(r)	Variables			(r)
DNO_3	vs	NO_2 T2	.14	FSO_4	vs	NO_2 T2	- .05
DNO_3	vs	NO_3 T1	.99	FSO_4	vs	NO_3 T1	.50
DNO_3	vs	NO_3 T2	.42	FSO_4	vs	NO_3 T2	.09
DNO_3	vs	SO_4 T1	.43	FSO_4	vs	SO_4 T1	.25
DNO_3	vs	SO_4 T2	.59	FSO_4	vs	SO_4 T2	.25
DSO_4	vs	NO_3 NY	.27	NH_3 T3	vs	DNO_2	.37
DSO_4	vs	NO_3 TF	.58	NH_3 T3	vs	DNO_3	.36
DSO_4	vs	SO_4 NY	.04	NH_3 T3	vs	DSO_4	- .02
DSO_4	vs	SO_4 TF	.27	NH_3 T3	vs	FNO_3	.42
DSO_4	vs	NO_2 T1	.57	NH_3 T3	vs	FSO_4	.35
DSO_4	vs	NO_2 T2	.26	DNO_2	vs	DNO_3	.64
DSO_4	vs	NO_3 T1	.48	DNO_2	vs	DSO_4	.60
DSO_4	vs	NO_3 T2	.09	DNO_2	vs	FNO_3	.64
DSO_4	vs	SO_4 T1	1.00	DNO_2	vs	FSO_4	.21
DSO_4	vs	SO_4 T2	.58	DNO_3	vs	DSO_4	.46
FNO_3	vs	NO_3 NY	.88	DNO_3	vs	FNO_3	.57
FNO_3	vs	NO_3 TF	.75	DNO_3	vs	FSO_4	.49
FNO_3	vs	SO_4 NY	.14	DSO_4	vs	FNO_3	.46
FNO_3	vs	SO_4 TF	.58	DSO_4	vs	FSO_4	.25
FNO_3	vs	NO_2 T1	.59	FNO_3	vs	FSO_4	.60

* NH_3 was quantified as NH_4^+.

Table 3.65 Summary of Pearson's correlation coefficients presented in Table 3.64.

Variable Pair*			(r)
SO_2(T1)	vs	SO_2(Cor)	1.00
SO_2(T1)	vs	HNO_2(Cor)	0.56
SO_2(T1)	vs	HNO_2(T1)	0.52
SO_2(T1)	vs	HNO_3(Cor)	0.43
SO_2(T1)	vs	HNO_3(T1)	0.50
SO_2(Cor)	vs	HNO_3(T1)	0.48
SO_2(T2)	vs	HNO_2(Cor)	0.78
SO_2(T2)	vs	HNO_2(T1)	0.76
SO_2(T2)	vs	HNO_3(T1)	0.58
SO_2(T2)	vs	NH_3	0.57
HNO_2(T1)	vs	HNO_2(Cor)	0.91

Continued...

Table 3.65 (Concluded).

Variable Pair*			(r)
HNO_2(T1)	vs	HNO_2(T2)	0.71
HNO_3(Cor)	vs	HNO_2(Cor)	0.64
SO_4^{2-}(TF)	vs	SO_4^{2-}(Tot)	0.93
SO_4^{2-}(TF)	vs	SO_4^{2-}(NY)	-0.06
SO_4^{2-}(Tot)	vs	SO_2(Cor)	0.62
NO_3^-(TF)	vs	HNO_3(Cor)	0.52
NO_3^-(TF)	vs	NO_3^-(Tot)	0.78
NO_3^-(TF)	vs	HNO_2(Cor)	0.68
NO_3^-(NY)	vs	NO_3^-(Tot)	0.79
NH_3	vs	SO_2(Cor)	0.02
NH_3	vs	NO_3^-(Tot)	0.424

* Cor = Corrected (T1+T2)
T1 = Denuder Tube #1
T2 = Denuder Tube #2
(Nos. 1 and 2 coated with Na_2CO_3)
Tot = Total from Teflon and Nylon filters
TF = Teflon filter
NY = Nylon filter

a. The nitrite concentrations in denuder tubes T1 and T2 were well correlated (r=0.71). This is consistent with the conversion of a small portion of the ambient atmospheric NO/NO_2 to NO_2^- in each of the annular denuder tubes, as was found by Sjödin and Fern (1985). The resulting interference in the nitrous acid, HNO_2, on T1 was corrected by the subtraction of NO_2T2 from NO_2T1, thus in Table 3.61 DNO_2 (HNO_2)= NO_2T1-NO_2T2.
b. The concentrations of SO_4^{2-} on the second denuder tube, SO_4T2, was well correlated with NO_2T1 and DNO_2, (r=0.75 and 0.78, respectively) with nitrate on the first tube, NO_3T1, (r=0.58) and with nitric acid, DNO_3, (r=0.59). These observations suggest that at the low SO_2 concentrations encountered at Fortress Mountain, the collection efficiency of T1 was affected by ambient concentrations of NO/NO_2, HNO_2, and HNO_3.
c. The sulphate concentrations on the first two denuder tubes,

SO_4T1 and SO_4T2, did not correlate very highly with each other ($r=0.47$). It has been common practice by investigators using the annular denuder system to subtract the value obtained in Tube 2, SO_4T2, from the value obtained in Tube 1, SO_4T1, to derive the corrected atmospheric SO_2 concentrations. This approach has been used to correct for the deposition of fine particle sulphate on the tubes and for possible sulphur contamination of the reagents used for the denuder coating. This method assumed that almost all of the SO_2 in the sampled air is trapped by the first denuder tube (T1) and is analyzed as SO_4^{2-}. In the present case, on an average basis at Fortress Mountain, the SO_4^{2-} concentration in T2 was 15% of SO_4T1. On a few occasions, the SO_4T2 concentrations were equal to the blank, but were mostly 4.8 $\pm$ 4.7 times higher than the blank. The concentration of fine particle SO_4^{2-} collected by the Teflon filter was low, and therefore, the amount of fine particle SO_4^{2-} deposition in T1 and T2 would be small. These findings suggest that a portion of the SO_2, up to 15% on average, passed through T1 and was collected by T2. In calculating the ambient atmospheric concentration of SO_2 at Fortress Mountain, fine particle SO_4^{2-} deposition was considered to be negligible and DSO_2 (in Table 3.61) was calculated from SO_4T1 + SO_4T2.

3.6.4.2 Dichotomous Sampler

Fine and coarse particle samples were collected at Fortress Mountain with a dichotomous sampler as 96-h integrated samples. These samples were analyzed by x-ray fluorescence spectroscopy (XRF). Summary statistics on the concentrations of various elements in the fine and coarse particle fraction are presented in Table 3.66. Crustal elements such as Al, Si, Ca, Ti, Fe, Mn, and K were present at higher concentrations in the coarse compared to the fine particle fraction. Conversely elements of anthropogenic origin such as S, As, and Pb were in higher concentrations in the fine compared to the coarse fraction. Table 3.67 provides a comparison of the S, Pb, and Br concentrations measured at Fortress Mountain with values reported for the Canadian Arctic and southern Sweden.

The SO_4^{2-} concentrations, smoothed as annual means and their similarity between Fortress Mountain and the three sites in the Canadian Arctic combined, suggest homogeneity of the regional-

Table 3.66 Summary statistics of chemical constituents in particulate matter (ng m^{-3}). Sampling location: Fortress Mountain. Sampling period: 1986 and 1987, 12 months. Sampling duration: 96 h.

				Range			Percentile							
Variable	(n)	Mean	S.D.	Min.	Max.	Median	10	30	70	90	95	99	Skewness	Kurtosis
Al F*	71	37.1	50.2	.84	276.7	22.2	6.1	15.8	32.8	63.3	159.0	--	3.5	13.2
Al C*	72	186.0	178.4	2.0	810.1	119.0	21.6	81.2	226.7	455.4	591.0	--	1.5	1.8
Si F	71	100.2	133.2	1.6	733.6	61.0	16.6	36.6	92.1	187.0	404.2	--	3.3	12.0
Si C	72	690.2	654.2	3.0	3260.1	439.5	85.0	247.8	880.6	1777.8	2031.0	--	1.5	2.4
P F	71	1.4	2.2	BD	8.2	.40	BD	BD	1.2	5.2	7.5	--	1.9	2.7
P C	72	2.2	2.4	BD	12.5	1.6	BD	.39	3.4	5.6	6.8	--	1.6	3.7
S F	71	246.5	163.1	BD	803.4	219.1	61.5	153.0	296.5	463.0	622.5	--	1.3	2.2
S C	72	51.1	33.2	BD	185.6	45.9	18.7	30.0	57.5	97.1	130.4	--	1.6	3.4
Cl F	71	.27	.66	BD	3.3	BD	BD	BD	.01	.83	2.2	--	3.1	9.6
Cl C	72	5.0	5.5	BD	31.9	3.6	.63	1.8	5.7	11.0	16.5	--	2.7	9.6
K F	71	21.9	20.4	.28	82.2	15.2	4.4	9.6	25.0	60.9	76.2	--	1.7	2.2
K C	72	44.5	38.9	.28	197.5	29.5	8.7	19.4	53.8	106.6	126.3	--	1.6	2.8
Ca F	71	14.8	20.2	.19	110.8	8.1	2.3	5.4	13.0	36.3	65.8	--	3.1	10.5
Ca C	72	143.9	143.1	.72	782.2	97.1	20.7	49.3	187.2	371.8	441.1	--	1.9	4.8
Ti F	71	1.4	1.9	BD	11.1	.74	.31	.52	1.2	2.7	5.2	--	3.6	15.0
Ti C	72	9.3	8.9	BD	42.2	6.0	1.1	3.6	11.8	24.5	28.5	--	1.4	1.9
V F	71	.14	.11	BD	.54	.12	.01	.08	.16	.26	.38	--	1.4	2.5
V C	72	.56	.53	BD	3.3	.36	.12	.26	.67	1.3	1.6	--	2.5	9.3
Cr F	71	.10	.09	BD	.50	.08	BD	.04	.12	.21	.26	--	1.7	5.6
Cr C	72	.37	.31	.03	1.9	.30	.06	.19	.42	.75	.99	--	2.2	7.1

Continued . . .

Table 3.66 (Continued).

Variable	(n)	Mean	S.D.	Range		Median	Percentile						Skewness	Kurtosis
				Min.	Max.		10	30	70	90	95	99		
Sr F	71	.26	.29	BD	1.2	.20	BD	.05	.31	.80	.91	--	1.4	1.1
Sr C	72	.68	.61	BD	2.8	.50	.05	.29	.84	1.6	2.0	--	1.4	1.9
Y F	71	.11	.19	BD	.87	.01	BD	BD	.12	.39	.64	--	2.5	6.4
Y C	72	.15	.20	BD	1.2	.08	BD	BD	.20	.40	.58	--	2.5	8.4
Zr F	71	.15	.32	BD	1.2	BD	BD	BD	BD	.83	1.00	--	2.0	2.6
Zr C	72	.25	.41	BD	1.4	BD	BD	BD	.40	.90	1.1	--	1.3	.15
Mo F	71	.10	.19	BD	.83	BD	BD	BD	BD	.46	.54	--	1.9	2.8
Mo C	72	.12	.21	BD	.86	BD	BD	BD	.13	.47	.58	--	1.6	1.7
Pd F	71	.16	.18	BD	.62	.08	BD	BD	.29	.42	.51	--	.75	- .71
Pd C	72	.20	.24	BD	.84	.09	BD	BD	.34	.55	.67	--	.90	- .46
Ag F	71	.33	.31	BD	1.1	.32	BD	BD	.52	.78	.88	--	.47	- .95
Ag C	72	.35	.33	BD	1.2	.29	BD	.05	.50	.79	.99	--	.71	- .31
Cd F	71	.26	.32	BD	1.2	.08	BD	BD	.37	.76	1.00	--	1.2	.51
Cd C	72	.44	1.2	BD	9.7	.13	BD	BD	.49	.99	1.1	--	7.2	57.8
In F	71	.49	.60	BD	2.5	.27	BD	BD	.81	1.4	1.7	--	1.1	.58
In C	72	.33	.42	BD	1.8	.16	BD	BD	.45	1.00	1.3	--	1.4	1.2
Sn F	71	.61	.68	BD	2.9	.50	BD	BD	.94	1.5	1.9	--	1.1	1.00
Sn C	72	.58	.68	BD	2.5	.32	BD	BD	.93	1.6	2.1	--	.95	- .14
Sb F	71	1.8	1.5	BD	6.0	1.6	BD	.34	2.8	3.8	4.6	--	.50	- .55
Sb C	72	1.2	1.3	BD	4.2	.97	BD	BD	1.8	3.6	3.9	--	.80	- .46

Continued . . .

Table 3.66 (Continued).

Variable	(n)	Mean	S.D.	Range		Median	Percentile						Skewness	Kurtosis
				Min.	Max.		10	30	70	90	95	99		
Mn F	71	.48	.46	.01	2.3	.36	.09	.26	.46	.98	1.8	--	2.4	6.4
Mn C	72	2.3	2.3	.12	13.8	1.6	.42	.91	2.9	5.5	6.4	--	2.3	8.2
Fe F	71	12.8	18.2	.33	98.9	7.3	2.1	4.6	11.22	5.9	52.6	--	3.4	12.9
Fe C	72	77.2	74.7	.79	370.7	47.8	10.9	28.7	99.8	200.5	238.7	--	1.6	2.5
Ni F	71	.08	.07	BD	.30	.05	BD	.03	.10	.16	.24	--	1.3	2.1
Ni C	72	.23	.18	BD	.88	.19	.04	.12	.27	.44	.59	--	1.3	2.1
Cu F	71	.19	.40	BD	3.3	.12	BD	.05	.21	.34	.41	--	7.1	56.4
Cu C	72	.39	.49	BD	3.6	.26	.04	.19	.39	.94	1.2	--	4.2	24.1
Zn F	71	1.6	1.6	BD	9.9	1.2	.33	.76	1.9	3.4	4.3	--	2.8	11.1
Zn C	72	1.1	.70	.16	2.8	.94	.32	.62	1.5	2.3	2.5	--	.68	- .60
Ga F	71	.02	.02	BD	.10	.01	BD	BD	.03	.05	.08	--	1.3	1.2
Ga C	72	.05	.05	BD	.22	.04	BD	.02	.08	.14	.14	--	.94	.58
As F	71	.11	.22	BD	1.00	BD	BD	BD	.12	.35	.69	--	2.6	7.0
As C	72	.08	.11	BD	.52	.01	BD	BD	.12	.25	.31	--	1.6	2.7
Se F	71	.08	.05	BD	.19	.08	BD	.05	.11	.15	.16	--	.01	- .96
Se C	72	.06	.05	BD	.19	.05	BD	.01	.09	.13	.15	--	.48	- .65
Br F	71	1.1	.62	.10	3.0	1.00	.33	.78	1.3	1.9	2.1	--	.65	.58
Br C	72	.31	.21	BD	1.1	.27	.08	.17	.39	.53	.66	--	1.5	3.8
Rb F	71	.17	.19	BD	.94	.10	BD	.02	.22	.44	.57	--	1.6	3.0
Rb C	72	.30	.27	BD	1.2	.25	BD	.10	.38	.68	.80	--	1.00	.90

Continued . . .

Table 3.66 (Concluded)

Variable	(n)	Mean	S.D.	Range		Median	Percentile						Skewness	Kurtosis
				Min.	Max.		10	30	70	90	95	99		
Ba F	71	1.6	2.4	BD	8.1	BD	BD	BD	2.4	5.9	6.9	--	1.3	.14
Ba C	72	2.2	3.0	BD	10.3	BD	BD	BD	3.9	7.0	8.0	--	1.00	- .28
La F	71	2.8	4.0	BD	14.7	BD	BD	BD	4.4	10.0	10.3	--	1.2	.13
La C	72	3.1	4.3	BD	14.7	BD	BD	BD	4.2	10.7	12.4	--	1.2	.35
Hg F	71	.04	.04	BD	.16	.03	BD	BD	.05	.10	.15	--	1.2	1.00
Hg C	72	.05	.05	BD	.21	.04	BD	.01	.07	.12	.14	--	1.1	1.1
Pb F	71	1.9	1.00	.58	4.8	1.7	.76	1.2	2.2	3.5	4.3	--	1.00	.66
Pb C	72	.89	.58	BD	2.7	.78	.21	.55	1.2	1.8	2.0	--	.76	.31

* F = fine particles (<2.5 μm); C = coarse particles (>2.5 μm). Chemical analysis was performed by x-ray fluorescence spectroscopy.

BD: below the limit of detection.

Table 3.67 Comparison of the annual means of S, Pb, and Br concentrations at Fortress Mountain with values from the Canadian Arctic and southern Sweden.

Sampling Location	Mean Concentration ng m^{-3}		
	S	Pb	Br
Fortress Mountain*	297	2.8	1.4
Canadian Arctic**	297	2.7	7.9
S. Sweden	800	21	--

* Fine and coarse particles, as the other two studies did not report actual particle sizes.

** Average of three sampling sites. Data extracted from Tables 3.7, 3.8, and 3.66.

scale air quality of the Canadian background sites. This characteristic was also similar for the Pb^{4+} concentrations. However, the mean Br^- concentration was much higher in the Canadian Arctic compared to Fortress Mountain. Br^- is considered to be an indicator of mobile sources (Stevens 1982). Given the geographic locations of the Canadian Arctic sites, Barrie and Hoff (1985) concluded that the Br^- at those sites was contributed by both sea salt and anthropogenic sources. At Fortress Mountain sea salt did not appear to be a factor.

Tables 3.68 through 3.70 provide summary statistics on the correlations between the chemical constituents in the coarse particles, in the fine particles, and in both fractions combined. An attempt was made to clarify the results presented in these tables by the application of principal component analysis (PCA) (Table 3.71). The crustal elements, Al, Si, K, Ca, Ti, Mn, Fe, and Sr in the coarse particles loaded high in Factor 1. This factor accounted for roughly 42% of the total variability of the chemical composition of both coarse and fine particles combined (Table 3.71). The crustal elements in the fine particle fraction (present at low concentrations compared to the coarse particles) also loaded high in Factor 1. Known tracers of anthropogenic pollution such as As, Br, and Pb in the fine particle fraction also loaded high in Factor 1. While Pb and Br are considered to originate from transportation activities, As is considered to be an indicator of coal combustion and non-ferrous metal processing. Fine particle Ag, In, and Sb were the only three elements which showed high values (>0.5) for loading in Factor 2. This factor accounted for 13.5% of the total variability of the chemical composition of coarse and fine particles combined. These elements (Ag, In, and Sb) are known to be emitted during coal combustion (Sheffield and Gordon 1985). Surprisingly, fine particle S and Zn were the only elements with values of loading >0.5 in Factor 8 (percent variability accounted by this factor was 2.9). S and Zn are known tracers for both metal production and coal combustion.

Table 3.68 Pearson's correlation coefficient (r) between variables in coarse particles (>2.5 μm). Dichotomous filter samples analyzed through x-ray fluorescence spectroscopy. Sampling location: Fortress Mountain. Sampling period: 1986-87. Sampling duration: 96 h.

Variable	(r)	Variable	(r)
Al vs Si	.99	Mn vs Ti	.91
Al vs K	.98	Mn vs V	.90
Al vs Ca	.94	Mn vs Cr	.75
Al vs Ti	.98	Fe vs Al	.99
Al vs V	.83	Fe vs Si	.99
Si vs K	.99	Fe vs K	.99
Si vs Ca	.95	Fe vs Ca	.94
Si vs Ti	.99	Fe vs Ti	.99
Si vs V	.89	Fe vs V	.86
K vs Ca	.96	Rb vs V	.70
K vs Ti	.99	Sr vs Al	.84
K vs V	.87	Sr vs Si	.80
Ca vs Ti	.94	Sr vs K	.82
Ca vs V	.86	Sr vs Ca	.79
Ti vs V	.87	Sr vs Ti	.81
V vs Cr	.78	Mn vs Fe	.91
Mn vs Al	.90	Sr vs Mn	.81
Mn vs Si	.92	Sr vs Fe	.82
Mn vs K	.93	Sr vs Rb	.75
Mn vs Ca	.92		

Table 3.69 Pearson's correlation coefficient (r) between variables in fine particles (<2.5μm). Dichotomous filter samples analyzed through x-ray fluorescence spectroscopy. Sampling location: Fortress Mountain. Sampling period: 1986-87. Sampling duration: 96 h.

Variable	(r)	Variable	(r)
Al vs Si	.99	Mn vs K	.88
Al vs K	.73	Mn vs Ca	.95
Al vs Ca	.95	Mn vs Ti	.91
Al vs Ti	.97	Mn vs V	.85

Continued...

Table 3.69 (Concluded).

Variable	(r)	Variable	(r)
Al vs V	.78	Fe vs Al	.98
Si vs K	.76	Fe vs Si	1.00
Si vs Ca	.97	Fe vs K	.76
Si vs Ti	.98	Fe vs Ca	.97
Si vs V	.80	Fe vs Ti	.98
K vs Ca	.80	Fe vs V	.79
K vs Ti	.74	As vs Ti	.71
K vs V	.82	Br vs K	.77
Ca vs Ti	.95	Mn vs Fe	.93
Ca vs V	.78	Mn vs Br	.70
Ti vs V	.81	Fe vs As	.70
Ti vs Cr	.38	Sr vs Rb	.79
Mn vs Al	.90	Y vs Rb	.81
Mn vs Si	.93	Pb vs Br	.69

Table 3.70 Pearson's correlation coefficient (r) between variables in coarse and fine particles. Dichotomous filter samples analyzed through x-ray fluorescence spectroscopy. Sampling location: Fortress Mountain. Sampling period: 1986-87. Time of data resolution: 96 h. F = fine; C = coarse particles.

Variable	(r)	Variable	(r)
Al C vs Si F	.75	Sr C vs Si F	.83
Al C vs K F	.91	Sr C vs K F	.88
Al C vs Ca F	.80	Sr C vs Ca F	.85
Al C vs Ti F	.76	Sr C vs Ti F	.80
Si C vs K F	.86	Al C vs Mn F	.84
Si C vs Ca F	.71	Al C vs Fe F	.75
K C vs K F	.88	Al C vs Br F	.71
K C vs Ca F	.75	Si C vs Mn F	.76
Ca C vs K F	.84	K C vs Mn F	.79
Ca C vs Ca F	.71	Ca C vs Mn F	.73
Ti C vs K F	.86	Ti C vs Mn F	.77
Ti C vs Ca F	.73	Mn C vs Mn F	.73
Mn C vs K F	.82	Fe C vs Mn F	.80
Fe C vs Si F	.70	Fe C vs Fe F	.70
Fe C vs K F	.88	Sr C vs Mn F	.87
Fe C vs Ca F	.76	Sr C vs Fe F	.82
Fe C vs Ti F	.70	Sr C vs Br F	.76
Rb C vs K F	.70	Pb F vs Al C	.70
Sr C vs Al F	.78	Al C vs Pb F	.70

Table 3.71 Varimax rotated factor matrix for XRF aerosol data for both fine (1) and coarse (2) particles (factor loading >0.5). Sampling location: Fortress Mountain. Sampling period: 1986-87. Sampling duration: 96 h.

	Factor							
Variable	1	2	3	4	5	6	7	8
Al 1	.83	--	--	--	--	--	--	--
Si 1	.88	--	--	--	--	--	--	--
P 1	--	--	--	--	.79	--	--	--
S 1	--	--	--	--	--	--	--	.69
Cl 1	--	--	--	.59	--	--	--	--
K 1	.93	--	--	--	--	--	--	--
Ca 1	.90	--	--	--	--	--	--	--
Ti 1	.89	--	--	--	--	--	--	--
V 1	.88	--	--	--	--	--	--	--
Mn 1	.93	--	--	--	--	--	--	--
Fe 1	.88	--	--	--	--	--	--	--
Zn 1	--	--	--	--	--	--	--	.53
As 1	.61	--	--	--	--	--	--	--
Br 1	.72	--	--	--	--	--	--	--
Rb 1	.52	--	--	--	--	--	--	--
Sr 1	.62	--	--	--	--	--	--	--
Zr 1	--	--	.61	--	--	--	--	--
Pd 1	--	--	--	--	--	--	.53	--
Ag 1	--	.71	--	--	--	--	--	--
In 1	--	.64	--	--	--	--	--	--
Sb 1	--	.71	--	--	--	--	--	--
Pb 1	.69	--	--	--	--	--	--	--
Al 2	.96	--	--	--	--	--	--	--
Si 2	.92	--	--	--	--	--	--	--
P 2	--	--	--	--	.55	--	--	--
S 2	--	--	--	--	--	.69	--	--
K 2	.93	--	--	--	--	--	--	--
Ca 2	.88	--	--	--	--	--	--	--
Ti 2	.93	--	--	--	--	--	--	--
V 2	.68	--	--	--	--	--	--	--
Cr 2	.59	--	--	--	--	--	--	--
Mn 2	.87	--	--	--	--	--	--	--
Fe 2	.94	--	--	--	--	--	--	--
Ni 2	.54	--	--	--	--	--	--	--
Zn 2	.60	--	--	--	--	--	--	--
Ga 2	.60	--	.50	--	--	--	--	--
Se 2	--	--	--	.68	--	--	--	--
Br 2	--	--	--	--	--	.51	--	--

Continued...

Table 3.71 (Concluded).

	Factor							
Variable	1	2	3	4	5	6	7	8
Rb 2	.70	--	--	--	--	--	--	--
Sr 2	.91	--	--	--	--	--	--	--
Y 2	--	--	--	--	--	--	.54	--
Zr 2	--	--	.59	--	--	--	--	--
Pd 2	--	--	--	--	--	--	.73	--
Sb 2	--	--	.53	--	--	--	--	--
Pb 2	--	--	--	.60	--	--	--	--
Pct of Var.	40.7	13.5	8.4	6.1	5.1	3.7	3.5	2.9
Eigenvalue	20.3	6.7	4.2	3.0	2.5	1.9	1.7	1.5

3.6.4.3 The Chemistry of Rain

During approximately April-October in the two years of study, rain samples were collected as 0.32 cm sequentially incremental fractions of each rain event using a refrigerated sampler (Coscio et al. 1982). The chemical analyses of these samples were according to the procedures described in Table 3.14.

Data on chemical composition were available for the sub-event rain samples. However, only summary statistics on averaged daily (24 hours) precipitation chemistry are presented in Table 3.72. The frequency distributions of almost all the constituents in the rain samples did not exhibit a normal distribution (refer to Chapter 2). Under these conditions, computations of mean values are inappropriate. The use of median values would be the correct approach since such values are free from the influence of the type of distribution. Nevertheless, to facilitate comparison with the results in published literature, both mean and median values are presented in Table 3.72.

The mean pH of rain at Fortress Mountain (pH 4.73) was quite comparable to the mean values (pH 4.78-4.96) reported by Galloway et al. (1982) for five remote sites (Table 3.12). While the mean SO_4^{2-} concentration (22.8 μeq L^{-1}) at Fortress Mountain was comparable to the corresponding values (2.9-36.3 μeq L^{-1}) for the remote sites, the NO_3 mean (13.2 μeq L^{-1}) was much higher than the values reported (1.7 to 5.5 μeq L^{-1}) by Galloway et al. (1982). This was also true for the mean NH_4^+ value (20.3 μeq L^{-1}) reported at Fortress Mountain. Given the non-normal frequency distribution,

Table 3.72 Summary statistics of daily precipitation chemistry at Fortress. Sampling period: approximately April-October, 1986 and 1987. Ion concentrations in $\mu eq.L^{-1}$.

				Range			Percentile						
Variable	(n)	Mean	S.D.	Min.	Max.	Median	10	30	70	90	95	Skewness	Kurtosis
H^+	50	18.4	15.8	.09	72.4	17.0	.46	3.3	28.3	36.0	54.7	1.2	2.1
Cl^-	37	3.7	2.4	1.00	10.3	2.8	1.3	1.8	5.1	7.4	9.5	1.1	.48
NO_3^-	52	13.2	7.8	3.3	42.4	10.8	6.1	8.3	15.8	24.4	30.6	1.7	3.3
SO_4^{2-}	52	22.8	14.1	5.6	63.1	19.0	8.1	11.9	30.2	43.9	54.3	1.0	.41
Ca^{2+}	40	30.4	24.7	2.2	100.2	22.3	8.0	11.1	39.1	72.0	90.6	1.3	1.2
Mg^{2+}	40	6.1	6.2	1.1	26.5	4.1	1.1	1.3	9.6	15.1	22.3	1.6	2.4
Na^+	40	4.6	6.3	.13	26.4	2.1	.13	1.2	4.8	13.2	23.9	2.6	5.2
K^+	31	1.4	1.4	.28	6.8	.90	.28	.28	2.0	3.4	5.0	2.2	5.8
NH_4^+	45	20.3	14.1	1.6	49.5	17.4	3.8	9.1	29.2	47.1	47.2	.69	-.57

Median pH 4.77;
Mean pH 4.73 and this value should be viewed with caution since it was derived from non-normally distributed data.

the median values for all rain parameters at Fortress Mountain were lower than their corresponding mean values (Table 3.72).

Table 3.73 provides comparative summary statistics of the results of correlation analysis between different variables at the three sites in the present study. While in this section only the results from Fortress Mountain are discussed, comparative analyses from the three sites are provided in Chapter 4.

Table 3.73 Precipitation chemistry. Correlation coefficients (r) among the important chemical components of daily rain samples, by sampling site (combined 1986 and 1987 daily samples).

	Sampling Location		
Variable Pairs	Fortress Mountain (r)	Crossfield West (r)	Crossfield East (r)
H^+ vs SO_4^{2-}	.13	.32	.06
pH^+ vs NO_3^-	.18	.13	-.12
NH_4^+ vs SO_4^{2-}	.58	.55	.34
NH_4^+ vs NO_3^-	.68	.57	.22
Ca^{2+} vs SO_4^{2-}	.63	.61	.59
Ca^{2+} vs NO_3^-	.50	.54	.66
Mg^{2+} vs SO_4^{2-}	.69	.55	.52
Mg^{2+} vs NO_3^-	.61	.44	.40
Na^+ vs Cl^-	.37	.001	.36
Na^+ vs NO_3^-	.27	.26	.42
Ca^{2+} vs Mg^{2+}	.85	.84	.70
Ca^{2+} vs Na^+	.65	.21	.27
Mg^{2+} vs Na^+	.43	.12	.31
SO_4^{2-} vs NO_3^-	.65	.77	.84
K^+ vs SO_4^{2-}	.73	.21	.40
K^+ vs NO_3^-	.64	.05	.32

The correlation coefficient (r) between H^+ and SO_4^{2-} and H^+ and NO_3^- was very low (0.13 and 0.18, respectively) at Fortress Mountain. In this respect, Fortress Mountain was very similar to another high altitude site, Alamosa, Colorado (Table 3.10). At Fortress Mountain, high r values for SO_4^{2-} were obtained with K^+ (0.73), Mg^{2+} (0.69), Ca^{2+} (0.63), and NH_4^+ (0.58). Similarly high r values for NO_3^- were obtained with NH_4^+ (0.68), K^+ (0.64), Mg^{2+}

(0.61), and Ca^{2+} (0.50). These results suggest that a major portion of rain SO_4^{2-} and NO_3^- at Fortress Mountain was not associated with free acidity and the rain at that site was regulated by well-aged air masses. The mean and median concentrations of excess cations (all measured cations minus anions) were equal to the corresponding values of total anion concentrations, suggesting the presence of high concentrations of anions (e.g., organic compounds) other than what was measured. Keene and Galloway (1984) estimated, based on measurements and indirect calculations, that organic acids may contribute 16-35% of the free acidity in precipitation of North America. In the present case, a large portion of the measured total anions (SO_4^{2-} + NO_3^- + Cl^-) most likely were associated with cations other than H^+.

Table 3.74 provides the summary statistics of principal component analysis. All the measured anions and Ca^{2+}, Mg^{2+}, and NH_4^+ loaded high (>0.60) in Factor 1, which accounted for approximately 69% of the variability of the total rain composition. The H^+ loaded high in Factor 2, but the correlation coefficient was negative in relation to most of the other ions. These results were further verified by cluster analysis (Table 3.75) where reasonable linkages or similarities were found between Ca^{2+} and Mg^{2+}, NO_3^- and SO_4^{2-}, and Ca^{2+} and Na^+, with relatively weak relationships between other ion pairs. All these results substantiate what has been described previously, at Fortress Mountain the pH of rain was not well regulated by SO_4^{2-} and/or NO_3^- and possibly other factors, such as weak acids, were involved.

Table 3.74 Precipitation chemistry. Varimax rotated factor matrix. The loadings or numbers in a given row, represent correlation coefficients of factors to describe a given rain variable. Data are from daily rain samples. Location: Fortress Mountain (combined 1985 and 1987).

Variable	Factor 1	Factor 2
H^+	.04	-.74
NO^{2-}	.10	-.10
SO_4^{2-}	.84	.17
NO_3^-	.86	.02
Cl^-	.63	.13
Ca^{2+}	.74	.68

Continued...

Table 3.74 (Concluded).

Variable	Factor 1	Factor 2
Mg^{2+}	.79	.42
Na^{+}	.50	.61
K^{+}	--	--
NH_4^{+}	.76	-.32
Volume	.25	.14
Duration	.15	-.05
Pct of Variability	68.6	22.8
Eigenvalue	4.4	1.5

Table 3.75 Summary results of cluster analysis of daily precipitation chemistry data, 1986 and 1987. Sampling location: Fortress Mountain.

Ion Pairs	Similarity Level
NO_3^{-} - SO_4^{2-}	0.59
Ca^{2+} - Mg^{2+}	0.83
NO_3^{-} - NH_4^{+}	0.48
Ca^{2+} - Na^{+}	0.55
H^{+} - NO_3^{-}	0.31
NO_2^{-} - Ca^{2+}	0.34
Volume - Duration	0.13
Cl^{-} - NO_2^{-}	0.14
Volume - Cl^{-}	0.07
Volume - H^{+}	0.03

3.6.4.4 Relationships Between Gaseous Air Pollutants and the Chemistry of Fine Aerosols and Rain

Correlation analysis was performed between the variables characterized through the KAPS serial annular denuder-filter pack system and the total of 10-minute average concentrations, individually of SO_2, NO_2, and NO for the 48 hours prior to the start of the KAPS sampler. The correlation coefficients were very low ($r < 0.3$) in all cases. The only exceptions were the relationships between NO_2 concentrations in the air and the NO_3^{-} concentration on the nylon filter ($r=0.60$) (possibly NH_4NO_3) and the NO_2 concentrations in the air and the total NO_3^{-} concentration (Teflon

+ nylon filter). Overall, these results were not particularly surprising since the ambient concentrations of SO_2, NO, and NO_2 at Fortress Mountain were below the detection limit (4 ppb) of the continuous monitoring methods for >98% of the total monitored time.

The summary results of the correlation analysis between SO_4^{2-} and NO_3^- in rain and the variables characterized through the KAPS annular denuder system for the 48 hours prior to the onset of the rain are presented in Table 3.76. The SO_4^{2-} and NO_3^- in rain showed reasonably high correlation (r=0.66). The SO_2 concentration in denuder tube #1 correlated reasonably well (r=0.66) with the SO_4^{2-} in rain. This was also the case between total SO_2 concentration (T1 + T2) and SO_4^{2-} in rain (r=0.60). According to Garland (1978) both SO_2 and SO_4^{2-} can contribute significantly to the dissolved sulphur in rain. Based on his assumptions, air concentrations of 10 $\mu g\ m^{-3}$ (approximately 4 ppb) SO_2 can contribute up to 3 mg L^{-1} sulphate in rain, with rain falling at 1 mm h^{-1} as 1 mm drops from a cloud containing 100 drops cm^{-1}, each drop with a radius of 10 μm. The ambient SO_2 concentrations at Fortress Mountain, as characterized by the annular denuder method, ranged from 0.08 up to 28 $\mu g\ m^{-3}$ (0.03 to 6.5 ppb).

Table 3.76 Relationships (correlation coefficients) between variables measured by the annular denuder system and variables in rainfall.[1] Sampling location: Fortress Mountain. Sampling period: 1985-87.

Variable Pairs	(r)
RNO_3^- vs D $NO_3^-(HNO_3)$	0.51
RNO_3^- vs NY NO_3^-	-0.26
RNO_3^- vs TF NO_3^-	-0.12
RNO_3^- vs F NO_3^-	-0.25
RNO_3^- vs T1 $NO_3^-(HNO_3)$	0.56
RNO_3^- vs T2 $NO_3^-(HNO_3)$	0.18
RSO_4^{2-} vs D SO_4^{2-} (2/3 SO_2)	0.60
RSO_4^{2-} vs T1 SO_4^{2-} (2/3 SO_2)	0.66
RSO_4^{2-} vs T2 SO_4^{2-} (2/3 SO_2)	0.23
RSO_4^{2-} vs NY SO_4^{2-}	0.17
RSO_4^{2-} vs TF SO_4^{2-}	0.13
RSO_4^{2-} vs F SO_4^{2-}	0.28
RSO_4^{2-} vs R NO_3^-	0.66

Continued...

Table 3.76 (Concluded).

R	=	rain;	D	=	Denuder Tube #1+#2;
NY	=	Nylon filter;	TF	=	Teflon filter.
F	=	total (Teflon + Nylon filter);	T1	=	Denuder Tube #1;
T2	=	Denuder Tube #2.			

1 Values from the annular denuder system included in the statistical analysis are from the 48 hours immediately preceding the onset of each daily rain event.

Even though fine particle scavenging by falling rain drops is not considered to be a major contributor to the sulphate in rain (Garland, 1978), positive, but weak correlations between fine particle SO_4^{2-} (Teflon filter, nylon filter, and total) and rain SO_4^{2-} were suggestive of common mechanisms or origin of the aerosol and rain SO_4^{2-}. The slightly higher correlation coefficient (r) between SO_4^{2-} (NY) and rain SO_4^{2-}, in comparison to SO_4^{2-} (TF) indicates a minor, but possible role of $(NH_4)_2SO_4$. Ammonium sulphate is well known for its properties as a cloud condensation nucleus (Husar et al. 1978).

The HNO_3 concentration in denuder Tube #1 (T1, uncorrected) was reasonably well correlated (r=0.56) with NO_3^- in rain. The r for the corrected (T1 + T2) HNO_3 concentration was slightly lower (0.51). As opposed to the relationship between T2 SO_2 and SO_4^{2-} in rain, T2 HNO_3 did not correlate well with NO_3^- in rain (r=0.18). This was not surprising since the T2 mean HNO_3 concentration was only 27% of the T1 mean HNO_3 concentration.

The negative correlation coefficients between fine particle nitrate species (Teflon filter, nylon filter) and NO_3^- in rain may be suggestive of poor below-cloud removal of fine particle nitrate by rain. It appears that the NO_3^- in rain is predominantly regulated by NO, NO_2, and HNO_3. As previously stated, the correlation coefficient between HNO_3(T1) and NO_3^- in rain was 0.56.

In contrast to SO_4^{2-}, much less information has been published regarding the removal mechanisms of nitrogen species by precipitation. There is some evidence for the formation of HNO_3 in clouds and rain-water. Recently, both theory and experimental evidence suggest that HNO_3 may be formed rapidly from a combined gas-phase/liquid-phase process (U.S. National Research Council 1983). Although significant uncertainty remains concerning the source of HNO_3 in clouds and rainwater, the limited evidence currently available favours the probable importance of the formation of dinitrogen pentoxide (N_2O_5) from nitrogen dioxide (NO_2), followed by its reaction with water droplets to form HNO_3. HNO_3 can be effectively scavenged by precipitation.

Correlation analysis was also performed between SO_4^{2-} and NO_3^- in rain and ambient SO_2, NO_2, and NO measured by continuous monitors over 48 hours prior to the onset of rain. While SO_2 and NO_2 and NO_2 and NO were highly correlated (0.72 and 0.76, respectively) with each other, no reasonable correlations could be established between these variables and SO_4^{2-} and NO_3^- in rain. Again, this was considered to be due to the huge percentage of measurement values below the detection limit of the continuous monitors.

3.7 CONCLUDING REMARKS

Two year mean hourly concentrations of SO_2, H_2S, NO, NO_2, and O_3 at Fortress Mountain were very comparable to the corresponding published values for remote locations. Thus, Fortress Mountain can be considered as a true background air quality site for Alberta and western Canada. However, periodically during late spring and early summer, up to 20-25% of the time, multiple-day O_3 episodes can occur at Fortress Mountain. During 1987, these episodes were generally regulated by air masses from the west-southwest (parts of Washington State and British Columbia) and by long range transport. Hourly O_3 concentrations during these episodes were as high as 124 ppb and the contribution of stratospheric O_3 intrusion appeared to be minimal, to these episodes.

Mean fine particle SO_4^{2-} concentration at Fortress Mountain was identical to the mean value reported for the Canadian Arctic. Similarly, mean concentrations of HNO_3, HNO_2, and NH_3 obtained through the KAPS serial annular denuder-filter pack method were in the low end of the range of concentrations reported by others for these pollutants. While the KAPS system performed up to expectations, some fundamental questions were raised concerning the principle by which chemical speciation is achieved in the system. More importantly, the utility of the KAPS system in our understanding of the relationships between SO_2, NO_x, fine particle SO_4^{2-} and NO_3^-, and rain SO_4^{2-} and NO_3^- was demonstrated.

Reasonable correlations were obtained between integrated air concentrations of SO_2 and HNO_3 and rain SO_4^{2-} and NO_3^-.

However, SO_4^{2-} and NO_3^- in rain were not well correlated with the acidity. While the mean pH of rain at Fortress Mountain was reported to be 4.73, caution is advisable in the interpretation of this value, based on statistical reasons. Weak acids may have an important role in regulating the pH of rain at Fortress Mountain.

3.8 LITERATURE CITED

Alberta Government/Industry Acid Deposition Research Program. Volume 1. 1984. *Researching Acid Deposition: Workshop Proposal.* Acid Deposition Research Program, Calgary, Alberta, Canada. 59 pp.

Angle, R.P. and H.S. Sandhu. 1986. Rural ozone concentrations in Alberta, Canada. *Atmospheric Environment* 20: 1221-1228.

Arnts, R.R. and S.A. Meeks. 1981. Biogenic hydrocarbon contribution to the ambient air of selected areas. *Atmospheric Environment* 15: 1643-1651.

Barrie, L.A. 1986. Background pollution in the arctic air mass and its relevance to North American acid rain studies. *Water, Air, and Soil Pollution* 30: 765-777.

Barrie, L.A. and R.M. Hoff. 1985. Five years of air chemistry observations in the Canadian arctic. *Atmospheric Environment* 19: 1995-2010.

Brown, R.H. and C.J. Purnell. 1979. Collection and analysis of trace organic vapor pollutants in ambient atmospheres. The performance of a Tenax-GC adsorbent tube. *Journal of Chromatography* 178: 79-90.

Chameides, W.L. and J.C. Walker. 1973. A photochemical theory of tropospheric ozone. *Journal of Geophysical Research* 78: 8751-8760.

Chameides, W.L. and J.C. Walker. 1976. A time-dependent photochemical model for ozone near the ground. *Journal of Geophysical Research* 81: 413-420.

Charlson, R.J. and H. Rodhe. 1982. Factors controlling the acidity of natural rainwater. *Nature* 295: 683-685.

Chatfield, R. and H. Harrison. 1977. Tropospheric ozone. 2. Variations along a meridional band. *Journal of Geophysical Research* 82: 5959-5976.

Chung, Y.S. and T. Dann. 1985. Observations of stratospheric ozone at the ground level in Regina, Canada. *Atmospheric Environment* 19: 157-162.

Coscio, M.R., G.C. Pratt, and S.V. Krupa. 1982. An automated refrigerated, sequential precipitation sampler. *Atmospheric Environment* 16: 1939-1944.

Crutzen, P.J. 1974. Photochemical reactions by and influencing ozone in the troposphere. *Tellus* 26: 47-57.

Crutzen, P.J. and L.T. Gidel. 1983. A two-dimensional photochemical model of the atmosphere. 2. The tropospheric budgets of the anthropogenic chlorocarbons, CO, CH_4, CH_3Cl, and the effect of various NO_x sources on tropospheric ozone. *Journal of Geophysical Research* 88: 6641-6661.

Danielsen, E.F. and V.A. Mohnen. 1977. Project Dustorm report: Ozone transport, in situ measurements, and meteorological analyses of tropopause folding. *Journal of Geophysical Research* 82: 5867-6877.

Demerjian, K.L. 1986. Atmospheric chemistry of ozone and nitrogen oxides. In: **Air Pollutants and Their Effects on the Terrestrial Ecosystem,** eds. A.H. Legge and S.V. Krupa, New York: John Wiley & Sons, pp. 105-127.

Driedger III, A.R., D.C. Thornton, M. Lalevic, and A.R. Bandy. 1987. Determination of part-per-trillion levels of atmospheric sulfur dioxide by isotope dilution gas chromatography/mass spectrometry. *Analytical Chemistry* 59: 1196-1200.

Dutkiewicz, V.A. and L. Husain. 1985. Stratospheric and tropospheric components of ^{7}Be in surface air. *Journal of Geophysical Research* 90: 5783-5788.

Ferm, M. 1979. Method for determination of atmospheric ammonia. *Atmospheric Environment* 13: 1385-1393.

Finlayson-Pitts, B.J. and J.N. Pitts. 1986. *Atmospheric Chemistry: Fundamentals and Experimental Techniques*. New York: John Wiley and Sons. 1098 pp.

Fishman, J., and P.J. Crutzen. 1977. A numerical study of tropospheric photochemistry using a one-dimensional model. *Journal of Geophysical Research* 82: 5897-5906.

Fishman, J. and P.J. Crutzen. 1978. The origin of ozone in the troposphere. *Nature* 274: 855-857.

Fishman, J. and W. Seiler. 1983. Correlative nature of ozone and carbon monoxide in the troposphere: Implications for the tropospheric ozone budget. *Journal of Geophysical Research* 88(C6): 3662-3670.

Forrest, J., D.J. Spandau, R.L. Tanner, and L. Newman. 1982. Determination of atmospheric nitrate and nitric acid employing a diffusion denuder with a filter pack. *Atmospheric Environment* 16: 1473-1485.

Galloway, J.N., G.E. Likens, and E.S. Edgerton. 1976. Acid precipitation in northeastern United States: pH and acidity. *Science* 194: 722-724.

Galloway, J.N., G.E. Likens, W.C. Keene, and J.M. Miller. 1982. The composition of precipitation in remote areas of the world. *Journal of Geophysical Research* 87: 8771-8786.

Garland, J.A. 1978. Dry and wet removal of sulfur from the atmosphere. *Atmospheric Environment* 12: 349-362.

Garrels, R.M. and F.T. MacKenzie. 1971. *Evolution of Sedimentary Rocks*. New York: Norton. 397 pp.

Gidel, L.T. and M. Shapiro. 1980. General circulation model estimates of the net vertical flux of ozone in the lower stratosphere and the implications for the tropospheric ozone budget. *Journal of Geophysical Research* 85: 4049-4058.

Harriss, R. and H. Niki. 1984. Sulfur cycle. In: **Global Tropospheric Chemistry, A Plan for Action.** Washington, D.C.: National Academy Press.

Husar, R.B., J.P. Lodge, and D.J. Moore. 1978. *Sulphur in the Atmosphere.* New York: Pergamon Press. 816 pp.

Keene, W.C. and J.N. Galloway. 1984. Organic acidity in precipitation of North America. *Atmospheric Environment* 18: 2491-2497.

Khalil, M.A. and R.A. Rasmussen. 1984. Sources of atmospheric carbonyl sulfide and carbon disulfide. In: **Transactions of Environmental Impact of Natural Emissions**, ed. V.P. Aneja, Pittsburgh: Air Pollution Control Association, pp. 32-40.

Krupa, S.V. and M. Nosal. 1984. Aerosol and Rainfall Chemistry in Minnesota. Prepared for Minnesota-Wisconsin Power Suppliers Group, Minneapolis, Minnesota. 99 pp.

Lefohn, A.S. and S.V. Krupa. 1988. The relationship between hydrogen and sulfate ions in precipitation - a numerical analysis of rain and snowfall chemistry. *Environmental Pollution* 49: 289-311.

Leighton, P.A. 1961. *Photochemistry of Air Pollution.* New York: Academic Press. 300 pp.

Levy, H. II. 1971. Normal atmosphere: Large radical and formaldehyde concentrations predicted. *Science* 173: 141-143.

Levy, H. II., J.D. Mahlman, W.J. Moxim, and S.C. Liu. 1985. Tropospheric ozone: The role of transport. *Journal of Geophysical Research* 90, 3753-3772.

Lewin, E.E. and D. Klockow. 1982. Application of the TCM denuder for SO_2 collection. In: **Proceedings of the 2nd European Symposium on Physico-chemical Behaviour of Atmospheric Pollutants.** Varese. 29 Sept.-1 Oct., 1981, eds. B. Versino and H. Ott, eds., pp. 54-61.

Likens, G.E. and F.H. Bormann. 1974. Acid rain: A serious environmental problem. *Science* 184: 1176-1179.

Likens, G.E., F.H. Bormann, and N.M. Johnson. 1981. Interactions between major biogeochemical cycles in terrestrial ecosystems. In: **Some Perspectives of the Major Biogeochemical Cycles**, ed. G.E. Likens, SCOPE Rep. 17, Wiley, Chichester.

Liu, S.C., D. Kley, M. McFarland, J.D. Mahlman, and H. Levy II. 1980. On the origin of tropospheric ozone. *Journal of Geophysical Research* 85(C12): 7546-7552.

Logan, J.A. 1983. Nitrogen oxides in the troposphere: Global and regional budgets. *Journal of Geophysical Research* 88: 10,785-10,807.

Logan, J.A. 1985. Tropospheric ozone: Seasonal behavior, trends, and atmospheric influence. *Journal of Geophysical Research* 90: 463-482.

Logan, J.A., M.J. Prather, S.C. Wofsy, and M.B. McElroy. 1981. Tropospheric chemistry: A global perspective. *Journal of Geophysical Research* 86: 7210-7254.

Mahlman, J.D., H. Levy II, and W.J. Moxim. 1980. Three-dimensional tracer structure and behaviour as simulated in two ozone precursor experiments. *Journal of Atmospheric Science* 37: 655-685.

Pankow, J.F. 1983. Cold trapping of volatile organic compounds on fused silica capillary columns. *Journal of High Resolution Chromatography and Chromatography Communications* 6: 293-299.

Pellizzari, E.D., W.F. Gutknecht, S. Cooper, and D. Hardison. 1984. Evaluation of sampling methods for gaseous atmospheric samples. EPA Report No. 600/3-84-062.

Pleil, J.D., K.D. Oliver, and W.A. McClenny. 1987. Enhanced performance of Nafion dryers in removing water from air samples prior to gas chromatography analysis. *Journal of the Air Pollution Control Association* 37: 244-248.

Possanzini, M., A. Febo, and A. Liberti. 1983. New design of a high-performance denuder for the sampling of atmospheric pollutants. *Atmospheric Environment* 17: 2605-2610.

Pruchniewicz, P.G. 1973. The average tropospheric ozone content and its variation with season and latitude as a result of the global ozone circulation. *Pure and Applied Geophysics* 106-108: 1058-1073.

Rahn, K.A. 1971. Sources of trace elements in aerosols - an approach to clean air. University of Michigan, Ann Arbor, Michigan. Xerox University Microfilms (order No. 72-4956), 325 pp. Ph.D. Thesis.

Riggin, R.M. and R.A. Markle. 1985. Comparison of solid adsorbent sampling techniques for volatile organic compounds in ambient air. EPA Report No. 600/4-85-077.

Schjoldager, J. 1981. Ambient ozone measurements in Norway, 1975-1979. *Journal of the Air Pollution Control Association* 31: 1187-1190.

Schjoldager, J., B. Sivertsen, and J.E. Hanssen. 1978. On the occurrence of photochemical oxidants at high latitudes. *Atmospheric Environment* 12: 2461-2467.

Sequeira, R.A. 1981. Chemistry of precipitation at high altitudes: interrelation of acid-base components. *Atmospheric Environment* 16: 329-335.

Sequeira, R.A. 1982. Acid rain: An assessment based on acid-base considerations. *Journal of the Air Pollution Control Association* 32: 241-245.

Shaw, R.W., T.G. Dzubay, and R.K. Stevens. 1978. The denuder difference experiment. In: **Current Methods to Measure Atmospheric Nitric Acid and Nitrate Artifacts,** ed. R.K. Stevens, EPA Report NO. 600/2/79/051. U.S. Environmental Protection Agency, Research Triangle Park, NC. pp. 79-84

Shaw, R.W., R.K. Stevens, and J. Bowermaster. 1982. Measurements of atmospheric nitrate and nitric acid: the denuder difference experiment. *Atmospheric Environment* 16: 845-853.

Sheffield, A.E. and G.E. Gordon. 1985. Variability of particle composition from ubiquitous sources: Results from a new source-composition library. In: **Receptor Methods for Source Apportionment-Real World Issues and Applications,** March 1985; ed. T.G. Pace, Williamsburg, Virginia, p. 14.

Singh, H.B., F.L. Ludwig, and W.B. Johnson. 1977. Ozone in clean remote atmospheres: Concentrations and variabilities. CRC-APRAC CAPA-15-76. Coord. Research Council Inc. June, 1977.

Singh, H.B., F.L. Ludwig, and W.B. Johnson. 1978. Tropospheric ozone: Concentrations and variabilities in clean remote atmospheres. *Atmospheric Environment* 12: 2185-2196.

Singh, H.B., W. Viezee, W.B. Johnson, and F.L. Ludwig. 1980. The impact of stratospheric ozone on tropospheric air quality. *Journal of Air Pollution Control Association* 30: 1009-1017.

Sjödin, A. and M. Fern. 1985. Measurements of nitrous acid in an urban area. *Atmospheric Environment* 19: 985-992.

Stedman, D.H. and R.E. Shetter. 1983. The global budget of atmospheric nitrogen species. *Advances in Environmental Science and Technology* 12: 411-454.

Stevens, R.K. 1986. Review of methods to measure chemical species that contribute to acid dry deposition. In: **Proceedings: Methods for Acidic Deposition Measurement.** EPA Report 600/9-86/014. p. 10.

Stevens, R.K. 1982. Status of source apportionment methods: Quail Roost II. In: **Proceedings: Receptor Models Applied to Contemporary Pollution Problems,** 17-20 Oct 1982, Danvers, Massachusetts, ed. Air Pollution Control Association, pp. 14.

Stevens, R.K., T.G. Dzubay, G. Russwurm, and D. Richel. 1978. Sampling and analysis of atmospheric sulfates and related species. *Atmospheric Environment* 12: 55-68.

Stewart, R.W., S. Hameed, and J.P. Pinto. 1977. Photochemistry of tropospheric ozone. *Journal of Geophysical Research* 82: 3134-3140.

Sze, N.D. and K.W. Ko. 1980. Photochemistry of COS, CS_2, CH_3SCH_3, and H_2S: Implications for the atmospheric sulfur cycle. *Atmospheric Environment* 14: 1223-1239.

Thornton, D.C., A.R. Driedger III, and A.R. Bandy. 1986. Determination of part-per-trillion levels of sulfur dioxide in humid air. *Analytical Chemistry* 58: 2688-2691.

U.S. Environmental Protection Agency, 1984. Air Quality Criteria for Ozone and Other Photochemical Oxidants. Volume II. Research Triangle Park, NC: U.S. Environmental Protection Agency, Environmental Criteria and Assessment Office, EPA Report No. EPA-600/8-84-020.

U.S. Environmental Protection Agency, 1986. Proceedings: Methods for Acidic Deposition Measurement. 1985. April 30-May 3, Raleigh, North Carolina. EPA Report No. 600/9-86/014. 212 pp.

U.S. Environmental Protection Agency Method TO-1, 1984. Method for the determination of volatile organic compounds in ambient air using Tenax adsorption and gas chromatography/mass spectrometry. In: **Compendium of Methods for the Determination of Toxic Organic Compounds in Ambient Air.** EPA-600/4-84-041.

U.S. National Academy of Sciences, 1977. *Ozone and Other Photochemical Oxidants.* Committee on Medical and Biological Effects of Environmental Pollutants. Washington, D.C. 719 pp.

U.S. NAPAP (The National Acid Precipitation Assessment Program), 1987. Interim Assessment Report, Refer to Volumes 1 through 4. Washington, D.C.

U.S. National Research Council, 1983. *Acidic Deposition. Atmospheric Processes in Eastern North America.* National Academy Press, Washington, D.C. 375 pp.

Viezee, W., W.B. Johnson, and H.B. Singh. 1983. Stratospheric ozone in the lower troposphere-II. Assessment of downwind flux and ground-level impact. *Atmospheric Environment* 17: 1979-1993.

Vukovich, F.M., J. Fishman, and E.V. Browell. 1985. The reservoir of ozone in the boundary layer of the eastern United States and its potential impact on the global tropospheric ozone budget. *Journal of Geophysical Research* 90: 5687-5698.

Whitby, K.T. 1978. The physical characteristics of sulfur aerosols. *Atmospheric Environment* 12: 135-159.

Wofsy, S.C., J.C. McConnell, and M.B. McElroy. 1972. Atmospheric CH_4, CO, CO_2. *Journal of Geophysical Research* 77: 4477-4493.

CHAPTER 4

AIR QUALITY OF AN AREA PROXIMAL TO ANTHROPOGENIC EMISSIONS

A. H. Legge, M. Nosal, E. Peake, M. Strosher, M. Hansen, and A. S. Lefohn

4.1 INTRODUCTION

Industrialization produces rapidly growing urban areas resulting in high density automotive traffic, enhanced power generation needs, expanding commercial and industrial activity and, consequently, the phenomenon of air pollution. Comparisons of pollutant concentrations in the clean troposphere or remote locations with moderately or heavily polluted areas were presented in Tables 3.1 through 3.4 of Chapter 3.

The high density of emissions originating from geographically concentrated sources introduces a pollutant mass flux burden that the natural atmospheric ventilating and cleansing processes cannot cope with. The result is high concentrations of combustion products such as CO, NO, NO_2, SO_2, and volatile organic compounds (VOCs). The VOCs participate in a complex series of chemical reactions driven by sunlight, leading to the phenomenon of photochemical smog (Figure 2.2 in Chapter 2).

Initially sulphur pollution and photochemical air pollution were viewed as rather independent atmospheric phenomena. Subsequently, however, during the 1970's, it had become clear that the major atmospheric phenomena, air pollution, acidic rain and other airborne toxic compounds, as well as the chemistry of the pristine troposphere, were inextricably linked in one continuous cycle (Chapters 2 and 3).

4.2 MONITORING SITE LOCATION AND DESCRIPTION

"There is a need to consider questions of air quality at several different levels. Clearly, there is a need to relate Alberta industrial emissions to natural emissions and the background air quality of the geographic region in which the province is located. Moreover, the environmental consequences of this modified air quality must be considered through effects of both primary and secondary pollutants. Studies of air chemistry and its effects are not necessarily sufficient since precipitation quality will also be modified (by a largely unknown amount) by Alberta emissions; hence precipitation-related studies must also be considered." (*Researching Acid Deposition: Workshop Proposals*, Acid Deposition Research Program, Calgary, Alberta, June 1984).

Thus, two air quality monitoring stations were established at Crossfield, Alberta (Table 4.1, and Figures 3.8, 4.1, 4.2 and 4.3). Air pollutants (both dry and wet, rain only) and meteorological parameters were measured (Table 3.13 in Chapter 3) at the two sites for two full years during 1985-87.

Table 4.1 Description of sampling sites at Crossfield West and Crossfield East, Alberta.

Site Name:	Crossfield West; Crossfield East
Coordinates:	51°23´1" 114°7´30"; 51°22´35" 114°1´0"
Elevation:	Crossfield West: 1158 m MSL Crossfield East: 1098 m MSL
Site Description:	
a) Location:	The monitoring sites are located 40 km north of Calgary near a sour gas processing plant at Crossfield. Crossfield West: Upwind monitoring site, located at about 6 km west-southwest of the sour gas plant. Crossfield East: Downwind monitoring site, located at about 2.5 km southwest of the sour gas plant.

Continued...

Table 4.1 (Concluded).

b) Access:	Access to the two monitoring sites from Calgary, is via Highway 2 north to the Crossfield junction. Crossfield West: Access is via a secondary gravel road. The final 0.5 km is a trail through a farmer's field. Crossfield East: Access is via a private road providing service to a gas well operated by the management of the gas plant. This road was extended by ADRP by about 0.5 km beyond the well to allow access to the monitoring site.
c) Topography:	Crossfield West: Terrain is flat. About 40 m above the plane of the sour gas plant. Crossfield East: Terrain is generally flat, featuring low, rolling hills with relative relief of about 30 m.
d) Vegetation:	Crossfield West and Crossfield East are located within the Fescue Grass ecoregion. Rough fescue dominates the native vegetation with secondary populations of Parry oat grass. June grass, chickweed, sticky geranium, lupines, yellow bean, sage, yarrow, bedstraw, and fleabane are amongst the more common herbs found in this ecoregion.
e) Sources in the Vicinity:	The land surrounding Crossfield West and East is under agricultural cultivation. Crossfield East is closer to sources of emissions: vehicles on Highway 2, the cities of Calgary and Airdrie, the town of Crossfield, two mushroom growing and processing plants, and several sour gas processing plants located at various distances.
f) Snowfall:	$\bar{x}$ 132.8 mm $\pm$ 37.9 mm y^{-1}
g) Rainfall:	$\bar{x}$ 375.5 mm $\pm$ 99.0 mm y^{-1}
h) Soil:	Shallow Black Chernozem (less than 30 cm). High organic matter.
i) Emissions:	▪ The total NO_x and SO_2 emissions within this ecoregion constitute 7.8% of the total provincial NO_x and SO_2 emissions. ▪ The total NO_x and SO_2 emissions within this ecoregion is 152.7 t d^{-1}. ▪ 2.2% of the total area of Alberta is occupied by this ecoregion.

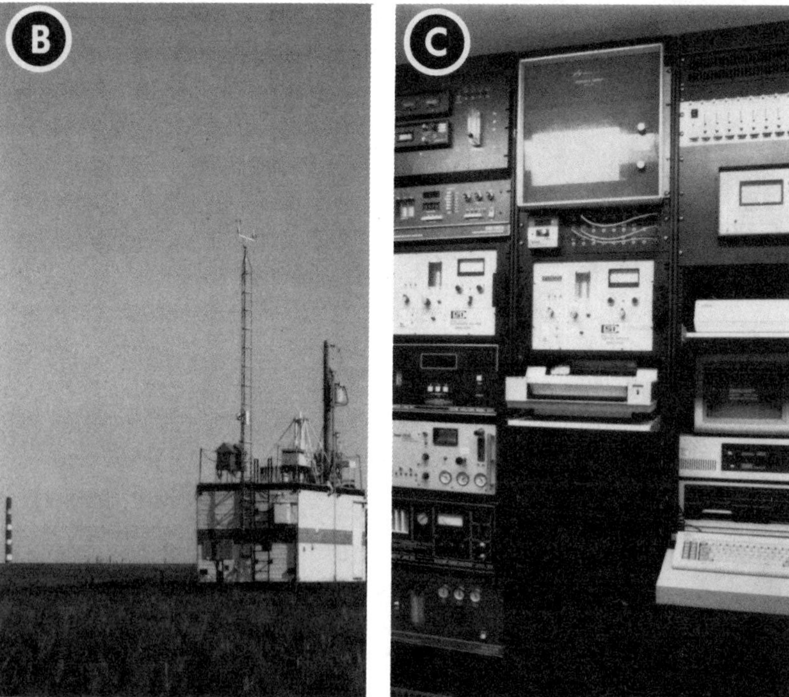

Figure 4.1 (A) View from a helicopter, looking northeast, with the Crossfield West air quality monitoring station in the foreground and a gas plant in the background. (B) Close-up view of the Crossfield West air quality monitoring station. (C) View of rack-mounted air quality and meteorological instrumentation inside the Crossfield West air quality monitoring station.

Figure 4.2 View from a helicopter, looking southeast, of the Crossfield East air quality monitoring station. Highway 2 is seen in the background.

Figure 4.3 (A) View from a helicopter, looking northwest, of the Crossfield East air quality monitoring station. A gas plant is seen in the centre background with the town of Crossfield to the right. (B) View on the ground of the Crossfield air quality monitoring trailer looking northwest with gas plant incinerator stack in the background. (C) View of rack-mounted air quality and meteorological instrumentation inside the Crossfield air quality monitoring station.

4.3 AUTOMATED MEASUREMENTS OF GASEOUS AIR POLLUTANTS AND METEOROLOGICAL PARAMETERS

A brief discussion of the considerations relevant to the evaluation of the frequency distributions of the pollutant concentrations was presented in Chapter 3.6.2. As with the data from the background site at Fortress Mountain, both at Crossfield West and Crossfield East gaseous pollutant concentrations and meteorological parameters were evaluated as 10-minute, as well as hourly means. Such means were computed from 2-minute data points or observations.

At Fortress Mountain (background site), with the exception of O_3 and CO_2, >99% of the 10-minute averages of all other pollutants were below the minimum detection limit during the two years of study. While at Fortress Mountain the frequency of the total observations for 10-minute means of SO_2 below the minimum detection limit was 99.3%, it was 96.8% and 93.9%, respectively, at Crossfield West (less frequently influenced by the plume from the sour gas plant) and Crossfield East (more frequently influenced by the plume from the sour gas plant). Similarly at Fortress Mountain, the frequency of 10-minute average NO_2 concentrations below the minimum detection limit was 99.2% while it was 79.8% and 58.4%, respectively, at Crossfield West and Crossfield East. However, with O_3, the percent of values below the minimum detection limit was the lowest at Fortress Mountain (0.01%) as opposed to Crossfield West (2.0%) and Crossfield East (10.7%). Fortress Mountain was a high elevation (2103 m MSL) remote site and Crossfield West (1158 m MSL) and Crossfield East (1098 m MSL) were located in an urban airshed, downwind (summer months) from the city of Calgary. Crossfield East, in particular, was considered to be influenced, in addition to the local point source, by a number of other sources in the vicinity. It appeared possible that the comparatively higher frequency of low O_3 (10-minute mean) values at Crossfield East may be due to the greater interaction of multiple pollutants leading to some reduction of O_3 at that site.

For the purpose of uniformity, as with the data from Fortress Mountain, 10-minute mean values below the minimum detection limit for SO_2, H_2S, NO and NO_2 at both Crossfield sites were substituted with values obtained by the application of the Weibull probability generator (Chapter 3). Summary statistics of the

Weibull distribution of the gaseous pollutant concentrations and the results of the test of the accuracy of the probability generator are presented in Table 3.15 of Chapter 3. Similarly, the results of the comparison of the computation of mean pollutant concentrations by three different approaches are presented in Tables 3.16 through 3.23. The lowest 10-minute and hourly mean values for SO_2 at all three sites were produced when all values below the minimum detection limit were substituted with "zero." The highest values were produced when one-half of the value for the minimum detection limit (1/2 DL) was used, while the use of the Weibull probability generator produced intermediate values that may be considered as reasonable. Nevertheless, as one would expect, both 10-minute and hourly SO_2 concentrations were the lowest at Fortress Mountain, intermediate at Crossfield West (less frequently influenced by the local point source), and highest at Crossfield East, independent of the computational method used. This was also true for the NO_2 concentrations at the three sites (Table 3.22).

As opposed to these results, use of the Weibull probability generator resulted in relatively higher 10-minute and hourly mean H_2S and NO concentrations at Fortress Mountain in comparison to the two Crossfield sites. This was considered to be due to a mathematical bias caused by the greater percent of values below the minimum detection limit at Fortress Mountain in comparison to the other two sites, and by the relative proportion of the percentage of values below the detection limit to the percentage of values representing the intermediate concentrations (Tables 4.2 through 4.7). Nevertheless, the 10-minute and hourly mean values of H_2S and NO derived by the application of the Weibull method were intermediate between the values obtained by the 1/2 DL or "zero" substitution methods.

The summary statistics on the reconstructed Weibull frequency distributions of the 10-minute pollutant concentrations are presented in Tables 4.2 through 4.7. Highest 10-minute SO_2 concentrations at Fortress Mountain were in the 20-30 ppb range, in the 130-140 ppb range at Crossfield West, and in the 160-170 ppb range at Crossfield East (Table 4.2). The percent frequency of occurrence of 10-minute SO_2 concentrations in the range of 4.0-140 ppb were quite different between the two Crossfield sites. In this concentration range, there were more frequent occurrences of SO_2 at Crossfield East (the site considered to be more frequently influenced by the point source) in comparison to Crossfield West.

Table 4.2 Best fit Weibull distribution parameters and empirical frequency distribution of 10-min. SO_2 concentrations (ppb) 1985-87.

	Fortress Mountain	Crossfield West	Crossfield East
Shape	0.396	0.447	1.141
Scale	0.042	0.145	0.509
Range			
0 - 4	99.271	96.786	93.883
4 - 10	0.629	1.997	3.327
10 - 20	0.099	0.870	1.514
20 - 30	0.001	0.185	0.562
30 - 40		0.070	0.299
40 - 50		0.032	0.158
50 - 60		0.017	0.104
60 - 70		0.013	0.040
70 - 80		0.010	0.027
80 - 90		0.010	0.034
90 - 00		0.006	0.014
100 - 110		0.001	0.011
110 - 120		0.001	0.008
120 - 130		0.001	0.005
130 - 140		0.001	0.004
140 - 150			0.005
150 - 160			0.003
160 - 170			0.002

Minimum detection limit for SO_2: 4 ppb.

Table 4.3 Best fit Weibull distribution parameters and empirical frequency distribution of 10-min. H_2S concentrations (ppb) 1985-87.

	Fortress Mountain	Crossfield West	Crossfield East
Shape	3.222	0.793	0.324
Scale	0.594	0.269	0.049
Range			
0 - 4	99.985	99.657	97.826
4 - 8	0.011	0.241	1.337
8 - 12	0.002	0.027	0.417
12 - 16	0.001	0.021	0.198
16 - 20	0.001	0.014	0.115

Continued...

Table 4.3 (Concluded).

	Fortress Mountain	Crossfield West	Crossfield East
20 - 24		0.009	0.042
24 - 28		0.007	0.024
28 - 32		0.009	0.013
32 - 36		0.006	0.010
36 - 40		0.000	0.005
40 - 44		0.001	0.003
44 - 48		0.004	0.003
48 - 52		0.000	0.002
52 - 56		0.001	0.002
56 - 60		0.000	0.001
60 - 64		0.002	0.001
64 - 68		0.000	0.000
68 - 72		0.001	0.001

Minimum detection limit for H_2S: 4 ppb.

Table 4.4 Best fit Weibull distribution parameters and empirical frequency distribution of 10-min. NO concentrations (ppb) 1985-87.

	Fortress Mountain	Crossfield West	Crossfield East
Shape	2.216	0.365	0.611
Scale	0.552	0.061	0.616
Range			
0 - 4	99.842	98.009	88.333
4 - 20	0.155	1.656	6.917
20 - 40	0.003	0.286	2.407
40 - 60		0.026	1.007
60 - 80		0.012	0.512
80 - 100		0.009	0.283
100 - 120		0.002	0.192
120 - 140			0.129
140 - 160			0.072
160 - 180			0.063
180 - 200			0.033
200 - 220			0.029
220 - 240			0.015
240 - 260			0.006
260 - 280			0.002

Minimum detection limit for NO: 4 ppb.

Table 4.5 Best fit Weibull distribution parameters and empirical frequency distribution of 10-min. NO_2 concentrations (ppb) 1985-87.

	Fortress Mountain	Crossfield West	Crossfield East
Shape	3.358	1.188	1.017
Scale	1.262	1.992	4.569
Range			
0 - 4	99.191	79.835	58.433
4 - 10	0.783	14.792	21.045
10 - 15	0.025	2.776	7.011
15 - 20	0.001	1.098	4.429
20 - 25		0.646	3.419
25 - 30		0.401	2.664
30 - 35		0.300	1.458
35 - 40		0.094	0.810
40 - 45		0.026	0.413
45 - 50		0.011	0.210
50 - 55		0.017	0.055
55 - 60		0.004	0.035
60 - 65			0.013
65 - 70			0.003
70 - 75			0.001
75 - 80			0.000
80 - 85			0.001

Minimum detection limit for NO_2: 4 ppb.

Table 4.6 Best fit Weibull distribution parameters and empirical frequency distribution of 10-min. O_3 concentrations (ppb) 1985-87.

	Fortress Mountain	Crossfield West	Crossfield East
Shape	4.375	2.755	1.692
Scale	45.723	31.310	28.939
Range			
0 - 5	0.010	2.024	10.651
5 - 10	0.006	2.865	5.775
10 - 20	1.126	18.027	18.986
20 - 30	8.339	37.575	27.775
30 - 40	34.242	25.569	24.747
40 - 50	33.565	8.989	8.698
50 - 60	13.114	3.879	2.864

Continued...

Table 4.6 (Concluded).

	Fortress Mountain	Crossfield West	Crossfield East
60 - 70	5.727	0.934	0.452
70 - 80	2.341	0.122	0.049
80 - 90	0.915	0.016	0.001
90 - 100	0.422		0.000
100 - 110	0.156		0.001
110 - 120	0.032		0.000
120 - 130	0.005		0.000
130 - 140			0.001

Minimum detection limit for O_3: 5 ppb.

Table 4.7 Best fit Weibull distribution parameters and empirical frequency distribution of 10-min. CO_2 concentrations (ppm) 1985-87.

	Fortress Mountain	Crossfield West	Crossfield East
Shape	50.743	37.527	32.791
Scale	348.872	351.049	352.623
Range			
285 - 300	0.000	0.020	0.000
300 - 315	0.188	0.193	0.093
315 - 330	2.318	5.053	5.929
330 - 345	37.871	36.160	33.653
345 - 360	57.859	46.910	46.335
360 - 375	1.608	10.146	9.902
375 - 390	0.156	1.193	2.298
390 - 405		0.220	0.815
405 - 420		0.063	0.429
420 - 435		0.025	0.243
435 - 450		0.010	0.138
450 - 465		0.004	0.071
465 - 480		0.002	0.057
480 - 495		0.001	0.036
495 - 500			0.001

Minimum detection limit for CO_2: none at known ambient concentrations.

The patterns for H_2S, NO, and NO_2 were very similar to what was described for SO_2 (Tables 4.3 through 4.5).

Contrary to the four primary pollutants, O_3 behaved differently. Fortress Mountain (high elevation remote site) had the lowest percent of 10-minute values below the detection limit; Crossfield East (more frequently influenced by the local point source) had the highest (Table 4.6). Crossfield East was also closer to a number of other sources (e.g., highway, two mushroom processing plants) in comparison to Crossfield West. Pollutant emissions from these sources might have interacted along with the local point source plume leading to some reduction in O_3 concentrations. Interestingly, however, the single highest 10-minute O_3 concentration (136.8 ppb) was recorded at Crossfield East (the corresponding highest hourly O_3 concentration, however, was only 73.8 ppb) (Figures 4.4 and 4.5).

The Fortress Mountain site exhibited the highest frequency of occurrence of 10-minute O_3 concentrations in the 30-130 ppb range among all three sites (Table 4.6). As opposed to SO_2, the frequency distribution statistics provided higher 10-minute O_3 values for Crossfield West (among the two urban sites less frequently influenced by the local point source and farther away from other sources) in comparison to Crossfield East.

Tables 4.8 through 4.19 provide summary statistics of Weibull substituted 10-minute and 1-hour pollutant and meteorological data at Crossfield West and East during 1985-86, 1986-87, and the combined years. The 10-minute and hourly mean SO_2, H_2S, NO, NO_2, and O_3 concentrations were almost identical to each other during a given year at Crossfield West (compare results in Table 4.8 with 4.9, and Table 4.10 with 4.11) and Crossfield East (compare results in Table 4.14 with 4.15, and Table 4.16 with 4.17). In this context neither the 10-minute nor the hourly averages were derived from single observations, but from 2-minute readings.

Table 4.20 provides a comparison of the mean and median hourly pollutant concentrations observed over two years (1985-87) at Fortress Mountain with the corresponding values for Crossfield West and East.

The mean and median hourly concentrations of SO_2 and NO_2 were at Fortress Mountain $<$ Crossfield West $<$ Crossfield East. As previously stated, the values of the mean and median H_2S and NO concentrations at Fortress Mountain appeared to be influenced by the mathematical bias of the computational method (Weibull

CROSSFIELD WEST

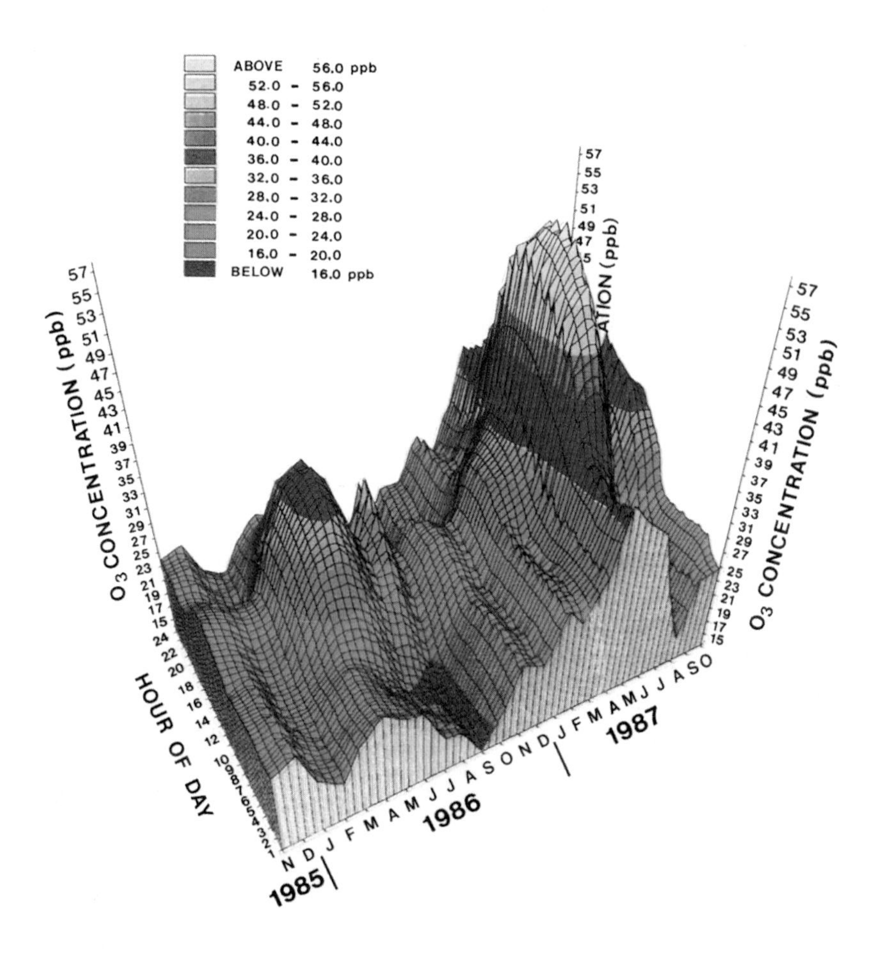

Figure 4.4 Crossfield West: Ozone concentration as a function of hour of the day and the month of the year (November 1, 1985 - October 31, 1987). The seasonal episodic and diurnal course of the O_3 concentration is clearly seen in this three dimensional plot.

CROSSFIELD EAST

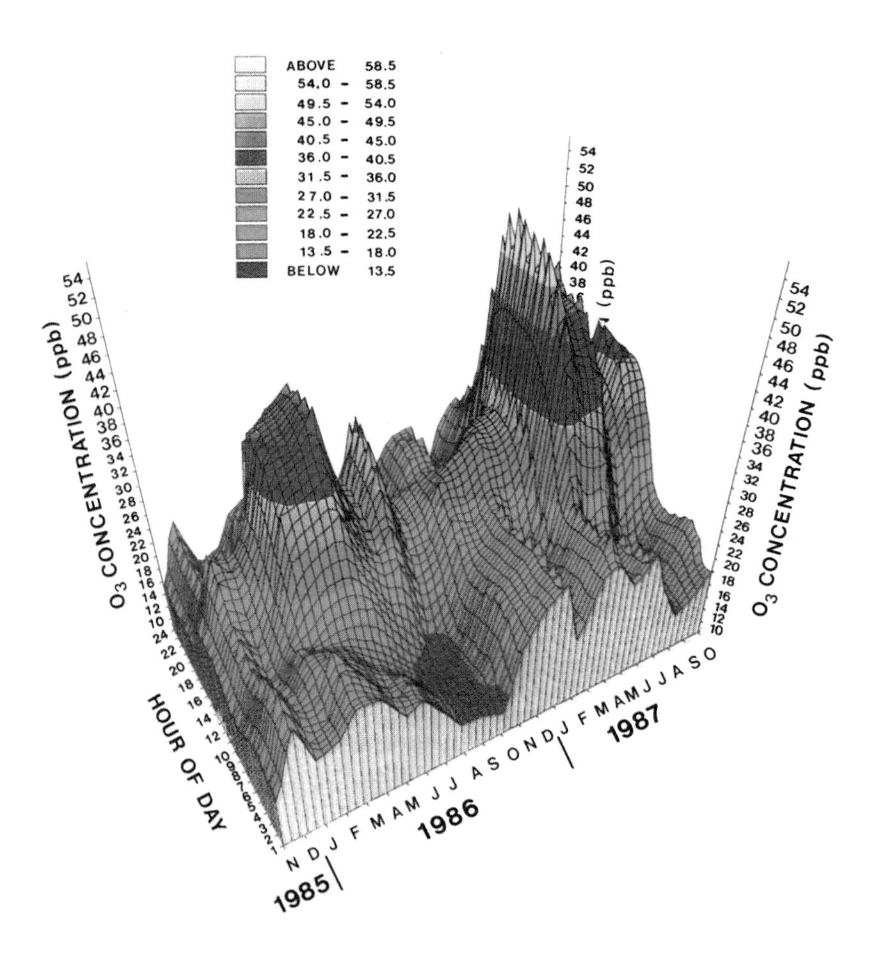

Figure 4.5 Crossfield East: Ozone concentration as a function of hour of the day and the month of the year (November 1, 1985 - October 31, 1987). The seasonal episodic and diurnal course of the O_3 concentration is clearly seen in this three dimensional plot.

Table 4.8 Summary statistics of automated gaseous pollutant easurements and meteorological variables*. Sampling location: Crossfield West. Sampling period: 1985-86. Time resolution of data: 10 minutes.

Variable	(n)	Mean	S.D.	Range		Median	Percentile						Skewness	Kurtosis
				Min.	Max.		10	30	70	90	95	99		
1. SO_2	46527	.60	3.1	.00	130.0	.10	.00	00	.20	.60	1.0	12.6	15.1	350.4
2. H_2S	45133	.31	.1	.00	70.3	.20	.00	.10	.30	.60	.70	.3	29.9	1209.4
3. NO	47188	.47	2.9	.00	102.8	.00	.00	.00	.10	.50	.70	11.0	15.6	350.1
4. NO_2	47125	2.9	4.6	.00	56.4	1.3	.50	.90	2.0	7.0	11.2	24.6	3.9	19.8
5. CO_2	47795	349.1	11.9	288.8	435.0	349.0	335.4	343.6	355.0	363.2	368.0	379.0	.07	1.5
6. O_3	46852	24.0	9.3	.00	86.8	24.2	11.6	19.4	28.8	35.8	39.0	45.6	.003	- .01
7. Wind Speed	49309	13.6	9.2	.00	77.6	12.4	2.4	8.2	17.2	25.6	30.4	41.6	.98	1.7
8. Wind Direction	46571	215.7	86.4	.80	60.0	235.8	84.6	169.8	267.6	323.0	337.0	350.4	- .53	- .50
9. Temperature	49312	4.1	10.9	-34.9	33.5	5.5	- 9.6	.60	10.1	16.5	19.6	25.6	- .82	1.1
10. Barometric Pressure	49319	880.2	6.7	850.0	900.0	881.0	871.0	877.0	884.0	888.0	890.0	893.0	- .44	.39
11. Precipitation	40099	.001	.02	.00	.80	.00	.00	.00	.00	.00	.00	.00	20.8	641.1
12. Humidity	45373	46.1	12.0	20.0	98.5	46.8	29.2	38.6	54.6	61.6	63.4	66.4	- .20	- .91
13. Radiation	32472	386.8	518.8	- 2.0	2075.0	102.6	- 1.0	.00	507.6	1280.4	1512.9	1825.3	1.3	.45

* All gas concentrations in ppb, except CO_2 (ppm); wind speed: km h^{-1}; wind direction: degrees; temperature: ^oC; pressure: millibars; precipitation: mm; humidity: %; radiation: Langleys min^{-1}.

Table 4.9 Summary statistics of automated gaseous pollutant measurements and meteorological variables*. Sampling location: Crossfield West. Sampling period: 1985-86. Time resolution of data: 1 hour.

Variable	(n)	Mean	S.D.	Range		Median	Percentile						Skewness	Kurtosis
				Min.	Max.		10	30	70	90	95	99		
1. SO_2	7991	.59	2.6	.00	78.6	.20	.10	.10	.20	.40	2.0	10.8	13.1	245.9
2. H_2S	7753	.30	.92	.00	34.4	.20	.10	.20	.30	.40	.40	1.3	24.5	729.2
3. NO	8105	.47	2.7	.00	89.5	.10	.00	.10	.20	.30	.50	10.7	15.1	333.4
4. NO_2	8097	2.9	4.4	.10	43.3	1.2	.80	1.0	1.9	7.0	10.9	24.3	3.9	19.9
5. CO_2	8178	349.2	11.7	291.1	409.1	349.0	335.6	343.7	354.9	363.4	368.0	377.9	.01	1.2
6. O_3	8201	23.9	9.2	2.3	64.9	24.1	11.6	19.2	28.6	35.5	38.8	45.2	.01	- .08
7. Wind Speed	8244	13.6	9.0	.00	72.4	12.3	3.1	8.3	17.1	25.3	30.0	40.6	1.00	1.8
8. Wind Direction	7786	215.7	80.8	6.9	359.6	233.5	98.8	171.0	263.7	317.2	330.3	344.8	- .48	- .54
9. Temperature	8244	4.2	10.9	-34.8	33.2	5.5	- 9.6	.70	10.1	16.5	19.6	25.7	- .82	1.1
10. Barometric Pressure	8246	880.2	6.7	850.0	900.0	881.0	871.1	877.0	884.0	888.0	890.08	93.1	- .44	.38
11. Precipitation	6705	.001	.01	.00	.30	.00	.00	.00	.00	.00	.00	.10	13.0	202.6
12. Humidity	7612	46.0	12.1	20.0	98.5	46.7	29.1	38.6	54.5	61.5	63.3	66.2	- .21	- .88
13. Radiation	5503	383.7	508.5	- 2.0	2058.1	106.3	- 1.0	.00	508.9	1258.1	1479.4	1776.7	1.2	.32

* All gas concentrations in ppb, except CO_2 (ppm); wind speed: km h^{-1}; wind direction: degrees; temperature: ^oC; pressure: millibars; precipitation: mm; humidity: %; radiation: Langleys min^{-1}.

Table 4.10 Summary statistics of automated gaseous pollutant measurements and meteorological variables*. Sampling location: Crossfield West. Sampling period: 1986-87. Time resolution of data: 10 minutes.

				Range			Percentile							
Variable	(n)	Mean	S.D.	Min.	Max.	Median	10	30	70	90	95	99	Skewness	Kurtosis
1. SO_2	45476	.52	2.6	.00	111.0	.10	.00	.00	.20	.60	.1	10.0	16.8	442.3
2. H_2S	42369	.25	.25	.00	7.4	.20	.00	.10	.30	.50	.70	1.1	4.7	67.2
3. NO	48168	.28	1.6	.00	51.4	.00	.00	.00	.10	.40	.70	5.8	14.4	269.3
4. NO_2	48168	2.60	4.1	.00	59.2	1.2	.50	.90	1.8	6.4	9.6	22.6	4.4	27.4
5. CO_2	48084	345.6	10.5	311.4	489.4	345.0	332.6	341.0	350.0	357.0	362.0	377.8	1.4	8.0
6. O_3	46914	32.2	12.3	2.3	82.8	30.8	18.0	25.0	37.6	49.6	54.8	64.4	.45	.19
7. Wind Speed	50003	15.5	9.4	.00	72.2	14.0	5.0	10.0	18.8	28.4	33.4	44.0	.98	1.4
8. Wind Direction	49169	220.4	82.3	1.0	360.0	241.0	91.6	183.2	266.4	320.6	334.4	350.0	- .72	- .09
9. Temperature	49006	5.6	9.3	-23.9	28.2	6.1	- 6.2	.70	10.9	17.3	20.1	25.0	- .27	- .14
10. Barometric Pressure	49873	880.4	6.9	854.8	901.6	881.0	871.0	877.0	884.0	889.0	891.0	895.0	- .27	.02
11. Precipitation	50019	.001	.01	.00	.50	.00	.00	.00	.00	.00	.00	.00	23.8	785.9
12. Humidity	47664	41.6	11.9	20.0	68.8	40.0	26.5	33.4	49.5	58.4	60.8	64.0	.18	-1.11
13. Radiation	40750	329.9	488.0	- 2.0	2104.0	31.0	- 1.0	- 1.0	401.7	1137.0	1483.5	1803.7	1.5	1.4

* All gas concentrations in ppb, except CO_2 (ppm); wind speed: km h^{-1}; wind direction: degrees; temperature: ^{o}C; pressure: millibars; precipitation: mm; humidity: %; radiation: Langleys min^{-1}.

Table 4.11 Summary statistics of automated gaseous pollutant measurements and meteorological variables*. Sampling location: Crossfield West. Sampling period: 1986-87. Time resolution of data: 1 hour.

Variable	(n)	Mean	S.D.	Range		Median	Percentile						Skewness	Kurtosis
				Min.	Max.		10	30	70	90	95	99		
1. SO_2	7868	.52	2.2	.00	84.3	.20	.10	.10	.20	.40	1.7	8.6	16.9	450.8
2. H_2S	7336	.25	.13	.00	4.0	.20	.10	.20	.30	.40	.40	.50	8.4	195.9
3. NO	8319	.29	1.4	.00	34.2	.10	.00	.10	.20	.20	.40	5.5	13.0	207.6
4. NO_2	8319	2.7	3.9	.105	1.3	1.2	.80	1.0	1.6	6.4	9.4	21.5	4.4	27.6
5. CO_2	8264	345.7	10.4	318.9	462.2	345.0	332.6	341.0	350.0	357.0	362.5	377.1	1.3	6.8
6. O_3	8294	32.0	12.2	3.0	80.5	30.6	17.9	25.0	37.3	49.1	54.3	63.9	.46	.21
7. Wind Speed	8388	15.5	9.1	.00	68.7	14.0	5.2	10.0	18.8	28.1	32.8	42.8	.99	1.4
8. Wind Direction	8251	220.2	77.5	5.2	360.0	239.6	108.7	183.3	263.3	313.4	328.2	342.1	- .68	- .14
9. Temperature	8222	5.79	.2	-23.4	28.0	6.2	- 6.2	.70	11.0	17.3	20.2	25.1	- .27	- .14
10. Barometric Pressure	8367	880.4	6.9	855.0	901.2	881.0	871.3	877.0	884.0	889.0	891.0	895.0	- .26	.01
11. Precipitation	8390	.001	.01	.00	.30	.00	.00	.00	.00	.00	.00	.00	17.7	405.4
12. Humidity	8033	41.5	11.9	20.0	67.4	39.8	26.3	33.3	49.3	58.4	60.6	63.7	.17	-1.11
13. Radiation	6884	330.8	479.9	- 2.0	1935.8	35.5	- 1.0	- .60	411.7	1140.7	1441.1	1743.5	.5	1.1

* All gas concentrations in ppb, except CO_2 (ppm); wind speed: km h^{-1}; wind direction: degrees; temperature: ^{o}C; pressure: millibars; precipitation: mm; humidity: %; radiation: Langleys min^{-1}.

Table 4.12 Summary statistics of automated gaseous pollutant measurements and meteorological variables*. Sampling location: Crossfield West. Sampling period: 1985-87. Time resolution of data: 10 minutes.

				Range			Percentile							
Variable	(n)	Mean	S.D.	Min.	Max.	Median	10	30	70	90	95	99	Skewness	Kurtosis
1. SO_2	92003	.56	2.9	.00	130.0	.10	.00	.00	.20	.60	1.1	11.2	15.9	393.7
2. H_2S	87502	.28	.84	.00	70.3	.20	.00	.10	.30	.60	.70	1.1	39.1	2148.6
3. NO	95356	.37	2.4	.00	102.8	.00	.00	.00	.10	.50	.70	8.0	17.2	447.0
4. NO_2	95293	2.8	4.3	.00	59.2	1.3	.50	.90	1.9	6.8	10.4	23.6	4.1	23.2
5. CO_2	95879	347.3	11.3	288.8	489.4	347.0	333.8	342.0	352.0	261.0	366.0	378.4	.64	3.6
6. O_3	93766	28.1	11.7	.00	86.8	27.0	14.2	22.0	33.0	43.2	49.8	60.4	.52	.63
7. Wind Speed	99312	14.6	9.4	.00	77.6	13.2	3.8	9.0	18.0	27.0	32.0	42.8	.97	1.6
8. Wind Direction	95740	218.1	84.9	.80	360.0	239.0	88.0	176.2	267.0	321.8	335.6	350.2	- .63	- .31
9. Temperature	98318	4.9	10.2	-34.9	33.5	5.8	- 7.8	.60	10.5	16.9	19.9	25.3	- .65	.87
10. Barometric Pressure	99192	880.3	6.8	850.0	901.6	881.0	871.0	877.0	884.0	889.0	891.0	894.2	- .34	.20
11. Precipitation	90118	.001	.01	.00	.80	.00	.00	.00	.00	.00	.00	.00	23.6	842.8
12. Humidity	93037	43.8	12.2	20.0	98.5	43.4	27.4	35.6	52.6	60.2	62.4	65.6	- .004	-1.1
13. Radiation	73222	355.1	502.6	- 2.0	2104.0	59.3	- 1.0	.00	449.0	1207.3	1498.8	1814.6	1.4	.91

* All gas concentrations in ppb, except CO_2 (ppm); wind speed: km h^{-1}; wind direction: degrees; temperature: ^{o}C; pressure: millibars; precipitation: mm; humidity: %; radiation: Langleys min^{-1}.

Table 4.13 Summary statistics of automated gaseous pollutant measurements and meteorological variables*. Sampling location: Crossfield West. Sampling period: 1985-87. Time resolution of data: 1 hour.

				Range			Percentile							
Variable	(n)	Mean	S.D.	Min.	Max.	Median	10	30	70	90	95	99	Skewness	Kurtosis
1. SO_2	15859	.56	2.4	.00	84.3	.20	.10	.10	.20	.40	1.8	9.6	14.7	322.1
2. H_2S	15089	.28	.66	.00	34.4	.20	.10	.20	.30	.40	.40	.60	33.2	1368.1
3. NO	16424	.38	2.2	.00	89.5	.10	.00	.10	.20	.30	.40	8.1	16.7	432.1
4. NO_2	16416	2.8	4.2	.10	51.3	1.2	.80	1.0	1.8	6.7	10.2	22.9	4.2	23.2
5. CO_2	16442	347.4	11.2	291.1	462.2	346.9	333.9	342.0	352.2	361.0	366.2	377.7	.57	2.9
6. O_3	16495	28.0	11.5	2.3	80.5	27.0	14.4	21.9	32.8	42.9	49.4	59.7	.53	.64
7. Wind Speed	16632	14.6	9.1	.00	72.4	13.2	4.2	9.2	18.0	26.6	31.6	41.8	.98	1.6
8. Wind Direction	16037	218.0	79.2	5.2	360.0	237.5	102.7	177.0	263.5	315.6	329.2	343.6	- .58	- .35
9. Temperature	16466	4.9	10.2	-34.8	33.2	5.8	- 7.7	.70	10.5	16.9	19.9	25.2	- .65	.87
10. Barometric Pressure	16613	880.3	6.8	850.0	901.2	881.0	871.2	877.0	884.0	888.8	891.0	894.5	- .34	.19
11. Precipitation	15095	.001	.01	.00	.30	.00	.00	.00	.00	.00	.00	.00	15.3	289.0
12. Humidity	15645	43.7	12.2	20.0	98.5	43.3	27.3	35.6	52.5	60.1	62.3	65.4	- .006	- 1.10
13. Radiation	12387	354.3	493.5	- 2.0	2058.1	65.4	- 1.0	.00	456.7	1202.1	1464.3	1761.0	1.4	.71

* All gas concentrations in ppb, except CO_2 (ppm); wind speed: km h^{-1}; wind direction: degrees; temperature: ^{o}C; pressure: millibars; precipitation: mm; humidity: %; radiation: Langleys min^{-1}.

Table 4.14 Summary statistics of automated gaseous pollutant measurements and meteorological variables*. Sampling location: Crossfield East. Sampling period: 1985-86. Time resolution of data: 10 minutes.

				Range			Percentile							
Variable	(n)	Mean	S.D.	Min.	Max.	Median	10	30	70	90	95	99	Skewness	Kurtosis
1. SO_2	47388	1.4	5.0	.00	161.6	.40	.10	.20	.70	1.8	5.9	20.8	11.9	222.0
2. H_2S	44627	.30	1.5	.00	68.0	.00	.00	.00	.10	.40	.80	5.8	16.6	457.8
3. NO	44754	4.2	16.6	.00	278.6	.40	.00	.10	.70	6.6	22.2	87.0	7.2	65.3
4. NO_2	45964	6.6	8.6	.00	67.0	2.6	.90	1.5	6.4	18.0	25.8	40.8	2.4	6.4
5. CO_2	47715	346.6	14.5	306.0	492.0	345.8	331.0	338.4	351.4	362.0	370.0	399.0	1.9	9.0
6. O_3	46788	22.7	12.0	.50	136.8	22.8	4.0	15.8	30.0	38.6	42.8	49.8	.12	- .46
7. Wind Speed	49479	12.1	9.3	.00	125.0	10.6	.40	6.6	15.4	24.6	29.6	40.5	1.1	2.3
8. Wind Direction	49481	205.2	87.5	1.2	358.6	197.4	81.2	163.6	257.4	323.6	338.2	350.4	- .22	- .71
9. Temperature	47402	4.9	9.7	-32.1	31.0	5.4	- 7.9	.60	10.2	16.7	19.8	25.1	- .46	.48
10. Barometric Pressure	49499	886.5	6.8	856.6	906.0	887.0	877.0	883.8	890.0	895.0	897.0	900.0	- .39	.36
11. Precipitation	37635	.001	.02	.00	1.2	.00	.00	.00	.00	.00	.00	.00	25.8	1036.3
12. Humidity	45395	60.5	20.1	20.0	100.0	61.0	32.6	48.0	74.4	87.4	90.8	95.0	- .13	-1.02
13. Radiation	44124	325.9	488.5	- 1.3	2202.0	25.0	- 1.0	.00	378.4	1142.7	1475.4	1805.1	1.5	1.4

* All gas concentrations in ppb, except CO_2 (ppm); wind speed: km h^{-1}; wind direction: degrees; temperature: ^oC; pressure: millibars; precipitation: mm; humidity: %; radiation: Langleys min^{-1}.

Table 4.15 Summary statistics of automated gaseous pollutant measurements and meteorological variables*. Sampling location: Crossfield East. Sampling period: 1985-86. Time resolution of data: 1 hour.

Variable	(n)	Mean	S.D.	Range		Median	Percentile						Skewness	Kurtosis
				Min.	Max.		10	30	70	90	95	99		
1. SO_2	8148	1.4	3.7	.00	69.8	.40	.30	.40	.60	3.1	6.1	17.1	7.9	91.5
2. H_2S	7670	.30	1.1	.00	49.3	.10	.00	.10	.10	.30	1.1	4.7	17.6	566.6
3. NO	7703	4.2	15.4	.00	247.6	.40	.20	.30	.50	7.3	22.4	82.0	6.8	60.4
4. NO_2	7919	6.6	8.3	.00	61.9	2.7	1.2	1.6	6.7	17.8	24.7	39.4	2.4	6.5
5. CO_2	8221	346.8	14.5	310.3	486.3	346.0	331.3	338.8	351.7	362.3	369.8	400.3	1.9	8.7
6. O_3	8221	22.5	12.0	2.0	66.1	22.6	4.8	15.7	29.6	38.1	42.4	49.5	.14	- .63
7. Wind Speed	8275	12.1	9.0	.00	70.4	10.6	1.6	6.6	15.3	24.3	29.1	39.7	1.1	2.0
8. Wind Direction	8276	205.2	80.5	6.4	353.5	196.8	93.9	167.1	252.7	318.4	329.5	344.1	- .16	- .70
9. Temperature	7922	4.9	9.7	-31.2	30.8	5.5	- 7.9	.60	10.2	16.6	19.8	25.0	- .47	.49
10. Barometric Pressure	8277	886.5	6.8	861.0	906.0	887.0	877.3	883.5	890.0	895.0	897.0	900.0	- .39	.30
11. Precipitation	6287	.001	.02	.00	.60	.00	.00	.00	.00	.00	.00	.10	18.5	535.0
12. Humidity	7626	60.3	20.1	20.0	100.0	61.0	32.4	47.7	74.1	87.4	90.5	94.5	- .14	- 1.02
13. Radiation	7371	327.1	479.8	-1.0	1935.3	29.6	-1.0	- .10	392.8	1129.9	1449.4	1759.0	1.5	1.1

* All gas concentrations in ppb, except CO_2 (ppm); wind speed: km h^{-1}; wind direction: degrees; temperature: ^{o}C; pressure: millibars; precipitation: mm; humidity: %; radiation: Langleys min^{-1}.

Table 4.16 Summary statistics of automated gaseous pollutant measurements and meteorological variables*. Sampling location: Crossfield East. Sampling period: 1986-87. Time resolution of data: 10 minutes.

				Range			Percentile							
Variable	(n)	Mean	S.D.	Min.	Max.	Median	10	30	70	90	95	99	Skewness	Kurtosis
1. SO_2	48039	1.3	5.8	.00	168.4	.40	.10	.20	.60	1.1	2.2	27.4	11.1	171.3
2. H_2S	46623	.36	1.7	.00	58.2	.00	.00	.00	.10	.50	.80	8.0	10.6	160.0
3. NO	48460	3.0	11.8	.00	219.6	.30	.00	.10	.70	4.8	16.0	56.4	8.0	84.8
4. NO_2	48408	6.4	8.3	.00	81.4	2.3	.90	1.5	6.2	19.2	26.0	35.8	2.0	3.8
5. CO_2	47641	350.1	15.6	311.8	496.0	349.0	334.0	344.0	353.6	365.2	374.0	410.5	2.3	12.7
6. O_3	46530	27.0	13.9	.50	75.4	28.0	5.4	19.8	33.8	45.0	51.0	59.6	.10	- .35
7. Wind Speed	50642	13.3	10.0	.00	70.0	11.6	1.6	7.2	16.8	26.8	32.4	44.2	1.1	1.5
8. Wind Direction	50638	211.7	85.9	1.00	358.8	211.3	85.4	173.8	261.0	326.2	338.4	351.2	- .40	- .43
9. Temperature	50653	5.6	9.6	-26.4	28.9	6.0	-6.9	.20	11.3	18.0	21.0	25.6	- .20	- .33
10. Barometric Pressure	50653	887.0	7.0	861.0	908.0	887.0	878.0	883.8	891.0	896.0	898.0	902.0	- .24	- .04
11. Precipitation	50668	.001	.01	.00	.60	.00	.00	.00	.00	.00	.00	.00	22.7	747.1
12. Humidity	48800	55.3	19.7	20.0	100.0	53.2	29.6	41.8	69.1	83.0	85.8	90.0	.09	- 1.2
13. Radiation	44161	374.4	502.1	- 2.0	2140.6	95.8	- .60	1.0	497.0	1224.2	1484.6	1779.9	1.3	.53

* All gas concentrations in ppb, except CO_2 (ppm); wind speed: km h^{-1}; wind direction: degrees; temperature: °C; pressure: millibars; precipitation: mm; humidity: %; radiation: Langleys min^{-1}.

Table 4.17 Summary statistics of automated gaseous pollutant measurements and meteorological variables*. Sampling location: Crossfield East. Sampling period: 1986-87. Time resolution of data: 1 hour.

Variable	(n)	Mean	S.D.	Range		Median	Percentile						Skewness	Kurtosis
				Min.	Max.		10	30	70	90	95	99		
1. SO_2	8288	1.3	4.6	.00	92.3	.40	.30	.40	.50	.90	4.9	22.8	9.1	108.7
2. H_2S	8043	.37	1.3	.00	25.7	.10	.00	.10	.20	.30	1.6	6.5	8.4	97.4
3. NO	8349	3.0	11.0	.00	179.3	.40	.20	.30	.50	5.7	16.1	49.9	8.1	89.4
4. NO_2	8339	6.4	7.9	.30	47.3	2.5	1.2	1.5	6.40	19.0	25.0	34.5	2.0	3.5
5. CO_2	8209	350.4	15.8	315.3	483.8	349.0	334.4	344.4	353.8	365.7	374.9	413.2	2.3	11.9
6. O_3	8345	26.8	13.6	1.9	73.8	27.7	6.5	19.7	33.5	44.7	50.5	59.3	.13	- .33
7. Wind Speed	8460	13.3	9.7	.00	66.1	11.4	2.2	7.3	16.7	26.4	31.7	43.4	1.1	1.6
8. Wind Direction	8460	211.7	78.3	5.1	354.4	208.1	104.8	176.5	257.4	319.7	330.6	344.6	- .34	- .38
9. Temperature	8462	5.6	9.6	-25.3	28.7	6.0	- 7.0	.30	11.3	17.9	21.0	25.6	- .20	- .33
10. Barometric Pressure	8462	887.0	7.0	861.0	908.0	887.0	878.0	883.6	891.0	896.0	898.0	901.9	- .24	- .04
11. Precipitation	8465	.001	.01	.00	.20	.00	.00	.00	.00	.00	.00	.00	12.6	177.3
12. Humidity	8212	55.0	19.7	20.0	97.6	53.0	29.2	41.6	68.8	82.8	85.5	89.7	.09	-1.17
13. Radiation	7473	370.2	490.9	- 2.0	1922.2	97.7	- .60	1.0	497.6	1199.1	1438.3	1730.1	1.2	.38

* All gas concentrations in ppb, except CO_2 (ppm); wind speed: km h^{-1}; wind direction: degrees; temperature: ^oC; pressure: millibars; precipitation: mm; humidity: %; radiation: Langleys min^{-1}.

Table 4.18 Summary statistics of automated gaseous pollutant measurements and meteorological variables*. Sampling location: Crossfield East. Sampling period: 1985-87. Time resolution of data: 10 minutes.

Variable	(n)	Mean	S.D.	Range		Median	Percentile						Skewness	Kurtosis
				Min.	Max.		10	30	70	90	95	99		
1. SO_2	95427	1.3	5.4	.00	168.4	.40	.10	.20	.60	1.3	5.0	23.8	11.5	193.2
2. H_2S	91250	.33	1.6	.00	68.0	.00	.00	.00	.10	.5	.80	7.0	12.9	266.9
3. NO	93214	3.6	14.3	.00	279.6	.30	.00	.10	.70	5.8	8.4	72.6	7.7	77.7
4. NO_2	94372	6.5	8.4	.00	81.4	2.4	.90	1.5	6.4	18.8	26.0	37.8	2.2	5.2
5. CO_2	95356	348.4	15.2	306.0	496.0	347.6	332.0	341.3	352.6	363.8	371.8	404.4	2.1	11.1
6. O_3	93318	24.8	13.3	.50	136.8	25.2	4.4	17.6	32.0	41.4	47.0	56.8	.16	- .33
7. Wind Speed	100121	12.7	9.7	.00	125.0	11.0	1.0	6.8	16.2	25.8	31.0	42.6	1.1	1.9
8. Wind Direction	100119	208.5	86.7	1.0	358.8	204.4	83.2	169.0	260.0	325.0	338.2	350.8	- .31	- .59
9. Temperature	98055	5.3	9.7	-32.1	31.0	5.7	- 7.4	.40	10.8	17.4	20.5	25.4	- .33	.09
10. Barometric Pressure	100152	886.7	6.9	856.6	908.0	887.0	878.0	883.8	891.0	895.0	897.8	901.0	- .31	.13
11. Precipitation	88303	.001	.02	.00	1.2	.00	.00	.00	.00	.00	.00	.00	26.7	1158.1
12. Humidity	94195	57.8	20.0	20.0	100.0	57.0	30.8	44.6	72.2	84.8	88.6	93.8	.01	-1.11
13. Radiation	88285	350.2	495.9	- 2.0	2202.0	57.0	- 1.0	.00	441.0	1185.8	1480.5	1792.7	1.4	.91

* All gas concentrations in ppb, except CO_2 (ppm); wind speed: km h^{-1}; wind direction: degrees; temperature: °C; pressure: millibars; precipitation: mm; humidity: %; radiation: Langleys min^{-1}.

Table 4.19 Summary statistics of automated gaseous pollutant measurements and meteorological variables*. Sampling location: Crossfield East. Sampling period: 1985-87. Time resolution of data: 1 hour.

Variable	(n)	Mean	S.D.	Range		Median	Percentile						Skewness	Kurtosis
				Min.	Max.		10	30	70	90	95	99		
1. SO_2	16436	1.3	4.2	.00	92.3	.40	.30	.40	.50	1.9	5.8	20.4	8.8	108.9
2. H_2S	15713	.33	1.2	.00	49.3	.10	.00	.10	.20	.30	1.3	5.5	12.1	269.9
3. NO	16052	3.6	13.3	.00	247.6	.40	.20	.30	.50	6.5	18.9	65.9	7.5	74.5
4. NO_2	16258	6.5	8.1	.00	61.9	2.6	1.2	1.5	6.6	18.3	24.9	36.5	2.2	5.1
5. CO_2	16430	348.6	15.3	310.3	486.3	347.9	332.0	341.6	352.8	364.2	372.3	405.8	2.1	10.5
6. O_3	16566	24.7	13.0	1.9	73.8	25.0	5.5	17.5	31.7	41.1	46.8	56.5	.19	- .37
7. Wind Speed	16735	12.7	9.4	.00	70.4	11.0	1.9	7.0	16.0	25.4	30.3	41.6	1.1	1.8
8. Wind Direction	16736	208.5	79.5	5.1	354.4	202.6	98.7	171.9	255.4	319.0	330.1	344.3	- .25	- .65
9. Temperature	16384	5.3	9.7	-31.2	30.8	5.8	- 7.3	.40	10.8	17.4	20.4	25.4	- .33	.10
10. Barometric Pressure	16739	886.7	6.9	861.0	908.0	887.0	877.7	883.6	890.7	895.2	897.6	901.0	- .31	.12
11. Precipitation	14752	.001	.01	.00	.60	.00	.00	.00	.00	.00	.00	.00	17.8	536.9
12. Humidity	15838	57.5	20.0	20.0	100.0	56.8	30.4	44.4	71.9	84.6	88.3	93.2	- .02	- 1.12
13. Radiation	14844	348.8	485.9	- 2.0	1935.3	64.0	- 1.0	.00	448.6	1167.1	1444.1	1739.6	1.4	.71

* All gas concentrations in ppb, except CO_2 (ppm); wind speed: km h^{-1}; wind direction: degrees; temperature: ^oC; pressure: millibars; precipitation: mm; humidity: %; radiation: Langleys min^{-1}.

Table 4.20 Comparison of the mean and median hourly pollutant concentrations (ppb) at Fortress Mountain, Crossfield West, and Crossfield East (1985-87).

	Fortress* Mountain		Crossfield West*		Crossfield East*	
Pollutant	Mean	Median	Mean	Median	Mean	Median
SO_2	0.16	0.10	0.56	0.20	1.30	0.40
H_2S	0.55**	0.50**	0.28	0.20	0.33	0.10*
NO	0.52**	0.50**	0.38	0.10	3.60	0.40
NO_2	1.20	1.10	2.80	1.20	6.50	2.60
CO_2	345.00	347.00	347.00	347.00	349.00	348.00
O_3	43.40	41.20	28.00	27.00	24.70	25.00

* Fortress Mountain: High elevation, remote or background site.
Crossfield West: Within urban airshed, but generally not influenced by a local point source plume.
Crossfield East: Within urban airshed, but generally influenced by a local point source plume and closer than Crossfield West to a line source and some other point sources.
** Value may be biased by the computational method used.

Model) used. Comparison of the values at the other two sites, however, showed slightly higher hourly mean H_2S concentrations at Crossfield East compared with Crossfield West. The opposite was true for the median concentrations. The higher mean value at Crossfield East was due to the fewer occurrences of 10-minute concentrations below the detection limit and higher occurrences of intermediate to high concentrations in comparison to Crossfield West. These features also provide the basis for the differences between the two sites in their median values.

The mean and median 10-minute NO concentrations were lower, as expected, at Crossfield West than at Crossfield East. There were no significant differences in the CO_2 concentrations between the two sites and with Fortress Mountain. In contrast, the mean and median hourly concentrations of O_3 at Fortress Mountain > Crossfield West > Crossfield East. Elevated O_3 concentrations at high altitude sites have been previously reported by a number of investigators (U.S. EPA 1984). In Chapter 3 a detailed discussion of the nature of the occurrence of high O_3 concentrations at Fortress Mountain has been presented. While both Crossfield sites are known to be influenced by the Calgary urban plume during appropriate months, the 40 km proximity of the two sites to Calgary may not have reflected the maximal O_3 production during the transport and arrival of the urban plume at the monitoring sites. This discussion concerns long term, smoothed data. For example, at Crossfield West, while the mean hourly O_3 concentrations ranged from 23.9 ppb to 32.0 ppb during 1985-86 and 1986-87, corresponding maximum concentrations ranged from 64.9 ppb to 80.5 ppb (Tables 4.9 and 4.11). These high hourly O_3 concentrations occurred during $<1\%$ of the total monitored time each year. In comparison, at Crossfield East, mean hourly O_3 concentrations ranged from 22.5 ppb to 26.8 ppb during the two years and the corresponding maximum concentrations were 66.1 ppb and 73.8 ppb (Tables 4.15 and 4.17), again occurring during $<1\%$ of the total monitored time each year. At the high elevation Fortress Mountain site, the maximum hourly O_3 concentrations were 70.9 ppb and 122.4 ppb, respectively, during 1985-86 and 1986-87 (Tables 3.31 and 3.33).

The O_3 episodes at Fortress Mountain occurred during April-June, 1987 (Table 4.21). High hourly O_3 concentrations also occurred at Crossfield West and Crossfield East during the same period. Table 4.22 provides values of hourly O_3 maxima at Crossfield West and East on the same days identified as episode or non-episode days at Fortress Mountain (located at approximately 90 km southwest of the Crossfield sites). The differences in the highest hourly O_3 concentrations on episode versus non-episode days at the two Crossfield sites were not as dramatic (15-21%) as at Fortress Mountain (38%). Table 4.23 provides a comparative summary of average daily 1-hour pollutant concentrations and meteorological parameters at Crossfield West and East on O_3 episode and non-episode days identified at Fortress Mountain during 1987. There were some differences in the primary pollutant concentrations (NO, NO_2) on episode versus non-episode days. However, the correlation coefficient between NO_2/NO versus O_3 concentrations was very poor ($r=0.05$) at both Crossfield sites. On episode days, mean wind speeds, temperature, and humidity were lower than on non-episode days. To the contrary, radiation on episode days was higher than on non-episode days. The correlation coefficients between these and other surface meteorological variables and O_3 were also poor. No consistent pattern could be established between peak hourly O_3 concentration on episode and non-episode days and surface wind direction at either Crossfield site.

Table 4.21 Characteristics of representative O_3 episode versus non- episode days at Fortress Mountain, 1987.

Episode			Non-Episode		
Date	Highest $\bar{x}$ Hourly O_3 (ppb)	Hour	Date	Highest $\bar{x}$ Hourly O_3 (ppb)	Hour
Apr 24	93.5	1800	Apr 23	66.5	0300
Apr 25	89.3	0100	Apr 28	68.1	1300
Apr 26	86.8	1200	Apr 29	63.5	2400
May 12	105.8	2400	May 05	81.1	2400
May 13	122.4	0100	May 10	72.6	0100
May 25	113.4	2200	May 21	63.2	0400
May 26	101.1	0100	Jun 04	67.6	1500
May 27	108.3	1200	Jun 08	69.7	0200
May 28	107.6	2000	Jun 09	49.2	2400
May 29	107.3	1000	Jun 11	56.4	2400
May 30	110.3	0400	Jun 13	66.0	1700
May 31	100.3	1100	Jun 16	68.8	1400
Jun 06	102.7	0800	Jun 22	52.8	0100
Jun 07	91.4	0100	Jun 30	53.3	0700

Based on the aforementioned considerations, it appeared highly probable that on episode days, O_3 in the air masses arriving at Fortress Mountain was subsequently carried along the down slope of the mountain and carried aloft to the two Crossfield sites. During the daytime vertical mixing, some portion of this O_3 was carried downwards, resulting in higher surface O_3 concentrations at the Crossfield sites. Evidence supporting this argument is presented in Table 4.24.

As previously stated in Chapter 3 and also in Figure 4.6, highest hourly O_3 concentrations on episode days at Fortress Mountain were observed during the midnight period (2300-2400 hours). This was followed by a steep decrease in hourly O_3 concentrations until 0800 hours after which concentrations increased again with a daytime peak at 1300 hours. In contrast, at the two Crossfield sites, the diurnal O_3 profiles (Figures 4.7 and 4.8) were very similar to the typical patterns observed at most monitoring sites (refer to U.S. National Academy of Sciences 1977). The O_3 concentrations increased from 0600 hours, reached peak concentrations between 1500 and 1700 hours (2 hours after Fortress Mountain), and decreased during the subsequent hours.

Table 4.22 Characteristics of highest hourly O_3 (ppb) at Crossfield West and Crossfield East on days identified as "episode" or "non-episode" at Fortress Mountain during 1987.*

Crossfield West						Crossfield East					
Episode			Non-Episode			Episode			Non-Episode		
Date	Highest $\bar{x}$ Hourly O_3 (ppb)	Hour	Date	Highest $\bar{x}$ Hourly O_3 (ppb)	Hour	Date	Highest $\bar{x}$ Hourly O_3 (ppb)	Hour	Date	Highest $\bar{x}$ Hourly O_3 (ppb)	Hour
Apr 24	69.8	1800	Apr 23	44.8	1800	Apr 24	68.0	1800	Apr 23	44.5	1100
Apr 25	71.6	1400	Apr 28	69.1	1400	Apr 25	70.9	1400	Apr 28	60.4	1400
Apr 26	64.0	1400	Apr 29	68.6	1300	Apr 26	61.9	1500	Apr 29	62.2	1800
May 12	39.2	1200	May 05	58.1	1500	May 12	50.8	1600	May 05	51.6	1500
May 13	56.3	1800	May 10	61.7	1500	May 13	51.1	1800	May 10	28.5	1800
May 25	69.2	1700	May 21	49.8	1200	May 25	65.3	1700	May 21	44.1	1100
May 26	64.8	1300	Jun 04	53.5	1700	May 26	60.0	1700	Jun 04	49.1	1300
May 27	64.4	1500	Jun 08	48.6	1600	May 27	60.8	1700	Jun 08	46.3	1600
May 28	76.5	1400	Jun 09	46.6	1600	May 28	73.8	1400	Jun 09	48.7	1500
May 29	59.4	1400	Jun 11	49.6	1700	May 29	58.1	1400	Jun 11	46.9	1700
May 30	55.7	0900	Jun 13	51.2	1600	May 30	52.7	1600	Jun 13	53.7	1600
May 31	54.6	1600	Jun 16	43.3	0100	May 31	48.3	1500	Jun 16	42.8	0100
Jun 06	53.8	1800	Jun 22	44.6	1100	Jun 06	50.4	1900	Jun 22	42.9	1100
Jun 07	64.1	1600	Jun 30	40.0	1700	Jun 07	59.3	1600	Jun 30	34.1	1700

* The episode and non-episode days are the same as those identified for Fortress Mountain.

Table 4.23 A comparison of average daily one hour pollutant concentrations and meteorological parameters at Crossfield West and East on O_3 episode and non-episode days identified at Fortress Mountain during April 15-June 30, 1987.

Pollutant or Meteorological Parameter	Crossfield West		Crossfield East	
	Episode*	Non-Episode*	Episode*	Non-Episode*
O_3 (max)**	61.7	52.2	59.4	46.8
O_3	46.5	41.0	39.0	33.4
NO_2	2.5	2.5	5.2	4.3
NO	.13	.21	.78	.58
SO_2	.51	.45	1.2	1.1
H_2S	.25	.24	.37	.26
Wind Speed	15.7	17.4	13.9	15.5
Temperature	12.0	13.0	11.8	13.1
Humidity	34.4	39.8	42.0	49.0
Radiation	542	500	546	488
Barometric Pressure	878	882	884	888
No. of Days	14	14	14	14

* Days of O_3 episode and non-episode were, respectively, the same as those identified at Fortress Mountain (refer to Table 4.21).

** Mean of daily maximum hourly concentrations.

All pollutant concentrations in ppb; wind speed: km h^{-1}; temperature: °C; humidity: %; radiation: Langleys min^{-1}; barometric pressure: mm.

Table 4.24 The relationships between the hourly O_3 concentrations during 2400-0700 hours at Fortress Mountain on episode (epis.) and non-episode (non-epis.) days, and the hourly O_3 concentrations on the same days during 0900-1600 hours at Crossfield West or Crossfield East.

Variables	(r)
Fortress Mountain epis. vs Crossfield West epis.	-0.94
Fortress Mountain epis. vs Crossfield East epis.	-0.96
Fortress Mountain non-epis. vs Crossfield West non-epis.	-0.42

Continued...

Table 4.24 (Concluded).

Variables	(r)
Fortress Mountain non-epis. vs Crossfield East non-epis.	-0.56
Ratio epis: non-epis. Fortress Mountain vs Ratio epis: non-epis. Crossfield West	-0.84
Ratio epis: non-epis. Fortress Mountain vs Ratio epis: non-epis. Crossfield East	-0.91
Ratio epis: non-epis. Crossfield West vs Ratio epis: non-epis. Crossfield East	0.94

On episode days there was a clear, inverse relationship (r=-0.94 to -0.96) between the late night (2400-0700 hours) decrease in the hourly O_3 concentrations at Fortress Mountain and mid-morning (0900 hours) rise in hourly O_3 concentrations at the two Crossfield sites (Table 4.24). The 'r' values for such a relationship on non-

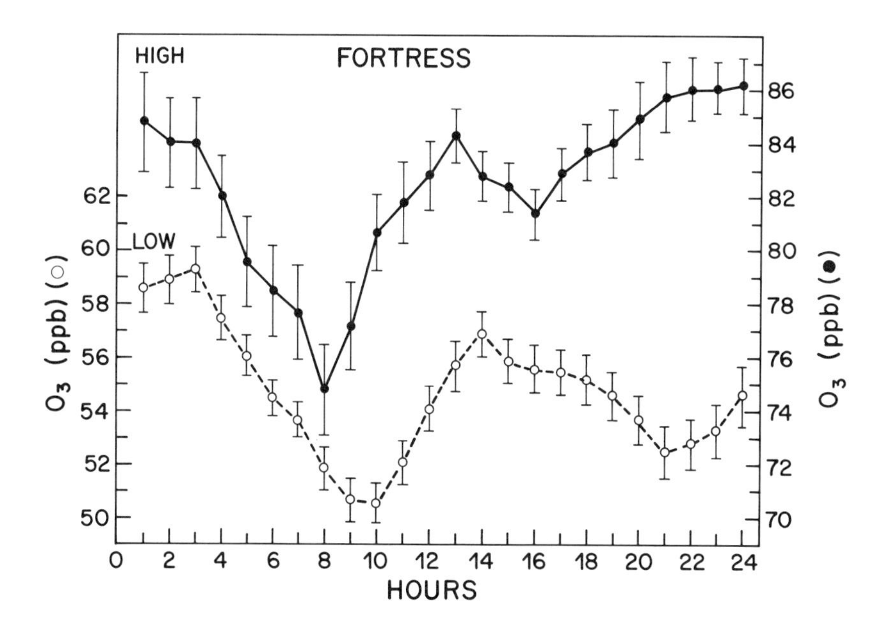

Figure 4.6 Diurnal O_3 profiles on episode versus non-episode days at Fortress Mountain. The vertical bars represent S.D. of the mean for each hour.

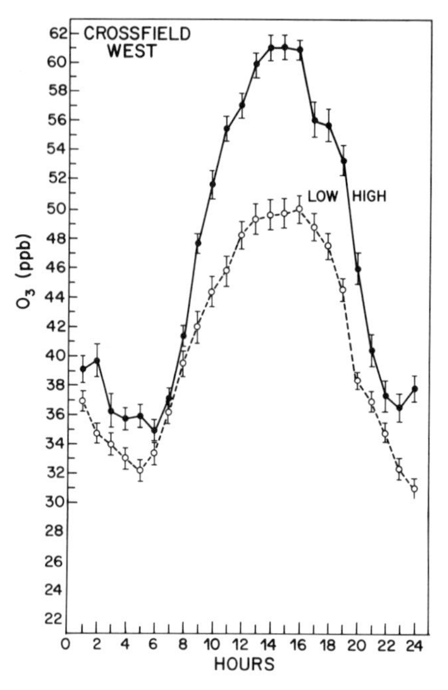

Figure 4.7 Diurnal O_3 profiles on episode versus non-episode days at Crossfield West. The vertical bars represent S.D. of the mean for each hour.

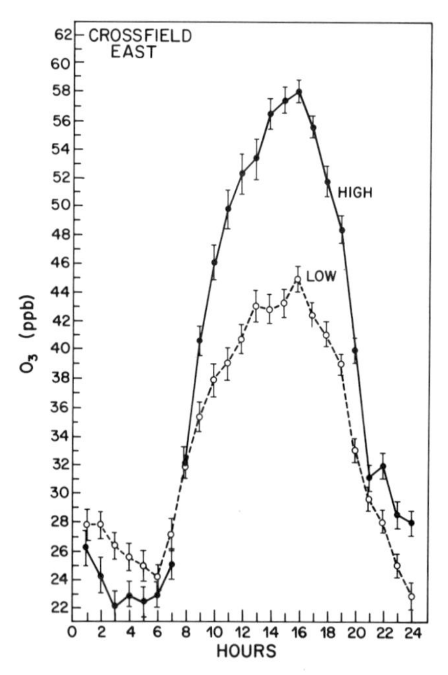

Figure 4.8 Diurnal O_3 profiles on episode versus non-episode days at Crossfield East. The vertical bars represent S.D. of the mean for each hour.

episode days were comparatively low (-0.42 to -0.56).

The ratios of hourly O_3 concentrations during 2400-0700 hours on episode days to the corresponding values on non-episode days at Fortress Mountain were well correlated to similar values on the same days, but at 0900-1600 hours, at both Crossfield sites (r=-0.84 and -0.91). Further, O_3 values for the ratios of episode to non-episode days at the two Crossfield sites were also strongly correlated (r=0.94). Thus, the results obtained offer a strong argument for the transport of O_3 on episode days from the Fortress Mountain area to Crossfield.

On an annual basis, values for O_3, NO, NO_2, and selected meteorological variables were plotted against the hour of the day (Figures 4.9 and 4.10). As opposed to the flat O_3 profile at Fortress Mountain, the typical diurnal pattern for O_3 and its inverse relationship to NO and NO_2 were observed at the two Crossfield sites. This is consistent with other published results (U.S. NAPAP 1987). There were no major deviations, from what is generally known, in the diurnal relationships between the meteorological parameters and the three air pollutants.

Figure 4.11 provides a graphic representation of the relationships between selected meteorological parameters and O_3, NO, and NO_2 on a monthly basis. As previously stated in Chapter 3, variations in any or all of these parameters is to be expected due to the inherent variability in climatology. As one would expect, maximum O_3 concentrations occurred during spring, with a decrease in the months following May. This was consistent with the changes in air temperature, with the variations in both parameters preceding similar decreases in solar radiation and dew point temperature.

Additional data on seasonal variations in the concentrations of specific air pollutants are presented in Tables 4.25 through 4.36. The SO_2 concentrations at Crossfield West exhibited monthly maxima in March during both 1986 and 1987. A gas well in the vicinity of this site was not considered to be a pollutant source of significance. However, this gas well was most likely serviced (clean-up of residue on the walls and frequent flaring) during March following the season of peak demand for natural gas.

In contrast, at Crossfield East the concentrations of both SO_2 and H_2S were somewhat higher in the warm season months compared with the cool season months. Independent of these considerations, it is important to note that at both Crossfield West and East, monthly mean concentrations of SO_2 and H_2S were very low.

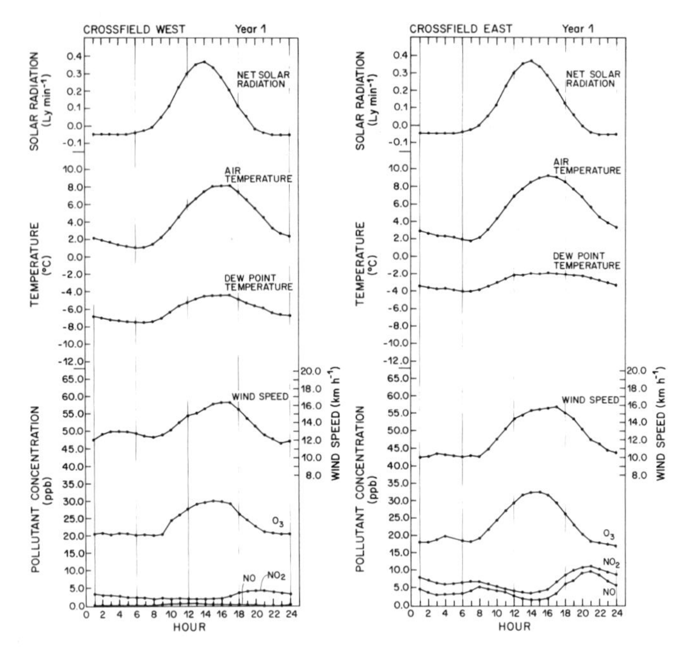

Figure 4.9 Mean diurnal patterns of O_3, NO, NO_2, and selected meteorological parameters at Crossfield West and Crossfield East. Year 1: November 1985 to October 1986.

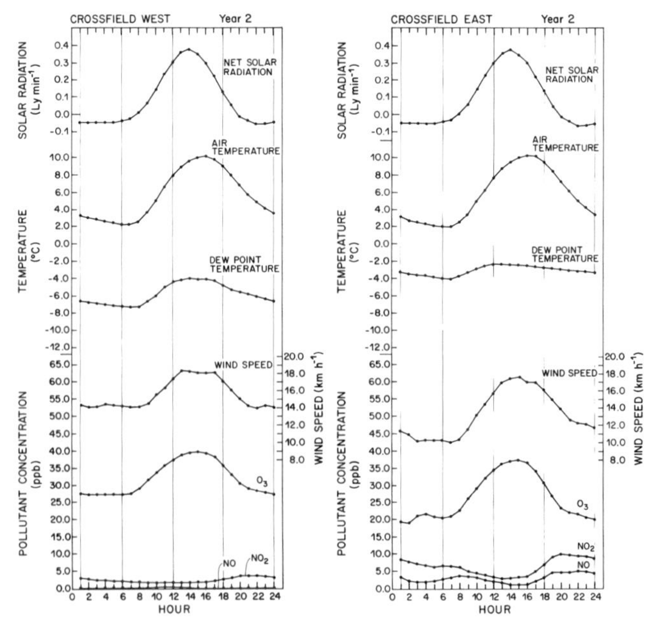

Figure 4.10 Mean diurnal patterns of O_3, NO, NO_2, and selected meteorological parameters at Crossfield West and Crossfield East. Year 2: November 1986 to October 1987.

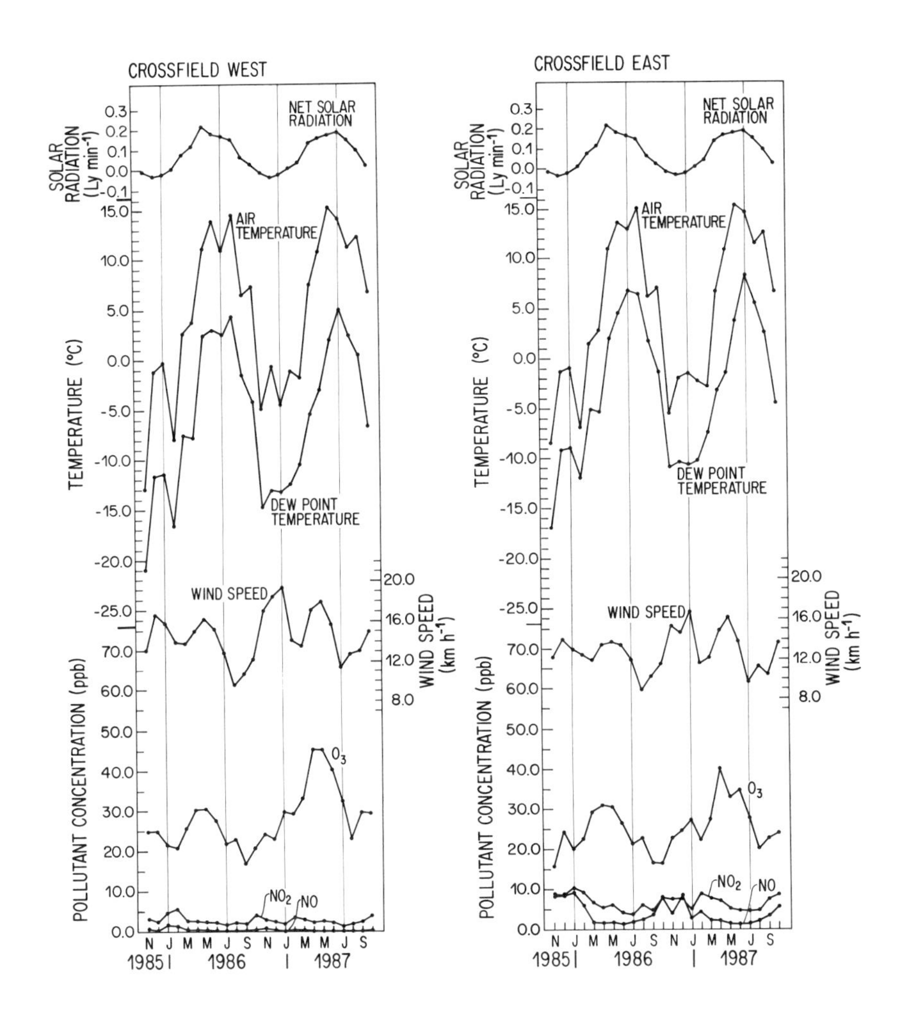

Figure 4.11 Mean monthly patterns of O_2, NO, NO_2 and selected meteorological parameters at Crossfield West and Crossfield East.

Table 4.25 Summary statistics of 10-min. SO_2 concentrations (ppb) by month. Sampling location: Crossfield West.

Month	Mean	Percentile				
		75th	90th	95th	99th	99.5th
Nov 85	.62	.30	.60	1.60	13.9	18.7
Dec 85	.41	.20	.50	.80	11.7	16.2
Jan 86	.34	.20	.50	.70	7.2	13.6
Feb 86	.83	.30	.80	5.60	13.0	16.9
Mar 86	1.33	.30	.80	6.50	34.2	44.6
Apr 86	.58	.30	.70	1.40	13.1	19.2
May 86	.65	.30	.70	1.60	11.2	19.3
Jun 86	.51	.30	.60	1.10	9.6	16.9
Jul 86	.43	.30	.60	1.00	9.1	12.4
Aug 86	.47	.30	.60	.90	9.4	14.9
Sep 86	.50	.30	.50	.80	9.4	20.8
Oct 86	.39	.20	.60	.80	9.0	13.4
Nov 86	.48	.20	.50	.80	11.6	17.7
Dec 86	.29	.20	.50	.70	4.2	12.3
Jan 87	.29	.20	.50	.70	5.1	8.9
Feb 87	.36	.30	.60	.90	6.8	9.1
Mar 87	1.67	.50	5.20	9.00	22.4	33.9
Apr 87	.42	.30	.70	1.10	6.9	10.7
May 87	.47	.30	.70	1.40	8.2	11.3
Jun 87	.51	.30	.60	.90	10.4	14.0
Jul 87	.56	.30	.60	.90	10.7	17.6
Aug 87	.33	.30	.60	.80	6.6	8.8
Sep 87	.32	.30	.60	.90	5.2	6.8
Oct 87	.39	.30	.60	.90	8.0	10.2

Table 4.26 Summary statistics of 10-min. SO_2 concentrations (ppb) by month. Sampling location: Crossfield East.

Month	Mean	Percentile				
		75th	90th	95th	99th	99.5th
Nov 85	1.70	.90	4.6	7.2	21.6	29.8
Dec 85	2.00	1.00	5.8	8.5	25.8	38.1
Jan 86	.60	.60	1.0	1.2	6.5	11.3
Feb 86	.89	.70	1.0	1.3	15.9	27.0
Mar 86	1.10	.70	1.1	2.8	17.2	30.7
Apr 86	1.80	.80	2.7	8.6	30.1	41.0
May 86	2.00	1.00	5.0	6.9	28.3	37.0
Jun 86	1.40	.70	1.2	3.2	29.4	45.7
Jul 86	1.90	1.00	5.0	7.6	23.4	35.3
Aug 86	1.00	.70	1.1	2.9	16.0	23.6
Sep 86	1.20	.70	1.1	2.7	21.1	31.0

Continued...

Table 4.26 (Concluded).

Month	Mean	Percentile 75th	90th	95th	99th	99.5th
Oct 86	1.20	.80	1.6	5.2	17.2	25.5
Nov 86	1.20	.70	1.1	3.3	20.1	30.1
Dec 86	1.60	.70	1.2	7.0	29.9	43.5
Jan 87	.80	.60	.9	1.2	14.2	30.7
Feb 87	.64	.60	.9	1.2	6.0	13.4
Mar 87	1.60	.70	1.1	1.7	37.3	67.2
Apr 87	1.00	.70	1.0	1.5	17.6	25.8
May 87	1.60	.70	1.2	4.8	33.7	45.9
Jun 87	1.10	.70	1.0	1.7	24.7	31.4
Jul 87	1.50	.70	1.1	2.6	31.9	49.2
Aug 87	1.50	.70	1.2	4.9	30.6	39.8
Sep 87	2.10	.70	1.5	11.4	42.1	52.7
Oct 87	1.00	.70	1.0	1.6	19.5	31.7

Table 4.27 Summary statistics of 10-min. H_2S concentrations (ppb) by month. Sampling location: Crossfield West.

Month	Mean	Percentile 75th	90th	95th	99th	99.5th
Nov 85	.25	.30	.60	.70	1.1	1.3
Dec 85	.25	.30	.60	.70	1.0	1.1
Jan 86	.27	.30	.50	.70	1.0	1.2
Feb 86	.29	.30	.60	.70	1.3	4.0
Mar 86	.68	.40	.70	1.00	15.6	24.5
Apr 86	.26	.30	.60	.70	1.1	1.3
May 86	.30	.30	.60	.80	1.4	4.0
Jun 86	.25	.30	.50	.70	1.0	1.2
Jul 86	.25	.30	.60	.70	1.1	1.3
Aug 86	.32	.30	.60	.80	4.8	5.0
Sep 86	.27	.30	.60	.70	1.1	1.3
Oct 86	.25	.30	.59	.70	1.0	1.1
Nov 86	.25	.30	.50	.70	1.1	1.2
Dec 86	.24	.30	.50	.70	1.0	1.2
Jan 87	.24	.30	.50	.70	1.1	1.2
Feb 87	.25	.30	.60	.70	1.1	1.2
Mar 87	.25	.30	.50	.70	1.0	1.1
Apr 87	.25	.30	.50	.70	1.0	1.1
May 87	.24	.30	.50	.70	1.1	1.1
Jun 87	.24	.30	.50	.70	1.1	1.2
Jul 87	.25	.30	.60	.70	1.0	1.2
Aug 87	.25	.30	.50	.70	1.0	1.2
Sep 87	.24	.30	.50	.70	1.0	1.1
Oct 87	.26	.30	.60	.80	1.2	2.4

Table 4.28 Summary statistics of 10-min. H_2S concentrations (ppb) by month. Sampling location: Crossfield East.

Month	Mean	Percentile				
		75th	90th	95th	99th	99.5th
Nov 85	.15	.10	.40	.60	1.3	3.8
Dec 85	.20	.10	.40	.60	3.1	6.8
Jan 86	.27	.10	.50	.80	4.3	5.0
Feb 86	.31	.10	.50	.80	5.1	10.7
Mar 86	.34	.10	.50	.80	7.2	11.3
Apr 86	.54	.20	.70	3.20	11.6	15.3
May 86	.20	.10	.40	.60	3.2	6.2
Jun 86	.23	.10	.40	.70	4.5	7.6
Jul 86	.55	.20	.60	2.00	10.0	16.2
Aug 86	.29	.10	.50	.80	6.0	8.5
Sep 86	.23	.10	.40	.70	4.3	7.0
Oct 86	.22	.10	.40	.70	4.8	6.2
Nov 86	.26	.10	.40	.70	5.9	8.6
Dec 86	.22	.10	.40	.60	3.9	7.2
Jan 87	.13	.10	.30	.50	.9	1.8
Feb 87	.14	.10	.40	.60	1.1	2.4
Mar 87	.12	.10	.30	.50	.8	1.1
Apr 87	.17	.10	.40	.60	2.4	4.2
May 87	.38	.20	.50	1.30	8.5	12.3
Jun 87	.22	.10	.40	.70	4.2	6.9
Jul 87	.41	.20	.60	2.40	7.2	10.8
Aug 87	.65	.20	.80	3.90	12.8	17.7
Sep 87	.83	.20	.90	5.50	16.0	20.3
Oct 87	.73	.20	.80	4.60	14.6	19.4

Table 4.29 Summary statistics of hourly NO concentrations (ppb) by month. Sampling location: Crossfield West.

Month	Mean	Percentile				
		75th	90th	95th	99th	99.5th
Nov 85	.45	.20	.50	.80	10.8	21.2
Dec 85	.19	.10	.40	.60	1.2	8.8
Jan 86	1.50	.20	.80	9.40	33.9	47.1
Feb 86	1.40	.20	2.40	9.80	25.3	33.8
Mar 86	.31	.10	.40	.70	7.8	15.2
Apr 86	.30	.20	.50	.70	6.0	8.9
May 86	.24	.20	.40	.70	3.3	5.4
Jun 86	.15	.10	.40	.60	1.3	3.5
Jul 86	.17	.10	.40	.60	1.5	4.9
Aug 86	.17	.10	.40	.60	2.4	4.4
Sep 86	.18	.10	.40	.60	3.7	4.8

Continued...

Table 4.29 (Concluded).

Month	Mean	Percentile				
		75th	90th	95th	99th	99.5th
Oct 86	.31	.20	.50	.80	6.00	7.6
Nov 86	.52	.20	.50	.80	15.90	23.8
Dec 86	.32	.20	.40	.70	7.20	15.2
Jan 87	.14	.10	.40	.60	.90	1.1
Feb 87	.46	.20	.50	1.00	10.40	16.1
Mar 87	.43	.20	.60	1.50	8.80	12.0
Apr 87	.29	.10	.40	.60	4.20	10.6
May 87	.18	.20	.40	.70	1.70	4.2
Jun 87	.14	.10	.40	.60	1.20	3.2
Jul 87	.13	.10	.40	.60	.96	1.4
Aug 87	.15	.10	.40	.60	1.20	2.4
Sep 87	.17	.10	.40	.60	1.60	3.4
Oct 87	.43	.20	.50	.70	12.50	17.1

Table 4.30 Summary statistics of hourly NO concentrations (ppb) by month. Sampling location: Crossfield East.

Month	Mean	Percentile				
		75th	90th	95th	99th	99.5th
Nov 85	7.7	1.4	12.00	41.1	176.0	225.0
Dec 85	8.3	.9	12.60	61.3	162.0	197.0
Jan 86	8.9	2.6	26.30	58.6	131.0	146.0
Feb 86	6.0	1.2	16.20	41.3	81.8	93.6
Mar 86	1.6	.7	1.90	7.6	29.4	40.2
Apr 86	1.3	.7	1.20	4.8	24.0	39.7
May 86	1.5	.8	2.80	8.0	24.5	30.6
Jun 86	.9	.6	.96	1.4	20.2	34.2
Jul 86	1.7	.7	1.20	5.2	35.0	64.4
Aug 86	2.1	.8	4.60	12.3	33.2	42.0
Sep 86	3.1	.8	4.60	9.3	58.2	73.8
Oct 86	7.6	1.7	21.80	49.2	117.0	157.0
Nov 86	3.3	.9	8.80	20.8	54.7	71.4
Dec 86	8.0	.9	17.25	1.4	156.2	173.0
Jan 87	2.7	.7	1.50	11.2	73.5	99.1
Feb 87	4.2	1.3	14.10	25.7	50.9	66.8
Mar 87	2.2	.8	5.40	12.0	33.5	41.0
Apr 87	2.0	.7	2.20	10.6	41.8	52.9
May 87	1.0	.6	1.00	2.2	18.6	30.2
Jun 87	1.0	.6	1.00	2.0	22.4	38.2
Jul 87	1.1	.7	1.20	4.4	20.9	28.4
Aug 87	1.9	.7	2.00	10.3	36.7	44.7
Sep 87	3.0	.8	7.00	20.3	48.7	58.0
Oct 87	5.7	1.2	16.00	35.9	82.1	104.0

Table 4.31 Summary statistics of hourly NO_2 concentrations (ppb) by month. Sampling location: Crossfield West.

Month	Mean	Percentile				
		75th	90th	95th	99th	99.5th
Nov 85	3.3	4.4	7.6	10.6	23.3	25.8
Dec 85	2.3	1.7	4.2	6.8	29.6	32.4
Jan 86	4.6	5.8	14.0	18.4	28.4	31.3
Feb 86	5.5	6.4	16.6	26.0	36.0	8.1
Mar 86	2.5	2.6	6.0	8.6	17.4	21.0
Apr 86	2.4	2.1	5.6	8.4	19.6	24.0
May 86	2.2	2.0	5.6	8.4	13.5	15.3
Jun 86	2.3	2.4	5.4	7.8	13.8	15.4
Jul 86	1.6	1.6	2.7	5.0	9.9	11.2
Aug 86	2.1	2.0	5.0	7.2	10.4	12.0
Sep 86	1.9	1.7	4.6	6.8	13.0	16.6
Oct 86	4.1	5.0	11.0	17.1	28.8	34.2
Nov 86	3.1	2.2	7.0	12.0	31.1	33.6
Dec 86	2.7	1.8	6.0	11.2	31.0	33.8
Jan 87	1.8	1.7	4.0	6.2	12.0	14.2
Feb 87	3.6	4.3	9.2	14.8	26.3	28.5
Mar 87	3.3	4.8	7.4	9.7	21.0	27.2
Apr 87	2.5	2.0	5.8	8.8	20.7	28.6
May 87	2.6	3.4	6.4	8.1	13.5	16.1
Jun 87	2.4	2.4	5.6	8.2	14.8	19.6
Jul 87	1.4	1.5	2.2	4.2	7.4	9.3
Aug 87	1.8	1.6	3.4	6.6	16.0	20.5
Sep 87	2.6	3.0	7.2	9.8	14.5	16.6
Oct 87	3.9	4.6	10.4	15.6	29.4	32.0

Table 4.32 Summary statistics of hourly NO_2 concentrations (ppb) by month. Sampling location: Crossfield East.

Month	Mean	Percentile				
		75th	90th	95th	99th	99.5th
Nov 85	8.4	10.0	24.2	30.0	41.0	43.0
Dec 85	8.2	8.2	28.2	41.1	51.0	55.6
Jan 86	10.1	14.4	27.8	34.6	46.2	51.0
Feb 86	9.1	12.4	25.6	32.1	48.3	51.6
Mar 86	6.4	7.8	16.2	24.3	33.0	36.3
Apr 86	5.2	6.2	12.8	19.2	30.8	34.0
May 86	5.9	7.6	17.8	22.0	28.0	29.7
Jun 86	3.9	5.0	8.6	12.7	20.4	22.6
Jul 86	3.3	3.2	8.8	13.2	20.2	21.4
Aug 86	5.8	7.8	16.2	20.8	29.4	34.2
Sep 86	4.2	5.4	12.2	15.9	22.4	24.6
Oct 86	7.7	11.8	22.2	27.4	35.6	38.4

Continued...

Table 4.32 (Concluded).

Month	Mean	Percentile 75th	90th	95th	99th	99.5th
Nov 86	7.2	8.5	23.8	29.6	39.6	40.6
Dec 86	7.4	9.4	24.6	29.8	40.2	42.1
Jan 87	4.7	4.4	13.2	23.0	35.0	38.2
Feb 87	8.7	13.8	23.4	27.6	32.8	34.8
Mar 87	7.7	8.6	20.0	27.8	39.0	41.6
Apr 87	7.0	8.2	22.6	30.8	40.5	42.0
May 87	5.0	6.3	12.4	19.7	29.8	32.2
Jun 87	4.5	5.4	11.0	17.0	31.2	34.0
Jul 87	4.3	5.2	11.0	16.0	26.6	30.7
Aug 87	4.4	5.2	13.4	18.2	23.8	26.0
Sep 87	7.5	11.0	21.8	27.5	34.1	36.4
Oct 87	8.5	13.1	25.0	30.2	35.4	36.2

Table 4.33 Summary statistics of hourly O_3 concentrations (ppb) by month. Sampling location: Crossfield West.

Month	Mean	Percentile 75th	90th	95th	99th	99.5th
Nov 85	24.8	30.0	34.8	37.9	40.9	42.0
Dec 85	24.9	27.9	30.6	32.6	35.2	36.2
Jan 86	21.6	29.1	33.0	34.4	36.6	37.2
Feb 86	20.9	26.6	29.2	30.4	32.0	32.4
Mar 86	25.7	32.8	39.0	40.8	45.3	46.0
Apr 86	30.0	36.0	39.8	42.0	45.6	46.2
May 86	30.1	36.2	41.2	45.2	50.6	52.2
Jun 86	27.8	34.2	39.2	42.8	50.8	52.4
Jul 86	21.9	27.4	32.4	34.6	38.2	39.2
Aug 86	22.9	30.0	35.2	39.2	50.2	53.2
Sep 86	16.9	20.8	25.8	29.5	35.4	36.4
Oct 86	20.9	27.4	32.4	34.6	39.4	40.0
Nov 86	24.1	28.4	30.2	31.0	32.0	32.4
Dec 86	23.0	26.2	28.6	30.5	35.6	36.0
Jan 87	29.9	35.0	37.2	38.4	40.0	40.6
Feb 87	29.1	36.2	38.2	39.2	41.4	41.8
Mar 87	33.6	42.4	45.6	47.6	54.2	57.3
Apr 87	45.6	52.6	58.8	64.2	73.6	78.8
May 87	45.5	55.0	60.0	63.8	72.6	75.7
Jun 87	40.6	48.0	55.2	58.8	65.2	66.4
Jul 87	32.6	40.9	50.0	52.8	62.2	66.1
Aug 87	23.1	29.0	36.2	40.4	49.7	54.7
Sep 87	29.8	36.4	44.6	48.6	55.4	57.7
Oct 87	29.6	36.2	40.2	43.4	53.8	56.0

Table 4.34 Summary statistics of hourly O_3 concentrations (ppb) by month. Sampling location: Crossfield East.

Month	Mean	Percentile				
		75th	90th	95th	99th	99.5th
Nov 85	15.6	20.4	23.8	26.8	31.6	32.2
Dec 85	24.1	32.8	36.6	38.6	41.0	41.4
Jan 86	19.8	31.6	35.2	36.4	38.8	39.2
Feb 86	22.0	30.2	36.4	39.6	47.2	48.0
Mar 86	29.3	39.4	44.6	47.6	51.2	52.0
Apr 86	30.7	40.0	44.6	46.8	50.0	51.4
May 86	30.4	40.0	46.8	50.4	58.2	59.4
Jun 86	26.1	33.4	39.4	41.6	50.2	52.2
Jul 86	21.1	28.2	32.6	34.4	38.4	41.9
Aug 86	22.6	32.4	38.6	43.2	54.1	62.2
Sep 86	16.3	22.4	27.6	31.8	39.6	42.4
Oct 86	16.1	24.0	31.0	34.6	39.5	41.6
Nov 86	22.3	29.6	33.2	34.2	36.8	37.1
Dec 86	24.3	32.4	36.4	38.0	39.4	39.6
Jan 87	27.0	32.8	34.8	35.8	38.0	38.6
Feb 87	21.8	31.6	35.6	38.0	41.2	42.0
Mar 87	27.1	36.6	40.8	43.6	46.3	48.2
Apr 87	39.9	51.8	57.4	61.4	68.0	69.8
May 87	32.8	46.8	52.8	57.0	64.7	71.1
Jun 87	34.7	44.8	51.0	54.6	59.5	60.9
Jul 87	28.0	38.5	46.4	48.9	57.2	59.8
Aug 87	19.9	26.6	34.0	37.5	45.9	49.1
Sep 87	22.7	32.4	40.4	44.2	54.4	58.2
Oct 87	23.9	33.8	39.0	41.6	49.1	54.5

Table 4.35 Summary statistics of hourly CO_2 concentrations (ppm) by month. Sampling location: Crossfield West.

Month	Mean	Percentile				
		75th	90th	95th	99th	99.5th
Nov 85	341	343	345	346	353	359
Dec 85	337	342	344	346	352	360
Jan 86	354	358	363	366	377	387
Feb 86	355	358	365	371	385	390
Mar 86	348	350	352	354	359	362
Apr 86	351	355	361	364	369	371
May 86	355	360	366	369	379	383
Jun 86	350	356	365	370	380	384
Jul 86	342	352	365	374	389	399
Aug 86	343	356	366	375	391	397
Sep 86	348	353	361	366	377	386
Oct 86	361	364	370	371	375	378

Continued...

Table 4.35 (Concluded).

Month	Mean	Percentile				
		75th	90th	95th	99th	99.5th
Nov 86	351	353	358	361	372	375
Dec 86	342	350	355	358	364	366
Jan 87	340	346	349	353	357	359
Feb 87	349	353	357	360	367	368
Mar 87	342	345	349	352	357	359
Apr 87	351	354	357	358	364	366
May 87	344	347	351	353	359	360
Jun 87	346	351	360	367	379	383
Jul 87	347	356	374	384	420	433
Aug 87	343	351	364	372	392	400
Sep 87	347	354	362	369	378	380
Oct 87	345	348	352	355	363	365

Table 4.36 Summary statistics of hourly CO_2 concentrations (ppm) by month. Sampling location: Crossfield East.

Month	Mean	Percentile				
		75th	90th	95th	99th	99.5th
Nov 85	334	336	340	344	351	357
Dec 85	335	335	342	351	368	373
Jan 86	349	352	359	364	380	387
Feb 86	352	354	359	364	384	391
Mar 86	354	356	361	366	375	384
Apr 86	346	351	356	360	369	372
May 86	353	357	367	373	405	420
Jun 86	346	352	361	372	396	406
Jul 86	346	356	374	391	437	447
Aug 86	350	361	378	395	435	450
Sep 86	347	350	364	373	397	410
Oct 86	349	357	365	368	387	397
Nov 86	352	354	362	367	389	402
Dec 86	349	351	356	362	393	404
Jan 87	351	354	359	362	375	377
Feb 87	355	357	362	367	384	389
Mar 87	354	364	368	370	378	381
Apr 87	350	355	360	364	372	377
May 87	350	353	362	369	383	393
Jun 87	351	356	369	386	424	435
Jul 87	349	356	400	427	472	483
Aug 87	342	350	368	379	395	401
Sep 87	350	358	375	385	403	410
Oct 87	348	353	361	366	379	385

The monthly mean concentration of NO_2 (particularly in the winter) was higher at Crossfield East than at West (Tables 4.31 and 4.32). The Crossfield East site was considered to be more frequently impacted by the gas plant plume than was the West site. In addition, NO_x compressors are generally in peak use during the cold months. As one would expect, the opposite (summer) patterns were observed for the seasonal O_3 profiles (Tables 4.33 and 4.34).

In contrast to the seasonal patterns, daily pollutant patterns are presented in Tables 4.37 through 4.48. Highest SO_2 concentrations were observed during the daytime at both Crossfield sites (Tables 4.36 and 4.37). This was also true for H_2S concentrations at Crossfield East (Table 4.40), but not at West (Table 4.39). As previously noted, Crossfield East was the site downwind from the gas plant. Higher concentrations during hours of increasing radiation and temperature (roughly 0800-2000 hours), may simply be due to the better vertical mixing of the air during that period. Crossfield East exhibited a pronounced peak SO_2 concentration slightly preceding the hours of highest daily radiation and temperature. This was also true for H_2S. Interestingly, NO_2 concentrations peaked at night at both sites (Tables 4.43 and 4.44). The local sources for SO_2 and NO_2 were quite different. While SO_2 was produced through the incinerator and flare stacks, NO_2 was primarily generated through the compressor operation. This NO_2 profile was similar to the CO_2 profile. Conversely, the O_3 concentrations were high during the day and low at night (Tables 4.45 and 4.46).

Table 4.37 Summary statistics of 10 minute SO_2 concentrations (ppb) by daily hour. Sampling location: Crossfield West.

		Percentile				
Hour	Mean	75th	90th	95th	99th	99.5th
1	.31	.30	.50	.80	5.8	8.3
2	.32	.20	.50	.80	5.2	9.9
3	.31	.20	.60	.80	4.5	8.9
4	.41	.20	.60	.90	6.3	10.2
5	.31	.20	.50	.70	6.2	10.2
6	.37	.30	.60	.80	6.7	11.0
7	.33	.30	.60	.80	5.2	9.2
8	.56	.30	.60	1.00	11.2	15.7
9	.85	.30	.70	3.1	14.1	23.9
10	.92	.30	.80	5.0	15.0	23.6

Continued...

Table 4.37 (Concluded).

Hour	Mean	Percentile				
		75th	90th	95th	99th	99.5th
12	.86	.30	.80	4.4	18.2	23.6
13	.70	.30	.70	3.2	12.8	20.9
14	.59	.30	.70	1.6	11.8	17.6
15	.81	.30	.70	3.7	15.0	21.6
16	.73	.30	.70	2.8	14.0	22.0
17	.84	.30	.70	3.2	17.6	26.8
18	.64	.30	.70	1.6	13.4	18.0
19	.51	.30	.60	1.0	10.8	15.4
20	.37	.30	.60	.8	8.4	12.0
21	.36	.20	.50	.8	8.1	11.2
22	.39	.30	.60	.8	9.0	12.6
23	.51	.30	.60	.9	12.4	18.5
24	.39	.30	.50	.8	9.8	13.5

Table 4.38 Summary statistics of 10 minute SO_2 concentrations (ppb) by daily hour. Sampling location: Crossfield East.

Hour	Mean	Percentile				
		75th	90th	95th	99th	99.5th
1	.76	.70	1.0	1.5	10.2	15.1
2	.91	.70	1.1	2.4	8.0	17.9
3	.91	.70	1.1	2.2	8.9	22.2
4	.75	.70	1.1	2.3	7.0	11.4
5	.76	.70	1.1	1.7	10.1	17.2
6	.91	.70	1.1	2.6	14.4	22.4
7	.98	.70	1.1	3.4	16.0	23.4
8	1.20	.70	1.2	4.8	20.9	28.2
9	1.80	.80	2.0	7.4	33.5	44.0
10	2.10	.80	3.9	11.5	36.1	49.5
11	2.60	.80	4.4	12.8	51.6	75.8
12	2.70	.90	6.0	14.2	46.0	58.8
13	2.50	.80	5.4	14.0	42.7	54.3
14	2.50	.90	5.2	13.0	41.9	50.2
15	1.00	.80	3.2	9.7	36.7	50.4
16	1.60	.80	1.4	6.4	27.6	39.4
17	1.30	.70	1.2	5.0	21.6	33.8
18	1.10	.70	1.1	2.8	19.1	34.9
19	.92	.70	1.0	1.8	14.4	26.9
20	.80	.70	1.0	1.6	11.2	19.6
21	.72	.70	1.0	1.5	10.0	15.5
22	.78	.70	1.0	1.4	10.6	17.0
23	.81	.70	1.0	1.5	10.4	15.3
24	.73	.70	1.0	1.5	8.6	14.3

Table 4.39 Summary statistics of 10 minute H_2S concentrations (ppb) by the daily hour. Sampling location: Crossfield West.

Hour	Mean	Percentile				
		75th	90th	95th	99th	99.5th
1	.26	.30	.60	.70	1.1	2.0
2	.25	.30	.60	.70	1.1	1.3
3	.25	.30	.60	.70	1.1	1.3
4	.26	.30	.60	.80	1.3	2.9
5	.26	.30	.60	.76	1.2	1.4
6	.25	.30	.50	.70	1.1	1.5
7	.26	.30	.60	.70	1.1	1.5
8	.26	.30	.60	.70	1.2	1.4
9	.30	.30	.60	.80	1.3	4.0
10	.31	.30	.60	.70	1.2	3.7
11	.31	.30	.60	.70	1.1	1.4
12	.28	.30	.50	.70	1.2	1.6
13	.27	.30	.60	.70	1.0	1.2
14	.32	.30	.60	.70	1.2	1.9
15	.33	.30	.60	.70	1.3	2.3
16	.30	.30	.60	.70	1.3	2.7
17	.30	.30	.60	.80	1.6	4.6
18	.26	.30	.50	.70	1.1	1.3
19	.24	.30	.60	.70	1.1	1.2
20	.25	.30	.50	.70	1.1	1.3
21	.26	.30	.50	.70	1.0	1.2
22	.28	.30	.60	.70	1.1	1.4
23	.28	.30	.60	.70	1.2	3.7
24	.28	.30	.60	.70	1.1	4.0

Table 4.40 Summary statistics of 10 minute H_2S concentrations (ppb) by the daily hour. Sampling location: Crossfield East.

Hour	Mean	Percentile				
		75th	90th	95th	99th	99.5th
1	.23	.10	.40	.70	5.0	6.9
2	.27	.10	.40	.70	5.6	8.0
3	.31	.10	.40	.80	6.5	11.3
4	.21	.10	.40	.60	4.0	6.0
5	.26	.10	.40	.80	5.3	7.4
6	.24	.10	.40	.70	5.6	7.5
7	.25	.10	.40	.70	5.0	6.6
8	.31	.20	.50	.80	6.0	10.6
9	.41	.20	.50	1.10	9.8	14.3
10	.45	.20	.60	2.00	10.4	12.5
11	.58	.20	.60	2.80	13.1	17.9
12	.60	.20	.70	3.40	12.6	17.2

Continued...

Table 4.40 (Concluded).

Hour	Mean	Percentile				
		75th	90th	95th	99th	99.5th
13	.58	.20	.70	3.30	11.4	17.6
14	.51	.20	.60	2.70	10.8	13.8
15	.45	.20	.50	1.40	8.4	13.5
16	.40	.20	.50	.80	8.3	13.5
17	.26	.10	.40	.70	6.1	8.5
18	.24	.10	.40	.60	4.2	8.1
19	.19	.10	.40	.60	2.2	4.2
20	.18	.10	.40	.60	3.2	6.0
21	.20	.10	.40	.70	4.2	6.1
22	.24	.10	.40	.70	4.2	7.0
23	.28	.10	.40	.70	6.2	9.6
24	.21	.10	.40	.60	4.2	7.0

Table 4.41 Summary statistics of hourly NO concentrations (ppb) by the daily hour. Sampling location: Crossfield West.

Hour	Mean	Percentile				
		75th	90th	95th	99th	99.5th
1	.27	.10	.40	.60	1.2	12.2
2	.23	.10	.40	.60	1.2	8.7
3	.30	.10	.40	.70	1.7	11.7
4	.27	.10	.40	.60	1.5	11.9
5	.19	.10	.40	.60	1.3	5.9
6	.20	.10	.40	.60	1.0	6.9
7	.21	.10	.40	.60	2.6	7.5
8	.27	.20	.40	.70	4.8	7.8
9	.40	.20	.50	.84	8.6	14.1
10	.54	.20	.60	2.80	10.2	16.2
11	.63	.20	.60	2.70	14.2	23.8
12	.67	.20	.60	2.90	14.6	20.8
13	.63	.20	.60	2.40	11.0	17.1
14	.55	.20	.50	1.30	9.7	16.4
15	.52	.20	.50	.90	9.8	23.0
16	.39	.20	.50	.80	9.1	14.0
17	.40	.20	.50	.70	10.0	17.4
18	.40	.20	.50	.70	12.6	18.7
19	.41	.20	.50	.70	10.4	21.1
20	.45	.20	.50	.70	16.3	22.5
21	.30	.20	.40	.60	5.6	15.9
22	.24	.10	.40	.60	2.4	8.4
23	.19	.10	.40	.60	1.0	8.0
24	.25	.20	.40	.60	1.0	7.2

Table 4.42 Summary statistics of hourly NO concentrations (ppb) by the daily hour. Sampling location: Crossfield East.

Hour	Mean	Percentile				
		75th	90th	95th	99th	99.5th
1	3.8	.80	7.7	21.0	75.4	102.0
2	2.5	.80	4.8	13.8	42.8	60.1
3	2.1	.70	1.8	10.5	40.2	58.5
4	2.2	.70	2.2	10.6	39.5	73.7
5	2.4	.70	2.7	12.6	49.4	66.7
6	2.7	.80	6.0	16.1	46.8	57.2
7	3.3	.90	7.8	22.1	53.4	67.4
8	4.3	.90	11.0	24.8	72.9	94.3
9	3.9	.90	8.4	22.6	66.2	93.7
10	3.6	.80	6.4	17.3	75.0	92.3
11	2.8	.80	5.2	12.0	65.4	86.6
12	2.2	.70	2.7	8.0	44.9	96.2
13	1.6	.70	1.6	5.6	28.2	40.6
14	1.2	.70	1.4	4.8	21.2	31.7
15	1.0	.70	1.3	4.4	16.9	25.6
16	1.4	.70	1.3	4.6	28.1	49.3
17	2.6	.70	1.7	9.2	58.8	96.6
18	4.5	.80	6.0	20.0	102.0	164.0
19	5.8	.90	8.8	29.2	135.0	172.0
20	6.9	.90	13.8	42.0	134.0	160.0
21	7.0	1.00	17.8	42.6	119.0	172.0
22	6.8	.90	21.5	44.4	105.0	143.0
23	5.8	.90	17.4	37.9	84.8	115.0
24	5.1	.90	13.2	30.0	86.4	114.0

Table 4.43 Summary statistics of hourly NO_2 concentrations (ppb) by the daily hour. Sampling location: Crossfield West.

Hour	Mean	Percentile				
		75th	90th	95th	99th	99.5th
1	3.3	4.0	8.4	12.0	24.3	29.2
2	2.9	3.3	7.3	10.4	22.0	30.7
3	2.8	2.7	6.6	10.0	25.2	30.8
4	2.6	2.3	6.0	8.8	25.0	31.1
5	2.5	2.2	5.6	8.0	23.6	27.9
6	2.3	2.1	5.2	7.6	21.8	26.7
7	2.3	2.0	5.0	7.2	19.4	23.9
8	2.1	1.9	5.0	7.4	14.6	20.3
9	2.1	1.9	4.8	7.1	14.0	17.6
10	2.0	1.8	4.8	7.2	15.4	18.9
11	2.1	1.7	4.4	7.6	17.6	20.6

Continued...

Table 4.43 (Concluded).

Hour	Mean	Percentile				
		75th	90th	95th	99th	99.5th
12	2.0	1.6	4.2	7.0	18.2	21.8
13	1.9	1.6	4.0	7.0	16.6	23.4
14	1.9	1.6	4.2	6.8	16.8	20.2
15	2.0	1.7	4.2	7.0	20.0	25.6
16	2.2	1.7	4.6	7.0	20.5	25.0
17	2.7	2.0	6.0	10.6	28.6	34.6
18	3.4	3.2	8.0	13.0	33.4	38.6
19	3.8	4.6	9.6	14.7	32.9	36.2
20	4.2	5.4	10.2	15.8	31.6	33.5
21	4.2	5.6	10.2	14.8	28.0	30.3
22	4.1	5.6	9.6	13.8	25.0	30.7
23	3.8	5.2	9.2	13.6	24.8	28.2
24	3.5	4.6	9.1	13.5	23.6	26.8

Table 4.44 Summary statistics of hourly NO_2 concentrations (ppb) by the daily hour. Sampling location: Crossfield East.

Hour	Mean	Percentile				
		75th	90th	95th	99th	99.5th
1	7.9	11.2	22.2	27.2	37.2	44.4
2	7.3	10.6	19.4	25.2	36.9	43.4
3	6.4	8.4	17.8	23.4	34.0	39.7
4	6.1	8.0	17.0	23.2	33.2	36.9
5	6.0	7.6	16.0	22.8	34.0	36.7
6	6.2	7.8	17.2	23.8	33.9	36.2
7	6.4	8.2	17.8	23.4	33.1	36.3
8	6.2	7.8	17.6	23.6	33.4	35.8
9	5.4	6.2	15.4	22.0	30.6	33.4
10	4.7	5.0	12.4	20.0	31.5	35.3
11	4.1	4.0	9.8	17.4	36.2	39.8
12	3.6	3.0	7.8	13.6	32.7	41.5
13	3.2	2.7	6.6	11.2	28.5	36.1
14	3.1	2.7	6.6	10.8	25.4	34.1
15	3.3	3.3	7.2	11.8	27.8	31.1
16	4.0	4.2	8.6	14.0	34.5	41.7
17	5.6	6.0	13.8	25.2	41.2	48.3
18	7.7	9.2	23.4	31.2	42.8	50.0
19	9.4	13.0	26.4	32.2	41.2	48.6
20	10.3	15.8	26.9	32.0	43.0	46.3
21	10.5	16.2	26.6	32.4	42.6	46.0
22	9.9	13.8	26.6	32.1	42.2	44.8
23	9.5	14.4	25.8	30.4	41.5	44.0
24	8.8	12.6	24.2	28.8	40.3	44.3

Table 4.45 Summary statistics of hourly O_3 concentrations (ppb) by the daily hour. Sampling location: Crossfield West.

Hour	Mean	Percentile				
		75th	90th	95th	99th	99.5th
1	24.1	30.2	36.6	41.2	47.1	49.0
2	24.0	29.8	35.5	37.8	44.5	46.3
3	23.5	29.6	35.6	38.4	44.8	47.8
4	24.1	30.4	36.6	40.2	44.8	45.6
5	24.1	30.6	36.6	40.2	45.0	46.1
6	23.8	29.8	36.6	40.0	45.2	47.0
7	24.1	30.0	36.8	40.8	47.0	48.6
8	25.2	30.8	38.6	44.0	50.8	52.5
9	27.1	33.0	42.2	48.0	54.8	56.6
10	29.0	35.2	43.8	50.6	58.2	60.6
11	31.1	37.2	47.4	52.9	61.4	63.5
12	32.8	38.8	49.2	54.2	63.8	67.6
13	34.2	40.0	50.8	55.8	67.2	69.9
14	34.7	40.8	51.4	56.8	67.8	71.5
15	35.1	41.4	52.2	57.6	67.3	71.1
16	34.7	41.6	52.6	57.8	67.1	69.1
17	33.3	40.0	52.0	57.2	66.7	69.2
18	31.3	38.4	49.8	55.8	65.4	68.7
19	29.0	36.2	46.0	52.8	60.8	63.9
20	26.7	33.6	41.4	47.4	54.2	57.4
21	25.4	32.0	38.8	43.0	51.4	53.0
22	24.8	31.0	37.8	42.6	50.0	51.9
23	24.5	30.6	37.4	42.0	49.0	51.6
24	24.2	30.4	37.0	41.2	48.0	50.6

Table 4.46 Summary statistics of hourly O_3 concentrations (ppb) by the daily hour. Sampling location: Crossfield East.

Hour	Mean	Percentile				
		75th	90th	95th	99th	99.5th
1	18.4	27.6	33.2	35.5	40.9	44.3
2	18.5	27.8	32.8	35.0	40.7	42.0
3	19.7	28.6	33.8	36.8	42.3	46.1
4	20.6	29.4	34.4	37.2	42.4	43.6
5	20.6	28.8	34.4	36.8	41.9	44.0
6	19.4	28.0	33.9	36.6	41.6	44.8
7	19.4	28.2	34.2	37.0	43.0	44.8
8	20.8	29.0	36.0	39.2	45.4	46.5
9	23.7	31.6	38.8	44.2	50.0	51.4
10	26.7	34.2	42.2	47.8	53.6	55.0
11	29.6	37.0	45.0	50.0	56.4	59.7

Continued...

Table 4.46 (Concluded).

Hour	Mean	Percentile				
		75th	90th	95th	99th	99.5th
12	31.9	39.0	47.4	51.8	59.9	64.2
13	33.7	40.4	49.2	53.5	62.2	64.8
14	34.5	41.8	49.8	53.8	61.4	64.6
15	34.9	42.3	50.6	55.0	62.7	67.6
16	34.0	42.0	50.6	56.0	63.4	66.9
17	31.6	40.8	49.2	54.0	61.4	64.2
18	28.3	37.8	46.2	51.0	59.4	62.2
19	24.9	34.0	42.0	47.2	54.6	56.6
20	21.8	31.6	37.8	42.4	49.0	51.1
21	20.3	29.9	35.8	39.2	46.2	47.6
22	19.8	29.4	34.8	38.4	44.8	46.4
23	19.1	29.0	34.4	37.8	44.0	45.7
24	18.8	28.4	34.0	36.8	44.3	46.4

Table 4.47 Summary statistics of hourly CO_2 concentrations (ppm) by the daily hour. Sampling location: Crossfield West.

Hour	Mean	Percentile				
		75th	90th	95th	99th	99.5th
1	352	357	367	372	387	397
2	352	358	366	372	391	396
3	353	359	367	373	392	403
4	353	359	368	374	388	400
5	353	360	369	375	396	407
6	352	358	366	373	387	397
7	350	356	363	368	380	387
8	348	354	360	364	372	377
9	346	352	359	362	370	371
10	345	351	358	361	370	372
11	345	351	357	361	371	375
12	345	351	357	362	371	376
13	344	350	356	361	370	372
14	343	350	356	360	369	374
15	343	349	355	359	369	375
16	342	349	355	359	367	369
17	342	349	355	359	366	368
18	342	349	355	360	366	369
19	343	350	356	360	369	370
20	345	351	357	361	370	373
21	347	353	359	363	372	375
22	350	355	362	366	375	381
23	351	356	364	369	381	385
24	351	357	365	371	385	391

Table 4.48 Summary statistics of hourly CO_2 concentrations (ppm) by the daily hour. Sampling location: Crossfield East.

Hour	Mean	Percentile				
		75th	90th	95th	99th	99.5th
1	355	360	372	387	430	447
2	356	361	375	392	428	439
3	356	361	377	392	446	461
4	357	363	378	393	440	469
5	357	363	379	398	453	470
6	355	360	374	393	440	466
7	353	358	369	384	411	428
8	350	355	365	372	393	402
9	347	353	360	365	378	387
10	346	352	358	365	381	389
11	345	352	357	364	385	402
12	345	351	358	364	384	394
13	344	351	357	362	375	392
14	343	350	355	360	370	383
15	342	350	355	359	369	375
16	342	349	354	359	369	375
17	341	349	354	358	368	376
18	341	349	354	359	368	374
19	343	350	355	359	371	379
20	344	351	356	360	372	381
21	347	353	359	364	374	382
22	350	356	363	369	381	389
23	352	358	367	375	402	409
24	354	359	371	382	409	427

4.4 MEASUREMENT OF SULPHUR GASES BY INSTANTANEOUS OR INTEGRATED SAMPLING AND GAS CHROMATOGRAPHY

In addition to the near-continuous measurements discussed in the previous section, sulphur gases were also quantified at discrete points in time (every 8 minutes) by gas chromatography using sulphur specific flame photometric detection. To examine what the values may be below the detection limits of continuous monitoring methods and conventional gas chromatography, the concentrations of these gases were also determined through the use of cryofocusing techniques described previously in Chapter 3.

Table 4.49 provides summary statistics on the sulphur gas concentrations measured by conventional sampling and gas chromatography at Crossfield West during 1985-87. Table 4.50 provides the corresponding information for Crossfield East.

Table 4.49 Summary statistics on the ambient concentrations of sulphur gases measured by conventional sampling and gas chromatography at Crossfield West (1985-87).

		Concentration (ppb)					
Pollutant	(n)	Mean	S.D.	Median	75%	99%	Range
COS	(2505)	1.14	1.51	1.0	1.0	3.0	1.0 - 71
CS_2	(5285)	1.22	1.20	1.0	1.0	7.0	1.0 - 28
H_2S	(8025)	1.22	1.04	1.0	1.0	5.0	1.0 - 41
SO_2	(6183)	2.14	3.66	1.0	1.0	2.0	1.0 - 82

Table 4.50 Summary statistics on the ambient concentrations of sulphur gases measured by conventional sampling and gas chromatography at Crossfield East (1985-87).

		Concentration (ppb)					
Pollutant	(n)	Mean	S.D.	Median	75%	99%	Range
COS	(3051)	1.73	1.82	1.0	2.0	9.0	1.0 - 33
CS_2	(7293)	1.74	2.18	1.0	2.0	11.0	1.0 - 50
H_2S	(8373)	2.56	4.02	1.0	2.0	21.0	1.0 - 57
SO_2	(16390)	4.49	7.84	2.0	4.0	41.0	1.0 - 108

The summary statistics on the sulphur gas concentrations at Crossfield West measured through cryogenic sampling are provided in Table 4.51. Similar measurements were not carried out at Crossfield East, since sulphur gas concentrations at this site were considered to be higher than at Crossfield West.

Mean and range of sulphur gas concentrations measured by conventional sampling and gas chromatography were higher at both Crossfield sites in comparison to Fortress Mountain (compare results presented in Tables 4.49 and 4.50 with 3.54). The mean values for all four pollutants were higher at Crossfield East compared to the West site (COS by 52%, CS_2 by 43%, H_2S by 210%, and SO_2 by 210%). While the mean values of H_2S and SO_2 at Crossfield East were higher than the corresponding values for Crossfield West by identical magnitude, the correlation coefficient (r) between H_2S and SO_2 at the two sites was 0.43 and 0.52.

Table 4.51 Summary statistics on the ambient concentrations of sulphur gases at Crossfield West measured by cryogenic sampling and gas chromatography.

		Concentration (ppt)					
Pollutant	(n)	Mean	S.D.	Median	75%	99%	Range
COS	(81)	244	35	244	271	--	178 - 346
CS_2	(81)	15	9	13	16	--	7 - 64
H_2S	(81)	71	131	56	65	--	27 - 1220
SO_2	(81)	1375	1231	850	1785	--	366 - 5560

Table 4.52 provides a comparison of the mean and the range of sulphur gas concentrations measured by cryogenic sampling at Fortress Mountain with the corresponding values for Crossfield West.

Table 4.52 Comparison of the mean and the range of sulphur gas concentrations measured by cryogenic sampling at Fortress Mountain and at Crossfield West.

	Fortress Mountain		Crossfield West	
Pollutant	Mean*	Range*	Mean*	Range*
COS	315	230 - 423	244	178 - 346
CS_2	18	12 - 26	15	7 - 64
H_2S	73	44 - 99	71	27 - 1220
SO_2	530	251 - 955	1375	366 - 5560

* Concentration in ppt; (n): Fortress Mountain (32)
Crossfield West (81)

The sample number for Fortress Mountain was 32 in comparison to the 82 samples analyzed for Crossfield West. Nevertheless, the mean values of COS and CS_2 were higher at Fortress Mountain in comparison to Crossfield West. Values obtained at both sites were, however, within the range of background values reported for the two pollutants by Sze and Ko (1980) (Table 3.5). While the mean values for H_2S were comparable at the two sites, maximum concentration of H_2S measured at Crossfield West was roughly 103 times higher than the maximum concentration measured at Fortress Mountain. Similarly, the mean SO_2 concentration measured at Crossfield West was

259% higher than at Fortress Mountain, and the range, 146-582% higher.

Table 4.53 provides a summary of the correlation analysis between the measured sulphur species by cryogenic sampling at Crossfield West.

Table 4.53 Correlation coefficients between sulphur gases collected by cryogenic sampling at Crossfield West.

Variables	(r)
COS vs H_2S	-0.10
COS vs CS_2	0.24
COS vs SO_2	0.35*
H_2S vs SO_2	0.47*
H_2S vs CS_2	0.36*
CS_2 vs SO_2	0.45*

* Significant at 0.001 level.

While at Fortress Mountain the r value between COS and H_2S was 0.53, at Crossfield West these two variables were negatively correlated. CS_2 and SO_2 were negatively correlated (r=-0.14) at Fortress Mountain but were positively correlated (r=0.45) at Crossfield West. These results suggest that the sources of COS, H_2S, CS_2, and SO_2 may be different at the two monitoring sites. This was confirmed by the results of the principal component analyses of the data from Crossfield West (Table 4.54).

Table 4.54 Results of the principal component analysis of sulphur gases characterized by cryogenic sampling at Crossfield West.

Variable	Factor Loading	
	1	2
COS	0.11	0.94
H_2S	0.84	-0.34
CS_2	0.71	0.28
SO_2	0.78	0.35
Pct. of variability	49.0	27.0
Eigenvalue	1.95	1.1

At Fortress Mountain the similarity in the factor loadings of CS_2, COS, and H_2S (Table 4.55, and Table 3.57) suggests the involvement of a common pathway of conversion of the three compounds (Figure 3.7) and their possible origin from natural sources. Both CS_2 and COS are produced from oceans and soils and during biomass burning. In addition, the conversion of CS_2 to COS constitutes a major portion in the CS_2 flux (Figures 3.5 and 3.6).

Table 4.55 Comparison of the results of principal component analysis of the data from Fortress Mountain and Crossfield West.

Variable	Factor Showing the Highest Value of Loading	
	Fortress Mountain	Crossfield West
COS	1	2
H_2S	1	1
CS_2	1	1
SO_2	2	1

CS_2, COS, H_2S, and SO_2 are also produced by anthropogenic activities (automobiles, fossil fuel combustion, wood pulping, and other industrial activities as appropriate, Graedel et al. 1986). The similarity in the loadings of CS_2, H_2S, and SO_2 at Crossfield West suggests an anthropogenic source(s) for these compounds. All three compounds are emitted during natural gas processing. Maximum concentrations of CS_2, H_2S, and SO_2 measured at Crossfield West were far higher than the corresponding values for Fortress Mountain. This was not true for COS (Table 4.52).

4.5 INTEGRATED MEASUREMENTS OF GASEOUS AIR POLLUTANTS AND AEROSOLS

4.5.1 KAPS Annular Denuder-Filter Pack System

Using a serial annular denuder-filter pack system (Kananaskis Air Pollutant Measurement System, KAPS) 24-hour integrated samples of SO_2, HNO_2, HNO_3, NH_3, and fine particle SO_4^{2-} and NO_3^- were collected at both Crossfield West and East. These samples were analyzed by the appropriate method described in

Table 3.14. The summary statistics of the observed concentrations of the various chemical constituents at the two monitoring sites are presented in Tables 4.57 and 4.58. The mean and maximum concentrations of HNO_3, SO_2, fine particulate SO_4^{2-}, and NO_3^- were generally higher at Crossfield East compared to the West. However, in some cases, these differences were not all that great. The mean concentrations of all the measured pollutants at both the Crossfield sites were higher than the corresponding values for Fortress Mountain (Table 3.61).

Table 4.56 KAPS annular denuder system. Explanation of the abbreviations used in Tables 4.57 through 4.62, as appropriate.

NY	=	Nylon filter
TF	=	Teflon filter
T1	=	Denuder Tube #1
T2	=	Denuder Tube #2
		(T1 and T2 were coated with sodium carbonate)
T3	=	Denuder Tube #3
		(T3 was coated with citric acid)
DNO_2	=	Corrected denuder NO_2^- concentration (HNO_2)
DNO_3	=	Corrected denuder NO_3^- concentration (HNO_3)
DSO_4	=	Corrected denuder SO_4^{2-} concentration (2/3 SO_4^{2-} = SO_2)
FNO_3	=	Total fine particulate NO_3^- concentration
FSO_4	=	Total fine particulate SO_4^{2-} concentration

The results of the correlation analysis between the variables sampled by the KAPS are presented in Tables 4.59 and 4.60. A summary of the information contained in these tables is presented in Table 4.61.

Both at Crossfield West and East the SO_2 (2/3 SO_4^{2-}) concentrations in denuder Tube #1 (T1) correlated reasonably well with their corresponding values for denuder Tube #2 (T2) (Table 4.61). This was also the case for NO_2^- on T1 and T2 at Crossfield East but the r value of 0.53 at Crossfield West was not high. The NO_3^- on T1 and T2 did not correlate well at either site. A good correlation was found between the total NO_3^- on the Teflon and nylon filters and the total SO_4^{2-} on these filters.

The correlation between HNO_2 (T1) and (T2) at Crossfield East may be due to the greater influence of the point source at that site. In comparison, good correlation between NO_3^- (TF) and SO_4^{2-} (TF) at the West site may be indicative of regional scale processes.

Table 4.57 Summary statistics of integrated (annular denuder system, KAPS) measurements of gaseous and particulate pollutants. Sampling location: Crossfield West. Sampling period: 1985-87. Sampling duration: 24 hours. All units, except (n), skewness and kurtosis in $\mu g\ m^{-3}$. For convenience, valence signs have been omitted.

				Range			Percentile							
Variable	(n)	Mean	S.D.	Min.	Max.	Median	10	30	70	90	95	99	Skewness	Kurtosis
NO_2 NY	169	.03	.04	.001	.21	.02	.01	.01	.04	.07	.10	.19	2.5	7.9
NO_2 TF	79	.01	.01	.001	.07	.01	.002	.004	.01	.1	.02	--	4.8	28.3
NO_3 NY	178	.26	.21	.03	1.2	.20	.09	.15	.30	.47	.70	1.2	2.3	6.7
NO_3 TF	184	.34	.67	.01	5.7	.10	.03	.05	.22	.91	1.6	4.1	4.5	26.8
SO_4 NY	177	.11	.21	.01	1.4	.05	.02	.04	.07	.19	.57	1.3	4.2	19.0
SO_4 TF	184	.89	.95	.01	6.3	.62	.11	.35	.95	2.0	2.9	4.9	2.4	7.5
NO_2 T1	173	.29	.31	.02	2.5	.20	.10	.15	.30	.55	.73	2.4	4.4	25.6
NO_2 T2	166	.09	.06	.003	.36	.07	.02	.05	.10	.17	.21	.30	1.4	2.1
NO_3 T1	175	.37	.31	.04	2.4	.28	.13	.20	.42	.68	.79	2.2	3.4	17.1
NO_3 T2	175	.11	.11	.01	.86	.08	.05	.06	.12	.18	.38	.61	3.5	16.7
SO_4 T1	175	6.4	4.8	.29	39.1	5.4	2.1	4.0	6.9	11.8	15.5	27.8	2.9	13.9
SO_4 T2	175	.50	.44	.07	3.7	.35	.21	.28	.52	1.0	1.3	2.9	3.5	19.5
NH_3 T3	176	1.5	1.3	.22	11.91	.2	.50	.89	1.6	2.8	3.6	7.3	3.9	26.1
DNO_2[a]	161	.21	.29	.003	2.3	.14	.04	.10	.21	.43	.53	2.2	5.1	32.3

Continued...

Table 4.57 (Concluded).

Variable	(n)	Mean	S.D.	Range		Median	Percentile						Skewness	Kurtosis
				Min.	Max.		10	30	70	90	95	99		
DNO_3[b]	172	.47	.37	.06	2.7	.38	.16	.26	.54	.82	1.1	2.6	2.9	12.8
DSO_4[c]	172	6.6	5.1	.26	42.9	5.5	1.9	4.1	7.2	12.5	16.2	28.4	4.4	22.1
FNO_3[d]	178	.60	.74	.05	5.8	.37	.13	.22	.55	1.4	2.0	4.6	3.6	18.2
FSO_4[e]	177	.99	.97	.04	6.3	.71	.18	.45	1.1	2.0	3.0	5.0	2.3	6.8

[a] $DNO_2 = NO_2T1 - NO_2T2$
[b] $DNO_3 = NO_3T1 + NO_3T2$
[c] $DSO_4 = SO_4T1 + SO_4T2$
[d] $FNO_3 = NO_3TF + NO_3NY$
[e] $FSO_4 = SO_4TF + SO_4NY$

Table 4.58 Summary statistics of integrated (annular denuder system, KAPS) measurements of gaseous and particulate pollutants. Sampling location: Crossfield East. Sampling period: 1985-87. Sampling duration: 24 hours. All units, except (n), skewness and kurtosis in $\mu g\ m^{-3}$. For convenience, valence signs have been omitted.

				Range			Percentile							
Variable	(n)	Mean	S.D.	Min.	Max.	Median	10	30	70	90	95	99	Skewness	Kurtosis
NO_2 NY	175	.04	.04	.002	.21	.10	.01	.01	.04	.09	.11	.20	2.0	5.2
NO_2 TF	800	.01	.01	.005	.07	.01	.003	.004	.01	.02	.05	--	3.4	12.6
NO_3 NY	182	.28	.22	.03	1.5	.22	.08	.16	.34	.51	.68	1.5	2.5	10.0
NO_3 TF	187	.42	1.1	.01	9.3	.11	.04	.07	.20	.88	2.0	8.1	5.6	36.5
SO_4 NY	180	.12	.20	.01	1.3	.05	.03	.04	.08	.27	.51	1.2	3.7	15.0
SO_4 TF	184	.96	1.3	.02	10.0	.64	.10	.37	.95	2.0	3.2	8.4	3.9	19.8
NO_2 T1	174	.52	.47	.02	3.0	.37	.15	.27	.55	1.0	1.6	2.5	2.4	7.3
NO_2 T2	167	.16	.13	.01	.89	.12	.04	.08	.19	.32	.42	.66	2.1	6.9
NO_3 T1	175	.40	.38	.03	3.8	.30	.11	.21	.47	.77	.97	2.3	4.7	36.6
NO_3 T2	175	.12	.10	.03	.74	.09	.05	.07	.12	.22	.36	.65	3.1	12.3
SO_4 T1	176	8.0	5.7	.10	35.7	6.8	2.4	5.1	9.2	14.9	18.5	33.6	02.0	6.1
SO_4 T2	176	.65	.51	.08	3.0	.51	.23	.38	.69	1.2	1.4	3.0	2.4	7.5
NH_3 T3	178	2.0	1.4	.24	8.3	1.7	.60	1.2	2.2	3.8	5.0	7.6	1.7	3.8

Continued...

Table 4.58 (Concluded).

Variable	(n)	Mean	S.D.	Range		Median	Percentile						Skewness	Kurtosis
				Min.	Max.		10	30	70	90	95	99		
DNO_2[a]	163	.38	.38	.003	2.1	.27	.07	.17	.42	.82	1.2	2.1	2.4	7.0
DNO_3[b]	171	.51	.43	.07	3.9	.39	.16	.28	.58	.93	1.2	2.3	3.7	23.5
DSO_4[c]	172	8.4	6.2	.18	38.3	7.0	2.3	5.1	9.7	16.0	20.1	35.6	3.0	9.2
FNO_3[d]	182	.71	1.2	.05	9.8	.42	.14	.27	.62	1.3	2.2	8.4	5.3	33.2
FSO_4[e]	177	1.1	1.3	.04	10.0	.80	.20	.48	1.1	2.2	3.3	8.6	3.8	18.6

[a] $DNO_2 = NO_2T1 - NO_2T2$
[b] $DNO_3 = NO_3T1 + NO_3T2$
[c] $DSO_4 = SO_4T1 + SO_4T2$
[d] $FNO_3 = NO_3TF + NO_3NY$
[e] $FSO_4 = SO_4TF + SO_4NY$

Table 4.59 Pearson's correlation coefficients (r) between variables measured through the annular denuder system (KAPS). Sampling location: Crossfield West. Sampling period: 1985-87. Sampling duration: 24 hours. For convenience, valence signs have been omitted.*

Variables			(r)	Variables			(r)
NO_3 NY	vs	NO_3 TF	.10	NO_2 T1	vs	NO_2 T2	.53
NO_3 NY	vs	SO_4 NY	.49	NO_2 T1	vs	NO_3 T1	.37
NO_3 NY	vs	SO_4 TF	.26	NO_2 T1	vs	NO_3 T2	.13
NO_3 NY	vs	NO_2 T1	.22	NO_2 T1	vs	SO_4 T1	.51
NO_3 NY	vs	NO_2 T2	.22	NO_2 T1	vs	SO_4 T2	.41
NO_3 NY	vs	NO_3 T1	.29	NO_2 T1	vs	NO_3 T1	.24
NO_3 NY	vs	NO_3 T2	.20	NO_2 T2	vs	NO_3 T2	.27
NO_3 NY	vs	SO_4 T1	.13	NO_2 T2	vs	SO_4 T1	.29
NO_3 NY	vs	SO_4 T2	.06	NO_2 T2	vs	SO_4 T2	.38
NO_3 TF	vs	SO_4 NY	-.07	NO_3 T1	vs	NO_3 T2	.43
NO_3 TF	vs	SO_4 TF	.79	NO_3 T1	vs	SO_4 T1	.34
NO_3 TF	vs	NO_2 T1	.30	NO_3 T1	vs	SO_4 T2	.16
NO_3 TF	vs	NO_2 T2	.38	NO_3 T2	vs	SO_4 T1	.07
NO_3 TF	vs	NO_3 T1	-.03	NO_3 T2	vs	SO_4 T2	.20
NO_3 TF	vs	NO_3 T2	.01	SO_4 T1	vs	SO_4 T2	.76
NO_3 TF	vs	SO_4 T1	.28	NH_3 T3	vs	NO_3 NY	.42
NO_3 TF	vs	SO_4 T2	.30	NH_3 T3	vs	NO_3 TF	.06
SO_4 NY	vs	SO_4 TF	-.06	NH_3 T3	vs	SO_4 NY	.08
SO_4 NY	vs	NO_2 T1	-.01	NH_3 T3	vs	SO_4 TF	.23
SO_4 NY	vs	NO_2 T2	.09	NH_3 T3	vs	NO_2 T1	.46
SO_4 NY	vs	NO_3 T1	.07	NH_3 T3	vs	NO_2 T2	.32
SO_4 NY	vs	NO_3 T2	.06	NH_3 T3	vs	NO_3 T1	.56
SO_4 NY	vs	SO_4 T1	-.06	NH_3 T3	vs	NO_3 T2	.28
SO_4 NY	vs	SO_4 T2	-.05	NH_3 T3	vs	SO_4 T1	.30
SO_4 TF	vs	NO_2 T1	.25	NH_3 T3	vs	SO_4 T2	.29
SO_4 TF	vs	NO_2 T2	.30	DNO_2	vs	NO_3 NY	.21
SO_4 TF	vs	NO_3 T1	.08	DNO_2	vs	NO_3 TF	.24
SO_4 TF	vs	NO_3 T2	.13	DNO_2	vs	SO_4 NY	-.03
SO_4 TF	vs	SO_4 T1	.22	DNO_2	vs	SO_4 TF	.21
SO_4 TF	vs	SO_4 T2	.32	DNO_2	vs	NO_2 T1	.98
DNO_2	vs	NO_2 T2	.37	FNO_3	vs	NO_2 T2	.43
DNO_2	vs	NO_3 T1	.32	FNO_3	vs	NO_3 T1	.06
DNO_2	vs	NO_3 T2	.05	FNO_3	vs	NO_3 T2	.07
DNO_2	vs	SO_4 T1	.48	FNO_3	vs	SO_4 T1	.30
DNO_2	vs	SO_4 T2	.35	FNO_3	vs	SO_4 T2	.20
DNO_3	vs	NO_3 NY	.29	FSO_4	vs	NO_3 NY	.40

Continued...

Table 4.59 (Concluded).

Variables			(r)	Variables			(r)
DNO_3	vs	NO_3 TF	.00	FSO_4	vs	NO_3 TF	.77
DNO_3	vs	SO_4 NY	.08	FSO_4	vs	SO_4 NY	.18
DNO_3	vs	SO_4 TF	.13	FSO_4	vs	SO_4 TF	.98
DNO_3	vs	NO_2 T1	.35	FSO_4	vs	NO_2 T1	.25
DNO_3	vs	NO_2 T2	.28	FSO_4	vs	NO_2 T2	.29
DNO_3	vs	NO_3 T1	.97	FSO_4	vs	NO_3 T1	.12
DNO_3	vs	NO_3 T2	.67	FSO_4	vs	NO_3 T2	.15
DNO_3	vs	SO_4 T1	.33	FSO_4	vs	SO_4 T1	.25
DNO_3	vs	SO_4 T2	.23	FSO_4	vs	SO_4 T2	.33
DSO_4	vs	NO_3 NY	.13	NH_3 T3	vs	DNO_2	.46
DSO_4	vs	NO_3 TF	.30	NH_3 T3	vs	DNO_3	.54
DSO_4	vs	SO_4 NY	-.06	NH_3 T3	vs	SO_4	.30
DSO_4	vs	SO_4 TF	.26	NH_3 T3	vs	FNO_3	.17
DSO_4	vs	NO_2 T1	.52	NH_3 T3	vs	FSO_4	.26
DSO_4	vs	NO_2 T2	.33	DNO_2	vs	DNO_3	.33
DSO_4	vs	NO_3 T1	.36	DNO_2	vs	DSO_4	.48
DSO_4	vs	NO_3 T2	.10	DNO_2	vs	FNO_3	.28
DSO_4	vs	SO_4 T1	1.0	DNO_2	vs	FSO_4	.21
DSO_4	vs	SO_4 T2	.80	DNO_3	vs	DSO_4	.33
FNO_3	vs	NO_3 NY	.38	DNO_3	vs	FNO_3	.25
FNO_3	vs	NO_3 TF	.96	DNO_3	vs	FSO_4	.13
FNO_3	vs	SO_4 NY	.07	DSO_4	vs	FNO_3	.34
FNO_3	vs	SO_4 TF	.81	DSO_4	vs	FSO_4	.26
FNO_3	vs	NO_2 T1	.34	FNO_3	vs	FSO_4	.83

* NH_3 was quantified as NH_4^+.

Table 4.60 Pearson's correlation coefficients (r) between variables measured through the annular denuder system (KAPS). Sampling location: Crossfield East. Sampling period: 1985-87. Sampling duration: 24 hours. For convenience, valence signs have been omitted.*

Variables			(r)	Variables			(r)
NO_3 NY	vs	NO_3 TF	.02	NO_2 T1	vs	NO_2 T2	.80
NO_3 NY	vs	SO_4 NY	.55	NO_2 T1	vs	NO_3 T1	.07
NO_3 NY	vs	SO_4 TF	.12	NO_2 T1	vs	NO_3 T2	.19
NO_3 NY	vs	NO_2 T1	.03	NO_2 T1	vs	SO_4 T1	.42
NO_3 NY	vs	NO_2 T2	.08	NO_2 T1	vs	SO_4 T2	.41

Continued...

Table 4.60 (Continued).

Variables			(r)	Variables			(r)
NO_3 NY	vs	NO_3 T1	.21	NO_2 T1	vs	NO_3 T1	.03
NO_3 NY	vs	NO_3 T2	.10	NO_2 T2	vs	NO_3 T2	.29
NO_3 NY	vs	SO_4 T1	.07	NO_2 T2	vs	SO_4 T1	.23
NO_3 NY	vs	SO_4 T2	-.005	NO_2 T2	vs	SO_4 T2	.40
NO_3 TF	vs	SO_4 NY	-.07	NO_3 T1	vs	NO_3 T2	.32
NO_3 TF	vs	SO_4 TF	.89	NO_3 T1	vs	SO_4 T1	.15
NO_3 TF	vs	NO_2 T1	.39	NO_3 T1	vs	SO_4 T	.11
NO_3 TF	vs	NO_2 T2	.35	NO_3 T2	vs	SO_4 T1	.09
NO_3 TF	vs	NO_3 T1	.05	NO_3 T2	vs	SO_4 T2	.26
NO_3 TF	vs	NO_3 T2	.23	SO_4 T1	vs	SO_4 T2	.78
NO_3 TF	vs	SO_4 T1	.35	NH_3 T3	vs	NO_3 NY	.31
NO_3 TF	vs	SO_4 T2	.36	NH_3 T3	vs	NO_3 TF	.36
SO_4 NY	vs	SO_4 TF	-.07	NH_3 T3	vs	SO_4 NY	-.006
SO_4 NY	vs	NO_2 T1	-.03	NH_3 T3	vs	SO_4 TF	.28
SO_4 NY	vs	NO_2 T2	-.04	NH_3 T3	vs	NO_2T1	.57
SO_4 NY	vs	NO_3 T1	.16	NH_3 T3	vs	NO_2 T2	.50
SO_4 NY	vs	NO_3 T2	.06	NH_3 T3	vs	NO_3 T1	.15
SO_4 NY	vs	SO_4 T1	.01	NH_3 T3	vs	NO_3 T2	.24
SO_4 NY	vs	SO_4 T2	-.06	NH_3 T3	vs	SO_4 T1	.13
SO_4 TF	vs	NO_2 T1	.31	NH_3 T3	vs	SO_4 T2	.15
SO_4 TF	vs	NO_2 T2	.28	DNO_2	vs	NO_3 NY	-.002
SO_4 TF	vs	NO_3 T1	.10	DNO_2	vs	NO_3 TF	.37
SO_4 TF	vs	NO_3 T2	.28	DNO_2	vs	SO_4 NY	-.009
SO_4 TF	vs	SO_4 T1	.40	DNO_2	vs	SO_4 TF	.28
SO_4 TF	vs	SO_4 T2	.38	DNO_2	vs	NO_2 T1	.98
DNO_2	vs	NO_2 T2	.68	FNO_3	vs	NO_2 T2	.36
DNO_2	vs	NO_3 T1	.08	FNO_3	vs	NO_3 T1	.08
DNO_2	vs	NO_3 T2	.15	FNO_3	vs	NO_3 T2	.27
DNO_2	vs	SO_4 T1	.45	FNO_3	vs	SO_4 T1	.35
DNO_2	vs	SO_4 T2	.39	FNO_3	vs	SO_4 T2	.35
DNO_3	vs	NO_3 NY	.23	FSO_4	vs	NO_3 NY	.24
DNO_3	vs	NO_3 TF	.11	FSO_4	vs	NO_3 TF	.88
DNO_3	vs	SO_4 NY	.17	FSO_4	vs	SO_4 NY	.09
DNO_3	vs	SO_4 TF	.17	FSO_4	vs	SO_4 TF	.99
DNO_3	vs	NO_2 T1	.13	FSO_4	vs	NO_2 T1	.32
DNO_3	vs	NO_2 T2	.11	FSO_4	vs	NO_2 T2	.30
DNO_3	vs	NO_3 T1	.97	FSO_4	vs	NO_3 T1	.14
DNO_3	vs	NO_3 T2	.53	FSO_4	vs	NO_3 T2	.32

Continued...

Table 4.60 (Concluded).

Variables			(r)	Variables			(r)
DNO_3	vs	SO_4 T1	.18	FSO_4	vs	SO_4 T1	.41
DNO_3	vs	SO_4 T2	.18	FSO_4	vs	SO_4 T2	.40
DSO_4	vs	NO_3 NY	.08	NH_3 T	vs	DNO_2	.54
DSO_4	vs	NO_3 TF	.37	NH_3 T3	vs	DNO_3	.21
DSO_4	vs	SO_4 NY	.02	NH_3 T3	vs	SO_4	.17
DSO_4	vs	SO_4 TF	.42	NH_3 T3	vs	FNO_3	.40
DSO_4	vs	NO_2 T1	.45	NH_3 T3	vs	FSO_4	.30
DSO_4	vs	NO_2 T2	.26	DNO_2	vs	DNO_3	.12
DSO_4	vs	NO_3 T1	.16	DNO_2	vs	DSO_4	.47
DSO_4	vs	NO_3 T2	.12	DNO_2	vs	FNO_3	.36
DSO_4	vs	SO_4 T1	1.0	DNO_2	vs	FSO_4	.30
DSO_4	vs	SO_4 T2	.82	DNO_3	vs	DSO_4	.17
FNO_3	vs	NO_3 NY	.21	DNO_3	vs	FNO_3	.15
FNO_3	vs	NO_3 TF	.98	DNO_3	vs	FSO_4	.21
FNO_3	vs	SO_4 NY	.04	DSO_4	vs	FNO_3	.34
FNO_3	vs	SO_4 TF	.89	DSO_4	vs	FSO_4	.37
FNO_3	vs	NO_2 T1	.39	FNO_3	vs	FSO_4	.90

*NH_3 was quantified as NH_4^+.

Table 4.61 Summary of Pearson's correlation coefficients presented in Tables 4.59 and 4.60 for Crossfield West and Crossfield East respectively.

Variable Pairs			r	
			Crossfield West	Crossfield East
SO_2 (T1)	vs	SO_2 (T2)	0.76	0.78
HNO_3 (T1)	vs	HNO_3 (T2)	0.43	0.32
HNO_2 (T1)	vs	HNO_2 (T2)	0.53	0.80
HNO_3 (T1)	vs	NH_3	0.56	0.16
NO_3^- (TF)	vs	NO_3^- (NY)	0.10	0.02
HNO_3	vs	NO_3^- (NY)	0.29	0.23
NO_3^- (TF)	vs	SO_4^{2-} (TF)	0.79	0.89
NO_3^- (Tot)	vs	SO_4^{2-} (Tot)	.83	.90
NO_3^- (NY)	vs	NH_3	0.42	0.31

T1 = Denuder Tube #1
T2 = Denuder Tube #2
TF = Teflon filter
NY = Nylon filter

Tot = Total from Teflon + nylon filters

Table 4.62 provides a comparison of some of the characteristics at the two sites. The results can be summarized as:

a. Mean values of all three chemical species (HNO_2, HNO_3, and SO_2) collected by (T1) and (T2) were higher at Crossfield East compared to West.
b. The mean mass of NO_3^- collected on the nylon filter was nearly the same at both sites. This NO_3^- was most likely NH_4NO_3.
c. While at Crossfield West the mean HNO_3 value in (T1) was higher than the corresponding HNO_2 value, the inverse was true at Crossfield East (Tables 4.57 and 4.58). At Crossfield West the r value between HNO_3 (T1) and NH_3 was 0.57, and at Crossfield East the r value between HNO_2 (T1) and NH_3 was 0.50.

4.5.2 Dichotomous Sampler

Fine and coarse particle samples were collected at Crossfield West and East with a dichotomous sampler as 48-hour integrated samples. These samples were analyzed by x-ray fluorescence spectroscopy (XRF). Summary statistics on the concentrations of various elements in the fine and coarse particle fraction are presented in Tables 4.63 and 4.64.

Crustal elements such as Al, Si, K, Ca, Mn, and Fe were present at higher concentrations in the coarse compared to the fine particle fraction. In general, the coarse particle mean concentration of these compounds were higher at Crossfield East compared to Crossfield West (Tables 4.64 and 4.63). The mean concentrations of crustal elements at both the Crossfield sites were greater than at Fortress Mountain (Table 3.66).

Non-crustal elements such as S, Zn, Se, Br, and Pb concentrations were higher in the fine compared with the coarse fraction at both the Crossfield sites. The mean fine particle S and Se concentrations at both sites were comparable, while Br, Zn, Ni, and Pb concentrations were higher at Crossfield East compared to Crossfield West (Tables 4.64 and 4.63). However, the mean concentrations of all these elements were higher at both the Crossfield sites compared to Fortress Mountain (Table 3.66).

Table 4.65 provides a comparison of selected coarse and fine particle elemental mean concentrations at the two Crossfield sites using Fortress Mountain as the background site.

Table 4.62 Comparison of selected KAPS parameters at Crossfield West and Crossfield East.

Location	Mean Concentration ($\mu g\ m^{-3}$)				
	SO_2 (T1)	SO_2 (T2)	HNO_3 (T1)	HNO_3 (T2)	HNO_2 (T1)
Crossfield West	4.27	0.33	0.37	0.11	0.29
Crossfield East	5.33	0.43	0.40	0.12	0.52

Location	Mean Concentration ($\mu g\ m^{-3}$)			
	HNO_2 (T2)	HNO_3 (T1 + T2)	NO_3^- (NY)	[HNO_3 (T1) - NO_3^- (NY)]
Crossfield West	0.09	0.48	0.26	0.11
Crossfield East	0.16	0.52	0.28	0.12

Table 4.63 Summary statistics of chemical constituents in particulate matter (ng m^{-3}). Sampling location: Crossfield West. Sampling period: 1986-87, 12 months. Sampling duration: 48 hours.

				Range			Percentile							
Variable	(n)	Mean	S.D.	Min.	Max.	Median	10	30	70	90	95	99	Skewness	Kurtosis
Al F*	184	66.6	69.2	BD	322.5	44.6	10.1	23.7	72.1	159.7	237.7	312.3	1.9	3.5
Al C*	184	407.3	408.0	.91	1709.0	279.6	44.5	128.7	512.1	1031.5	1371.7	1651.0	1.4	1.4
Si F	184	233.0	273.5	4.2	1534.9	128.6	29.6	70.0	259.2	587.0	821.5	1496.5	2.3	6.4
Si C	184	1930.9	2032.5	4.1	9881.5	1230.0	195.7	591.4	2311.7	4688.8	6797.0	9117.7	1.7	2.6
P F	184	3.3	4.4	BD	20.2	1.5	BD	BD	3.9	10.7	12.4	19.9	1.6	2.3
P C	184	8.1	10.5	BD	66.3	5.8	BD	1.7	9.0	17.2	29.1	57.3	2.8	9.8
S F	184	357.8	274.3	5.5	1963.6	288.4	100.3	192.9	402.6	758.1	901.9	1366.1	2.1	6.8
S C	184	84.3	58.3	BD	471.1	69.9	29.3	53.9	92.6	149.0	195.9	296.7	2.5	10.8
Cl F	184	1.7	4.2	BD	37.4	BD	BD	BD	1.2	4.1	8.6	28.0	5.2	34.6
Cl C	184	16.0	18.9	BD	148.0	10.2	1.7	5.1	19.9	36.5	56.3	94.0	3.0	14.1
K F	184	32.0	28.6	1.0	147.0	21.3	8.5	14.5	35.1	79.5	92.2	141.3	1.8	3.3
K C	184	105.2	103.2	1.6	487.8	67.3	18.2	37.7	120.9	268.5	361.4	457.4	1.6	2.1
Ca F	184	57.8	58.8	1.6	298.3	38.2	7.5	17.6	69.3	141.6	196.6	291.5	1.8	3.4
Ca C	184	530.7	513.8	2.0	2520.0	382.5	50.1	191.2	624.8	1132.2	1722.8	2453.7	1.7	3.1
Ti F	184	2.7	3.1	BD	16.2	1.6	.40	1.0	2.9	7.1	10.0	15.2	2.2	5.0
Ti C	184	20.3	21.7	BD	103.0	12.4	1.7	6.2	24.3	48.4	76.0	96.8	1.7	2.7
V F	184	.28	.25	BD	1.2	.22	BD	.12	.35	.64	.79	1.2	1.3	1.9
V C	184	1.2	1.2	BD	5.9	.90	.07	.37	1.5	2.8	3.9	5.6	1.6	2.5

Continued...

Table 4.63 (Continued).

Variable	(n)	Mean	S.D.	Range		Median	Percentile						Skewness	Kurtosis
				Min.	Max.		10	30	70	90	95	99		
Cr F	184	.23	.20	BD	1.1	.20	BD	.09	.28	.50	.61	.85	1.2	1.9
Cr C	184	.71	.78	BD	4.8	.49	.11	.28	.77	1.5	2.4	4.1	2.5	7.7
Mn F	184	1.3	1.1	BD	5.9	.97	.26	.61	1.6	2.9	3.8	5.2	1.5	2.2
Mn C	184	6.8	6.9	.13	34.4	4.7	.88	2.2	8.6	16.3	24.2	31.1	1.7	2.9
Fe F	184	28.2	30.8	.65	157.3	16.7	4.6	8.7	32.2	71.0	105.4	149.4	2.0	4.3
Fe C	184	202.6	213.3	.09	986.6	124.1	23.0	59.5	241.9	481.2	707.4	971.4	1.7	2.6
Ni F	184	.16	.13	BD	.76	.15	BD	.09	.22	.34	.44	.59	1.2	2.0
Ni C	184	.48	.41	BD	1.8	.33	.10	.20	.61	1.1	1.4	1.8	1.3	1.1
Cu F	184	.42	1.00	BD	12.3	.24	BD	.05	.39	.76	1.3	5.1	8.8	95.2
Cu C	184	.46	.49	BD	3.5	.37	BD	.17	.58	.95	1.3	3.2	2.9	12.5
Zn F	184	2.7	2.2	.07	12.6	2.0	.65	1.3	3.1	6.0	7.4	12.0	1.8	3.8
Zn C	184	1.9	1.5	BD	9.0	1.4	.47	.87	2.4	3.8	4.6	7.0	1.5	3.2
Ga F	184	.05	.06	BD	.24	.02	BD	BD	.07	.13	.17	.22	1.4	1.3
Ga C	184	.11	.12	BD	.74	.07	BD	.03	.15	.28	.33	.57	1.8	4.9
As F	184	.13	.22	BD	1.0	BD	BD	BD	.14	.39	.61	1.00	2.2	5.0
As C	184	.09	.17	BD	.87	BD	BD	BD	.07	.34	.51	.85	2.4	5.7
Se F	184	.22	.21	BD	1.2	.19	BD	.09	.28	.46	.65	.99	1.6	3.4
Se C	184	.14	.13	BD	.61	.11	BD	.07	.17	.36	.41	.57	1.2	.97
Br F	184	2.0	1.3	.07	9.2	1.7	.60	1.2	2.5	4.0	4.5	6.3	1.4	3.7

Continued...

Table 4.63 (Continued).

Variable	(n)	Mean	S.D.	Range		Median	Percentile						Skewness	Kurtosis
				Min.	Max.		10	30	70	90	95	99		
Br C	184	.82	.66	BD	4.6	.69	.19	.40	1.0	1.7	2.1	3.7	1.9	6.4
Rb F	184	.36	.41	BD	1.8	.24	BD	.07	.40	1.1	1.3	1.7	1.5	1.5
Rb C	184	.73	.69	BD	3.2	.49	BD	.28	.99	1.8	2.1	3.0	1.2	.97
Sr F	184	.52	.59	BD	2.3	.32	BD	.19	.57	1.5	1.9	2.3	1.4	1.1
Sr C	184	1.6	1.4	BD	7.1	1.2	.19	.74	2.2	3.5	4.2	6.9	1.4	2.4
Y F	184	.27	.42	BD	1.9	.08	BD	BD	.28	.81	1.4	1.9	2.1	4.3
Y C	184	.34	.45	BD	2.1	.16	BD	BD	.45	.85	1.3	2.1	1.9	3.8
Zr F	184	.16	.44	BD	2.3	BD	BD	BD	BD	.78	1.3	2.1	3.0	8.1
Zr C	184	.27	.66	BD	4.8	BD	BD	BD	BD	1.3	2.0	3.0	3.3	14.0
Mo F	184	.22	.42	BD	2.2	BD	BD	BD	.01	.92	1.1	1.9	2.1	4.5
Mo C	184	.27	.56	BD	4.6	BD	BD	BD	.09	1.1	1.3	2.5	3.5	20.1
Pd F	184	.41	.45	BD	1.8	.24	BD	BD	.61	1.1	1.3	1.7	.96	.10
Pd C	184	.52	.52	BD	2.3	.39	BD	.05	.78	1.2	1.5	2.2	.86	.17
Ag F	184	.41	.45	BD	1.8	.24	BD	BD	.61	1.1	1.3	1.7	.96	.10
Ag C	184	.80	.68	BD	3.3	.78	BD	.31	1.2	1.7	2.1	2.8	.67	.23
Cd F	184	.47	.62	BD	2.6	.08	BD	BD	.61	1.4	1.8	2.6	1.5	1.5
Cd C	184	.71	.80	BD	2.8	.35	BD	BD	1.3	1.8	2.1	2.8	.79	- .53
In F	184	1.0	1.1	BD	4.5	.80	BD	BD	1.5	2.5	3.1	4.4	1.1	1.0
In C	184	1.3	1.2	BD	5.6	1.1	BD	.35	1.9	3.0	3.5	5.1	.91	.62

Continued...

Table 4.63 (Concluded).

Variable	(n)	Mean	S.D.	Range		Median	Percentile						Skewness	Kurtosis
				Min.	Max.		10	30	70	90	95	99		
Sn F	184	1.1	1.3	BD	4.7	.61	BD	BD	1.6	3.2	3.9	4.4	1.0	- .15
Sn C	184	1.0	1.2	BD	5.4	.67	BD	BD	1.6	2.7	3.3	5.3	1.1	.82
Sb F	184	2.6	2.8	BD	12.9	2.0	BD	BD	4.2	6.6	7.8	10.4	.85	.07
Sb C	184	2.7	3.0	BD	14.2	1.9	BD	BD	4.0	7.2	8.7	13.8	1.2	1.0
Ba F	184	3.1	4.3	BD	22.7	BD	BD	BD	4.5	9.7	11.4	16.4	1.4	1.8
Ba C	184	6.1	7.1	BD	31.0	3.3	BD	BD	9.7	15.6	18.6	28.7	1.1	.59
La F	184	5.1	8.0	BD	31.5	BD	BD	BD	6.1	17.9	23.0	29.4	1.4	.69
La C	184	5.5	8.9	BD	41.2	BD	BD	BD	5.8	20.3	25.5	40.3	1.7	2.5
Hg F	184	.08	.08	BD	.30	.04	BD	BD	.11	.20	.22	.29	.87	- .22
Hg C	184	.12	.11	BD	.54	.11	BD	.04	.17	.26	.37	.51	1.1	1.1
Pb F	184	7.6	5.9	.61	35.6	6.3	1.5	3.6	9.1	15.4	19.3	31.0	1.5	3.1
Pb C	184	3.0	2.5	BD	15.3	2.3	.66	1.4	3.8	6.2	8.0	14.3	1.7	4.5

* F = fine particles (<2.5 μm); C = coarse particles (>2.5 μm). Chemical analysis was performed by x-ray fluorescence spectroscopy.
BD = concentration below detection limit.

Table 4.64 Summary statistics of chemical constituents in particulate matter (ng m^{-3}). Sampling location: Crossfield East. Sampling period: 1986-87, 12 months. Sampling duration: 48 hours.

				Range			Percentile							
Variable	(n)	Mean	S.D.	Min.	Max.	Median	10	30	70	90	95	99	Skewness	Kurtosis
Al F*	142	78.2	69.4	BD	382.2	65.4	12.5	30.7	90.9	168.4	212.1	370.1	1.9	4.7
Al C*	142	520.3	435.7	BD	1899.0	397.7	44.5	221.6	699.1	1159.3	1381.5	1850.3	.92	.24
Si F	142	283.0	291.2	.48	1730.0	227.1	38.6	99.4	318.9	636.3	796.7	1670.9	2.3	7.2
Si C	142	2547.4	2221.1	BD	13173.7	1956.7	192.4	1084.3	3487.2	5772.5	6689.9	11387.5	1.4	2.9
P F	142	3.6	4.7	BD	21.9	1.8	BD	BD	3.8	10.9	15.8	20.4	1.7	2.6
P C	142	9.0	10.9	BD	63.0	6.3	BD	2.4	10.6	21.9	28.0	61.9	2.6	9.3
S F	142	368.9	252.0	BD	1827.9	329.4	133.2	226.8	416.2	666.6	763.5	1553.0	2.3	8.7
S C	142	107.8	141.8	BD	1634.8	83.9	40.1	65.6	107.1	183.8	229.8	1104.1	9.1	96.6
Cl F	142	1.6	4.3	BD	40.5	BD	BD	BD	.61	4.8	7.6	32.5	6.2	48.8
Cl C	142	21.3	30.0	BD	189.7	11.7	1.8	6.2	22.8	48.2	83.7	188.1	3.5	14.7
K F	142	34.5	31.3	.83	189.1	25.8	8.9	17.2	37.7	76.0	99.8	169.1	2.2	6.0
K C	142	133.5	112.3	.20	633.4	98.1	23.1	58.2	169.0	282.0	365.2	583.9	1.5	2.8
Ca F	142	81.9	109.5	.74	897.7	51.8	6.9	23.9	86.7	190.4	261.6	786.6	4.2	25.5
Ca C	142	682.3	613.6	.22	2687.0	545.1	48.7	257.4	916.4	1458.0	2056.8	2628.3	1.3	1.3
Ti F	142	3.2	3.4	BD	25.3	2.4	.41	1.2	3.8	7.0	9.7	21.6	2.9	13.5
Ti C	142	26.3	23.9	BD	140.5	19.2	2.1	10.8	35.6	58.9	70.7	126.3	1.5	3.4
V F	142	.33	.28	BD	1.7	.27	.04	.15	.39	.74	.89	1.5	1.5	3.3
V C	142	1.5	1.4	BD	6.2	1.1	.02	.59	1.8	3.3	4.9	6.1	1.4	1.8

Continued...

Table 4.64 (Continued).

Variable	(n)	Mean	S.D.	Range		Median	Percentile						Skewness	Kurtosis
				Min.	Max.		10	30	70	90	95	99		
Cr F	142	.32	.49	BD	5.4	.24	.04	.15	.37	.60	.72	3.6	8.4	87.3
Cr C	142	.86	.84	BD	4.0	.59	.15	.39	.85	2.0	2.7	4.0	2.0	4.4
Mn F	142	2.0	1.9	BD	14.2	1.3	.46	.89	2.2	4.5	5.7	11.1	2.7	12.9
Mn C	142	8.8	7.4	.15	33.4	6.8	.87	3.7	11.3	17.9	28.0	33.1	1.3	1.8
Fe F	142	38.9	39.6	BD	255.7	29.2	7.2	15.0	42.8	84.7	104.9	230.6	2.5	8.5
Fe C	142	266.4	231.6	BD	1170.2	196.8	24.9	107.2	350.1	583.0	720.0	1111.7	1.3	1.7
Ni F	142	.76	6.7	BD	80.1	.17	.07	.13	.24	.36	.43	46.1	11.9	141.9
Ni C	142	.57	.43	BD	3.0	.43	.15	.32	.74	1.1	1.4	2.5	1.8	6.1
Cu F	142	.53	.76	BD	7.3	.37	BD	.22	.57	1.2	1.6	5.4	5.6	45.2
Cu C	142	.86	1.5	BD	14.5	.54	.04	.35	.83	1.3	2.1	11.9	6.3	48.5
Zn F	142	3.2	4.2	BD	45.3	2.3	.86	1.5	3.6	5.7	9.2	31.1	7.3	69.5
Zn C	142	2.9	1.9	BD	9.3	2.4	.87	1.4	3.7	5.9	6.4	9.3	.95	.55
Ga F	142	.06	.07	BD	.30	.04	BD	BD	.09	.17	.20	.29	1.2	.85
Ga C	142	.14	.11	BD	.61	.11	BD	.07	.20	.30	.35	.53	.97	1.2
As F	142	.16	.30	BD	2.3	BD	BD	BD	.20	.50	.67	1.9	3.7	21.2
As C	142	.15	.23	BD	1.2	BD	BD	BD	.20	.47	.65	1.1	1.9	3.8
Se F	142	.23	.20	BD	1.0	.20	BD	.09	.28	.49	.71	.96	1.3	1.9
Se C	142	.17	.13	BD	.72	.15	BD	.09	.22	.35	.44	.66	1.1	1.6
Br F	142	4.8	4.5	.33	21.8	3.3	1.1	2.2	4.9	11.6	13.9	21.6	2.0	4.2

Continued...

Table 4.64 (Continued).

Variable	(n)	Mean	S.D.	Range		Median	Percentile						Skewness	Kurtosis
				Min.	Max.		10	30	70	90	95	99		
Br C	142	1.7	1.5	.07	8.2	1.3	.44	.87	2.0	3.6	5.0	7.9	1.9	4.5
Rb F	142	.34	.39	BD	1.9	.21	BD	.06	.44	.97	1.2	1.9	1.6	2.5
Rb C	142	.92	.81	BD	3.7	.69	BD	.39	1.2	2.2	2.5	3.6	1.2	1.1
Sr F	142	.57	.59	BD	3.6	.42	BD	.22	.72	1.3	1.5	3.2	2.0	6.0
Sr C	142	2.1	1.7	BD	8.0	1.7	.30	.91	2.7	4.5	5.5	7.8	1.2	1.4
Y F	142	.24	.32	BD	1.9	.11	BD	BD	.28	.65	.87	1.7	2.0	5.7
Y C	142	.39	.45	BD	1.8	.20	BD	BD	.61	1.00	1.3	1.7	1.1	.32
Zr F	142	.24	.49	BD	2.5	BD	BD	BD	BD	1.00	1.3	2.3	2.3	5.2
Zr C	142	.34	.65	BD	3.9	BD	BD	BD	.02	1.4	1.8	3.2	2.3	6.5
Mo F	142	.23	.40	BD	1.9	BD	BD	BD	.25	.80	1.00	1.8	2.0	3.9
Mo C	142	.24	.47	BD	2.6	BD	BD	BD	BD	.90	1.1	2.5	2.4	6.6
Pd F	142	.43	.44	BD	1.7	.34	BD	BD	.65	1.1	1.2	1.6	.72	- .45
Pd C	142	.52	.56	BD	3.6	.42	BD	BD	.81	1.2	1.5	2.8	1.5	4.7
Ag F	142	.68	.61	BD	2.6	.57	BD	.19	1.00	1.5	1.7	2.5	.66	- .26
Ag C	142	.81	.74	BD	3.5	.74	BD	.24	1.2	1.9	2.1	3.4	.94	.91
Cd F	142	.53	.60	BD	2.5	.32	BD	BD	.80	1.5	1.6	2.2	.90	- .18
Cd C	142	.68	.75	BD	4.2	.55	BD	BD	1.1	1.7	1.9	3.9	1.3	2.9
In F	142	1.1	1.1	BD	4.9	.90	BD	BD	1.5	2.7	3.1	4.6	1.00	.74
In C	142	1.3	1.4	BD	6.5	.88	BD	BD	1.8	3.7	4.3	5.9	1.2	.77

Continued...

Table 4.64 (Concluded).

Variable	(n)	Mean	S.D.	Range		Median	Percentile						Skewness	Kurtosis
				Min.	Max.		10	30	70	90	95	99		
Sn F	142	.85	1.0	BD	4.5	.48	BD	BD	1.3	2.3	3.0	4.1	1.2	.78
Sn C	142	1.2	1.4	BD	6.6	.82	BD	BD	2.0	3.2	4.0	6.0	1.1	.82
Sb F	142	2.7	2.7	BD	13.2	2.3	BD	BD	4.1	6.7	7.5	11.5	.85	.31
Sb C	142	2.8	3.2	BD	13.2	1.5	BD	BD	4.4	7.3	9.5	13.1	1.1	.60
Ba F	142	3.1	4.7	BD	17.4	BD	BD	BD	4.0	11.6	13.7	16.5	1.3	.45
Ba C	142	5.4	7.1	BD	32.8	.90	BD	BD	8.4	14.8	20.7	31.1	1.4	1.8
La F	142	5.6	8.6	BD	33.1	BD	BD	BD	8.1	19.3	25.4	31.5	1.3	.62
La C	142	4.1	7.0	BD	27.8	BD	BD	BD	4.2	16.5	19.4	26.1	1.6	1.4
Hg F	142	.06	.08	BD	.37	.02	BD	BD	.09	.17	.24	.37	1.6	2.3
Hg C	142	.13	.11	BD	.46	.11	BD	.07	.17	.28	.33	.46	.72	.03
Pb F	142	16.9	14.3	.85	67.7	12.0	3.4	8.1	20.0	38.4	50.8	65.8	1.5	2.1
Pb C	142	6.2	4.9	BD	24.7	4.7	1.4	3.2	7.3	12.5	16.6	24.4	1.5	2.4

* F = fine particles (<2.5 μm); C = coarse particles (>2.5 μm). Chemical analysis was performed by x-ray fluorescence spectroscopy.
BD = concentration below detection limit.

Table 4.65 Ratio of selected coarse and fine particle elemental mean concentrations at the two Crossfield sites and at Fortress Mountain.

Element*	Ratio Crossfield West: Fortress Mountain	Ratio Crossfield East: Fortress Mountain
Al (C)	2.2	2.8
Si (C)	2.8	3.7
Ca (C)	3.7	4.7
Mn (C)	3.0	3.8
Fe (C)	2.6	3.4
S (F)	1.4	1.5
Zn (F)	1.7	2.0
Se (F)	2.7	2.9
Br (F)	1.8	4.4
Pb (F)	4.0	8.9

* C: coarse (>2.5 μm) and F: fine (<2.5 μm) particles.

Tables 4.66 through 4.71 provide summary statistics on the correlations between the chemical constituents in the coarse particles, in the fine particles, and in both the fractions. An attempt was made to clarify the results presented in these tables by the application of principal component analysis (PCA) (Tables 4.72 and 4.73). The crustal elements Al, Si, K, Ca, Ti, Mn, Fe, and Sr in the coarse particles loaded high in Factor 1, both at Crossfield West and East. This factor accounted for roughly 46.6% and 34.5% of the total variability of the chemical composition of both the coarse and fine particles combined at Crossfield West and Crossfield East, respectively. The crustal elements in the fine particle fraction (present at low concentrations compared to the coarse particles) also loaded high in Factor 1 at both sites. These results were similar to what was observed at Fortress Mountain (Table 3.73). Known tracers of anthropogenic pollution such as fine particle Br and Pb, did not load in Factor 1 at either Crossfield West or at Crossfield East (Tables 4.72 and 4.73). This was in contrast to Fortress Mountain where As, Br, and Pb loaded in Factor 1 (Table 3.73). However, Br and Pb loaded high in Factor 2 at Crossfield West and in Factor 3 at Crossfield East. These two factors accounted for 13.9% and 9.9% of the total variability of the chemical composition of coarse and fine particles combined at Crossfield West and East, respectively. As previously stated in

Chapter 3, fine particle Br and Pb at inland sites are considered to be tracers for the influence of transportation activities or line sources.

Table 4.66 Pearson's correlation coefficient (r) between selected variables in coarse particles (>2.5 μm). Dichotomous filter samples analyzed through x-ray fluorescence spectroscopy. Sampling location: Crossfield West. Sampling period: 1986-87. Time of data resolution: 48 h.

Variable	(r)	Variable	(r)
Al vs Si	.99	Zn vs Ca	.72
Al vs K	.98	Zn vs Ti	.77
Al vs Ca	.88	Ga vs Al	.70
Al vs Ti	.94	Ga vs Si	.71
Al vs V	.84	Ga vs Ti	.71
Si vs K	.98	Rb vs Al	.76
Si vs Ca	.83	Rb vs Si	.73
Si vs Ti	.96	Rb vs K	.77
Si vs V	.83	Rb vs Ca	.72
K vs Ca	.86	Rb vs Ti	.71
K vs Ti	.95	Rb vs V	.71
K vs V	.84	Sr vs Al	.87
Ca vs Ti	.80	Sr vs Si	.83
Ca vs V	.78	Sr vs K	.87
Ti vs V	.86	Sr vs Ca	.87
Mn vs Al	.97	Sr vs T	.80
Mn vs Si	.95	Sr vs V	.78
Mn vs K	.97	Sr vs Cr	.73
Mn vs Ca	.92	Mn vs Fe	.97
Mn vs Ti	.92	Mn vs Ni	.78
Mn vs V	.86	Mn vs Zn	.80
Mn vs Cr	.75	Mn vs Rb	.80
Fe vs Al	.98	Fe vs Ni	.81
Fe vs Si	.98	Fe vs Zn	.82
Fe vs K	.98	Fe vs Ga	.70
Fe vs Ca	.90	Fe vs Rb	.76
Fe vs Ti	.95	Ni vs Zn	.79
Fe vs V	.85	Se vs Rb	.71
Ni vs Al	.78	Sr vs Mn	.91
Ni vs Si	.78	Sr vs Fe	.87
Ni vs K	.80	Sr vs Zn	.74
Ni vs Ti	.76	Sr vs Br	.71
Zn vs Al	.79	Sr vs Rb	.93
Zn vs Si	.78	Pb vs Br	.95
Zn vs K	.82		

Table 4.67 Pearson's correlation coefficient (r) between selected variables in fine particles (<2.5 μm). Dichotomous filter samples analyzed through x-ray fluorescence spectroscopy. Sampling location: Crossfield West. Sampling period: 1986-87. Time of data resolution: 48 h.

Variable	(r)	Variable	(r)
Al vs Si	.97	Mn vs Ca	.88
Al vs K	.77	Mn vs Ti	.83
Al vs Ca	.87	Mn vs V	.78
Al vs Ti	.94	Fe vs Al	.97
Al vs V	.76	Fe vs Si	.97
Si vs K	.75	Fe vs Ca	.91
Si vs Ca	.85	Fe vs Ti	.95
Si vs Ti	.96	Fe vs V	.75
Si vs V	.78	Cu vs V	.15
K vs Ca	.73	Mn vs Fe	.89
K vs Ti	.73	Mn vs Br	.74
Ca vs Ti	.81	Sr vs Rb	.91
Ti vs V	.80	Y vs Rb	.81
Mn vs Al	.85	Pb vs Br	.89
Mn vs Si	.86	Sr vs Y	.83
Mn vs K	.72		

Table 4.68 Pearson's correlation coefficient (r) between selected variables in fine and coarse particles. Dichotomous filter samples analyzed through x-ray fluorescence spectroscopy. Sampling location: Crossfield West. Sampling period: 1986-87. Time of data resolution: 48 h.

Variable	(r)	Variable	(r)
Al C vs Al F	.84	Rb C vs Ca F	.72
Al C vs Si F	.88	Sr C vs Al F	.73
Al C vs K F	.74	Si C vs Si F	.74
Al C vs Ca F	.86	Sr C vs K F	.72
Al C vs Ti F	.84	Sr C vs Ca F	.83
Si C vs Al F	.84	Sr C vs Ti F	.71
Si C vs Si F	.90	Al C vs Mn F	.84
Si C vs K F	.71	Al C vs Fe F	.88
Si C vs Ca F	.82	Si C vs Mn F	.83
Si C vs Ti F	.86	Si C vs Fe F	.88
S C vs S F	.85	K C vs Mn F	.83
K C vs Al F	.82	K C vs Fe F	.86
K C vs Si F	.87	Ca C vs Mn F	.79
K C vs K F	.73	Ca C vs Fe F	.75
K C vs Ca F	.83	Ti C vs Mn F	.81

Continued...

Table 4.68 (Concluded).

Variable	(r)	Variable	(r)
K C vs Ti F	.83	Ti C vs Fe F	.85
Ca C vs Ca F	.92	V C vs Mn F	.73
Ti C vs Al F	.80	V C vs Fe F	.73
Ti C vs Si F	.86	Mn C vs Mn F	.85
Ti C vs K F	.71	Mn C vs Fe F	.86
Ti C vs Ca F	.79	Fe C vs Mn F	.86
Ti C vs Ti F	.82	Fe C vs Fe F	.89
V C vs Si F	.71	Ni C vs Mn F	.72
V C vs Ca F	.75	Ni C vs Fe F	.70
Mn C vs Al F	.80	Zn C vs Mn F	.77
Mn C vs Si F	.84	Zn C vs Fe F	.70
Mn C vs K F	.74	Br C vs Br F	.83
Mn C vs Ca F	.88	Rb C vs Mn F	.76
Mn C vs Ti F	.79	Sr C vs Mn F	.83
Fe C vs Al F	.83	Sr C vs Fe F	.78
Fe C vs Si F	.87	Pb C vs Br F	.79
Fe C vs K F	.73	Br C vs Pb F	.82
Fe C vs Ca F	.87	Pb C vs Pb F	.89
Fe C vs Ti F	.83		

Table 4.69 Pearson's correlation coefficient (r) between selected variables in coarse particles (>2.5 μm). Dichotomous filter samples analyzed through x-ray fluorescence spectroscopy. Sampling location: Crossfield East. Sampling period: 1986-87. Time of data resolution: 48 h.

Variable	(r)	Variable	(r)
Al vs Si	.98	Ni vs Ti	.75
Al vs K	.97	Ga vs Al	.71
Al vs Ca	.90	Ga vs Si	.73
Al vs Ti	.96	Ga vs K	.72
Al vs V	.82	Ga vs Ti	.72
Si vs K	.98	Br vs P	.76
Si vs Ca	.84	Rb vs Al	.84
Si vs Ti	.98	Rb vs Si	.81
Si vs V	.82	Rb vs K	.85
P vs Cr	.79	Rb vs Ca	.82
K vs Ca	.86	Rb vs Ti	.80
K vs Ti	.98	Rb vs V	.72
K vs V	.84	Sr vs Al	.90
Ca vs Ti	.83	Sr vs Si	.85
Ca vs V	.77	Sr vs K	.89

Continued...

Table 4.69 (Concluded).

Variable	(r)	Variable	(r)
Ti vs V	.87	Sr vs Ca	.91
Mn vs Al	.96	Sr vs Ti	.84
Mn vs Si	.94	Sr vs V	.76
Mn vs K	.96	Pb vs P	.70
Mn vs Ca	.92	Mn vs Fe	.97
Mn vs Ti	.94	Mn vs Zn	.70
Mn vs V	.83	Mn vs Ga	.70
Fe vs Al	.99	Fe vs Ni	.77
Fe vs Si	.97	Fe vs Zn	.70
Fe vs K	.98	Fe vs Ga	.71
Fe vs Ca	.91	Fe vs Rb	.84
Fe vs Ti	.97	Ni vs Zn	.74
Fe vs V	.83	Se vs Rb	.70
Ni vs Al	.76	Sr vs Mn	.93
Ni vs Si	.75	Sr vs Fe	.90
Ni vs K	.76	Sr vs Rb	.95

Table 4.70 Pearson's correlation coefficient (r) between selected variables in fine particles (<2.5 μm). Dichotomous filter samples analyzed through x-ray fluorescence spectroscopy. Sampling location: Crossfield East. Sampling period: 1986-87. Time of data resolution: 48 h.

Variable	(r)	Variable	(r)
Al vs Si	.97	Mn vs Ti	.86
Al vs K	.82	Mn vs Cr	.76
Al vs Ca	.84	Fe vs Al	.92
Al vs Ti	.91	Fe vs Si	.97
Si vs K	.84	Fe vs K	.83
Si vs Ca	.92	Fe vs Ca	.92
Si vs Ti	.96	Fe vs Ti	.95
Cl vs Ca	.73	Ni vs Cr	.89
K vs Ca	.75	Cu vs Cr	.70
K vs Ti	.85	Zn vs Cr	.82
Ca vs Ti	.89	Mn vs Fe	.89
Ca vs Cr	.80	Mn vs Zn	.74
Ti vs V	.71	Ni vs Cu	.76
Ti vs Cr	.70	Ni vs Zn	.84
Mn vs Al	.79	Cu vs Zn	.74
Mn vs Si	.87	Sr vs Rb	.84
Mn vs K	.78	Pb vs Br	.96
Mn vs Ca	.91		

Table 4.71 Pearson's correlation coefficient (r) between selected variables in fine and coarse particles. Dichotomous filter samples analyzed through x-ray fluorescence spectroscopy. Sampling location: Crossfield East. Sampling period: 1986-87. Time of data resolution: 48 h.

Variable	(r)	Variable	(r)
Al C vs Al F	.64	Fe C vs Si F	.63
Al C vs Si F	.63	Fe C vs K F	.63
Al C vs K F	.63	Sr C vs K F	.60
K C vs Al F	.60	Al C vs Br F	.96
K C vs Si F	.60	P C vs Br F	.62
K C vs K F	.61	Fe C vs Fe F	.60
Ti C vs Al F	.60	Br C vs Br F	.92
Ti C vs K F	.60	Pb C vs Br F	.89
Mn C vs Al F	.62	Br C vs Pb F	.92
Mn C vs Si F	.60	Pb C vs Pb F	.92
Mn C vs K F	.61		

Table 4.72 Varimax rotated factor matrix for XRF aerosol data for both fine* and coarse** particles (factors >0.5). *Fine = 1; **Coarse = 2. Sampling location: Crossfield West. Sampling period: 1986-87. Sampling duration: 48 h.

	Factor							
Variable	1	2	3	4	5	6	7	8
Al 1	.91	--	--	--	--	--	--	--
Si 1	.95	--	--	--	--	--	--	--
S 1	--	--	--	--	--	.93	--	--
K 1	.78	--	--	--	--	--	--	--
Ca 1	.89	--	--	--	--	--	--	--
Ti 1	.92	--	--	--	--	--	--	--
V 1	.79	--	--	--	--	--	--	--
Cr 1	.60	--	--	--	--	--	--	--
Mn 1	.87	--	--	--	--	--	--	--
Fe 1	.95	---	---	--	--	--	--	--
Ni 1	.60	--	--	--	--	--	--	--
Cu 1	--	--	--	--	--	--	--	.74
Br 1	--	.71	--	--	--	--	--	--
Rb 1	--	--	.82	--	--	--	--	--
Sr 1	--	--	.86	--	--	--	--	--
Y 1	--	--	.78	--	--	--	--	--
Cd 1	--	--	--	--	.57	--	--	--
In 1	--	--	--	.51	--	--	--	--
Sn 1	--	--	--	--	.51	--	--	--
Sb 1	--	--	--	.67	--	--	--	--
La 1	--	--	--	--	.58	--	--	--

Continued...

Table 4.72 (Concluded).

Variable	Factor 1	2	3	4	5	6	7	8
Pb 1	--	.90	--	--	--	--	--	--
Al 2	.97	--	--	--	--	--	--	--
Si 2	.97	--	--	--	--	--	--	--
P 2	--	--	--	--	--	--	.57	--
S 2	--	--	--	--	--	.88	--	--
Cl 2	--	.51	--	--	--	--	--	--
K 2	.97	--	--	--	--	--	--	--
Ca 2	.82	--	--	--	--	--	--	--
Ti 2	.95	--	--	--	--	--	--	--
V 2	.84	--	--	--	--	--	--	--
Cr 2	.58	--	--	--	--	--	--	--
Mn 2	.95	--	--	--	--	--	--	--
Fe 2	.97	--	--	--	--	--	--	--
Ni 2	.78	--	--	--	--	--	--	--
Zn 2	.77	--	--	--	--	--	--	--
Ga 2	.71	--	--	--	--	--	--	--
Br 2	--	.72	--	--	--	--	--	--
Rb 2	.79	--	--	--	--	--	--	--
Sr 2	.86	--	--	--	--	--	--	--
Y 2	--	--	.57	--	--	--	--	--
In 2	--	--	--	.52	--	--	--	--
Sb 2	--	--	--	.59	--	--	--	--
Pb 2	--	.83	--	--	--	--	--	--
Pct of Var.	46.6	13.9	7.7	6.4	5.3	3.7	3.4	2.7
Eigenvalue	21.5	6.4	3.5	3.0	2.4	1.7	1.6	1.2

Table 4.73 Varimax rotated factor matrix for XRF aerosol data for both fine* and coarse** particles (factors >0.5). *Fine = 1; **Coarse = 2. Sampling location: Crossfield East. Sampling period: 1986-87. Sampling duration: 48 h.

Variable	Factor 1	2	3	4	5	6	7	8
Al 1	--	.98	--	--	--	--	--	--
Si 1	--	.99	--	--	--	--	--	--
P 1	--	--	.59	--	--	--	--	--
Cl 1	--	.87	--	--	--	--	--	--
K 1	--	.89	--	--	--	--	--	--
V 1	--	.91	--	--	--	--	--	--
Cr 1	--	.76	--	--	--	--	--	--

Continued...

Table 4.73 (Concluded).

Variable	Factor 1	2	3	4	5	6	7	8
Mn 1	--	.94	--	--	--	--	--	--
Fe 1	--	.98	--	--	--	--	--	--
Ni 1	--	--	--	.93	--	--	--	--
Cu 1	--	--	--	.75	--	--	--	--
Zn 1	--	--	--	.90	--	--	--	--
Br 1	--	--	.91	--	--	--	--	--
Rb 1	--	.66	--	--	--	--	--	--
Sr 1	--	.83	--	--	--	--	--	--
Pd 1	--	--	--	--	.53	--	--	--
In 1	--	--	--	--	--	.55	--	--
Sb 1	--	--	--	--	--	.68	--	--
Ba 1	--	--	--	--	.69	--	--	--
La 1	--	--	--	--	.67	--	--	--
Hg 1	--	--	--	--	.51	--	--	--
Pb 1	--	--	.88	--	--	--	--	--
Al 2	.96	--	--	--	--	--	--	--
Si 2	.97	--	--	--	--	--	--	--
P 2	--	--	.63	--	--	--	--	--
Cl 2	--	--	.66	--	--	--	--	--
K 2	.97	--	--	--	--	--	--	--
Ca 2	.82	--	--	--	--	--	--	--
Ti 2	.98	--	--	--	--	--	--	--
V 2	.85	--	--	--	--	--	--	--
Cr 2	--	--	.67	--	--	--	--	--
Mn 2	.94	--	--	--	--	--	--	--
Fe 2	.97	--	--	--	--	--	--	--
Ni 2	.72	--	--	--	--	--	--	.53
Cu 2	--	--	--	--	--	--	--	.92
Zn 2	.60	--	--	--	--	--	--	.56
Ga 2	.71	--	--	--	--	--	--	--
Br 2	--	--	.89	--	--	--	--	--
Rb 2	.85	--	--	--	--	--	--	--
Sr 2	.88	--	--	--	--	--	--	--
In 2	--	--	--	--	--	.53	--	--
Sb 2	--	--	--	--	--	.68	--	--
Ba 2	--	--	--	--	.50	--	--	--
Pb 2	--	--	.87	--	--	--	--	--
Pct of Var.	34.5	19.5	9.9	8.6	6.1	4.7	3.6	2.7
Eigenvalue	16.7	9.5	4.8	4.2	3.0	2.3	1.7	1.3

Interestingly, fine particle Ni loaded high, along with Cu and Zn, in Factor 4 at Crossfield East, but not in any of the 8 factors at Crossfield West, or at Fortress Mountain. Fine to coarse particle ratio of mean Ni concentrations at Crossfield East and West was, respectively, 1.33 and 0.33. Ratio of mean concentrations of Ni at Crossfield East to West was, respectively: fine particles 4.75 and coarse particles 1.0. Similarly, the ratio of fine particle mean Ni concentrations at Crossfield East and West to Fortress Mountain was: Crossfield East 9.5 and Crossfield West 2.0. These results suggest that Crossfield East is under some influence of a source(s) emitting primary particles enriched in Ni, Cu, and Zn. Source categories for these elements include metal smelting, incineration, use of metal alloys containing these three elements, and residual oil combustion. Factor 4 which grouped these three elements, accounted for only 8.6% of the variability of the total aerosol chemical composition.

Surprisingly, fine particle S at Crossfield East did not load in any of the first 8 factors. Similarly at Crossfield West, S was loaded in Factor 6 which accounted for only 3.7% of the variability of the chemical composition of the combined aerosols. These results are not surprising since a predominant portion of the fine particle S at the two Crossfield sites appeared to be in the form of ammonium salts (Table 4.74).

Table 4.74 Summary statistics of correlation coefficients between fine particle S and selected elements.

	r	
Variables	Crossfield West	Crossfield East
S vs Zn	0.41	0.25
S vs Se	0.34	0.24
S vs Br	0.37	0.08
S vs Pb	0.21	0.08
SO_4^{2-} vs NH_4^+	0.94*	0.83*

* values independently obtained in the correlation analysis of KAPS serial denuder-filter pack data.

4.6 THE CHEMISTRY OF RAIN

During approximately April-October in each of the two years of study, rain samples were collected as 0.32 cm sequentially incremental fractions of each rain event using a refrigerated sampler (Coscio et al. 1982). The chemical analyses of these samples were according to the procedures described in Table 3.14.

Data were available on the sub-event rain samples. However, in this report, only summary statistics on averaged daily (24-hour) precipitation chemistry are presented in Tables 4.75 and 4.76 for Crossfield West and Crossfield East. The frequency distributions of almost all the constituents in the rain samples did not exhibit a normal distribution. Under these conditions, computations of mean values are inappropriate. The use of median values would be the correct approach since such values are free from the influence of the type of distribution. Nevertheless, to facilitate comparison with the results in published literature, both mean and median values are presented in Tables 4.75 and 4.76.

The median pH of rain at Crossfield West and East were, respectively, 4.87 and 5.00. In comparison at Fortress Mountain the corresponding value was 4.77. The median H^+ concentrations at the three sites were: Crossfield West 13.5 μeq L^{-1}; Crossfield East 10.2 μeq L^{-1}; and Fortress Mountain 17.0 μeq L^{-1}. In Table 4.77 these results are compared to the median values of rain H^+ concentrations obtained at several other locations in North America. Even though, the median values for SO_4^{2-} and NO_3^- were relatively high at the two Crossfield sites, the H^+ concentrations were comparatively low. These results suggest that much of the SO_4^{2-} and NO_3^- at the two Crossfield sites was not present as their acid species in rain.

Table 4.78 provides comparative summary statistics on the results of the correlation analysis between the measured chemical constituents at the three ADRP sites. In general, SO_4^{2-} and NO_3^- were not well correlated with H^+ at any of the three sites, although the r between H^+ and SO_4^{2-} at Crossfield West was slightly higher than at the other two sites. The SO_4^{2-} and NO_3^- correlated much better with NH_4^+ at Fortress Mountain and Crossfield West. The SO_4^{2-} also correlated reasonably well with Ca^{2+} and Mg^{2+} at all three sites, while the NO_3^- showed a similar property with Ca^{2+}.

The results of the principal component analysis (Table 4.79) show that at Crossfield West, SO_4^{2-}, NO_3^-, and NH_4^+ exhibited

Table 4.75 Summary statistics of daily precipitation chemistry at Crossfield West. Sampling period: approximately April-October, 1986-87. Ion concentrations in $\mu eq\ L^{-1}$.

				Range			Percentile						
Variable	(n)	Mean	S.D.	Min.	Max.	Median	10	30	70	90	95	Skewness	Kurtosis
H^+	85	16.4	16.4	.11	72.2	13.5	.72	2.6	27.3	36.5	47.8	1.4	2.3
Cl^-	55	4.2	3.5	.59	15.3	2.9	1.2	1.9	5.1	8.7	13.7	1.8	3.0
NO_3^-	90	22.7	15.2	3.1	77.7	18.0	7.6	12.1	28.7	46.6	53.2	1.4	1.7
SO_4^{2-}	90	56.6	36.0	10.8	183.3	46.1	21.5	29.0	77.3	102.2	137.2	1.4	2.1
Ca^{2+}	70	35.1	37.9	.15	166.7	26.2	5.0	8.3	40.0	87.2	141.8	2.1	4.1
Mg^{2+}	70	9.2	9.0	1.1	42.8	6.9	1.1	3.3	10.9	23.8	33.4	2.0	3.8
Na^+	70	7.4	13.1	.13	93.0	3.0	.13	.87	8.8	21.1	27.5	4.5	26.6
K^+	50	2.3	3.0	.28	15.9	1.3	.29	.83	2.5	5.0	10.6	3.0	10.0
NH_4^+	75	53.1	35.7	5.9	177.6	42.3	14.4	26.82	80.3	100.9	121.4	1.0	.88

Median pH 4.87.
Mean pH 4.79 and this value should be viewed with caution since it was derived from non-normally distributed data.

Table 4.76 Summary statistics of daily precipitation chemistry at Crossfield East. Sampling period: approximately April-October, 1986-87. Ion concentrations in μeq L^{-1}.

				Range			Percentile						
Variable	(n)	Mean	S.D.	Min.	Max.	Median	10	30	70	90	95	Skewness	Kurtosis
H^+	78	16.2	16.6	.03	66.1	10.2	.14	1.3	25.9	41.8	52.5	1.0	.53
Cl^-	53	4.5	3.5	.16	17.0	3.4	1.1	1.7	7.0	8.2	11.9	1.5	3.0
NO_3^-	85	34.9	64.6	1.1	436.6	22.1	7.0	12.1	34.0	45.3	65.2	5.1	26.5
SO_4^{2-}	85	87.9	95.0	1.9	639.9	69.9	24.8	40.3	97.3	143.7	284.9	3.9	18.0
Ca^{2+}	67	51.7	50.0	.25	235.0	35.7	6.3	10.8	73.8	114.7	172.2	1.5	2.6
Mg^{2+}	68	16.4	17.2	1.1	115.2	10.9	2.4	6.6	20.6	34.9	48.4	3.3	15.7
Na^+	68	6.9	8.3	.13	39.8	3.0	.13	1.00	10.9	21.0	22.6	1.6	2.8
K^+	51	2.7	3.6	.28	18.2	1.4	.28	.83	2.5	8.8	11.2	2.6	7.3
NH_4^+	74	47.3	29.5	.06	135.9	43.0	14.2	22.8	63.7	86.8	110.0	.93	.74

Median pH 5.00.
Mean pH 4.79 and this value should be viewed with caution since it was derived from non-normally distributed data.

Table 4.77 Median ion concentrations in rain: A comparison of data from the three ADRP sites with data from selected locations in Canada and the United States.

Location	Range Median Ion Concentration (mg L^{-1})				
	H^+	SO_4^{2-}	NO_3^-	NH_4^+	Ca^{2+}
Fortress Mountain	0.013 - 0.019	0.67 - 1.18	0.55 - 0.80	0.16 - 0.49	0.60 - 0.65
Crossfield West	0.007 - 0.015	2.05 - 2.29	1.00 - 1.19	0.68 - 0.90	0.73
Crossfield East	0.009 - 0.015	2.83 - 3.58	1.31 - 1.48	0.65 - 0.86	0.83 - 1.46
Bonner Lake, Ontario	0.018 - 0.021	0.73 - 0.76	0.66 - 0.71	0.053 - 0.095	0.08 - 0.09
Chalk River, Ontario	0.052 - 0.063	1.67 - 2.05	2.04 - 2.17	0.194 - 0.302	0.09 - 0.11
Cormack B, Newfoundland	0.014 - 0.015	0.71 - 0.79	0.27 - 0.35	0.045 - 0.058	0.02 - 0.04
Cree Lake, Saskatchewan	0.011 - 0.015	0.43 - 0.49	0.18 - 0.40	0.053 - 0.058	0.03 - 0.04
Island Lake, Manitoba	0.009 - 0.011	0.43 - 0.46	0.35 - 0.53	0.084 - 0.087	0.07 - 0.08
Montmorency, Quebec	0.031 - 0.036	0.99 - 1.31	0.80 - 1.06	0.127 - 0.145	0.02 - 0.05
Port Cartier, Quebec	0.027 - 0.038	0.26 - 0.79	0.58 - 1.73	0.053 - 0.059	0.02 - 0.03
Marcell, Minnesota	0.003 - 0.006	0.95 - 1.01	1.30 - 1.53	0.170 - 0.200	0.17 - 0.22
Hubbard Brook, New Hampshire	0.041 - 0.046	1.73 - 1.77	1.28 - 1.40	0.120 - 0.130	0.04 - 0.08
Whiteface Mountain, New York	0.040 - 0.046	1.47 - 1.72	1.28 - 1.45	0.120 - 0.150	0.04 - 0.07
Caldwell, Ohio	0.076	3.20 - 2.87	1.89 - 2.07	0.230 - 0.340	0.17 - 0.18
Leading Ridge, Pennsylvania	0.056 - 0.076	2.53 - 3.16	1.75 - 2.43	0.230 - 0.350	0.10 - 0.14
Olympic National Park, Washington	0.004	0.24 - 0.32	0.01	0.010	0.03 - 0.05

Values for the Canadian monitoring sites are from the Environment Canada, Canadian Air and Precipitation Monitoring Network (CAPMoN).
Values for the U.S. monitoring sites are from the National Atmospheric Deposition Program (NADP).
In most cases, the values for 1985 and 1986 are used; ADRP values are for 1986 and 1987.

Table 4.78 Precipitation chemistry. Correlation coefficients (r) among the important chemical components of daily rain samples, by sampling site (combined 1986 and 1987 daily samples).

Variable Pairs			Sampling Location		
			Fortress Mountain (r)	Crossfield West (r)	Crossfield East (r)
H^+	vs	SO_4^{2-}	.13	.32	.06
H^+	vs	NO_3^-	.18	.13	-.12
NH_4^+	vs	SO_4^{2-}	.58	.55	.34
NH_4^+	vs	NO_3^-	.68	.57	.22
Ca^{2+}	vs	SO_4^{2-}	.63	.61	.59
Ca^{2+}	vs	NO_3^-	.50	.54	.66
Mg^{2+}	vs	SO_4^{2-}	.69	.55	.52
Mg^{2+}	vs	NO_3^-	.61	.44	.40
Na^+	vs	Cl^-	.37	.001	.36
Na^+	vs	NO_3^-	.27	.26	.42
Ca^{2+}	vs	Mg^{2+}	.85	.84	.70
Ca^{2+}	vs	Na^+	.65	.21	.27
Mg^{2+}	vs	Na^+	.43	.12	.31
SO_4^{2-}	vs	NO_3^-	.65	.77	.84
K^+	vs	SO_4^{2-}	.73	.21	.40
K^+	vs	NO_3^-	.64	.05	.32

Table 4.79 Precipitation chemistry. Varimax rotated factor matrix. The loadings or numbers in a given row, represent correlation coefficients of factors to describe a given rain variable. Data are from daily rain samples (combined 1986 and 1987).

Chemical	Crossfield West		Crossfield East	
Species	Factor 1	Factor 2	Factor 1	Factor 2
H^+	.13	-.58	-.18	-.02
NO_2^-	.03	.17	.27	.10
SO_4^{2-}	.82	.04	.83	.18
NO_3^-	.94	.10	.82	.42
Cl^-	.70	.35	.84	.06
Ca^{2+}	.69	.70	.72	.11
Mg^{2+}	.65	.73	.87	-.13
Na^+	.29	.12	.15	.28
K^+	--	--	--	--

Continued...

Table 4.79 Concluded.

Chemical Species	Crossfield West Factor 1	Crossfield West Factor 2	Crossfield East Factor 1	Crossfield East Factor 2
NH_4^+	.79	.01	.69	.39
Volume	.03	.01	-.19	.18
Duration	-.11	.03	.18	.74
Pct. of Var.	60.4	19.4	70.7	17.2
Eigenvalue	4.3	1.4	4.3	1.1

highest values for loading in Factor 1, which accounted for 60% of the variability of the rain chemistry as reflected by the measured variables. Ca^{2+} and Mg^{2+} also loaded well in Factor 1, but their values were somewhat higher in Factor 2 which accounted for 19% of the total variability. H^+ was poorly correlated with the variables in Factor 1 and was negatively correlated with the variables in Factor 2 (primarily Ca^{2+} and Mg^{2+}). The results of the cluster analysis (Figure 4.12) substantiated these findings.

At Crossfield West, Ca^{2+} and Mg^{2+} showed the best similarity (0.88). The similarity in the behaviour of SO_4^{2-} and NO_3^+ was related to the behaviour of Ca^{2+}-Mg^{2+}-NH_4^+ grouping. The balance of this interaction regulated the role of free H^+. Thus, rain pH at Crossfield West was dependent on the acid-base equilibria. The results of the principal component analysis at Crossfield East (Table 4.79) were similar to those at Crossfield West. Factor 1 loading were stronger in this case and Factor 1 accounted for 71% of the total variability of the rain composition. In addition, H^+ was negatively correlated with all chemical variables in Factor 1. These results were further substantiated by cluster analysis (Figure 4.13). While SO_4^{2-} and NO_3^- showed strong similarity, H^+ showed virtually no positive relationship with any of the measured variables.

4.7 RELATIONSHIPS BETWEEN GASEOUS AIR POLLUTANTS AND THE CHEMISTRY OF FINE AEROSOLS AND RAIN

Correlation analysis was performed between the variables characterized through the KAPS, serial annular denuder-filter pack system and the sum of 10-minute average concentrations, individually of SO_2, NO_2 and NO for the 48 hours prior to the start of

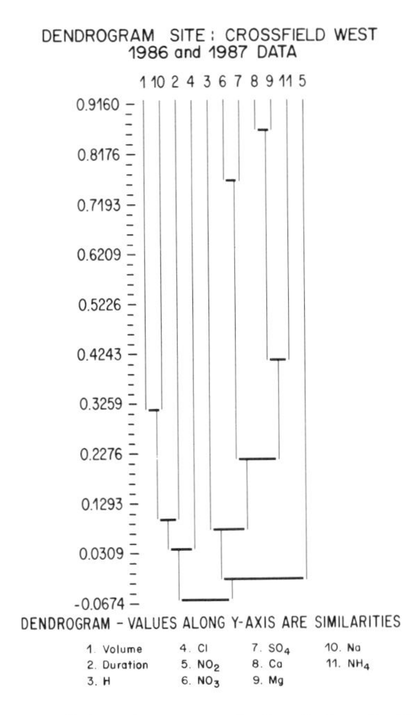

Figure 4.12 Dendrogram of the cluster analysis of the similarity between ion pairs in daily rain samples at Crossfield West.

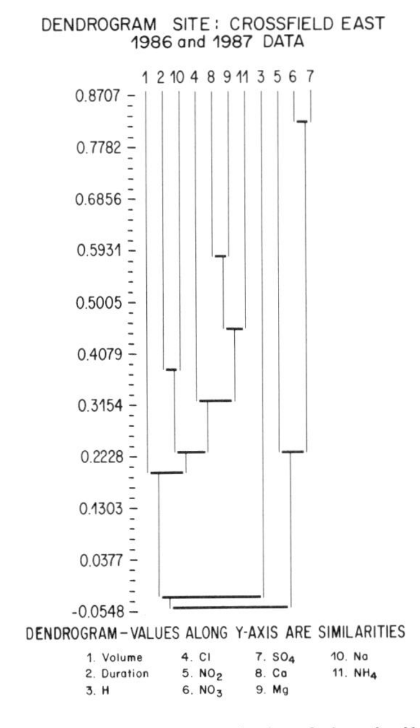

Figure 4.13 Dendrogram of the cluster analysis of the similarity between ion pairs in daily rain samples at Crossfield East.

the KAPS sampler. At Crossfield West NO and NO_2 correlated well (r=0.74). The correlation coefficient between NO and HNO_2 (T1) was 0.37, while it was 0.46 with HNO_2 (T2). These relatively low correlations may be due to the collection of both NO and HNO_2 on the denuder tubes. Both NO and NO_2 correlated weakly (r=0.32 and 0.48, respectively) with fine particle NO_3^- (TF). Fine particle NO_3^- (TF) and SO_4^{2-} (TF) showed an r of 0.43. Similarly r between SO_2 and SO_4^{2-} (TF) was 0.41. All these results are reasonable. The positive but weak correlations are most likely due to the minimum detection limits of the continuous monitors and the presence of the gases below that limit during most of the KAPS sampling duration.

At Crossfield East r between NO and NO_2 was 0.72 similar to the results obtained at Crossfield West, NO was correlated with HNO_2 (T1) and HNO_2 (T2) (r=0.37 in each case). SO_2 and NO_2 also correlated with each other, but weakly (r=0.32).

The summary results of the correlation analysis between SO_4^{2-} and NO_3^- in rain and the variables characterized through the KAPS annular denuder system for the 48 hours prior to the onset of the rain are presented in Table 4.80. At all three ADRP sites SO_4^{2-} and NO_3^- in rain were reasonably well correlated with each other. As opposed to Fortress Mountain (for a discussion refer to Chapter 3.6.4.4), at both Crossfield West and East, all KAPS variables evaluated were either poorly or negatively correlated with the rain variables. This could be interpreted to mean poor or lack of below cloud scavenging of SO_2, HNO_3 and fine particulate nitrate and sulphate. Alternatively, this could also mean that the precipitating clouds originated from regions far off and thus little or no local cloud incorporation of gaseous and fine particulate pollutants occurred in the Crossfield region. The results of the correlation analysis between rainfall SO_4^{2-} and NO_3^- and the sum of 10-minute average of continuously monitored gaseous pollutant concentrations (SO_2, NO and NO_2 individually) over the 48 hours prior to the onset of rain also provided poor results at both Crossfield sites, thus confirming the previous conclusions.

Table 4.80 Relationships between variables measured through the annular denuder system and variables in rainfall.[1] A comparison of the three ADRP monitoring sites.

Variable Pairs	r Fortress Mountain	r Crossfield West	r Crossfield East
RNO_3^- vs $D\ NO_3^-$ (HNO_3)	0.45	-0.02	0.25
RNO_3^- vs $NY\ NO_3^-$	-0.26	0.03	-0.21
RNO_3^- vs $TF\ NO_3^-$	-0.12	0.08	-0.14
RNO_3^- vs $F\ NO_3^-$	-0.25	0.05	-0.24
RNO_3^- vs $T1\ NO_3^-$ (HNO_3)	0.56	-0.06	-0.18
RNO_3^- vs $T2\ NO_3^-$ (HNO_3)	0.18	-0.12	0.15
RSO_4^{2-} vs $D\ SO_4^{2-}$ (2/3 SO_2)	0.66	-0.16	0.18
RSO_4^{2-} vs $T1\ SO_4^{2-}$ (2/3 SO_2)	0.66	-0.15	-0.12
RSO_4^{2-} vs $T2\ SO_4^{2-}$ (2/3 SO_2)	0.61	0.04	-0.15
RSO_4^{2-} vs $NY\ SO_4^{2-}$	0.17	-0.05	--
RSO_4^{2-} vs $TF\ SO_4^{2-}$	0.13	0.09	-0.25
RSO_4^{2-} vs $F\ SO_4^{2-}$	0.28	0.07	-0.34
RSO_4^{2-} vs $R\ NO_3^-$	0.66	0.79	0.59

R: Rain; D: denuder tube #1+#2; NY: Nylon filter; TF: Teflon filter.
F: Total (Teflon + Nylon filter); T1: Denuder Tube #1; T2: Denuder Tube #2.
1 Values from the annular denuder system included in the statistical analysis are from the 48 hours immediately preceding the onset of each daily rain event.

4.8 CONCLUDING REMARKS

Air quality was monitored over a two year period during 1985-87 at two sites located within an urban airshed. One of the two sites (Crossfield East) was downwind from a local point source, a gas plant, while the other site (Crossfield West) was upwind from the point source. Results obtained at these two locations were compared to the corresponding results from a high elevation, regional background site (Fortress Mountain).

Hourly O_3 concentrations were lower at the two Crossfield sites compared to the high elevation site. However, there was a relationship between the occurrences of O_3 episodes at Fortress Mountain and the occurrences of elevated O_3 concentrations on the corresponding days at the two Crossfield sites. Fortress Mountain is located approximately 90 km southwest of the two Crossfield sites.

On days of O_3 episodes at Fortress Mountain, peak O_3 concentrations were observed during 2300-2400 hours. However, during the following hours there was a steep decrease in hourly O_3 concentrations terminating at approximately 0900 hours. This hourly decrease at Fortress Mountain was inversely and highly correlated with relatively high hourly concentrations of O_3 at the two Crossfield sites during 0900-1600 hours of the same day. It was proposed that O_3 was transported aloft from Fortress Mountain to Crossfield, followed by downward mixing during midday, leading to peak O_3 concentrations at 1600 hours at the two Crossfield sites. The direction of the 500 m winds on these days of O_3 episode were consistent with the proposed explanation.

Highest 10-minute SO_2 concentrations were in the 130-140 ppb range at Crossfield West and in the 160-170 ppb range at Crossfield East. In the 4.0-140 ppb range of 10-minute SO_2 concentrations, there were more frequent occurrences at Crossfield East compared to Crossfield West. Similar patterns were also observed with 10-minute H_2S and hourly NO and NO_2 concentrations.

Mean concentrations of COS, CS_2, H_2S, and SO_2 at the two Crossfield sites were higher than at Fortress Mountain. At Crossfield East they were higher than at Crossfield West. The results of cryogenic sampling of the four pollutants suggested the possible origin of CS_2, COS, and H_2S from natural sources at Fortress Mountain and from regional scale anthropogenic sources at Crossfield West (data were not collected for these parameters at Crossfield East).

In general, the mean and maximum 48-hour concentrations of SO_2, HNO_2, HNO_3, and NH_3 as measured by the KAPS annular denuder system were higher at Crossfield East compared to Crossfield West.

The mean coarse particle elemental concentrations and the fine particle Br, Zn, Ni, and Pb concentrations were higher at Crossfield East compared to West. It appears that Ni, Zn, and Cu ratios may be used as a tracer for the local point source, the gas plant.

The median pH of rain was higher at Crossfield West (pH 4.87) and Crossfield East (pH 5.00) in comparison to Fortress Mountain (pH 4.77). It is suggested that the higher rain pH at Crossfield East compared to Crossfield West was due to the neutralization by the coarse particulate matter. Ambient particulate matter concentrations (sampled by Hi-Vol Sampler) at the three sites were: Fortress Mountain 8.19 ± 7.15; Crossfield West 14.60 ± 10.19; and

Crossfield East 18.59 ± 10.19. The SO_4^{2-} and NO_3^- in rain at the two Crossfield sites correlated poorly with the H^+. A predominant fraction of the two anions appeared to be associated with basic cations. It is suggested that weak acids may be involved and that the rain pH appeared to be regulated by acid-base equilibrium.

4.9 LITERATURE CITED

Alberta Government/Industry Acid Deposition Research Program. Volume 1. 1984. *Researching Acid Deposition: Workshop Proposals*. Acid Deposition Research Program: Calgary, Alberta. 59 pp.

Finlayson-Pitts, B.J. and J.N. Pitts. 1986. *Atmospheric Chemistry: Fundamentals and Experimental Techniques*. New York: John Wiley and Sons. 1098 pp.

Graedel, T. E., D. T. Hawkins, and L. D. Claxton. 1986. *Atmospheric Chemical Compounds*. New York: Academic Press. 732 pp.

Sze, N.D. and K.W. Ko. 1980. Photochemistry of COS, CS_2, CH_3SCH_3, and H_2S: Implications for the atmospheric sulfur cycle. *Atmospheric Environment* 14: 1223-1239.

U.S. Environmental Protection Agency. 1984. Air Quality Criteria for Ozone and Other Photochemical Oxidants. Volume II. Research Triangle Park, NC: U.S. Environmental Protection Agency, Environmental Criteria and Assessment Office, EPA Report No. EPA-600/8-84-020.

U.S. NAPAP. The National Acid Precipitation Assessment Program. 1987. Interim Assessment Report, Volumes 1 through 4.

U.S. National Academy of Sciences. 1977. *Ozone and Other Photochemical Oxidants*. Committee on Medical and Biological Effects of Environmental Pollutants. Washington, D.C. 719 pp.

CHAPTER 5

SOURCE APPORTIONMENT STUDIES NEAR CROSSFIELD, ALBERTA

J. A. Cooper and E. Peake

5.1 INTRODUCTION

Regional, area, line, and stationary point sources contribute to the anthropogenic pollutants in the atmosphere. The qualitative and quantitative characteristics of these pollutants vary with the type of source and in time and space. Once emitted into the atmosphere, many primary pollutants are transformed into secondary chemical products. Such transformation is dependent on the reactivity and the atmospheric residence time of the initially-emitted pollutant, presence of other chemical constituents in the atmosphere, and meteorological conditions.

In addition to the anthropogenic processes, many primary pollutants in the atmosphere are also contributed by natural emissions. The relative importance and magnitude of the contribution by the individual processes vary with the pollutant in question and in time and space. Nevertheless, the air quality at a given geographic location is the product of all these complex, but interrelated, processes.

There is a critical need to better understand the relationships between pollutant emissions and atmospheric processes in the context of environmental impact assessment and application of control strategies. Thus, over the past 10 years a number of investigators have developed methods for segregating or identifying the relative contributions of multiple, appropriate sources of interest to air quality at a given receptor site (APCA 1982; Hidy 1984; Rahn and Lowenthal 1985; Henry et al. 1984; Stevens and Pace 1984).

5.2 CHEMICAL MASS BALANCE METHOD OF SOURCE APPORTIONMENT

The relationship between particulate emissions and their ambient concentrations measured at a receptor site distant from the pollution source is a complicated one. Many variables, primarily meteorological, make the direct correlation between source emissions and ambient pollutant concentrations a poor one. Each of these variables is random in nature, will vary in space and time, and may combine with other variables in a nonlinear manner. Thus, any estimate of the effect of a given source on atmospheric pollutant concentrations, e.g., mass of particles collected on a Hi-Vol filter, using dispersion modelling, is approximate at best. The conceptualization of this complex and intractable "real-life" situation is a comparatively simple "model" based on physical principles which can be used to determine the average contribution of specific sources to particulate loadings at a geographic location.

It is possible to start with ambient air particulate samples collected at a receptor site by a representative sampling technique, determine some properties such as elemental composition of that sample which are unique to specific source or source types, and assign the origin of that fraction of the sample possessing the required property to its appropriate source.

The chemical mass balance (CMB) method is based on the conservation of the relative aerosol chemistry from the time a chemical species is emitted from its source to the time it is measured at a receptor site. That is, if p sources are emitting M_j mass of particles, then:

$$m = \sum_{j=1}^{p} Mj \tag{1}$$

where m is the total mass of the particulate matter collected on a filter at a receptor site. It assumes that the mass on the filter is a linear combination of the mass contributed by each of the sources.

The mass of a specific chemical species, m_i, is given by the following equation

$$m_i = \sum_{j=1}^{p} Mij = \sum_{j=1}^{p} F_{i'j} M_j \tag{2}$$

where M_{ij} is the mass of element i from source j and $F_{i'j}$ is the fraction of chemical species i in the mass from source j collected at the receptor. It is usually assumed that

$$F_{i'j} = F_{ij} \tag{3}$$

where F_{ij} is the fraction of chemical i emitted by source j as measured initially at the source. The degree of validity of this assumption depends on the chemical and physical properties of the species and its potential for atmospheric modification such as condensation, volatilization, chemical reaction, or sedimentation.

If we accept this equation, however, and divide both sides of equation (2) by the total mass of the material collected at the receptor site, it follows that:

$$\frac{m_i}{m} = \sum_{j=1}^{p} F_{ij} \frac{M_j}{m} \quad (4)$$

$$C_i = \sum_{j=1}^{p} F_{ij} S_j \quad (5)$$

where C_i is the concentration of the chemical component i measured at the receptor site and S_j is the source contribution, i.e., ratio or the mass contributed from source j to the total mass collected at the receptor site. In practice, it is this fraction of particulate pollution (S_j) measured at a receptor due to source j, which is of primary interest in receptor modelling calculations.

If the C_i and the F_{ij} at the receptor site for all sources (p) of the types suspected of affecting that receptor site are known, and $p < n$ (n=number of chemical species), a set of n simultaneous equations exists from which the source type contributions S_j may be calculated by least squares methods.

Implementation of a CMB analysis requires the collection of data sets on both source and ambient elemental composition. The approach considered in the present study was to use direct source measurements and values published in the literature to develop source elemental profiles.

5.3 MONITORING OF AMBIENT PARTICULATE MATTER

Fine (<2.5 μm) and coarse (2.5-10 μm) particles were collected at Crossfield West and Crossfield East using Sierra model 244 dichotomous virtual impactors with 10 μm inlets. The Crossfield West monitoring site, was about 7 kilometres west of the Crossfield natural gas processing plant and represented, in general, the upwind site from the point source. The Crossfield East monitoring site was located at 2.4 kilometres downwind from the Crossfield natural gas processing plant.

Ambient particulate samples were collected (sampling duration: 48 hours) on ring-mounted Teflon membrane filters at the two monitoring sites over a two year period. From these, one hundred filters were selected for intensive chemical analysis and CMB calculations. The filter samples were selected to represent days with high particulate loading or days when a specific pollutant source(s) was either generally influencing or not influencing the monitoring sites. Tables 5.1 and 5.2 provide a listing of the days for which filter samples were selected, their corresponding fine, coarse, and PM_{10} particle concentrations, and prevailing wind direction for the day.

Table 5.1 Summary of particle concentration and wind direction on days selected for source apportionment. Sampling Site: Crossfield West.

Sampling	Particle Concentration ($\mu g\ m^{-3}$)			Wind
Date	Fine	Coarse	PM_{10}	Direction
08/16/86	7.95			SSE,S
09/07/86	8.89	4.54	13.43	SSE,S
09/21/86	23.15			W,WSW,SW,SSW
10/05/86	1.87			W,WSW,SW,SSW
10/07/86	5.41			SSE,S
10/19/86	6.51			SW
11/02/86	2.24	3.70	5.94	SW
11/06/86	5.70			NW,NNW
11/08/86	2.26	3.58	5.84	NW,NNW
11/14/86	3.42			W,WSW,SW,SSW
11/20/86	1.80			SW
11/22/86	2.38			SW
11/24/86	1.34	3.05	4.39	W,WSW,SW,SSW
12/04/86	3.53			NW,NNW
12/06/86	1.18			SW
12/21/86	3.40			W,WSW,SW,SSW
01/14/87	2.75			SSE,S
01/18/87	0.95			NW,NNW
01/22/87	1.55	4.88	6.43	SW
01/30/87	0.97	3.06	4.03	W,WSW,SW,SSW
02/03/87	1.22			SW
02/12/87	7.16	3.59	10.75	SSE,S
03/01/87	9.86			SSE,S

Continued...

Table 5.1 (Concluded).

Sampling Date	Particle Concentration ($\mu g\ m^{-3}$) Fine	Coarse	PM_{10}	Wind Direction
03/19/87	4.09			NW,NNW
03/29/87	3.42			W,WSW,SW,SSW
04/09/87	3.14			SSE,S
04/23/87	8.59			SSE,S
05/03/87	1.71			SW
05/18/87	8.22			NW,NNW
05/30/87	2.91			W,WSW,SW,SSW
06/21/87	1.16	3.09	4.25	W,WSW,SW,SSW
07/03/87	4.83			NW,NNW
07/31/87	1.16			SW
08/02/87	3.80	7.06	10.86	NW,NNW
08/04/87	3.10			NW,NNW
08/08/87	11.37			SSE,S
09/02/87	3.88			NW,NNW
09/18/87	7.65	16.76	24.41	SSE,S
10/02/87	6.19			W,WSW,SW,SSW

Table 5.2 Summary of particle concentration and wind direction on days selected for source apportionment. Sampling Site: Crossfield East.

Sampling Date	Particle Concentration ($\mu g\ m^{-3}$) Fine	Coarse	PM_{10}	Wind Direction
08/14/86	5.98			SE,E,ESE
08/16/86	7.53			S
08/18/86	4.90			NNW
08/25/86	7.14			S
09/03/86	2.75			SE,E,ESE
09/07/86	9.08			SE,E,ESE
09/11/86	3.28	3.77	7.05	S
09/13/86	2.61			SSE
09/15/86	0.67			SE,E,ESE
12/04/86	4.64			NNW
12/21/86	5.80			S
02/12/87	9.08			SSE
03/09/87	12.08	32.47	4.55	SE,E,ESE
03/11/87	9.56	4.74	14.30	SSE
03/17/87	6.40	2.83	9.23	NNW
04/02/87	5.41			SE,E,ESE
04/04/87	9.15			SSE
04/09/87	1.41			S

Continued...

Table 5.2 (Concluded).

Sampling Date	Particle Concentration ($\mu g\ m^{-3}$)			Wind Direction
	Fine	Coarse	PM_{10}	
04/23/87	8.71			SSE
05/10/87	8.09	30.06	38.15	S
05/14/87	10.40	69.00	79.40	NNW
05/21/87	5.61	11.1	16.69	SSE
05/23/87	7.78			SE,E,ESE
05/29/87	3.70			S
06/13/87	5.98			SE,E,ESE
06/17/87	5.57			SSE
07/03/87	41.67			NNW
07/19/87	3.35			SE,E,ESE
07/21/87	5.24			NNW
07/27/87	10.93			SSE
08/02/87	3.65	10.61	14.26	NNW
08/08/87	4.07			SSE
08/14/87	3.10	2.81	5.91	SE,E,ESE
08/22/87	4.48			SSE
08/25/87	7.39			S
09/08/87	7.99			NNW
10/07/87	3.83			NNW
10/19/87	6.58			S

5.4 DEVELOPMENT OF THE SOURCE PROFILE

5.4.1 Overview

Source profiles or "fingerprints" of emissions in the vicinity of Crossfield (Figure 5.1) were developed through samples collected and analyzed from sources in the area: the Crossfield natural gas plant, local fugitive dust, and motor vehicles. In addition, information was also obtained from previously published source profiles (U.S. EPA 1984; NEA Inc. 1986).

5.4.2 Natural Gas Plant

At the Crossfield processing plant, natural gas from Elkton and D-1 geologic formations is treated to remove condensate, water vapour, and certain gases (CO_2 and H_2S). The resulting "sweet" gas is either piped from the plant for sale or used, to a limited extent,

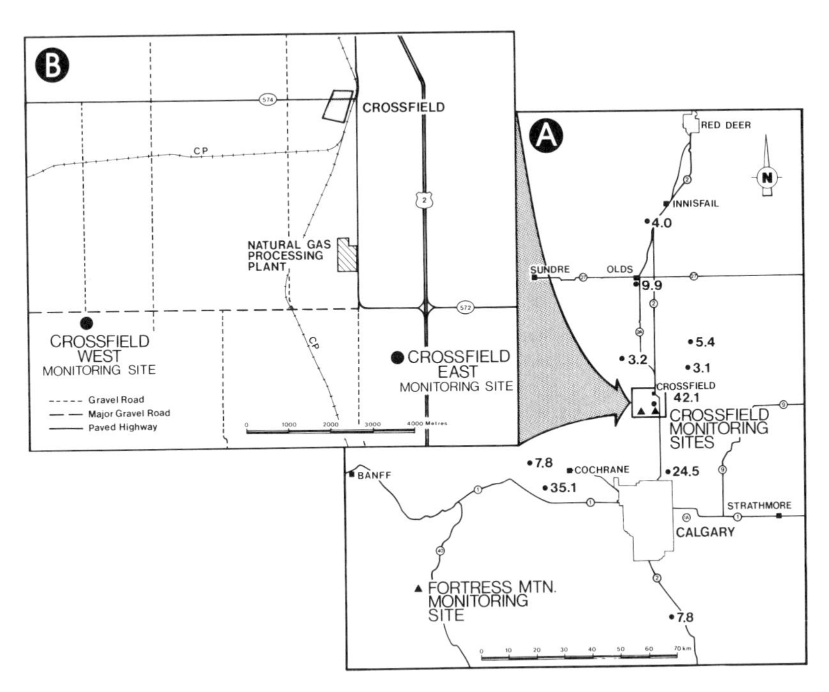

Figure 5.1 (A) Map showing the comparative geographic locations of the three ADRP air quality monitoring sites. Background air quality was monitored at Fortress Mountain. Sources of SO_2 ($t\ d^{-1}$) in the vicinity of the Crossfield monitoring sites are noted in the Figure. (B) A closer view of the geographic location of the two air quality monitoring sites in relation to Crossfield and the local point source of SO_2.

within the plant for such purposes as fueling compressors and providing flare stack pilot lights. The acid gas generated from the sweetening process is treated to remove the H_2S as elemental sulphur. Any remaining gases are sent to the incinerator where the H_2S in the mixture is burned before the release of emissions from a 91-metre stack.

Three 30-metre flare stacks are located on the north side of the plant. These flares are used infrequently to vent gases should problems develop at any point in the process. When used, the west stack carries sales gas, the middle stack vents gases ranging in composition from sales gas to pure H_2S, and the east stack vents sour inlet gas (field gas). A fourth, smaller flare, vents small amounts of gases from glycol and sulphinol regeneration processes and from steps in salt water recovery.

The liquid sulphur derived from the acid gas is either pumped to a number of on-plant storage pits or to a solid sulphur handling facility just west of the plant. Liquid sulphur is also trucked to the plant from other gas processing plants which do not have their own sulphur-handling facilities. The solid sulphur handling operation deals primarily with cooling and storage of the sulphur as large flakes and the subsequent loading of the slated sulphur into open boxcars. A small prilling operation is also used occasionally. Dust control consists of spraying conveyor belts with a water/detergent mixture.

Five large blocks of elemental sulphur exist on-site but have not been used as a means of storage for approximately six years. At present, this sulphur is being slowly remelted and added to the slating process, although at the time of air quality sampling the remelting project was not in operation due to inadequate fume controls.

5.4.3 Stack Sampling

The primary objective of source sampling was to collect particulate samples from sources of acidic material in a downwind local airshed, consistent with receptor modelling requirements. The tall, high temperature stack emissions were sampled with a size-segregating dilution sampler. Emissions from short stacks were sampled directly from the plume with special extension probes. Fugitive dust sources were sampled by collecting bulk grab samples followed by laboratory aerosolization.

A size-selective dilution sampler (SSDS) was used to collect fine and coarse particles from the incinerator stack, simultaneously on Teflon and quartz fibre filters with two dichotomous virtual impactors. The sampler is schematically illustrated in Figure 5.2. The sampler probe and dilution chamber were all stainless steel. In its normal operation, a pressure difference between the chamber and the stack is established by varying the flow rates of the inlet and outlet blowers. Particles less than 10 μm in diameter are drawn into the dilution chamber isokinetically through a cyclone and a heated transfer tube or probe. Filtered dilution air is mixed with the stack aerosol in the dilution chamber. This cools and dilutes the stack aerosol simulating the condensation and evaporation in a plume. The cooled aerosol is then sampled with two dichotomous virtual impactors. In this way, size-selective samples are collected that minimize modifications in the chemistry due to evaporation, condensation, and sedimentation during transport. In addition, the filter samples were analyzed with methods similar to those used for the ambient air samples to minimize systematic errors due to differences in analytical or sampling techniques.

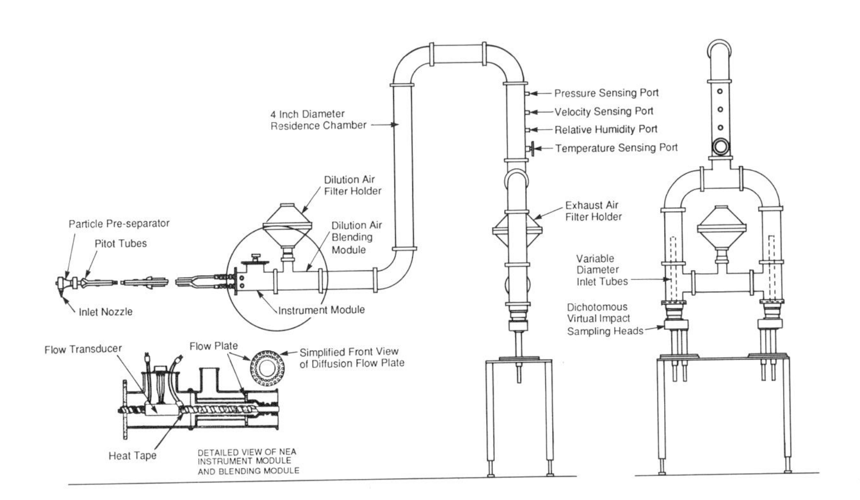

Figure 5.2 Schematic diagram of size-segregating dilution sampler used to collect representative samples from the gas plant incinerator stack.

The flare emissions were sampled using a 5 cm diameter stainless steel extension probe held in the plume with a crane. The emissions were drawn through the probe and sampled downstream using Teflon and quartz fibre filters. Cyclones were used to obtain 2.5 and 15 μm cut points. A similar sampling arrangement was used to collect compressor samples.

In the end, particulate samples of emissions were collected from the incinerator stack, three flare stacks, one inlet compressor stack, the prill stack, and motor vehicles. A total of 51 valid fine and coarse particle samples were collected from these sources on Teflon and quartz fibre filters.

Thirty-one bulk fugitive dust samples of soil and road dust, elemental sulphur, and other miscellaneous fugitive emissions from the gas plant were collected. These samples were sieved, composited, and suspended in a dust chamber from which fine and coarse particle samples were collected on filters for analysis.

5.5 ANALYSIS

5.5.1 Sample Selection

Source profile filter samples were selected for analysis based on the following criteria: (a) valid deposited particle mass (uniform deposit), (b) proper functioning of the sampler, (c) proper field and laboratory QA/QC, (d) deposited particle mass in the range of 1-10 mg per filter, (e) simultaneous sampling in other experiments, (f) sampling of special events, (g) expected significance of source impact, (h) selection of representative replicate samples, and (i) emphasis on size fractions that are expected to represent the largest impact.

All the selected ambient and source profile filters were analyzed by XRF (X-ray fluorescence spectroscopy), and 30 of the highest priority source profile filters were analyzed by neutron activation analysis (NAA).

Twelve of the highest priority bulk dust samples were selected for resuspension and analysis by XRF. Emphasis was given to those bulk dust samples most likely to make significant contributions to the particulate levels at the receptor site and also provide representative samples of the various potential sources of fugitive dust.

5.5.2 Deposited Particle Mass

The deposited particle mass was determined gravimetrically using a Cahn 27 electro-balance. The standard operating procedure was modified for those samples expected to have substantial sulphuric acid deposits, by making the measurements in a dry nitrogen-controlled environment to minimize artifacts induced by moisture absorption.

5.5.3 X-ray Fluorescence Analysis (XRF)

This analysis was performed with a TEFA III tube excited X-ray analyzer. Selected source profile samples were analyzed in air to minimize loss of sulphuric acid prior to additional analysis by ion chromatography. The remaining filters were analyzed in vacuum using an ORTEC TEFA III energy dispersive X-ray fluorescence analyzer. Each filter was analyzed under three different excitation conditions to optimize the sensitivity for specific elements. The elements determined with each of the excitation conditions were:

Al, Si, P, and Fe	-	Mo anode, No filter, 15KV, 200 μamps
S, Cl, K, Ca, Ti	-	W anode, Cu filter, 35KV, 200 μamps
V, Cr, Mn, Fe, Ni, Cu, An, Ga, As, Se, Br, Rb, Sr, Y, Zr, Nb, Mo, Pd, Ag, Cd, In, Sn, Sb, Ba, La, and Pb	-	Mo anode, Mo filter, 50KV, 200 μamps

A quality assurance sample was included with the analysis of every ten filters, and corrections were made for inter-element interferences, blanks, and for particle size.

5.5.4 Neutron Activation Analysis (NAA)

Thirty of the highest priority source profile samples were analyzed by NAA following the XRF analysis. The filters were separated from their supporting rings and sealed in polyethylene envelopes after folding the filters so that the particulate deposit was on the inside. The sealed envelopes were placed in 1.5 mL polyethylene vials, which in turn were placed in 8.5 mL polyethylene vials. These triple sealed samples, along with copper flux monitors and standards, were placed in irradiation tubes and irradiated at the Oregon State University Triga Reactor for five minutes at a flux of 5×10^{12} neutrons cm^{-2} sec^{-2}. The resulting short-lived radionuclides were determined after a 10 minute decay using a high purity germanium gamma-ray spectrometer with a resolution of 1.7 KeV at 1332 KeV and a relative efficiency of 17%. The gamma-ray spectra were analyzed using GELIGAM and the elemental concentrations determined by ORTEC, Inc. model A29 neutron activation analysis comparative software.

Subsequently, the samples still sealed in their polyethylene envelopes, were removed from their counting vials and placed in a second polyethylene envelope and sealed for long irradiation. The samples were irradiated at the Hanford Nuclear Reactor in Richland, Washington, to an integrated flux of 2.3×10^{18} neutrons cm^{-2}. The intermediate-lived activation products were determined after a 3 day decay period, and the long-lived products determined after a 30 day decay period.

5.5.5 Organic and Elemental Carbon Analysis

The organic (OC) and elemental (EC) carbon content of selected source profile samples was determined by a pyrolysis-flame ionization method (Huntzicker et al. 1982). About 1 cm^2 of each quartz fibre filter, which was used for sample collection at the same time as the Teflon filters, was heated in an argon atmosphere to remove the organic carbon. The volatile organic carbon was then combusted to carbon dioxide and converted to methane prior to measurement by a flame ionization detector. After the organic carbon had been removed, the argon was replaced with oxygen and the elemental carbon combusted and determined in a similar manner. The values of both elemental and organic carbon were corrected for the influence of pyrolysis during the analysis, using a laser absorption technique (Huntzicker et al. 1982).

5.6 DATA PROCESSING

5.6.1 Source Profile Data

The results of the elemental analyses, including the uncertainties attached to them, were processed to obtain average source composition profiles as indicated in Figure 5.3. It is important to note that the uncertainties related to the analysis of each element and sample were propagated through this procedure so as to yield source profile uncertainties that were representative of the variability of each source profile.

Three primary data bases were used as input to generate the composite average source profiles:

1. Field data, representing the filter identification code, source type, and particle size.
2. Data representing the mass of the particulate deposit on each filter.
3. Chemical analysis data files containing the results of the XRF, OC/EC, and NAA.

Appropriate portions of the files containing the field and particulate mass data were integrated with the chemistry data files prior to numerical analysis. The data from each chemical analysis were reviewed for the quality assurance, results of replicate sample analysis, and selected fine to coarse particle ratios. The average blank filter concentrations were then determined and subtracted from the results for each individual filter. This step was carried out automatically with the XRF analysis but was a completely separate step with the NAA and OC/EC determinations. Because of the large number of chemical species determined (almost 60) and the low concentrations of many of them, a large number of the elemental concentrations were below detection limits or negative in value after this first subtraction. Consequently in the data processing, a zero value was used where the results were below the detection limit or negative in value.

A second quality assurance check was performed after the blank subtraction step was completed. This consisted of a review of the results to see if the data were consistent with detection limits for lightly loaded filters.

Corrections were then made on the fine and coarse particle concentrations for the approximate 10% contamination of the fine particles on to the coarse particle filter at the time of sample

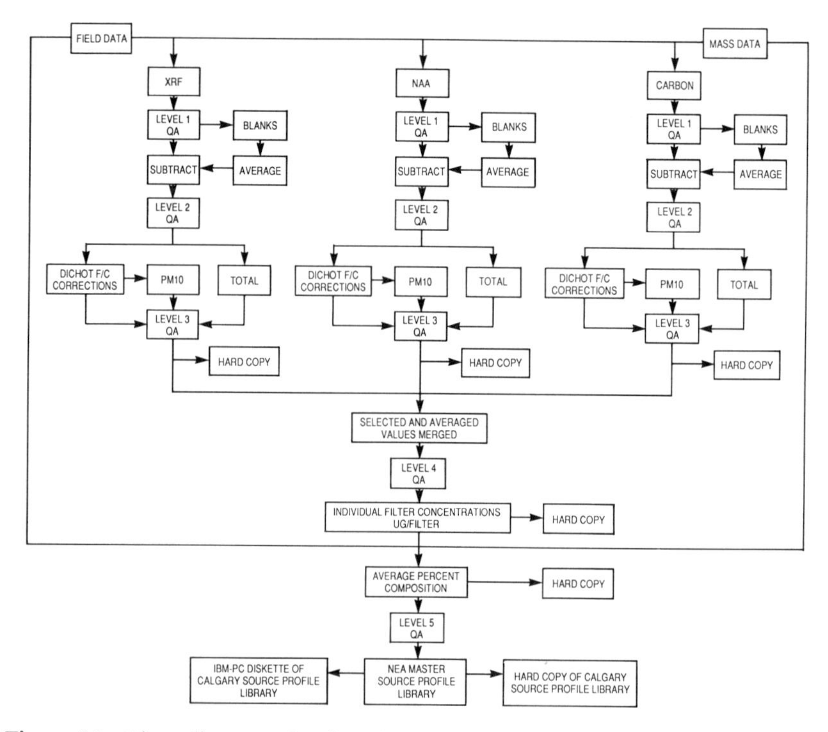

Figure 5.3 Flow diagram showing the major steps in data processing in the development of source profiles for the Crossfield-Calgary airshed.

collection. The value for this correction was quite large in many cases because of the large number of sources with emissions primarily consisting of fine particles. This correction also produced a substantial number of negative values which were replaced by zero. The corrected fine and coarse particle concentrations were summed to yield a calculated PM_{10} composition.

These results were then subjected to another QA where results from different analytical methods were compared for internal consistency.

The results from each chemical analysis were then selected and averaged to form a merged data set for each individual aerosol sample. The elemental data selection and measurement codes are listed in Table 5.3. The selected values represent the results reported by either XRF or NAA analysis technique. This selection was based on the uncertainties, detection limits, and potential inter-elemental interferences for each element which were generally quite different between the selected elements.

Table 5.3 Elemental data selection and measurement codes.

Element	Code[a]	Element	Code[a]	Element	Code[a]
He	0	Br	3	Er	0
Li	0	Kr	0	Tm	2
Be	0	Rb	3	Yb	2
B	0	Sr	1	Lu	2
C	0	Y	1	Hf	2
N	0	Zr	1	Ta	2
O	0	Nb	0	W	0
F	0	Mo	1	Re	0
Ne	0	Ag	0	Os	0
Na	2	Ru	0	Ir	0
Mg	2	Rh	0	Pt	0
Al	2	Pd	1	Au	2
Si	1	Ag	2	Hg	2
P	1	Cd	1	Tl	0
S	1	In	2	Pb	1
Cl	2	Sn	1	Bi	0
Ar	0	Sb	2	Po	0
K	1	Te	0	At	0
Ca	1	I	2	Rn	0
Sc	2	Xe	0	Fr	0
Ti	1	Cs	2	Ra	0
V	2	Ba	2	Ac	0

Continued...

Table 5.3 (Concluded).

Element	Code[a]	Element	Code[a]	Element	Code[a]
Cr	3	La	2	Th	2
Mn	2	Ce	2	U	2
Fe	1	Pr	0	OC (C)	4
Co	2	Nd	2	EC (C)	5
Ni	1	Pm	0	NO_3(N)	0
Cu	1	Sm	2	NH_4(N)	0
Cn	1	Eu	2	CO_2(C)	0
Ga	1	Gd	0	NO_2(N)	0
Ge	0	Tb	2	PO_4(P)	0
As	2	Dy	2	SO_4(S)	0
Se	2	Ho	0		

[a] 0. Not reported
1. x-ray fluorescence
2. Neutron Activation Analysis
3. Average of the results from (1) and (2)
4. Organic carbon.
5. Elemental carbon

The merged analytical data were then combined with the field and particle mass data and average source composition profiles calculated. Fine and coarse particle data were also combined to create PM_{10} source profiles. The uncertainties listed are either the standard deviation of the distribution of values included in the average or the propagated uncertainty, whichever is the larger value.

5.7 RESULTS AND DISCUSSION

5.7.1 Concentrations of Ambient Particulate Matter

The fine, coarse, and PM_{10} particulate matter concentrations on the days selected for intensive analysis are listed in Tables 5.1 and 5.2. The average fine particle mass at Crossfield West and East was 4.6 $\mu g\ m^{-3}$ and 7.2 $\mu g\ m^{-3}$, respectively. The average coarse particle mass was 5.3 $\mu g\ m^{-3}$ at Crossfield West and 18.6 $\mu g\ m^{-3}$ at Crossfield East.

The average values of particle mass in relation to the wind regimes are summarized in Table 5.4. The fine and coarse particle concentrations at Crossfield East were highest when the wind was from the north-north-westerly direction (the gas plant plume impacting the monitoring site). The highest fine and coarse particle concentrations at Crossfield West were measured when the wind was from the south and south-southeasterly direction. The fine particle mass was more sensitive to wind direction at Crossfield West, varying by more than a factor of 3, from 2.2-7.3 $\mu g\ m^{-3}$. The lowest fine particle concentrations were measured when the wind was from the southwest.

Table 5.4 Wind regimes and average concentrations of particles ($\mu g\ m^{-3}$).

	Crossfield East		Crossfield West	
Regime	Fine	Coarse	Fine	Coarse
NNW	9.2	27.5	--	--
SSE	7.0	7.9	--	--
SE		5.7	16.9	--
SE, E, ESE	6.2	17.6	--	--
SSE, S	--	--	7.3	8.3
SW	--	--	2.2	4.3
NW, NNW	--	--	4.0	5.3
W, WSW, SW, SSW	--	--	4.8	3.1
All	7.2	18.6	4.6	5.3

The concentrations of particulate matter at Crossfield East were about two times higher than at Crossfield West.

5.7.2 Chemistry of Ambient Particulate Matter

The most abundant chemical species in ambient fine particles were sulphate and nitrate followed by Si. The most abundant coarse particle species were Si and Al. About 90% of the PM_{10} Si and Al was in the coarse particle fraction, and about 90% of the sulphate and nitrate was in the fine particle fraction.

5.7.3 Source Profiles

5.7.3.1 Overview

An example of the source profiles is presented in Table 5.5. In many samples only a low percent of the particle mass was explained by the measured chemical species. This was considered, in part, to be due to the presence of some elements primarily associated with oxides or carbon species which were not measured. This was particularly the case for most of the road and soil dust samples. Another source of unexplained mass was associated with sulphur species. Most of the sulphur was in the form of sulphate, accounting for three times more mass than sulphur. Additional mass appeared to be associated with cations not measured, such as hydrogen or ammonium. Although these chemical species can account for a significant portion of the mass, the largest portion of unaccounted mass was probably associated with water. Most of the gas plant samples (e.g., incinerator and flares) consisted of visible liquid droplets which were presumed to be mostly sulphuric acid with varying amounts of water.

Table 5.5 Source description - incinerator stack (composite).

	Percent Composition		
Species	Fine	Coarse	Total (PM_{10})
Na	.0138 ± .0246	.0629 ± .1233	.0163 ± .0233
Mg	.0020 ± .0060	.2939 ± .3395	.0127 ± .0147
Al	.2820 ± .4535	2.2176 ± 1.8629	.3584 ± .4998
Si	.1886 ± .3501	8.8791 ± 6.9551	.5223 ± .5803
P	.0786 ± .1457	.0534 ± .7910	.0774 ± .1434
S	16.5683 ± 3.0711	4.4321 ± 4.1699	16.0902 ±2.9486
Cl	.0147 ± .0254	.2889 ± .4759	.0247 ± .0282
K	.0100 ± .0057	.5110 ± .2978	.0300 ± .0131
Ca	.0037 ± .0055	1.8303 ± 1.1376	.0757 ± .0476
Sc	.0000 ± .0001	.0002 ± .0001	.0000 ± .0001
Ti	.0019 ± .0007	.1287 ± .0606	.0073 ± .0033
V	.0002 ± .0001	.0087 ± .0056	.0006 ± .0005
Cr	.0085 ± .0140	1.0321 ± 1.3211	.0642 ± .1041
Mn	.0015 ± .0016	.1455 ± .0737	.0083 ± .0068
Fe	.0435 ± .0479	4.6884 ± 2.9949	.2670 ± .2515
Co	.0001 ± .0001	.0077 ± .0061	.0005 ± .0005
Ni	.0071 ± .0112	.5926 ± .6733	.0388 ± .0538
Cu	.0004 ± .0007	.0211 ± .0177	.0013 ± .0012
Zn	.0013 ± .0013	.0777 ± .0915	.0049 ± .0052
Ga	.0000 ± .0001	.0008 ± .0013	.0000 ± .0001
As	.0002 ± .0003	.0064 ± .0090	.0005 ± .0004

Continued...

Table 5.5 (Concluded).

Species	Percent Composition Fine	Coarse	Total (PM_{10})
Se	.0002 ± .0002	.0010 ± .0013	.0002 ± .0002
Br	.0003 ± .0005	.0019 ± .0026	.0003 ± .0004
Rb	.0006 ± .0007	.0007 ± .0066	.0006 ± .0007
Sr	.0004 ± .0008	.0028 ± .0036	.0005 ± .0008
Y	.0006 ± .0011	.0018 ± .0036	.0007 ± .0012
Zr	.0000 ± .0010	.0024 ± .0178	.0001 ± .0011
Mo	.0010 ± .0018	.0008 ± .0097	.0010 ± .0017
Pd	.0006 ± .0006	.0047 ± .0104	.0007 ± .0007
Ag	.0000 ± .0003	.0002 ± .0066	.0000 ± .0004
Cd	.0000 ± .0010	.0127 ± .0203	.0006 ± .0012
In	.0000 ± .0005	.0000 ± .0094	.0000 ± .0006
Sn	.0008 ± .0016	.0000 ± .0278	.0008 ± .0018
Sb	.0001 ± .0024	.0148 ± .0427	.0008 ± .0029
I	.0001 ± .0001	.0000 ± .0007	.0001 ± .0001
Cs	.0000 ± .0001	.0001 ± .0001	.0000 ± .0001
Ba	.0006 ± .0027	.0158 ± .0484	.0012 ± .0032
La	.0024 ± .0080	.0379 ± .1369	.0043 ± .0094
Ce	.0000 ± .0001	.0007 ± .0010	.0000 ± .0001
Nd	.0000 ± .0001	.0000 ± .0017	.0000 ± .0001
Hg	.0001 ± .0001	.0018 ± .0026	.0001 ± .0001
Pb	.0099 ± .0044	.0191 ± .0077	.0103 ± .0043
Th	.0000 ± .0001	.0002 ± .0002	.0000 ± .0001
U	.0000 ± .0001	.0000 ± .0004	.0000 ± .0001
OC	.9914 ± .8184	4.2823 ± 4.2072	1.1918 ± .6602
EC	.2139 ± .1729	.4608 ± .9938	.2230 ± .1947
Sum	18.4494	30.1416	19.0382

The average fine to coarse particle mass ratios were determined only for the incinerator stack, vehicle emissions, and road dust. The incinerator stack emissions consisted primarily of particles less than 2.5 μm (F/C=24.9), while the road and soil dust samples consisted of mostly coarse particles (F/C=0.041). Vehicle emissions were mostly fine particles (F/C=15.0). It was assumed that the emissions from the flare, compressor, and sulphur handling facilities were also mostly fine particles.

The major elemental features of the sources with the most common fit are listed in Table 5.6. The profiles are clearly distinguishable in most cases. The differences between vegetation and vegetation residue burning are rather minor, and the contribution of these two source categories could not be separated.

Table 5.6 Source profiles: elemental percent composition of the fine particle fraction.

Chemical Species	Soil	Road Dust	Mobile Source Emissions	Vegetation	Vegetation Residue Burning
Al	10.6843	9.4884	0.6415	0.0870	0.3400
Si	36.8305	33.4279	0.7856		0.3000
S	0.2809	0.6760	0.0000		1.5000
Cl	0.0788	0.2506	0.2624		3.1000
K	1.2131	1.4310	0.0511	3.5600	5.9000
Ca	1.6522	7.0099	0.0918	1.3500	0.4000
Ti	0.4110	0.3564	0.0040	0.0050	
V	0.0222	0.0232	0.0018		
Mn	0.0883	0.1054	0.9873	0.0165	0.0040
Fe	3.0968	4.1812	0.1113	0.0550	0.2000
Ni	0.0062	0.0114	0.0120	0.0006	
Cu	0.0107	0.0103	0.0264	0.0012	0.1100
Zn	0.0559	0.0657	0.1483	0.0050	
Br	0.0017	0.0092	6.1834	0.0054	0.0300
Pb	0.0033	0.1145	21.6486		0.0300
SO_4^{2-}					5.0000
NO_3					0.8900

5.7.3.2 Emissions from the Incinerator Stack

The incinerator stack emissions consisted primarily of fine particle sulphur species which were assumed to be mostly sulphuric acid, based on the liquid droplets on the filters and the reactivity of the deposited material with ammonia. Sulphur represented 16.6% ± 3.1% of the mass measured in a controlled nitrogen atmosphere. However, even under these conditions it was possible that the material collected on the filter still contained substantial quantities of water. The concentration of sulphur was relatively constant over the large number of samples collected.

The next most abundant fine particle species was organic carbon (OC) which accounted for 0.99% ± 0.82% of the deposited mass. Elemental carbon accounted for 0.21% ± 0.17% of the mass.

Aluminum and silicon concentrations were about 0.2%, but values for the standard deviation or propagated analytical uncertainty were almost twice the average. The concentration of iron was 0.044% ± 0.048%, while the concentration of most of the other species was less than 0.01%. The concentration of over half the measured species was less than 0.001%.

The average chemistry of the coarse particle species was substantially different from the fine particle species, but the concentration of even the most abundant species showed a wide range of variability after subtracting for the fine particle contamination from the coarse particle filter. Silicon was the most abundant species at 8.9% ± 7.0%, but Fe and Mn exhibited the lowest relative values for the standard deviation. The ratios of Cr, Fe, and Ni concentrations were similar to what would be expected from the stainless steel incinerator construction materials.

5.7.3.3 Emissions from the Prill, Flare, and Inlet Compressor Stacks

Emissions from these sources were similar to the incinerator stack; they were dominated by fine particle sulphur. Although the trace metal species were, in some cases, substantially different, the utility of these trace species in resolving the sources must still be evaluated. The quartz filters from which the OC/EC concentrations were determined were not matched by simultaneous sampling with Teflon filters for elemental analysis. Thus, comparable filter pairs were not available in all cases.

5.7.3.4 Fugitive Soil and Road Dust

The source characteristics of the soil and road dust were similar to the previous results. Over 90% of the PM_{10} mass was associated with coarse particles, and Si was the most abundant species, representing between 20% and 30% of the mass. The next most abundant species were generally Al, Ca, and Fe. Calcium was the more variable of the three elements, ranging from less than 1% in soil to more than 10% in unpaved road dust samples.

5.7.3.5 Emissions from Vehicles

The major components in the vehicular emissions were similar to what has been previously reported for diesel and leaded fuel-utilizing vehicles. The diesel emissions were dominated by carbon (about 60%), while the emissions from leaded gasoline fuel-utilizing vehicles were dominated by Pb, at about 36%, and Br at 18%.

The emissions from the use of unleaded fuel were quite low and the analytical uncertainties high. The most abundant inorganic species were Mn and Cl. The data on organic and elemental carbon were omitted from the final source profiles because of the high uncertainty.

5.7.4 Results of Source Apportionment Modelling

Chemical mass balance-source apportionment calculations were performed on the data from about 100 ambient particulate samples. Table 5.7 provides an example of the results of the calculations.

Table 5.7 An example of CMB-source apportionment results. Particle size: Fine (<2.5μm). Sampling Site: Crossfield East. Sampling duration: 48 h. Reduced chi square: 0.494. Degrees of freedom: 6.

Measured Concentration			Calculated Concentration	Ratio of Calculated and Measured Concentrations	
Species	$\mu g\ m^{-3}$	Percent	$\mu g\ m^{-3}$		
Al *	.082 ± .010	2.140	.092 ± .002	1.121 ± .142	Al
Si *	.347 ± .039	9.050	.331 ± .008	.954 ± .111	Si
P	<.002		.005 ± .000	.000 ± .000	P
S *	.367 ± .042	9.560	.359 ± .035	.979 ± .148	S
Cl	<.002	87	.004 ± .001	.000 ± .000	Cl
K *	.030 ± .004	.787	.030 ± .002	.999 ± .132	K
Ca *	.100 ± .011	2.613	.103 ± .007	1.028 ± .134	Ca
Ti *	.004 ± .001	.107	.003 ± .000	.848 ± .124	Ti
V	.000 ± .000	.008	.000 ± .000	.756 ± .442	V
Cr	.000 ± .000	.011	.000 ± .000	.513 ± .224	Cr
Mn *	.002 ± .000	.060	.002 ± .000	.833 ± .203	Mn
Fe *	.043 ± .005	1.131	.044 ± .003	1.026 ± .138	Fe
Ni	.000 ± .000	.007	.000 ± .000	.415 ± .257	Ni
Cu	<.000		.000 ± .000	1.288 ± 3.712	Cu
Zn *	.002 ± .000	.064	.002 ± .000	1.000 ± .163	Zn
A1	0.082 ±.002	2.140	.002 ± .008	1.121 ± .142	A1

Continued...

Table 5.7 (Concluded).

Measured Concentration			Calculated Concentration	Ratio of Calculated and Measured Concentrations	
Species	$\mu g\ m^{-3}$	Percent	$\mu g\ m^{-3}$		
Ga	.000		.000 ± .000	1.116 ± 5.686	Ga
As	.001		.000 ± .000	.000 ± .000	As
Se	.000		.000 ± .000	.176 ± 1.096	Se
Br *	.003 ± .000	.083	.004 ± .001	1.108 ± .397	Br
Rb	.000 ± .000	.011	.000 ± .000	.124 ± .117	Rb
Sr	.001 ± .000	.017	.000 ± .000	.352 ± .175	Sr
Y	.000		.000 ± .000	.147 ± .355	Y
Zr	.002		.000 ± .000	.000 ± .000	Zr
Mo	.001		.000 ± .000	.000 ± .000	Mo
Pd	.001		.000 ± .000	.000 ± .000	Pd
Ag	.001 ± .001	.036	.000 ± .000	.044 ± .063	Ag
Cd	.003 ± .002	.068	.000 ± .000	.051 ± .051	Cd
In	.002		.000 ± .000	.152 ± .264	In
Sn	.003		.000 ± .000	.150 ± .256	Sn
Sb	.006		.000 ± .000	.000 ± .000	Sb
Ba	.011		.000 ± .001	.000 ± .000	Ba
La	.019		.000 ± .001	.000 ± .000	La
Hg	.000		.000 ± .000	.472 ± 2.261	Hg
Pb *	.013 ± .002	.335	.013 ± .004	1.001 ± .356	Pb
SO_4^{2-} *	1.001 ± .224	26.111	1.054 ± .105	1.052 ± .258	SO_4^{2-}
NO_3^- *	.452 ± .101	11.790	.452 ± .045	1.000 ± .245	NO_3^-
Mass	3.8 ± 0.40				

* Fitting species

The CMB results presented represent the best solution as determined by an iterative procedure which optimized goodness-of-fit parameters. These parameters, in the order of their importance, are:

1. Source contributions should be positive and greater than their uncertainties.
2. Reduced chi square should be minimized, generally to less than 2.
3. The ratio of calculated to measured concentration for individual chemical species should approach 1.0 within the listed uncertainty.
4. The calculated total mass concentrations should approach the measured mass concentrations within the uncertainties.
5. The number of degrees of freedom should be maximized.

In the example shown in Table 5.7, 93.421% ± 13.962% of the mass was explained by the sources indicated. A reduced chi square of 0.494 with 6 degrees of freedom was obtained.

The annual average source contributions are listed by site and particle size in Table 5.8. The two largest primary source categories in both the fine and coarse particle size fractions were soil and road dust and vegetation or combustion of vegetation residue. The largest source categories in the fine particle size fraction at both monitoring sites were identified by sulphate and nitrate ions, presumably as their ammonium salts.

Table 5.8 Crossfield East and Crossfield West: summary of average annual contributions of various sources to primary particle concentrations.

	Fine		Coarse	
Source	$\mu g\ m^{-3}$	Percent	$\mu g\ m^{-3}$	Percent
Crossfield East				
Soil and Road Dust	0.90	12.75	8.52	45.81
Vehicle Exhaust	.09	1.27	0.00	0.00
Vegetation/Residue Burning	0.59	8.36	0.56	3.01
Ammonium Sulphate	1.46	20.68	0.00	0.00
Ammonium Nitrate	0.99	14.02	0.00	0.00
Sulphur	0.12	1.70	0.03	0.16
Other	0.01	0.14	0.00	0.00
Unexplained	2.90	41.08	9.49	51.02
Mass	7.06	100.00	18.60	100.00
Crossfield West				
Soil and Road Dust	0.51	11.02	4.16	78.20
Vehicle Exhaust	0.03	0.65	0.00	0.00
Vegetation/Residue Burning	0.58	12.53	0.26	4.89
Ammonium Sulphate	0.99	21.38	0.00	0.00
Ammonium Nitrate	0.61	13.17	0.00	0.00
Sulphur	0.08	1.73	0.02	0.38
Other	0.00	0.00	0.00	0.00
Unexplained	11.83	39.52	0.88	16.54
Mass	4.63	100.00	5.32	100.00

The road dust contribution to both the fine and coarse particle size fractions was well established based on the fitting of elements such as Al, Si, Ti, Mn, and Fe, which appeared to be relatively unique to this source category. In addition, the fine to coarse ratio for this category was similar to the ratio measured during the laboratory aerosolization studies. The vegetation category was much more uncertain because most of the fitting pressure was based on excess K which was not unique to this source. In addition, the amount of K associated with both vegetation and combustion of vegetation residues can vary by an order of magnitude. The contribution from this source category could have been better defined if the organic and elemental carbon content of the ambient air filter samples had been determined.

The average percent unexplained mass was substantial at both monitoring sites and in both size fractions. About 40% of the fine particle mass remained unexplained at both sites, while 51% of the coarse particle mass at Crossfield East and 16.5% at Crossfield West could not be explained. The large unexplained mass was possibly associated with an aerosol source with emissions that were composed mostly of carbon species. Carbon is one of the few potential major aerosol species that was not determined for the ambient particulate samples.

The contribution of vehicle exhaust was calculated for vehicles using unleaded fuel. The contribution from the entire vehicle fleet may be as much as twice this value depending on the characteristics of the fleet population and the concentration of lead in the fuel used in the airshed. Most of the particulate mass relevant to vehicle exhaust was generally associated with diesel emissions, primarily carbon.

The contributions to fine and coarse particles by soil and road dust were about twofold greater at Crossfield East as compared with West. The differences in the contributions by secondary particles were not as great, but still significant, with the highest values reported at Crossfield East.

The source contributions under varying wind regimes are summarized in Table 5.9. The influence of soil and road dust at Crossfield East was greatest when the wind was out of the north-northwest direction, down-wind direction of the gas plant. A major gravel road is located north of the site, between the sampling site and the gas plant (Figure 5.1). The contribution of vehicle exhaust at Crossfield East was highest when the wind was from the south. Similarly, the impact of vehicle exhaust at Crossfield West was highest when the wind was from the south-southeast, with both monitoring sites in the downwind direction from Calgary.

Table 5.9 Comparison of average source contributions ($\mu g\ m^{-3}$) under varying wind conditions.

Site	Wind Direction	Fine Particles Soil and Road Dust	Fine Particles Vehicle Exhaust	Vegetation	$(NH_4)_2SO_4$	NH_4NO_3	Coarse Particles Soil and Road Dust
1	NNW	1.33	0.07	0.74	1.26	0.72	12.95
1	SSE	0.68	0.08	0.50	1.89	1.36	3.52
1	S	0.88	0.13	0.55	1.03	0.90	11.71
1	SE,E,ESE	0.70	0.07	0.56	1.63	0.97	3.68
2	SSE,S	0.56	0.06	0.73	1.61	1.06	7.03
2	SW	0.39	0.03	0.47	0.42	0.40	3.72
2	NW,NNW	0.61	0.03	0.46	1.30	0.49	3.93
2	W,WSW,SW,SSW	0.46	0.02	0.65	0.57	0.49	1.72

The secondary sulphate and nitrate concentrations were also high during southerly wind flows when the sites were influenced by the Calgary plume. The concentrations at Crossfield East when it was influenced by the gas plant (north-northwest), were lower than when the winds were from the south-southeast or other easterly directions. This suggested that the gas plant did not make a significant contribution to the secondary particle concentrations at Crossfield East. This was not too surprising considering the height of the incinerator stack, the largest source, and the time required to convert the gaseous emissions to secondary particles. Another explanation for the lack of a clear dependence of the source impacts on the wind direction may be because, the aerosol data were from 48 hour samples, and the variability in the wind direction during each sampling period may have been significant. Unfortunately sufficient mass of the aerosols could not be obtained for the chemical analyses when the samples were collected over shorter duration.

5.8 COMPOSITE ANALYSIS

5.8.1 Overview

The objective of the composite analysis was to evaluate the source apportionment results in terms of a complete data set that included all of the available information as a distinct set from the CMB calculations performed on data from individual filters.

5.8.2 Quality of Individual CMB Regression Analysis

The quality of each individual CMB apportionment calculation was judged on the basis of three primary goodness-of-fit criteria: reduced chi square (should be less than 2.0), measured to calculated elemental ratios (should be 1.0 within the listed uncertainty, two-thirds of the time), and the percent of total mass explained should be 100% within the listed uncertainty, two-thirds of the time.

In this particular case, the reduced chi square values were less than 2.0 in every case and, in general, substantially less than 1.0. The ratios of calculated to measured elemental mass were within one standard deviation of 1.0 for almost all the major elements in the coarse fraction. The ratios of calculated to measured elemental mass for the primary fitting elements in the fine fraction were also close to 1.0, but a few minor elements, typically included in the regression analysis, were removed from the fitting because of the obvious limitations discussed below. The average mass explained was generally not close to 100%. This deviation from the ideal was of utmost concern as to the completeness of the modelling results.

In general, the quality of the individual CMB regression analyses was quite good. Deviations from the optimum are discussed in the following subsections, but such deviations were considered to have minimal impact on the conclusions regarding the major sources identified.

5.8.3 Elemental Ratios

The ratios of calculated to measured elemental mass in both the fine and coarse particle fractions were close to 1.0, within the uncertainties for most of the elements included in the fitting. Occasionally, Al deviated substantially from 1.0, and K would have been consistently underexplained if the vegetation source category was not included in the fit.

Almost all the major elements in the coarse particle size fraction fitted well with the soil and road dust profiles, suggesting that the laboratory-generated profiles were good representations of what was in the ambient aerosol. Even the coarse particle fits at Crossfield West, where the local soil and road dust samples were not collected, was quite good.

Most of the major crustal elements in the fine particle size fraction, such as Al, Si, Ti, Mn, and Fe, were explained with the soil and road dust and in the expected fine to coarse particle ratio if these elements were primarily from the same sources. The fine to coarse ratio for the other major crustal elements were consistent with the laboratory-generated source profiles, while the K fine to coarse ratio was higher than expected. However, this was consistent with the inclusion of the vegetation and burning of vegetation refuse category. Measurement of the organic and elemental carbon concentrations in the ambient aerosol would have helped to resolve the overall issue.

The total elemental S measured by XRF was generally in good agreement with the sulphate measured by ion chromatography. However, there were times when the sulphur concentration was in excess of what could be explained by sulphate. Presence of this excess sulphur suggested a small contribution from a source of reduced sulphur such as fugitive dust or one of the stacks. Because only a few coarse particle samples were analyzed, it was not clear whether this excess sulphur was associated primarily with fine or coarse particles.

5.8.4 Dependence on the Meteorology and the Site

Although there was a clear difference in the average particulate matter concentrations between the two monitoring sites, the sampling days selected at the two sites were not necessarily the same; this made it difficult to draw strong conclusions from intersite comparisons. The particulate matter concentrations were clearly low at Crossfield West as would be expected from its relatively more remote location. In addition, the annual average concentrations of vehicle exhaust components and secondary sulphate and nitrate were higher at Crossfield East as might be expected due to potential sources in its vicinity (Highways 1 and 1A, a major gravel road and a natural gas processing plant). The lack of a clear correlation with wind direction, and the potential influence of the Calgary metropolitan area on both monitoring sites, reduced the strength of conclusions drawn from meteorology and site-specific source contributions.

5.8.5 Dependence on the Particle Size

As discussed previously, the particle size dependence of the source CMB contributions was consistent with the known particle size characteristics of the sources. The soil and road dust fine to coarse particle ratio of about 1:10 was similar to the ratio measured in laboratory aerosolization studies. In addition, most of the mobile and secondary particulate source contributions were apportioned to the fine particle size fractions as expected from their source characteristics.

5.8.6 Impacts of Individual Sources

5.8.6.1 Soil and Road Dust

The soil and road dust contribution was clearly established with the large number of crustal elements that fitted to this source category. The elemental ratios in both the fine and coarse particle size fractions were in excellent agreement with the source profiles, and the fine to coarse particle ratios were also in agreement with laboratory results. In addition, the percent composition of the major elemental species such as Al, Si, Ca, and Fe were in good agreement with the expected crustal material. The contribution from this source was clear, and the relative uncertainty in its impact was expected to be less than 10%.

5.8.6.2 Secondary Sulphate and Nitrate

The sulphate and nitrate were assumed to occur as ammonium salts. In addition, these salts were assumed to be predominantly associated with secondary particle formation from gaseous precursors. This source contribution was based on the residual sulphate and nitrate after removing the portion associated with the sources of primary particles. This represented an upper limit to the contribution by the secondary particles because the sources of primary particles, composed mostly of these chemical species, could not be resolved. The lack of a strong meteorological dependence for these sources suggested that the influence of local sources was not dominant, and therefore, mainly a larger area source of secondary particles was involved.

5.8.6.3 Vegetation and Burning of Vegetation Residue

This source category fitted mainly to the excess K and was thought to have a very high degree of uncertainty both in its qualitative features as a source and in its magnitude of contribution. Because only one element (K) was used as the main indicator of the source type, and since that element could have also been contributed by another source, identification of a specific source type must be considered to be highly uncertain. There was, however, a consistent pattern of unexplained K when the vegetation source was not used.

Even if the source was correctly identified, its level of contribution was uncertain because the concentration of K in this source category varied by more than a factor of 10. Inclusion of organic and elemental carbon measurements and the use of K corrected for the soil derived fraction would have helped to establish the source contribution.

5.8.6.4 Mobile Source Emissions

As already noted, the contribution of the vehicle exhaust was based on the fitting of the leaded gasoline exhaust profile to the Br and Pb concentrations. The attribution of this source impact was clearly established by its relatively unique profile and expected high fine to coarse particle ratio. Since this represents only the contribution from exhaust of vehicles using leaded gasoline, the estimation of the contribution of the entire motor vehicle fleet would require either developing a composite source profile, based on emission inventory weighting of individual source profiles, or emissions inventory scaling of the attributed source contribution.

Other subcategories in the motor vehicle classification include unleaded and diesel exhaust, as well as tire and brake wear. Although there may be many more vehicles using unleaded than leaded gasoline, the emissions from vehicles using unleaded gasoline were much lower. Diesel exhaust was usually the largest subcategory. Tire and brake wear emissions were also significant. The total motor vehicle contribution to primary particulate concentrations could easily be two to four times greater than what was indicated by the contributions from leaded gasoline (Cooper, et al. 1987).

5.8.6.5 Unexplained Mass

The large unexplained mass represented one of the most significant limitations in the CMB results. The excellent CMB fits of the measured inorganic species, and the agreement of the fine to coarse ratio with the source characteristics, provided a high level of confidence in the source contributions identified, except for the category of vegetation/vegetation refuse burning. Thus, the major components associated with the unexplained mass must consist mostly of species not already included with the sources identified or measured, such as H, C, N, and O. Sources of these elements not identified by the species already measured include: secondary organic carbon, diesel exhaust, tire wear, artifact organic carbon, and water. The first four are primarily sources of carbon. Because of the relatively low particle concentrations at the two monitoring sites, these sources may have been responsible for the magnitude of unexplained mass. As already noted, another source of unexplained mass could be associated with the profile uncertainties used for the vegetation and vegetation refuse combustion. In the case of the coarse particle size fraction, loss of particles during shipping could also be a source of the unexplained mass.

5.9 CONCLUDING REMARKS

The major sources of primary particles at the Crossfield ADRP monitoring sites were soil and road dust. The soil erosion in ploughed fields, windy conditions, the generation of dust by vehicles travelling on gravel roads, low precipitation during the summer, and the lack of snow cover in the winter were considered to be responsible.

Secondary fine particles consisted largely of SO_4^{2-} and NO_3^- derived from the conversion of regional SO_2 and NO_x emissions. If the SO_4^{2-} and NO_3^- occurred as ammonium salts, these species would represent about 60% of the mass accounted for in the CMB analysis. The assignment of the SO_4^{2-} and NO_3^- to $(NH_4)_2SO_4$ and NH_4NO_3 was considered to be reasonable because the correlation coefficient (r) between SO_4^{2-} and NH_4^+ at Crossfield West was 0.94, and between NO_3^- and NH_4^+ was 0.77.

The CMB gave no evidence of a direct contribution of fine particle SO_4^{2-} from the Crossfield natural gas processing plant to the monitoring sites nor of the formation of secondary SO_4^{2-} from the SO_2 emissions from the gas plant. This was consistent with the short distance between the gas plant and the monitoring sites, lack of sufficient time for conversion of SO_2 to SO_4^{2-}, and the low atmospheric concentrations of SO_2 at the two monitoring sites.

Fine particle sulphate concentrations were highest at both monitoring sites when the wind was from the direction of the city of Calgary. Because the Balzac natural gas processing plant is located 6 km north of the city, it was not possible, at the Crossfield East site, to distinguish the contributions of the Balzac gas processing plant and the city of Calgary. The abundance of fine particle SO_4^{2-} at the Crossfield West monitoring site, when the wind was from the S and SSE, indicated that the city rather than the gas plant was the source of the fine particulate SO_4^{2-}.

In addition, the highest concentrations of fine particulate nitrate occurred at both sites when the wind was from the direction of Calgary, indicating that the city was the source of the fine particles or its precursors. There was no evidence from the CMB analysis of a direct contribution of fine particulate nitrate from the Crossfield natural gas processing plant to either of the monitoring sites.

5.10 LITERATURE CITED

APCA. Air Pollution Control Association. 1984. *Receptor Models Applied to Contemporary Pollution Problems*. SP-48. 368 pp.

Cooper, J.A., D.C. Redline, J.R. Sherman, L.M. Valvolinos, W.L. Pollard, L.C. Scavone, and C. Badgett-West. 1987. PM_{10} Source Composition Library for the South Coast Air Basin. Final Report to South Coast Air Quality Management District. Vol. 2. NEA, Inc., Beaverton, Oregon.

Henry, R.C., C.W. Lewis, P.K. Hopke, and H.J. Williamson. 1984. Review of receptor model fundamentals. *Atmospheric Environment* 18: 1507-1515.

Hidy, G.M. 1984. Source-receptor relationships for acid deposition: Pure and simple? *Journal of the Air Pollution Control Association* 34: 518-531.

Huntzicker, J.J., R.L. Johnson, J.J. Shah, and R.A. Cary. 1982. Analysis of organic and elemental carbon in ambient aerosols by a thermal-optical method. In: **Particulate Carbon: Atmospheric Life Cycle**, eds. G.T. Wolff and R.L. Klinisch. New York: Plenum Press. pp. 79.

NEA, Inc. 1986. QSAS III Source Profile Library. Beaverton, Oregon.

Rahn, K.A. and D.H. Lowenthal. 1985. Pollution aerosols in the northeast: Northeastern-midwestern contributions. *Science* 228: 275-284.

Stevens, R.K. and T.G. Pace. 1984. Overview of the Mathematical and Empirical Receptor Models Workshop. (Quail Roost II). *Atmospheric Environment* 18: 1499-1506.

U.S. EPA. Environmental Protection Agency. 1984. Receptor Model Source Composition Library, U.S. EPA, Research Triangle Park, North Carolina.

CHAPTER 6

WET AND DRY DEPOSITION OF AIR POLLUTANTS IN ALBERTA

E. Peake and C. I. Davidson

6.1 INTRODUCTION

The deposition of acidic and acidifying substances from the atmosphere on to receptor surfaces has been of concern in Europe and North America for the past 30 years. Plants, soils, and surface waters respond to chemical constituents which are deposited on their surfaces by wet deposition, in the form of rain, snow, and fog, and by dry deposition, in the form of gases and particles. The ecological effects of anthropogenic emissions are complex and difficult to understand, but a key parameter is the deposition of the hydrogen ion and other species derived from both natural sources and man-made emissions. A description of the major species of interest, the mechanisms of their deposition, and the relative importance of wet and dry deposition was provided in Chapter 2. The wet and dry deposition of major ions was estimated for each of five major geographic emission regions in Alberta. These regions, 3, 4, 6, 7, and 9 in Figure 6.1, were selected for modelling the effects of atmospheric deposition on crops, soils, and surface waters (Chapter 9).

Estimates of wet deposition for Region 9, which stretches east from the Great Divide of the Rocky Mountains to the plains of southern Alberta, 80 km east of Calgary, were based on measurements made at the Crossfield and Fortress Mountain monitoring sites (Chapters 3 and 4). In Regions 3, 4, 6, and 7 wet deposition was measured by Alberta Department of Environment at one site within each region, and this deposition value was taken to represent the region.

There are no simple, direct methods for measuring dry deposition (Hicks et al. 1987), but two approaches are generally used. The first approach is to carefully analyze the chemistry of exposed surfaces such as leaves or surrogate surfaces; the second is to infer dry deposition fluxes from the atmospheric concentrations of the

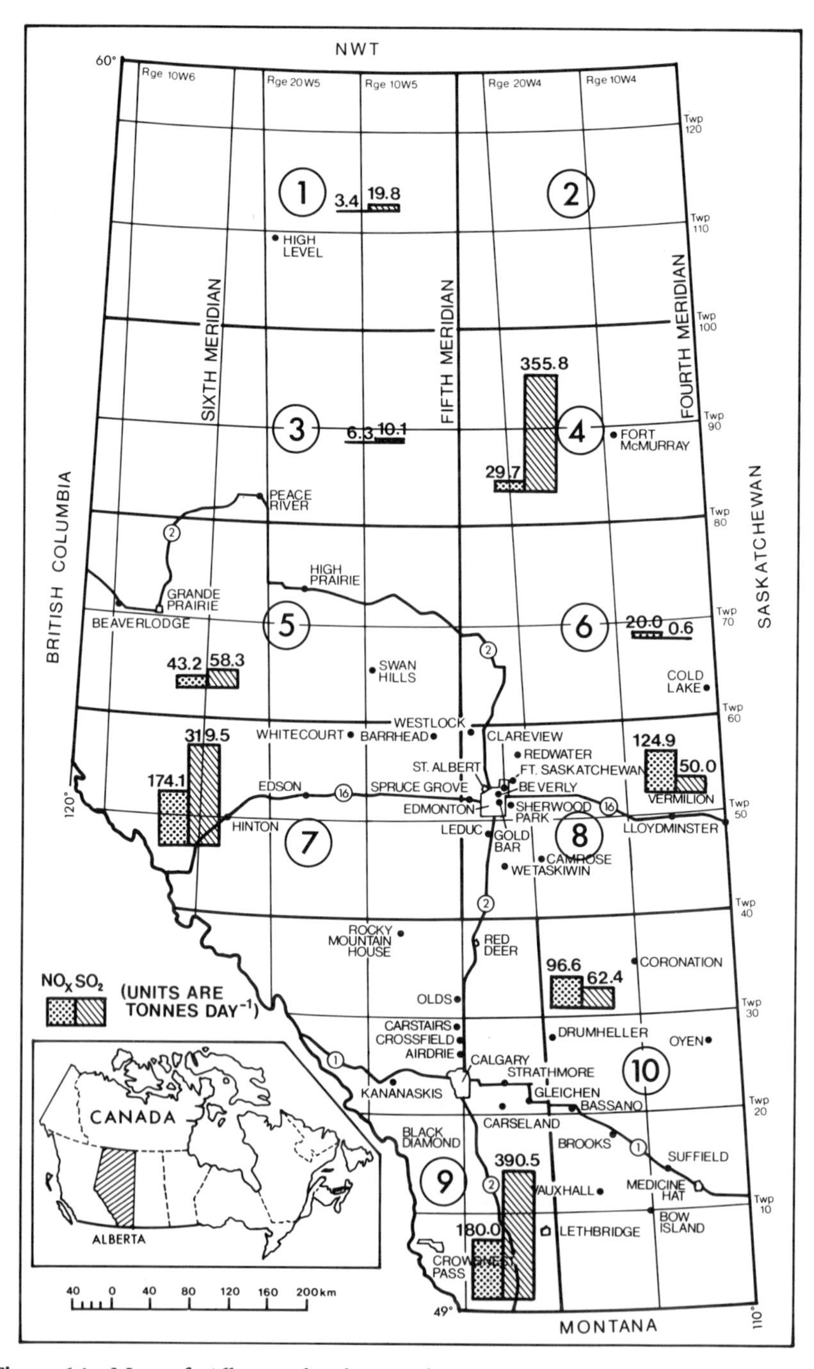

Figure 6.1 Map of Alberta showing total NO_x and total SO_2 emissions as histograms in each of the 10 geographic emission regions.

pollutants in question using the formula:

$$-F = C \bullet V_d$$

where F is the flux away from the surface (thus the flux toward the surface is -F), C is the atmospheric concentration of the chemical species of interest and V_d is its deposition velocity. Atmospheric concentrations of SO_2, NO_x, and fine particle SO_4^{2-} were obtained from the atmospheric dispersion modelling described in Chapter 8, together with measurements of fine and coarse particle concentrations at the three ADRP sampling sites.

The average deposition velocity of about 1.0 cm sec^{-1} for gases and 0.1 cm sec^{-1} for fine particles are typically used to estimate pollutant fluxes (Sheih et al. 1979). However, deposition velocities vary greatly between gases and as a function of meteorology and the nature of the surface. There are no direct means of measuring the deposition velocity of a gas, such as SO_2, over diverse topography such as that contained in Region 9. One indirect approach is to use a resistance model:

$$V_d = \frac{1}{r_a + r_b + r_c}$$

in which the resistance to transport is divided into three components: the aerodynamic resistance, r_a; the boundary layer resistance, r_b; and the canopy resistance, r_c. Approximate values of r_a and r_b can be calculated from knowledge of the wind speed and standard deviation in the wind direction (Hicks et al. 1985) and estimates of r_c have been made for different land-use types under a variety of conditions (Walcek et al. 1986). Calculations of r_a and r_b were made at each of the three monitoring sites for daytime and nighttime, summer and winter. The V_d of SO_2 was then calculated using r_c values from the literature for agricultural land at Crossfield and coniferous forest at Fortress Mountain. It was assumed that the landscape was dry 85% of the time and that r_c approached zero for a wet landscape.

Site specific measurements of deposition velocity can be made by a number of methods (Hicks 1987). The gradient method was applied to measure the V_d of SO_2 over a field of stubble near Crossfield. Deposition velocities for SO_2 in Alberta were then estimated from the experimental results and from published values.

Table 6.1 provides a listing of the symbols and their definitions as they relate to the following sections of this chapter.

6.2 WET DEPOSITION

6.2.1 Wet Deposition in Region 9

6.2.1.1 Methodology

Precipitation in Alberta generally occurs in the form of rain and snow with very little fog. Rainwater was collected using refrigerated sequential rain samplers (Coscio et al. 1982) at each of the three ADRP sampling sites. Samples of fresh snowfall were collected during winter visits to Fortress Mountain and occasionally at the Crossfield sampling sites. At Crossfield, the snow cover was scant and usually in the form of a mixture of wind-blown soil and snow. Only fresh, relatively uncontaminated, snow samples were collected.

The depth of rainfall at each site was measured during the summer months by a tipping bucket rain gauge, but the values for the depth of snowfall were not available. Therefore, the total yearly precipitation measured by the Atmospheric Environment Service of Canada at Madden, located about 13 km NW of the Crossfield West sampling site, was used in calculating annual wet deposition in the Crossfield area, and, similarly, precipitation amounts measured in the Kananaskis Valley were used to calculate wet deposition at the nearby Fortress Mountain. The annual average precipitation at Madden over the two-year period (1985-87) was 457 mm of which 360 mm was rain. This compared favourably with 349 mm of precipitation, mainly rain, measured at the ADRP Crossfield West site.

The depth of precipitation would be expected to vary greatly within Region 9 which includes both the Rocky Mountains (Fortress Mountain monitoring site), and the plains (Crossfield monitoring sites). The mean annual precipitation over the past 30 years at a Kananaskis Valley monitoring station (near to Fortress Mountain) was 657 mm and during the study period was 604 mm y^{-1}. In estimating pollutant deposition for Region 9 it was considered that the amount of precipitation for one-third of the region would more closely approximate that of the Kananaskis Valley and two-thirds

Table 6.1 Definition of symbols.*

Symbol		Definition
C	=	airborne concentration of contaminant (g cm^{-3})
C_o	=	airborne concentration of contaminant at z = 0 (g cm^{-3})
$C(z)$	=	airborne concentration of contaminant at height z above the surface (g cm^{-3})
D	=	Brownian diffusivity of the contaminant (cm^2 sec^{-1})
d	=	zero-plane displacement (cm)
F	=	flux or dry deposition rate of contaminant (g cm^{-2} sec^{-1})
F_a	=	flux of air momentum (g cm^{-2} sec^{-1})
K	=	eddy diffusivity of contaminant mass (cm^2 sec^{-1})
K_M	=	kinematic eddy viscosity of air (cm^2 sec^{-1})
k	=	von Karman's constant = 0.4
L	=	Monin-Obukhov length (cm)
r_a	=	aerodynamic resistance (s cm^{-1})
r_b	=	boundary layer resistance (s cm^{-1})
r_c	=	canopy resistance (or surface resistance) based on the flux to the horizontal area of the earth's surface (s cm^{-1})
r_t	=	total resistance to transport (s cm^{-1})
μ	=	air velocity in the direction of mean wind flow (cm sec^{-1})
μ_*	=	friction velocity (cm sec^{-1})
v_d	=	dry deposition velocity (cm sec^{-1})
z	=	height above the surface (cm)
z_o	=	roughness height or momentum sink (cm)
z_{oc}	=	contaminant sink (cm)
ζ	=	dimensionless height = (z-d)L^{-1} used in stability-dependent correction factors and in the model of Shreffler (1978)
P_a	=	bulk density of air (g cm^{-3})
σ_θ	=	standard deviation of the horizontal wind direction (degrees)
σ_μ	=	standard deviation of the horizontal wind direction at an air velocity of μ (degrees)
τ_o	=	shear stress at the surface (g cm s^2)$^{-1}$
ϕ_C	=	stability-dependent correction factor applied to eddy transport of contaminant mass
ϕ_H	=	stability-dependent correction factor applied to eddy transport of heat
Ψ_C	=	stability-dependent correction factor in the equation for r_a for contaminant mass
Ψ_M	=	stability-dependent correction factor in the equation for r_a for air momentum

* Source: Davidson and Wu (1989).

would resemble that at Madden. Using this ratio, the estimated annual precipitation for the region as a whole was 506 mm.

Wet deposition was calculated on a daily volume-weighted basis. All samples of rainwater collected by the refrigerated sampler on a given day were analyzed for SO_4^{2-}, NO_3^-, and Cl^- by ion chromatography, Ca^{2+}, Mg^{2+}, Na^+, and K^+ by atomic absorption spectrometry, and NH_4^+ by colorimetry. The pH was also determined. The mean chemical composition of the rainwater for a given day was multiplied by the amount of rainfall for that day to derive daily volume-weighted wet deposition. Daily volume-weighted values were summed to give annual wet deposition and corrected for days when rain or snow occurred but no sample was collected. For those days, the value for the mean chemical composition for all samples collected at that site was substituted for the missing value to calculate wet deposition.

For use in subsequent effects models, Chapter 9, a presumed "maximum" amount of annual wet deposition was calculated. The maximum was calculated using the value of the 90th percentile of the daily volume-weighted deposition obtained during the study at Crossfield. This value of the 90th percentile for daily wet deposition was applied to the annual deposition. These hypothetical values could only be reached if, at some future date, emissions dramatically increased. Present day emissions in Alberta are the equivalent of 21 kg ha^{-1} y^{-1} total wet and dry sulphate deposition over the entire province.

6.2.1.2 Results

6.2.1.2.1 Estimated Wet Deposition.

The values of wet deposition for Region 9 are presented in Table 6.2. Wet deposition of sulphate for Region 9 was 8.2 kg ha^{-1} y^{-1}, lower than the 12.3 kg ha^{-1} y^{-1} reported for Calgary for the period 1978-84 by Lau and Das (1985). The wet deposition of nitrate, 4.9 kg ha^{-1} y^{-1}, and ammonium ion, 2.4 kg ha^{-1} y^{-1}, were close to the values obtained by Alberta Department of Environment, 5.0 kg ha^{-1} y^{-1} and 2.6 kg ha^{-1} y^{-1}, respectively (Lau and Das 1985). Concentrations of these ions on an average were lower in the ADRP samples used in the present calculation of wet deposition in Region 9 compared to the Calgary samples analyzed by Alberta Department of Environment (see Tables 2.17 and 2.18, in Chapter 2). The estimated rainfall in the present study for Region 9 was 509 mm versus an average for the period 1978-84 of 398 mm in Calgary. Thus the two estimates of volume-weighted deposition were approximately the same. The wet deposition of sulphate and nitrate in Region 9 was low compared to most other regions in Canada (Table 2.18 in Chapter 2).

Table 6.2 Mean and maximum wet and dry deposition of air pollutants for Region 9 (kg ha^{-1} y^{-1}).

	Mean			Maximum		
	Wet	Dry	Total	Wet	Dry	Total
REGION 9						
SO_4^{2-}	8.2	14.2 *	22.4	26.8	27.4	54
NO_3^-	4.9	8.1 **	13.0	15.9	23	39
Ca^{2+}	2.8	1.9	4.7	5.4	7.8	13
Mg^{2+}	0.45	0.29	0.74	0.98	1.05	2.0
Na^+	0.51	0.14	0.65	1.34	0.12	1.6
K^+	0.37	0.50	0.87	0.82	1.48	2.3
H^+	0.09	--	0.09	0.28	--	0.28
Cl^-	0.76	0.01	0.77	1.34	0.12	1.5
NH_4^+	2.4	4.1	6.5	9.6	17.6	27.2

* Includes SO_2 and fine and coarse particle SO_4^{2-}

** Includes NO_x, HNO_2, HNO_3, and fine and coarse particle NO_3^-.

The hydrogen ion deposition for Region 9 was about 13 times greater than what was reported by Alberta Department of Environment for Calgary. This may be partially due to differences in the sampling methods. Comparisons were made of the chemical composition of rainfall collected with CCM Model 8100 refrigerated sampler and with the MIC Type B wet-only precipitation sampler. The mean volume weighted pH of the refrigerated samples collected at Crossfield was 4.77 as compared with 5.18 with the MIC sampler. Similarly, the calcium concentration in the refrigerated samples was 43 μeq L^{-1} as compared with 83 μeq L^{-1} with the MIC sampler. The ADRP used a refrigerated rain sampler which limited contamination by wind blown dust and reduced chemical changes which may occur in samples after collection. During 1978-84, samples were collected by Alberta Department of Environment using the Sangamo wet-only precipitation collector where samples are known to be subject to dust contamination. Prior to 1980, these samples were collected only on a monthly basis. After 1980, the collection buckets were emptied every few days and a monthly sample was accumulated. The differences in these sampling methods are more likely to affect the hydrogen ion rather than the sulphate or nitrate concentrations. The wet deposition of calcium ion, which may be indicative of dust contamination, was 7.6 kg ha^{-1} y^{-1} in Calgary during the period 1979-86 compared with 2.8 kg ha^{-1} y^{-1} calculated for Region 9 in the present study. Calcium deposition measured by Alberta Department of Environment, after new sampling procedures were initiated, were closer to the values calculated from the ADRP data. Thus, wet deposition of calcium in Calgary was reported as 3.4 kg ha^{-1} y^{-1} for the 1982-86 period (Alberta Department of Environment, personal communication).

6.2.2 Wet Deposition in Other Regions

Wet deposition was measured at one site in each of the Regions 3, 4, 6, and 7 (Figure 6.1) between 1978-86. Not all sites were monitored for the whole period; thus the wet deposition values reported in Table 6.4 are for the latest year when measurements were made, 1984 or 1986. In Region 3 the Alberta Department of Environment monitoring site, located at Beaverlodge, reported 463 mm of precipitation in 1986. In Region 4 the monitoring site was at Fort McMurray, and the last available precipitation data from that site were for 1984 when the total precipitation was 516 mm. At Cold Lake in Region 6, the 1986 precipitation was 417 mm and at Edson, in Region 7, the 1984 precipitation was 547 mm. The wet deposition values for each Region are given in Table 6.4.

6.2.3 "Maximum" Wet Deposition

The maximum wet deposition of sulphate for Region 9, as calculated from the 90th percentile of the volume-weighted daily wet deposition, was 26.8 kg ha^{-1} y^{-1}, about 2.5 times the arithmetic mean of the annual wet deposition at Crossfield (Tables 6.2 and 6.3). The corresponding maximum values for NO_3^-, Ca^{2+}, H^+, and NH_4^+ were 15.9, 5.4, 0.28, and 9.6 kg ha^{-1} y^{-1}, respectively (3.1, 1.9, 3.1, and 4 times the annual values).

The same ratios of "maximum" to the estimated annual deposition found in Region 9 were used to estimate maximum deposition

Table 6.3 Deposition of air pollutants at the Fortress Mountain, Crossfield West, and Crossfield East ADRP monitoring sites (kg ha^{-1} y^{-1}).

	SO_4^{2-}	NO_3^-	Ca^{2+}	Mg^{2+}	Na^+	K^+	H^+	NH_4^+
Fortress Mountain								
Wet deposition	7.1	4.6	3.1	0.38	0.47	0.35	0.10	1.9
Deposition of gases	4.3	5.1	--	--	--	--	--	1.2
Fine particle deposition	0.13	0.04	0.00	0.00	0.00	0.01	--	0.06
Coarse particle deposition	0.51	0.48	0.89	0.03	0.13	0.28	--	0.26
Total deposition	12.0	10.2	4.0	0.41	0.61	0.64	0.10	3.4
Crossfield West								
Wet deposition	10.0	5.5	2.4	0.39	0.86	0.37	0.069	3.2
Deposition of gases	14.6	15.9	--	--	--	--	--	7.2
Fine particle deposition	0.28	0.19	0.02	0.00	0.00	0.01	--	0.09
Coarse particle deposition	1.29	0.87	3.3	0.46	0.14	0.65	--	0.41
Total deposition	26.2	22.5	5.7	0.85	1.0	1.0	0.069	10.9
Crossfield East								
Wet deposition	10.9	5.6	3.1	0.57	0.63	0.34	0.067	3.1
Deposition of gases	18.4	18.4	--	--	--	--	--	9.4
Fine particle deposition	0.30	0.22	0.03	0.00	0.00	0.01	--	0.11
Coarse particle deposition	1.39	1.01	4.19	0.55	0.25	0.82	--	0.49
Total deposition	31.0	25.2	7.3	1.13	0.88	1.17	0.067	13.1

Table 6.4 Mean and maximum wet and dry deposition of air pollutants for Regions 3, 4, 6, and 7 (kg ha^{-1} y^{-1}).

	Mean			Maximum		
	Wet	Dry	Total	Wet	Dry	Total
REGION 3						
SO_4^{2-}	3.7	6.0 *	9.7	19	10	19
NO_3^-	2.1	0.9 **	3.0	5.7	1.4	7.1
Ca^{2+}	0.7	--	--	1.8	--	--

Continued...

Table 6.4 (Concluded).

	Mean			Maximum		
	Wet	Dry	Total	Wet	Dry	Total
H^+	0.036	--	--	0.09	--	--
REGION 4						
SO_4^{2-}	9.6	11	21	24	42	66
NO_3^-	1.8	0.9	2.7	4.9	2.9	7.8
Ca^{2+}	5.4	--	--	15	--	--
H^+	0.005	--	--	0.014	--	--
REGION 6						
SO_4^{2-}	2.5	7.6	10	6	12	18
NO_3^-	1.0	2.7	3.7	2.6	5.7	8.3
Ca^{2+}	1.1	--	--	3.0	--	--
H^+	0.024	--	--	0.074	--	--
REGION 7						
SO_4^{2-}	10	12	22	25	18	43
NO_3^-	4.3	3.3	7.6	11.6	10.0	22
Ca^{2+}	5.3	--	--	14.3	--	--
H^+	0.005	--	--	0.014	--	--

* Includes SO_2 and fine and coarse particle SO_4^{2-}.

** Includes NO_x, HNO_2, HNO_3, and fine and coarse particle NO_3^-.

in the other regions, as shown in Table 6.4.

6.3 DRY DEPOSITION

As previously stated a number of methods for measuring dry deposition have been described and evaluated by Hicks et al. (1987). They range from surface analysis methods, such as leaf washing and snow sampling, to atmospheric flux methods, such as vertical gradient and eddy accumulation measurements. The method most widely used to estimate dry deposition over a region is to measure the atmospheric concentration, C, of a given chemical species and to calculate the deposition from the formula:

$$-F = C \bullet V_d$$

where F is the flux away from the surface and V_d is the deposition velocity for that species. The concentration of a chemical species in the atmosphere may be estimated using dispersion models, as described in Chapter 8, to predict concentrations of SO_2, NO_x and sulphate particles in Alberta, or the actual concentrations may be measured, as with the ADRP air quality monitoring trailers near Crossfield and Fortress Mountain.

Published deposition velocities for SO_2 and other species are listed in Table 2.6 of Chapter 2. To determine whether deposition velocities in the ADRP study area are comparable with those in the published literature, measurements were made during the fall and winter of 1987-88 at the Crossfield West sampling site using the gradient method.

6.3.1 Theory

The transport of mass along a concentration gradient in a quiescent fluid is a function of the Brownian diffusion coefficient D:

$$F = -D \frac{dC}{dz} \qquad (1)$$

where F is the flux of contaminant mass (g cm^{-2} sec^{-1}), C is the contaminant mass concentration (g cm^{-3}), and z is a length scale, e.g. the distance above a surface (cm). For surfaces which are a perfect sink to the contaminant, the concentration will always be zero adjacent to the surface. Hence, there will be a perpetual flux of contaminant onto the surface as long as there is a non-zero concentration in the fluid.

Although D is important in a quiescent fluid or in laminar flow, a parameter describing transport by turbulent eddies must be applied to the free atmosphere. For such conditions, Equation (1) becomes:

$$F = -(D + K) \frac{dC}{dz} \qquad (2)$$

where K is the eddy diffusivity for contaminant mass (g cm^{-2} sec^{-1}). In general K is much greater than D except very close to the surface. D is usually considered to be a constant for a particular species at a given temperature and pressure. K, however, varies with distance from the surface because of the varying sizes of turbulent eddies responsible for transport. It is often assumed that the eddy diffusivity for mass is equal to the eddy kinematic viscosity for momentum transport K_M given by:

$$K_M = k\mu_* (z - d) \tag{3}$$

where k is von Karman's constant (equal to about 0.4), μ_* is the friction velocity (cm sec^{-1}), and d is the zero-plane displacement (cm). The friction velocity is equal to:

$$(\tau_o/\rho_a)^{0.5}$$

where τ_o is the shear stress at the surface (g cm^{-2} sec^{-1}) and ρ_a is the air density (g cm^{-3}). The shear stress at the surface is merely the deposition flux of air momentum F_a, analogous to F for contaminant mass. The zero plane displacement is an empirical parameter in the equation for windspeed versus height.

Substituting Equation (3) into Equation (2) with $K=K_M$ yields a solution for C(z):

$$C(z) - C_o = \frac{F}{k\mu_*} \ln \frac{z-d}{z_{oc}} \tag{4}$$

Here z_{oc} is a constant of integration representing the effective contaminant sink, where the concentration falls to a minimum value C_o. For a perfect sink surface, $C_o=0$.

Equation (4) shows that a straight line is obtained when C(z) is plotted against ln (z-d). A similar expression is obtained for the wind-speed $\mu(z)$ when an equation is written for momentum flux F_a (similar to Equation 2) and solved:

$$\mu(z) = \frac{\mu_*}{k} \ln \frac{z-d}{z_o} \tag{5}$$

where z_o is the momentum sink (cm), a hypothetical height where the windspeed reaches zero. It is important to note that z_{oc} and z_o are mathematical tools used to fix the shapes of the profiles well above the surface.

The aforementioned Equations indicate that both contaminant concentration and windspeed vary linearly with the logarithm of height above the surface. The contaminant concentration can be either zero or non-zero at the contaminant sink height z_{oc}, while the windspeed generally reaches zero at the momentum sink height z_o.

6.3.2 Dry Deposition Velocity and Resistance to Transport

The aforementioned relationships can be used to determine the deposition velocity V_d (cm sec^{-1}) defined as the ratio -F/C. Note that since C is a function of height, V_d is also a function of height. However, these variations are weak except very close to the surface.

The deposition velocity at any height z can be viewed as a parameter describing the overall transport between height z and the surface. This transport can be divided into three components, each representing a resistance to transport: the aerodynamic resistance, r_a; the boundary layer resistance, r_b; and the canopy resistance, r_c. r_a, which is a function of height z, describes resistance to transport from height z down to the momentum sink z_o. r_b describes transport across the quasi-laminar boundary layer just above the surface, represented mathematically as the height interval from z_o down to the contaminant sink z_{oc}. Finally, r_c refers to interaction with the surface; a perfect sink surface will have $r_c = 0$.

The relationship between deposition velocity and the component resistances is given by:

$$V_d = \frac{1}{r_t} = \frac{1}{r_a + r_b + r_c} \tag{6}$$

where r_t is the total resistance to transport (sec cm^{-1}). The expressions for r_a, r_b, and r_c can be written as:

$$r_a = \frac{1}{k\mu_*} \ln \frac{z - d}{z_o} \tag{7}$$

$$r_b = \frac{1}{k\mu_*} \ln \frac{z_o}{z_{oc}} \tag{8}$$

$$r_c = -\frac{C_o}{F} \tag{9}$$

Comparison of the Equations (5) and (7) shows that the aerodynamic resistance to transport can be expressed simply as:

$$r_a = \frac{\mu(z)}{\mu_*^2} \tag{10}$$

6.3.3 Corrections for a Non-adiabatic Atmosphere

The aforementioned analysis applies only when the atmosphere is adiabatic, characterized by neutral stability. A number of modifications have been developed to account for the influence of buoyancy in a non-adiabatic atmosphere. The expressions for eddy kinematic viscosity for momentum and eddy diffusivity for contaminant mass become:

$$K_M = \frac{k\mu_*(z-d)}{\phi_M} \qquad K = \frac{k\mu_*(z-d)}{\phi_C} \tag{11}$$

There is an analogous expression for the eddy diffusivity of heat K_H incorporating the function ϕ_H. The correction factors ϕ_M, ϕ_C, and ϕ_H are functions of $\zeta = (z - d)/L$, where L is the Monin-Obhukov length. L is positive in a stable atmosphere, negative in an unstable atmosphere, and approaches $+\infty$ or $-\infty$ when the atmosphere is neutral. The magnitude of L is a measure of the height above ground where the production of turbulence by mechanical forces equals the production of turbulence by buoyant forces.

Appropriate expressions for the correction factors must be determined by experimentation. Businger et al. (1971) analyzed field data from a variety of stability conditions and suggested the expressions for ϕ_M and ϕ_H given in Table 6.5. Some authors have

Table 6.5 Expressions for stability-dependent correction factors for aerodynamic transport.

Reference	Parameter	Atmospheric Condition		
		Stable	Neutral	Unstable
Businger et al. 1971	ϕ_M	$1 + 4.7\zeta$	1	$[1 - 15\zeta]^{-1/4}$
Businger et al. 1971	ζ_H	$0.74 + 4.7\zeta$	0.74	$0.74\,[1 - 9\zeta]^{-1/2}$
Wesely and Hicks 1977	Ψ_M	-5ζ	0	$\exp\ 0.032 + 0.448 \ln(-\zeta) - 0.132\,[\ln(-\zeta)]^2$
Wesely and Hicks 1977	Ψ_C	-5ζ	0	$\exp\ 0.598 + 0.390 \ln(-\zeta) -0.090\,[\ln(-\zeta)]^2$
Hicks et al. 1985	r_a for contaminants	4 — 2 $\mu\alpha$ θ	4 — 2 $\mu\alpha$ θ	9 — 2 $\mu\alpha$ θ

suggested that eddy transport of contaminant mass is similar to the eddy transport of heat (e.g., Galbally 1971). Therefore, the expressions for ϕ_H in Table 6.5 are sometimes used identically for ϕ_C. According to this table, the equations developed by Businger et al. (1971) show ϕ_H/ϕ_M=0.74 for neutral conditions, rather than unity as suggested by other authors (e.g. Lumley and Panofsky 1964; Munn 1966; Webb 1970; and Thom 1975). Both alternatives have been used in eddy diffusion models found in the literature.

Rather than substituting the stability-corrected diffusivities into Equation (2) and the analogous equation for F_a, one can develop correction factors for non-adiabatic conditions to be applied after those equations are integrated:

$$r_a \text{ for momentum} = \frac{1}{k\mu_*} \left[\ln \frac{z - d}{z_o} - \psi_M \right]$$

$$r_a \text{ for contaminants} = \frac{1}{k\mu_*} \left[\ln \frac{z - d}{z_o} - \psi_C \right] \qquad (12)$$

where Ψ_M and Ψ_C represent modifications applied directly to Equation (7). Wesely and Hicks (1977) give expressions for these functions, based on tabulations by Dyer and Hicks (1970). These expressions are shown in Table 6.5. Relationships between the Pasquill stability categories (Turner 1970) and values of ζ in the correction factors are discussed by Sheih et al. (1979).

As another way to correct for stability, Hicks et al. (1985) suggest the use of σ_θ, the standard deviation of the horizontal wind direction. Based on Equation (7) and the relation $\sigma_\theta = \sigma_\mu/\mu$, an expression for r_a can be written as:

$$r_a = \frac{\sigma_u^2}{\mu_*^2} \frac{1}{\mu \sigma_\theta^2} \qquad (13)$$

where σ_μ is the standard deviation of the windspeed. The authors note that σ_μ/μ_* has a value of about 2 for stable and neutral conditions, but increases asymptotically to about 3 in an unstable atmosphere, leading to the expressions shown in Table 6.5. The expression for r_a for an unstable atmosphere is used only if the net radiation is positive and if σ_θ exceeds about 10°. Hicks et al. (1985) also calculated the boundary layer resistance from the equation:

$$r_b \simeq \left\{\frac{2}{K\mu_*}\right\}\left\{\frac{Sc}{Pr}\right\}^{p} \tag{14}$$

in which the parameters K, Sc, Pr, and p are either known or can be estimated. Thus, by measuring μ and σ_θ, r_b can be calculated using Equations (13), (10), and (14).

A more detailed discussion of the mathematical equations and the physical processes involved in dry deposition is given by Davidson and Wu (1989).

6.3.4 Estimates of SO_2 Dry Deposition From ADRP Data

The aforementioned numerical expressions can be used to infer SO_2 transport from the atmosphere onto surfaces studied in the ADRP. Equations (2)-(4) show that the flux of SO_2 in an adiabatic atmosphere can be estimated from the concentration gradient if μ_*, z_o, and d are known. These last three micrometeorological parameters can be determined from the windspeed profile defined by Equation (5).

In the ADRP the windspeed was measured at z=0.5 m, 2 m and 10 m over a field of wheat stubble, allowing determination of the micro-meteorological parameters. Note that z_o and d are not expected to change significantly with mean windspeed as they are dependent mainly on surface characteristics. Therefore, these two variables can be determined for time periods when anemometers at two heights were operating and the atmosphere was adiabatic. Determination of μ_* is then possible even for time periods when windspeed at only one height is available, based on the fixed values of z_o and d.

Atmospheric SO_2 concentrations were measured at heights of 1 m and 10 m, providing estimates of dC/dz and z_{oc}. The flux F was calculated for several sets of measurements of the SO_2 gradient and micro-meteorological parameters during adiabatic atmospheric conditions.

It is important to consider atmospheric stability. Temperature data were available at 0.5 m, 2 m, and 10 m heights, allowing identification of those times when buoyancy effects must be accounted for. A temperature gradient dT/dz of - 0.0098°C m^{-1}, characteristic of an adiabatic atmosphere, implies a temperature difference of approximately - 0.1°C between the 0.5 m and 10 m temperature sensors. From the temperature data, the times when the atmosphere was either adiabatic, slightly stable, or slightly unstable, were identified for near-neutral conditions, and the procedure described in the preceding few paragraphs was used to estimate SO_2 flux and deposition velocity.

Examination of the meteorological and SO_2 data showed very few times when adiabatic conditions existed and concentrations of SO_2 were sufficiently high to allow an accurate measure of the difference in concentrations between the upper and lower monitoring points of the tower. Two examples are presented in Tables 6.6 and 6.7.

Table 6.6 Results of the SO_2 deposition velocity experiments, March 17, 1988 at Crossfield West.

Time	Average Wind Speed ($km\ h^{-1}$)			SO_2 Concentration (ppb)		μ_*	z_o	V_d
	10 m	2 m	0.5 m	10 m	1 m	($km\ h^{-1}$)	(mm)	($cm\ sec^{-1}$)
957	19.2	17.2	14.9	14.7		0.57	0.012	0.84
1002					10.2			
1004	15.8	13.4	12.2	11.3		0.47	0.018	0.46
1009					7.9			
1013	15.1	13.2	11.6	9.4		0.47	0.024	0.47
1018					7.0			
1022	14.6	12.6	16.7	8.7		0.55	0.047	0.53
1024					6.9			
1035	17.0	15.1	13.1	9.2		0.52	0.019	0.57
1039					7.3			
1042	17.8	15.6	13.4	9.1		0.59	0.050	0.65
1047					6.7			
1050	16.7	14.6	13.3	8.7		0.46	0.005	0.59
1056					6.1			
1059	14.2	12.7	11.1	7.6		0.42	0.012	0.47
1104					5.6			

On March 17, 1988, adiabatic conditions existed between 0957 and 1104 with SO_2 concentrations fairly stable and above 5 ppb. The friction velocity μ_* ranged from 0.42-0.57 km h^{-1}, and the momentum sink height z_o was low, 0.05-0.47 cm. SO_2 concentrations at the top of the tower ranged from 7.6-14.7 ppb and from 5.6-10.2 ppm at the bottom with measurements made every four minutes at each height. The SO_2 concentrations were measured using a calibrated pulsed fluorescence analyzer and a switching mechanism to draw samples alternately from the upper and lower measurement points of the tower through Teflon lines. Deposition velocities were calculated using the measurements of SO_2 concentrations at the top of the tower and the average SO_2 concentrations at the bottom of the tower during periods preceding and following the measurement at the top of the tower.

The results provided deposition velocities ranging from 0.46-0.65 cm sec^{-1} with one exception; the first set of calculations in the series provided a V_d of 0.84 cm sec^{-1}. The mean deposition velocity over the 67 minute adiabatic period was 0.57 cm sec^{-1}. This is well above the minimum deposition velocity which would be measurable under these conditions. Given a μ_* value of about 0.5 km h^{-1}, an SO_2 concentration of 10 ppb, and a minimum difference of 0.5 ppb,

Table 6.7 Results of the SO_2 deposition velocity experiments, February 28, 1988 at Crossfield West.

Time	Average Wind Speed (km h^{-1})			SO_2 Concentration (ppb)		μ_*	z_o	V_d
	10 m	2 m	0.5 m	10 m	1 m	(km h^{-1})	(mm)	(cm sec^{-1})
1615					3.4			
1619	11.4	.9	7.8	4.3		0.46	0.46	0.52
1624					3.2			
1634	13.2	11.0	8.9	4.5		0.58	1.0	0.84
1637					3.1			
1642	14.8	11.8	9.2	4.4		0.75	3.7	1.23
1645					2.7			
1649	16.6	13.0	9.8	3.1		0.98	7.0	1.53
1654					2.2			
1658	17.7	14.3	10.3	3.1		0.98	7.0	1.53
1701					2.0			
1712	15.8	12.7	9.0	3.9		0.90	8.3	1.84
1717					2.5			
1719	13.9	11.2	8.0	3.7		0.79	7.9	1.23
1725					2.5			
1727	13.7	10.6	7.2	3.1		0.86	17	1.01
1730					2.2			
1735	14.1	9.7	6.5	3.0		1.02	42	1.48
1739					2.0			

the minimum deposition velocity measurable would be 0.12 cm sec^{-1}.

A second example, for the period 1615-1739 hours on February 28, 1988, is shown in Table 6.7. The friction velocity during this period ranged from 0.46-1.02 km h^{-1} and z_o from 0.04-4.1 cm. SO_2 concentrations ranged from 3.0-4.5 ppb at the top of the tower and from 2.0-3.4 ppb at the bottom of the tower. The values of deposition velocity increased systematically from 0.52 cm sec^{-1} at 1619 to 1.84 cm sec^{-1} at 1712 before declining to 1.0 cm sec^{-1} at 1727. The mean deposition velocity over the 84 minute adiabatic period was 1.2 cm sec^{-1}.

The values are well within the range of deposition velocities for SO_2 reported by other investigators and summarized in Table 2.7 of Chapter 2. On the basis of the measurements made within the ADRP at Crossfield and previously published deposition velocities for SO_2, a value of 0.7 cm sec^{-1} was chosen as representative of the V_d for SO_2 in Alberta.

6.3.5 Regional Scale Dry Deposition Estimates for Alberta

6.3.5.1 Methodology

Estimates of dry deposition for each geographic region of the Province were developed using atmospheric concentrations of SO_2, NO_x, and particulate SO_4^{2-} as predicted by the model described in Chapter 8. Models of this type are generally considered to be accurate within a factor of two. The flux was then calculated using representative deposition velocities for SO_2, NO_x, HNO_2, HNO_3, and NH_3 as well as for fine and coarse particles (Tables 2.7 and 2.8 in Chapter 2) (Davidson and Wu 1989). The V_d values in the tables were obtained by different investigators making measurements over different surfaces using different methods; therefore the values vary greatly. Deposition velocities would be expected to differ according to the type of surface (e.g., forest, grassland, or crop; wet, snow-covered, or dry) and with the meteorological conditions. A range of V_d values for SO_2 was obtained from the resistance model, from the measurements at Crossfield, and from mass balance calculations. The selected V_d of SO_2 fell between that of the resistance model and the mass balance calculations. Similar criteria were used in the selection of V_d values for other gases and for particles. The V_d values were: for SO_2, 0.7 cm sec^{-1}; NO_x, 0.1 cm sec^{-1}; HNO_2 and HNO_3, 3 cm sec^{-1}; NH_3, 1.5 cm sec^{-1}, fine particles 0.1 cm sec^{-1}, and for coarse particles 2.0 cm sec^{-1}. These values should be considered as estimates and may vary at any given time and space by more than a factor of two. The 0.7 cm sec^{-1} value for the V_d of SO_2 is comparable to that used in the eastern United States but is lower than the 0.2-0.3 cm sec^{-1} applied by the Atmospheric Environment Service in studies of dry deposition in Ontario (Barrie and Sirois 1986). A map of dry deposition velocities of SO_2 over North America east of the Rocky Mountains (Voldner et al. 1986) indicated values of 0.5-1.2 cm sec^{-1} for south-central Alberta in April, declining to about 0.2 cm sec^{-1} for most of the year.

The deposition of SO_2 was calculated as sulphate to allow a direct comparison of wet and dry deposition in Table 6.2. The total dry deposition of SO_4^{2-} consisted of SO_2, fine particle SO_4^{2-}, and coarse particle sulphate. Coarse particles were not analyzed directly for sulphate but the sulphur content was determined by XRF and, for the purpose of these calculations, the S was assumed to occur as SO_4^{2-}. The atmospheric concentration of sulphur in coarse particles was 23% of that in fine particles at the Crossfield sites and 19% at Fortress Mountain.

6.3.5.2 Dry Deposition in Region 9

The annual dry deposition of sulphate, nitrate, hydrogen ion, and other species was calculated for Region 9 and the results are presented in Table 6.2.

The mean atmospheric concentration of SO_2 in the region was predicted by the model to be 3.7 $\mu g\ m^{-3}$. Applying a deposition velocity of 0.7 cm sec^{-1}, the annual SO_2 deposition was calculated to be 8.2 kg ha^{-1} y^{-1} or the equivalent of 12.2 kg ha^{-1} y^{-1} as SO_4^{2-}. Fine particle sulphate contributed a further 0.36 kg ha^{-1} y^{-1} and coarse particles 1.6 kg ha^{-1} y^{-1} for a total of 14.2 kg ha^{-1} y^{-1} dry sulphate deposition. This value is similar to those reported for eastern Canada but lower than the values in the range of 24-28 kg ha^{-1} y^{-1} reported for southern Ontario (Tang et al. 1986; Barrie and Sirois 1986). For some sites in the eastern United States, the estimated total dry deposition of sulphate for the period 1979-83 were as high as 37 kg ha^{-1} y^{-1} (Hicks et al. 1987). The annual dry deposition estimate for Region 9 is in the same range as that predicted by the long-range transport model of the Atmospheric Environment Service of Canada (AES) (Kociuba 1984). Predictions were made by the AES model for three locations within Region 9: Red Deer, Calgary, and Rocky Mountain House, and the results were 12, 21, and 20 kg ha^{-1} y^{-1}, respectively, as SO_4^{2-}. The AES model predicted an annual SO_4^{2-} dry deposition of 2.4 kg ha^{-1} y^{-1} for Cree Lake, Saskatchewan, the only western Canadian station with long-term air monitoring and precipitation data. Using the methods developed by Voldner et al. (1986) for determining the regional average dry deposition velocities for SO_2 and SO_4^{2-} in combination with the air concentrations measured by filter pack, Summers (1986) calculated the dry sulphate deposition at Cree Lake to be 1.5 kg ha^{-1} y^{-1} for the period July 1982 to December 1984. The mean deposition velocity of SO_2 used by Summers (1986) was 0.16 cm sec^{-1} as compared with 0.7 cm sec^{-1} used by Kociuba (1984) and in the ADRP estimates for Region 9.

Measurements of the atmospheric concentrations of SO_2, fine particulate SO_4^{2-}, and coarse particulate S were made using the KAPS annular denuder system and dichotomous samplers at Fortress Mountain and at the Crossfield East and West ADRP sampling sites. Deposition estimates based on these data are contained in Table 6.3.

The mean SO_2 concentration at Fortress Mountain, as determined using the KAPS system, was 1.31 μg m^{-3} or 1.96 μg m^{-3} as SO_4^{2-}. With an estimated deposition velocity of 0.7 cm sec^{-1} the rate of dry deposition of SO_2 would be 4.3 kg ha^{-1} y^{-1} as SO_4^{2-} and total dry deposition of SO_2, fine particle SO_4^{2-} and coarse particle SO_4^{2-} would be 3.7 kg ha^{-1} y^{-1}. At Crossfield West the corresponding values were 15, 0.3, and 1.3 kg ha^{-1} y^{-1} for SO_2 (as SO_4^{2-}), fine particle SO_4^{2-}, and coarse particle SO_4^{2-}, respectively, and at Crossfield East 18, 0.3, and 1.4 kg ha^{-1} y^{-1}, respectively. The model was expected to predict the atmospheric SO_2 concentrations accurately within a factor of two (Section 8). The predicted value for the Crossfield region of 5.0 μg m^{-3} was close to the 4.4 and 5.6 μg m^{-3} observed at the Crossfield West and Crossfield East sites, respectively. The predicted deposition in the Crossfield area was 17 kg ha^{-1} y^{-1} (in units of SO_4^{2-}) as compared to the 15 and 18 kg ha^{-1} y^{-1} estimated from observed SO_2 concentrations at the Crossfield sites.

The annual dry deposition of nitrate in Region 9 was calculated to be 8.1 kg ha^{-1} y^{-1} or about 50% of the SO_4^{2-} deposition. This is in keeping with the lower NO_x emissions as compared with SO_2 (Chapter 7). The NO_x and HNO_3 deposition contributed 2.6 kg ha^{-1} y^{-1} and 3.5 kg ha^{-1} y^{-1}, respectively, and HNO_2 1.7 kg ha^{-1} y^{-1}, when calculated as NO_3^-. Coarse particle nitrate deposition was 0.37 kg ha^{-1} y^{-1} and fine particle nitrate only 0.08 kg ha^{-1} y^{-1}. The regional scale atmospheric concentrations were obtained from the model described in Chapter 8, and the HNO_3 and fine particle concentrations from the KAPS measurements at Fortress Mountain and Crossfield East and West. No analysis of coarse particle NO_3^- was performed; therefore, the same coarse to fine particle ratio of 0.23 previously determined for SO_4^{2-} was used.

It should be emphasized that the total dry deposition of nitrate was influenced greatly by the deposition velocities of NO_x, HNO_2, and HNO_3 chosen to be representative of Region 9 which includes mountains and plains, forests, crops, grasslands and, in winter, snow-covered surfaces. As discussed earlier, the V_d of HNO_3 would vary greatly within the region and with changing surface characteristics. Similarly no measurements of the V_d of HNO_2 are available.

No published dry deposition estimates of nitrate exist for Alberta or for the western United States (Young et al. 1988). The dry deposition of nitrate is 30% of the total estimated deposition in eastern Canada (Summers et al. 1986).

Deposition of ammonium was calculated for Region 9 from atmospheric concentrations of ammonia and from fine particle ammonium concentrations, determined with the KAPS system at the three ADRP sampling sites. The ammonia concentration over Region 9 was estimated to be 0.79 $\mu g\ m^{-3}$ and the fine particle ammonium ion concentration 0.22 $\mu g\ m^{-3}$. There are but few measurements of ammonia deposition and fluxes are expected to be positive on occasion (a negative V_d), reflecting the release rather than the uptake of ammonia by surfaces (Davidson and Wu 1989). Applying a deposition velocity of 1.5 cm sec^{-1} for NH_3 and 0.1 cm sec^{-1} for NH_4^+ results in an annual deposition flux of 3.7 kg ha^{-1} y^{-1} of NH_3 and 0.07 kg ha^{-1} y^{-1} of fine particle NH_4^+. No direct measurement of the atmospheric concentrations of coarse particle NH_4^+ was available so the coarse to fine particle ratio of SO_4^{2-} was applied to give a value of 0.32 kg ha^{-1} y^{-1}.

The dry deposition flux of calcium was estimated for Region 9 using the results of the XRF analysis of fine and coarse particles collected by dichotomous samplers to give atmospheric concentrations at the Fortress Mountain, Crossfield West, and Crossfield East air quality monitoring sites. The estimated rate of fine particle calcium deposition was 0.01 kg ha^{-1} y^{-1}, and coarse particle deposition was 1.9 kg ha^{-1} y^{-1}. Some areas within the region are expected to experience much higher particulate Ca^{2+} deposition. At the Crossfield East site, the dry flux was 4.3 kg ha^{-1} y^{-1} of which coarse particle Ca^{2+}, particles in the size range of 2.5-10 μm, comprised over 98%. Similarly calcium contained in coarse particles comprised 98% of the 3.3 kg ha^{-1} y^{-1} dry deposition at the Crossfield West site. In comparison to these values, estimated calcium emission flux from windblown dust of an unspecified size range was 12.5 kg ha^{-1} y^{-1} for dry prairie regions (Summers 1982).

Dry deposition of Mg^{2+}, Na^{+}, and K^{+} were also calculated using atmospheric concentrations of these elements determined from the chemical analysis of coarse and fine particles collected with dichotomous samplers. The dry deposition of Mg^{2+}, Na^{+}, and K^{+} for Region 9 were 0.29, 0.14, and 0.50 kg ha^{-1} y^{-1}, respectively (Table 6.2).

6.3.5.3 Dry Deposition in Other Regions

Dry deposition in Regions 3, 4, 6, and 7 was estimated on the basis of predicted SO_2 and NO_x concentrations. Deposition of S and/or N from fine and coarse particles and HNO_3 were not included in these computations (Table 6.4).

6.3.6 "Maximum" Dry Deposition

For the purpose of modelling the potential effects on crops, soils, and surface waters of a major increase in emissions a "maximum" dry deposition for major anions and cations was calculated for Region 9. In the case of deposition of SO_4^{2-} equivalents, the maximum contribution from SO_2 deposition was calculated using the maximum concentration of 7.5 μg m^{-3} and 1.4 μg m^{-3} for fine particulate SO_4^{2-} from the model described in Chapter 8. The maximum coarse particle SO_4^{2-} deposition was calculated from the fine particle SO_4^{2-} deposition adjusted for the fine to coarse S ratio as determined by XRF analysis. The maximum dry SO_4^{2-} deposition for Region 9 was 27.4 kg ha^{-1} y^{-1}, of which 24.8 was due to SO_2 deposition, 2.2 kg ha^{-1} y^{-1} coarse particle SO_4^{2-}, and 0.4 kg ha^{-1} y^{-1} due to fine particle SO_4^{2-}.

The maximum dry deposition for NO_3^- in Region 9 was 23 kg ha^{-1} y^{-1}, of which 13.2 kg ha^{-1} y^{-1} was derived from the maximum NO_x concentration of 25 $\mu g/m^{-3}$ predicted by the atmospheric model. The maximum nitric acid deposition of 5.0 kg ha^{-1} y^{-1} was derived from the 90th percentile of the atmospheric nitric acid concentrations at the Crossfield monitoring sites. Similarly maximum deposition of fine and coarse particle nitrate of 0.2 and 0.9 kg ha^{-1} y^{-1}, respectively, were calculated from the 90th percentile values of the atmospheric concentrations at the Crossfield sites. The same procedure was followed in computing the maximum HNO_2 deposition of 3.6 kg ha^{-1} y^{-1}.

Maximum dry deposition of calcium, magnesium, sodium, and potassium were estimated from the 90th percentile value of their atmospheric concentrations in fine and coarse particles at Crossfield. Values for maximum dry ammonium ion deposition were estimated from the 90th percentile value of ammonia (NH_3) concentrations together with the 90th percentile values for the atmospheric concentrations of NH_4^+ associated with fine and coarse particles at Crossfield. The results are given in Table 6.2.

Estimates of the maximum deposition were made using the predicted maximum values of SO_2 and NO_x from the atmospheric dispersion model (Chapter 8) for Regions 3, 4, 6, and 7. The results, presented in Table 6.4, do not include fine and coarse particle deposition, other than fine SO_4^{2-}. These results also do not include HNO_3 deposition.

6.3.7 Total Wet and Dry Deposition

The total wet and dry deposition of SO_4^{2-} in Region 9 was estimated to be 22 kg ha^{-1} y^{-1} with a dry to wet ratio of 1.7.

The 22 kg ha^{-1} y^{-1} annual sulphate deposition in Region 9, an area with major SO_2 emissions, is much higher than the 5.1 kg ha^{-1} y^{-1} estimated for the CAPMoN station at Cree Lake, Saskatchewan, by Summers (1986) using a deposition velocity for SO_2 of 0.16 cm sec^{-1}. Estimates of total wet and dry sulphate deposition in southeastern Ontario and the eastern United States are in the range of 25-68 kg ha^{-1} y^{-1} (U.S. NAPAP 1987).

The dry to wet ratio of 1.7 in Region 9 should be expected close to major sources of SO_2, such as natural gas processing plants. Such a ratio should, however, decline with distance from the sources (Summers 1986). The ratio at Crossfield East (2.5 km from the gas plant) was 1.84; at Crossfield West (7 km away from the gas plant) the ratio was 1.62, and at the remote Fortress Mountain site, 0.69. The average dry to wet ratio over the period 1979-82 at six sites in eastern Canada ranged from 0.16-0.58 with the highest ratios occurring at Long Point in southern Ontario and the lowest ratios at locations more remote from major sources, Montmorency, Quebec and Algoma, Ontario (Barrie and Sirois 1986). These lower ratios result from the use of lower deposition velocities and higher rainfall in calculations for central Canada as compared with Alberta. Dry to wet ratios in the eastern United States exhibit values as high as 1.5 (U.S. NAPAP 1987).

Total wet and dry nitrate deposition for Region 9 was estimated to be 13 kg ha^{-1} y^{-1} with a dry to wet ratio of 1.6. Measurements of the atmospheric concentrations of NO_x and particulate nitrate have been made at monitoring sites in Canada and the United States for several years but comparable data are rare. Total wet and dry deposition at Long Point for the period 1979-82 was 35 kg ha^{-1} y^{-1} with a dry to wet ratio of 0.23 declining to 12 kg ha^{-1} y^{-1}, with a ratio of 0.22 at Kejimkujik, Nova Scotia. These values do not include the contribution of NO_2 to the total dry deposition of nitrate (Barrie and Sirois 1986). With the inclusion of the 50% contribution of NO_2, the dry to wet ratios were estimated to be about 0.4.

As in the case with SO_4^{2-}, the dry to wet ratio of NO_3^- would be expected to decline with distance from the source. The ratio at the Crossfield monitoring sites was 3.1 (excluding coarse particle deposition), declining to 1.1 at Fortress Mountain. These results are consistent since the City of Calgary may be considered as the major source of NO_x emissions in the region. Dry deposition is believed to equal wet over eastern North America as a whole (Summers et al. 1986). Wet deposition of nitrate in the north-eastern United States and the lower Great lakes region is estimated to be 19-20 kg ha^{-1} y^{-1} (Table 6.8) (Munger and Eisenreich 1983). Thus, the total wet and dry nitrate deposition in Region 9 is well below those of the Great Lakes region and the eastern United States. The total wet and dry deposition for calcium in Region 9 was estimated to be 4.7 kg ha^{-1} y^{-1} of which 60% was wet and 40% was dry deposition. Because of the contribution of windblown prairie soils to calcium deposition, the total calcium deposition flux would be expected to exceed that of eastern Canada. The annual emission of calcium to the atmosphere from agricultural soils is

Table 6.8 Wet deposition of major cations and anions (kg ha^{-1} y^{-1}).

	H_2O (cm)	H^+	Na^+	K^+	NH_4^+	Ca^{2+}	Mg^{2+}	Cl^-	NO_3^-	SO_4^{2-}
1. Northeastern U.S.	100	0.70	4.6	0.8	2.7	1.0	3.0	11	19	31
2. Maritimes	100	0.40	6.9	1.2	2.7	4.0	0.6	11	13	46
3. Lower Great Lakes	80	0.48	1.8	0.6	4.3	4.8	1.0	4.3	20	42
4. Upper Great Lakes	70	0.21	1.8	0.5	3.8	4.2	1.0	2.5	13	32
5. Southeastern U.S.	130	0.39	4.6	0.5	3.6	2.6	0.8	7.1	17	34
6. Western prairies	40	0.02	1.4	0.2	2.5	3.2	1.0	1.4	5	12
7. Eastern prairies	90	0.09	3.0	0.5	5.8	5.4	1.1	3.2	12	26
8. Pacific Northwest	200	0.30	14	2.3	3.6	8.0	2.4	21	13	52
9. Los Angeles area	40	0.10	2.3	0.6	1.8	0.8	0.5	4.3	8	10
10. San Francisco area	60	0.06	4.8	0.5	2.2	1.2	0.7	4.3	8	8

Source: Munger and Eisenreich (1983).

estimated to be 12.5 kg h^{-1} y^{-1} along the St. Lawrence River and 0.5 kg ha^{-1} y^{-1} in the Maritimes (Summers 1982). Thus, there is a high potential for the neutralization of acidic gases by basic cations in the Prairie region.

6.4 CONCLUDING REMARKS

The processes responsible for SO_4^{2-} deposition in southwestern Alberta, Region 9,were estimated to be: SO_2 deposition 54%, wet SO_4^{2-} deposition 37%, coarse particle deposition 7%, and fine particle deposition 2%. Wet deposition accounted for 38% of the total nitrate deposition, nitrogen oxides 20%, nitrous acid 13%, nitric acid 25%, fine particles 1%, and coarse particles 3%.

Wet sulphate and nitrate deposition in southwestern Alberta was about one-half of that in southern Ontario. The 8.2 kg ha^{-1} y^{-1} wet SO_4^{2-} deposition and 4.9 kg ha^{-1} y^{-1} NO_3^- deposition in Region 9 were well below the highest values of 43 kg ha^{-1} y^{-1} SO_4^{2-} and 28 kg ha^{-1} y^{-1} NO_3^- calculated for the period 1979-82 at Long Point, Ontario. They were comparable to the estimates made for a site near the Ontario-Manitoba border of 7.8 kg ha^{-1} y^{-1} SO_4^{2-} and 6.7 kg ha^{-1} y^{-1} NO_3^-.

Hydrogen ion deposition in southwestern Alberta was estimated to be 0.09 kg ha^{-1} y^{-1}, well below the 0.4-0.6 kg ha^{-1} y^{-1} found in southern Ontario and the eastern United States.

Dry deposition of SO_2 and fine particulate sulphate in southwestern Alberta was 12.6 kg ha^{-1} y^{-1} (expressed as SO_4^{2-}) based on an atmospheric concentration of 3.7 $\mu g\ m^{-3}$ SO_2, derived from an atmospheric model, and a deposition velocity of 0.7 cm sec^{-1}. This was less than the 18-28 kg ha^{-1} y^{-1} measured in southern Ontario, comparable to the deposition in the region from Sault St. Marie to Ottawa, and more than the 4.2 kg ha^{-1} y^{-1} measured near the Manitoba-Ontario border.

The deposition values in Ontario were based on an average deposition velocity of 0.19-0.30 cm sec^{-1}. Deposition velocities of SO_2 over a stubble field at Crossfield commonly fell in the range of 0.5-1.5 cm sec^{-1}.

The values for the total wet and dry deposition of SO_4^{2-} calculated at the Crossfield monitoring sites, 25 and 29 kg ha^{-1} y^{-1} at Crossfield West and East, respectively, were close to the SO_4^{2-} deposition predicted by the atmospheric model for southwestern

Alberta, whereas the calculated nitrate deposition, 22 kg ha^{-1} y^{-1}, was twice the 11 kg ha^{-1} y^{-1} predicted by the model.

The total wet and dry deposition of SO_4^{2-} and NO_3^- at the remote Fortress Mountain monitoring site was about 40% of the values at Crossfield, indicating that sources east of the Rocky Mountains contribute about 60% of the wet and dry deposition of these species in the Crossfield area.

6.5 LITERATURE CITED

Barrie, L.A. and A. Sirois. 1986. Wet and dry deposition of sulphates and nitrates in eastern Canada: 1979-1982. *Water, Air and Soil Pollution* 30: 303-310.

Businger, J.A., J.C. Wyngaard, Y. Izumi, and E.F. Bradley. 1971. Flux-profile relationships in the atmospheric surface layer. *Journal of Atmospheric Science* 28: 181-189.

Coscio, M.R., G.C. Pratt, and S.V. Krupa. 1982. An automated, refrigerated, sequential precipitation sampler. *Atmospheric Environment* 16: 1939-1944.

Davidson, C.I. and Y.L. Wu. 1989. Dry deposition of particles and vapors. In: **Acid Precipitation Volume 2. Sources, Emissions, and Mitigation.** ed., D.C. Adriano, Advances in Environmental Sciences Series, New York: Springer-Verlag, pp. 103-216.

Dyer, A.J. and B.B. Hicks. 1970. Flux-gradient relationships in the constant flux layer. *Quarterly Journal of the Royal Meteorological Society* 96: 715-721.

Galbally, I.E. 1971. Ozone profiles and ozone fluxes in the atmospheric surface layer. *Quarterly Journal of the Royal Meteorological Society* 97: 18-29.

Hicks, B.B. (ed.). 1987. Proceedings of the NAPAP Workshop on Dry Deposition, Harpers Ferry, West Virginia, March 25-27, 1986.

Hicks, B.B., D.D. Baldocchi, R.P. Hosker, B.A. Hutchinson, D.R. Matt, R.T. McMillen, and L.C. Satterfield. 1985. On the use of monitored air concentrations to infer dry deposition, NOAA Technical Memorandum ERL ARL-141, Silver Spring, Maryland.

Hicks, B.B., R.P. Hosker, and J.D. Womack. 1987. Comparisons of wet and dry deposition: The first year of trial dry deposition monitoring. In: **The Chemistry of Acid Rain**, eds. R.W. Russell and G.E. Gordon. Washington, D.C.: American Chemical Society. pp. 196-203.

Kociuba, P.J. 1984. Estimates of sulphate deposition in precipitation for Alberta. Atmospheric Environment Service, Environment Canada. Western Region. Report 84-2. 6 pp.

Lau, Y.K. and N.C. Das. 1985. Precipitation quality monitoring in Alberta. Presented at the PNWIS-APCA 1985 Annual Meeting. Calgary, Alberta, 1985 November 13-15.

Lumley, J.L. and H.A. Panofsky. 1964. *The structure of atmospheric turbulence.* Interscience Monographs and Texts in Physics and Astronomy, Vol. XII, New York: Wiley. 239 pp.

Munger, J.W. and S.J. Eisenreich. 1983. Continental-scale variations in precipitation chemistry. *Environmental Science and Technology* 17: 32A-42A.

Munn, R.E. 1966. *Descriptive Micrometeorology.* New York: Academic Press. 245 pp.

Sheih, C.M., M.L. Wesely, and B.B. Hicks. 1979. Estimated dry deposition velocities of sulfur over the eastern United States and surrounding regions. *Atmospheric Environment* 13: 1361-1368.

Shreffler, J.H. 1978. Factors affecting dry deposition of SO_2 on forests and grasslands. *Atmospheric Environment* 12: 1497-1503.

Summers, P.W. 1982. From emission to deposition: Processes and budgets. In: **A Specialty Conference on Acid Deposition**, Detroit, Michigan, 1982 November 7-10. The Air Pollution Control Association. pp. 15-45.

Summers, P.W. 1986. The role of acid and base substances in wet and dry deposition in Western Canada. In: **Acid Forming Emissions in Alberta and Their Ecological Effects**, Proceedings of the Second Symposium/Workshop, H.S. Sandhu, A.H. Legge, J.I. Pringle, and B.L. Magill, eds. Calgary, Alberta, 1986 May 12-15. pp. 85-116.

Summers, P.W., Van C. Bowersox, and G.J. Stensland. 1986. The geographical distribution and temporal variations of acidic deposition in eastern North America. *Water, Air, and Soil Pollution* 31: 523-535.

Tang, A.J.S., W.H. Chan, D.H.S. Chung, and M.A. Lusis. 1986. Precipitation and air concentration and wet and dry deposition fields of pollutants in Ontario. 1983. APIUS-001-86. Ontario Ministry of Environment. Toronto, Canada.

Thom, A.S. 1975. Momentum, mass and heat exchange of plant communities. In: **Vegetation and the Atmosphere**, J.L. Monteith, ed. London: Academic Press. pp. 57-109.

Turner, D.B. 1970. Workbook of atmospheric dispersion estimates. U.S. Environmental Protection Agency, Office of Air Programs, Publication No. AP-26, EPA, Research Triangle Park, North Carolina.

U.S. NAPAP. National Acidic Precipitation Assessment Program. 1987. Interim Assessment Report Vol. III. Atmospheric Processes. pp. 4.1-5.116.

Voldner, E.C., L.A. Barrie, and A. Sirois. 1986. A literature review of dry deposition of oxides of sulphur and nitrogen with emphasis on long-range transport modelling in North America. *Atmospheric Environment* 20: 2101-2123.

Walcek, C.J., R.A. Brost, J.S. Chang, and M.L. Wesely. 1986. SO_2, Sulfate and HNO_3 deposition velocities computed using regional landuse and meteorological data. *Atmospheric Environment* 20:949-964.

Webb, E.K. 1970. Profile relationships: The log-linear range and extension to strong stability. *Quarterly Journal of the Royal Meteorological Society* 96: 67-90.

Wesely, M.L. and B.B. Hicks. 1977. Some factors that affect the deposition rates of sulfur dioxide and similar gases on vegetation. *Journal of the Air Pollution Control Association* 27: 1110-1116.

Young, J.R., E.C. Ellis, and G.M. Hidy. 1988. Deposition of airborne acidifiers in the western environment. *Journal of Environmental Quality* 17: 1-26.

CHAPTER 7

ANTHROPOGENIC SOURCES OF ACIDIC AND ACIDIFYING AIR POLLUTANTS IN ALBERTA

D. J. Picard, D. G. Colley, and A. H. Legge

7.1 INTRODUCTION

As stated in Chapter 3, background air quality at any given location can be considered as a product of true natural processes and the influence of long-range transport of air pollutants on that location. Local man-made emissions of air pollutants are added to the background air resulting in a net pollutant burden of a given location or region. In evaluating the pollutant burden within a given region, a description of the location, type and size, characteristics and performance of sources, and the nature and rate of their air pollutant emissions are fundamental to an understanding of their influence on the air quality and the consequential observed or predicted environmental impacts in their vicinity and at larger distances from their locations.

While previous emission inventories in Alberta were considered to be either incomplete or did not provide sufficient detail to permit an evaluation of specific regions within the Province, Picard et al. (1987a,b,c,d) completed a detailed emission inventory for SO_x and NO_x in Alberta. What follows is a summary of Picard et al. (1987a).

7.2 THE SO_2 EMISSION INVENTORY

The SO_2 emission inventory identified all SO_2 emission sources in Alberta which were licensed by the provincial government agency (Alberta Department of Environment), and all sour oil batteries approved by the Alberta Energy Resources Conservation Board (ERCB) that emit SO_2 at a rate of 0.2 tonnes per day (t d^{-1}) or more.

As a first approximation, it could have been assumed that the actual emissions are, on an average, equal to the maximum design or licensed values. However, it was observed that the actual SO_2 emissions are, in general, significantly lower than the licensed values (typically between one-third and two-thirds of the licensed values).

Consequently, to estimate the average emission rate for each source, emission factors were employed. Where available, actual emission data were used to determine the emission factors. Otherwise, emission factors were estimated based on values determined for similar sources. To indicate the quality of the emission factors that were estimated or evaluated, the precision in each case was determined.

7.3 THE NO_x EMISSION INVENTORY

The NO_x emission inventory was based on three types of sources: industrial, urban centre, and highway. The NO_x emissions from each type of source were quantitatively expressed as equivalent amounts of NO_2. Since the characteristics of each emission source type are quite different, three different inventory formats were used.

The inventory of industrial NO_x sources included all NO_x emission sources currently licensed to operate under the Clean Air Act, and all NO_x emission sources approved by the ERCB that have a maximum rated power output of 100 kw or more. Industrial NO_x emission sources that have a maximum rate power output of less than 100 kw were excluded as detailed documentation of these sources was generally difficult to obtain, and the total NO_x emissions from these small sources were not considered significant relative to the total NO_x emissions from large industrial sources.

The NO_x emission inventory of urban centres included villages (population approximately 100-1000), towns (population approximately 1000-10,000), and cities (population approximately $\geq$10,000). The NO_x emissions from the urban centres were attributed to vehicular traffic and to the residential, commercial, and industrial consumption of natural gas.

In developing the NO_x inventory for highways, the three major sections of highways in the Province with major traffic were divided into a number of short segments and treated as line sources of NO_x. These sections consisted of all of Highways 1 and 16, and Highway 2 from Westlock, located roughly 70 km north of Edmonton, south to the U.S. border. The NO_x emissions due to vehicular traffic on all other roadways, excluding urban roadways which were included in the analysis of urban centres, were generally too low for these roadways to be treated as significant line sources (typically less than 0.005 t d^{-1} km^{-1} compared with an average value of 0.06 t d^{-1} km^{-1} for highways).

In summary, the emission sources excluded from the NO_x emission inventory were:

a. industrial sources with a rated total power output of <100 kw;
b. vehicular traffic outside urban centres on roadways other than Highways 1, 2, and 16;
c. off-road vehicular traffic;
d. air and rail traffic;
e. natural gas consumption by summer villages and rural residences; and
f. forest fires.

7.4 TOTAL SO_2 AND NO_x EMISSIONS IN ALBERTA

The emission data base comprised 565 sources of SO_2 and a total of 4025 industrial, urban, and highway sources of NO_x. A map of Alberta is presented in Figure 7.1 showing the major urban centres and highways, as well as the townships and ranges which form the basis of Alberta's legal land system. Superimposed on Figure 7.1 is a grid defining the geographic regions used for the summarization of emissions data.

Emission regions were selected to be approximately equivalent in geographic area, while allowing for:

a. variances to locate similar types of emission sources in the same region (e.g., oil sands plants, chemical plants, power plants); and
b. variances to accommodate the topographic grid system.

The emission sources were identified based on 1985-86 records available from government agencies and existing inventories (Alberta Environment 1984, 1985; Legge and Baker 1987). The emissions were estimated based on 1984 production, population, and traffic data (Alberta Municipal Affairs 1985; Alberta Department of Transportation 1985; and Energy Resources Conservation Board 1985a,b,c,d). Limited verification and updating of sources occurred in 1987.

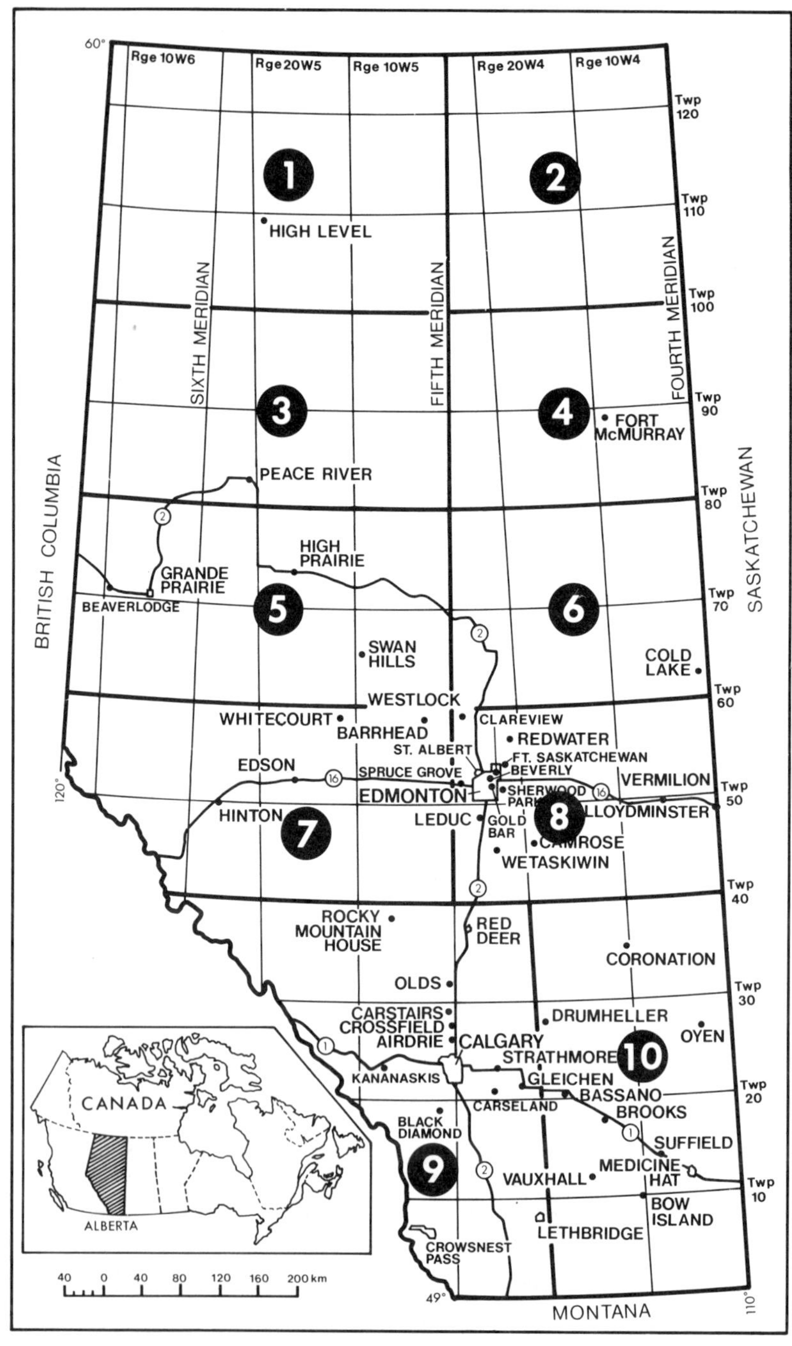

Figure 7.1 Map of Alberta showing major cities, and selected towns and villages, and three major highways (1, 2, and 16) with respect to geographic emission regions used in summarizing the SO_2 and NO_x emissions data.

Table 7.1 A comparison of annual SO_2 and NO_x emissions in Alberta with the corresponding emissions in selected geographic locations in the U.S. and Canada.[1]

Province or State	Annual Emissions (in 1000 Metric Tonnes)		% A/B	
	SO_2	NO_x	SO_2	NO_x
Alberta (A)	462.5	247.5		
Others (B)				
Ohio	2,401.1	1,038.4	19.3	23.8
Indiana	1,821.5	941.2	25.2	26.3
Illinois	1,334.1	912,0	34.7	27.1
Mississippi	1,180.4	258.8	39.2	95.6
Texas	1,158.2	2,307.7	40.0	10.7
Kentucky	1,016.7	482.0	45.5	51.3
Washington	246.7	262.1	187.5	94.4
Montana	148.6	114.0	311.2	217.1
Ontario	1,830.6	536.5	25.3	46.1
Quebec	1,157.9	331.4	39.9	74.7
Manitoba	489.9	78.1	94.4	316.9
British Columbia	192.9	199.0	239.8	124.4
Saskatchewan	57.8	148.3	800.2	166.9

[1] Source: For Alberta, Picard et al. (1987); for others, United States-Canada Memorandum of Intent on Transboundary Air Pollution, Final Report, Work Group 3B (1982). Emission estimates for all geographic areas (other than Alberta) were for 1980.

In total, 1945.2 t d^{-1} of SO_2 + NO_x emissions were inventoried. These total emissions were composed of: SO_2: 1267.0 t d^{-1}, and NO_x: 678.2 t d^{-1}.

Table 7.1 provides a comparison of these values with the corresponding SO_2 and NO_x emissions for selected geographic areas of the U.S. and Canada. On a relative basis, of the 15 states and provinces compared, Alberta ranked 11th lowest in both its annual SO_2 and NO_x emissions. Annual SO_2 emissions in Alberta were: 19% of the corresponding value for Ohio; 25% of the value for Indiana, and Ontario, and 40% of the value for Quebec (Table 7.1). Similarly annual NO_x emissions in Alberta were: 11% of the corresponding value for Texas; and 24%, 26%, and 27% of the values for Ohio, Indiana, and Illinois, respectively. Similarly, Alberta NO_x emissions were 46% and 75% of the corresponding values for Ontario and Quebec, respectively. As expected, Alberta's annual SO_2 and NO_x emissions were higher than the less industrialized or populated states or provinces such as Montana and Saskatchewan.

Total SO_2 and NO_x emissions by emission region are presented in Figure 7.2 as histograms superimposed on the Alberta map. Emission Region 9, which included the City of Calgary, had the highest total emissions (570.5 t d^{-1}). Other major sources in this region are two highways with major traffic, several major sour gas processing plants, four fertilizer plants, and a cement plant.

Emission Region 7, which included four major coal-fired power plants, several major sour gas processing plants, and a pulp and paper mill, had the second highest emissions (493.6 t d^{-1}). Emission Region 4, which included two major oil sands plants, had the third highest emissions (385.5 t d^{-1}). Together, Emission Regions 9, 7, and 4 accounted for 75% of all emissions in the Province.

Other regions with significant emissions were Emission Region 8 (the City of Edmonton, major highways, four refineries, and chemical and fertilizer plants), Emission Region 10 (power plants and oil and gas facilities), and Emission Region 5 (sour gas processing and a pulp and paper mill). Emission Regions 1, 3, and 6 had very low emissions (resulting primarily from oil and gas facilities), and Emission Region 2, which consisted primarily of Wood Buffalo National Park, had no significant emissions.

Sulphur dioxide emissions were significantly higher than NO_x emissions in most emission regions; this can generally be attributed to the extraction of sulphur from natural gas and the burning of sulphur-containing fuels in the Province. Only in three Emission Regions, 6, 8 and 10, were NO_x emissions higher than SO_2 emissions. This was primarily due to the lack of combustion of sulphur-containing fossil fuels in these regions.

It can be seen from Figure 7.2 that, except for Emission Region 4 which contained two major oil sands plants, both NO_x and SO_2 emissions in the northern half of the Province (Emission Regions 1-6) were considerably lower than in the southern half (Emission Regions 7-10). With respect to total provincial NO_x and SO_2 emissions, 28.1% or 547.2 t d^{-1} occurred in the northern half of the Province and 71.9% or 1398.0 t d^{-1} occurred in the southern half of the Province. This was due to the presence and use of sulphur-

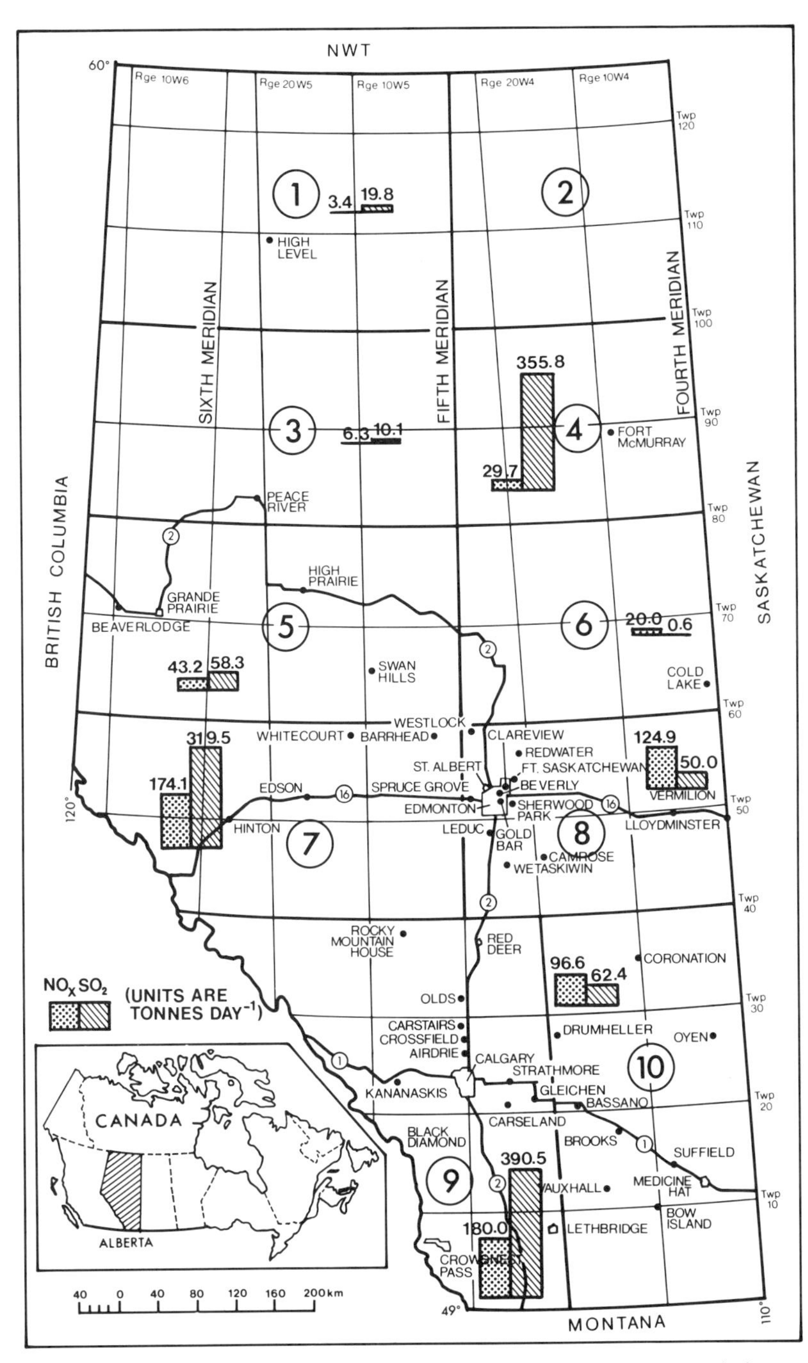

Figure 7.2 Map of Alberta showing total NO_x and total SO_2 emissions as histograms in each of the 10 geographic emission regions.

containing fossil fuels and the heavy concentration of major industry, major urban centres, and major highways in the southern half of the Province.

A pie chart depicting NO_x and SO_2 emissions by major source category for the Province is presented in Figure 7.3. The petroleum industry was the largest single contributor (as a source category) of both types of emissions, accounting for 38.0% of the NO_x emissions (257.5 $t\ d^{-1}$) and 81.5% of the SO_2 emissions (1032.6 $t\ d^{-1}$). For this reason, both NO_x and SO_2 emissions from various sectors of the petroleum industry are detailed in Figure 7.4. Electric power plants, highways, and urban centres were also important sources of NO_x emissions representing 22.4, 15.9 and 15.7%, respectively, of total NO_x emissions. Electric power plants were the second most important source of SO_2 emissions with 16.6% or 210.6 $t\ d^{-1}$ of total SO_2 emissions.

As an addition to the information presented in Figure 7.3, a histogram showing NO_x and SO_2 emissions for the Province by major source category in tonnes per day is presented in Figure 7.5. Information on the number of sources (or population or highway length, where applicable) for each major emission source category and emission type is shown in parentheses.

Figure 7.6 provides histograms showing total NO_x and SO_2 emissions by major source category as a function of each emission region. This figure is a breakdown by emission region of the total NO_x and SO_2 histograms shown in Figure 7.2 and depicts the geographic patterns of emissions throughout the Province as described previously.

7.5 SO_2 AND NO_x EMISSIONS BY ECOREGION

Since ecosystems vary in their sensitivity to air pollution stress, data for both SO_2 and NO_x in the emission inventory were separated according to source location in each of the 12 ecoregions of Alberta as characterized by Strong and Leggat (1981). The term "ecoregion" is derived from "ecosystem" and "region" and is defined as an area characterized by a distinctive regional climate expressed by the distribution of vegetation. The pattern of recurring vegetation, and soil and moisture sequences was the main basis for the recognition and delineation of the ecoregions. Secondary successional species such as spruce (*Picea*) and fir (*Abies*) were used as additional criteria to separate ecoregions. Figure 7.7 shows a map

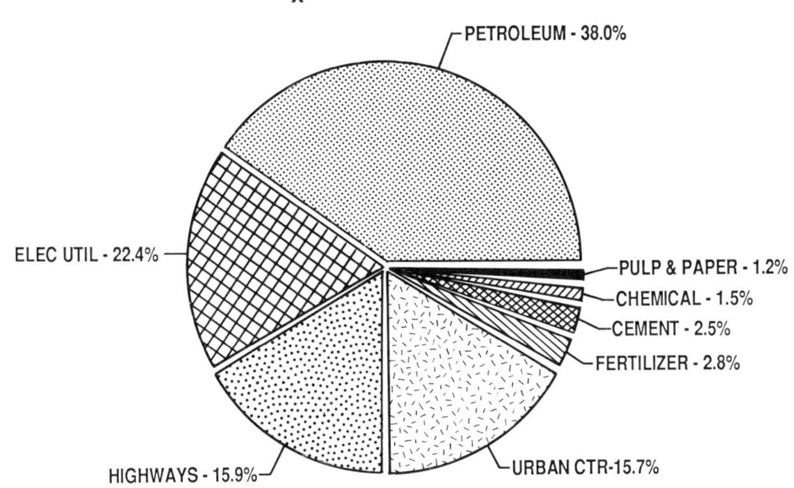

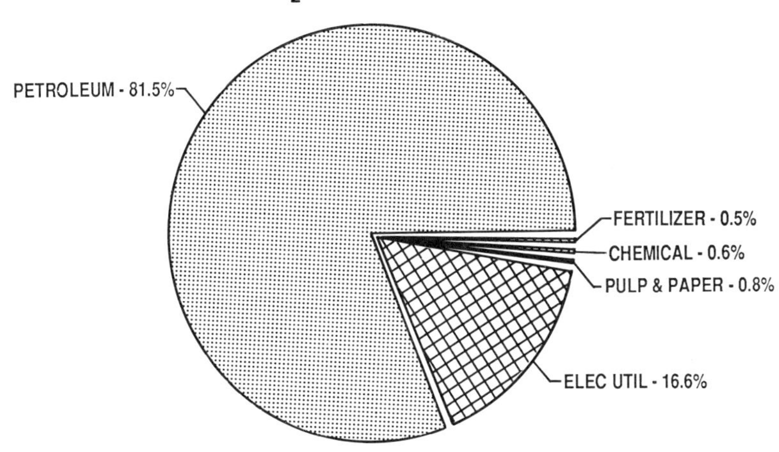

Figure 7.3 Total NO_x and total SO_2 emissions as a function of the percentage contribution by each major source category in Alberta.

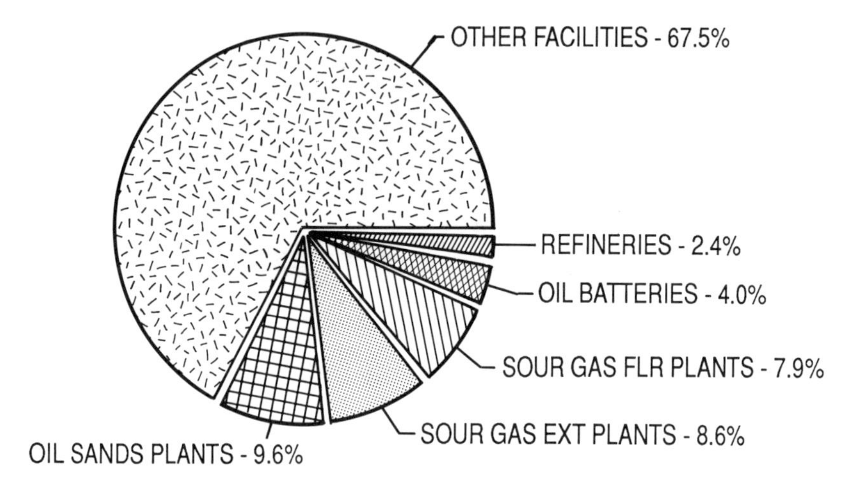

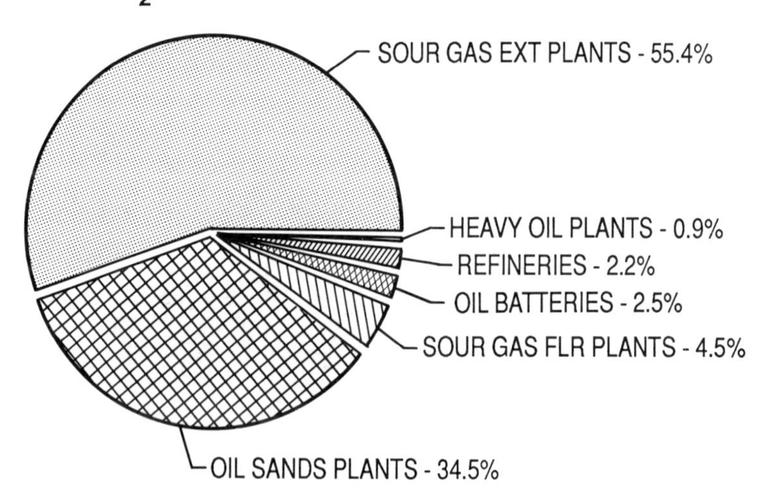

Figure 7.4 Total NO_x and total SO_2 emissions from the Alberta petroleum industry as a function of the percentage contribution by each facility type.

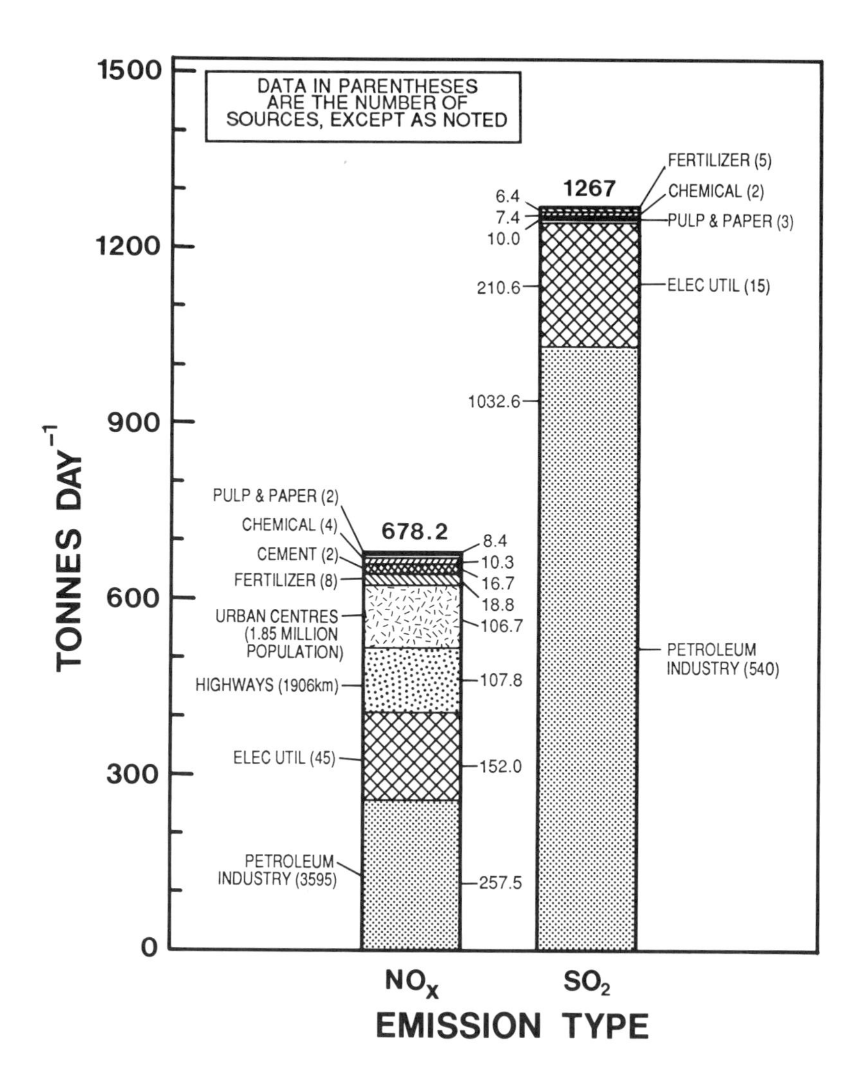

Figure 7.5 Total NO_x and total SO_2 emissions in Alberta by major source category.

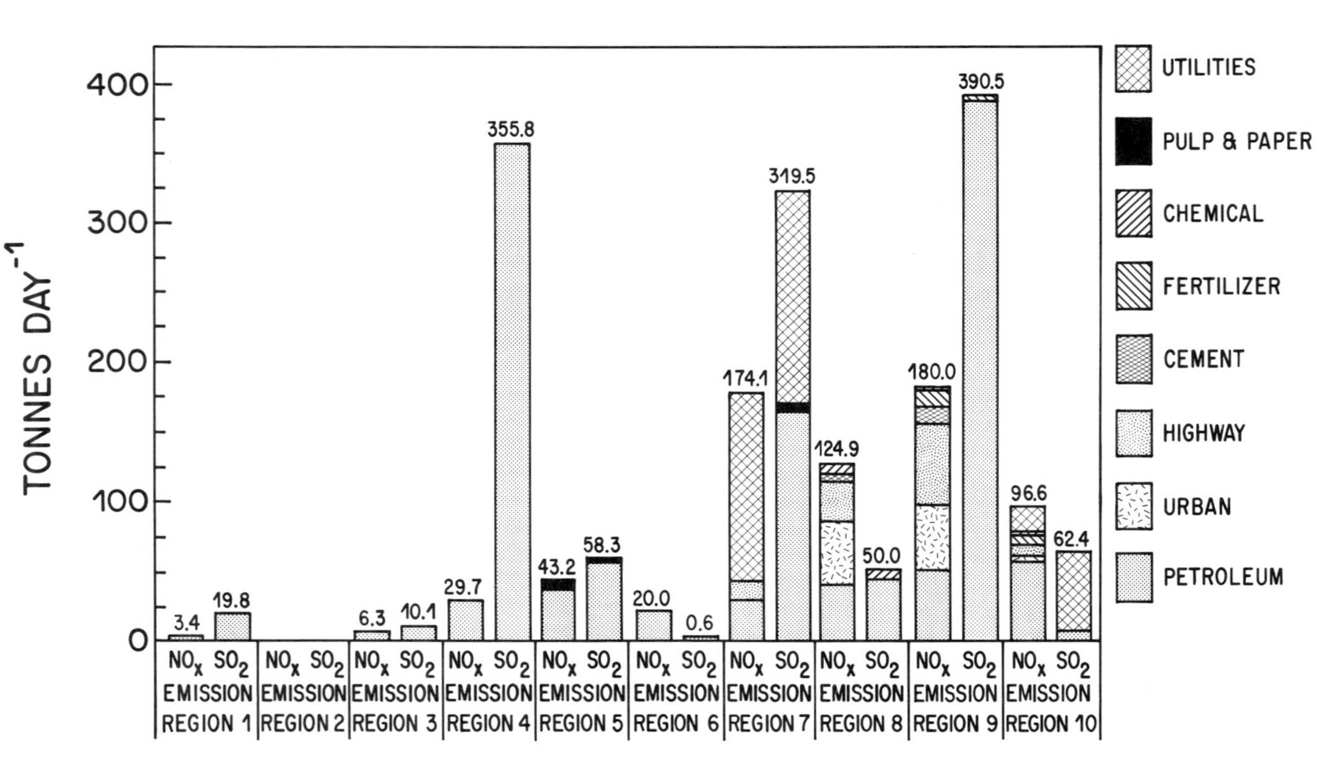

Figure 7.6 Total NO_x and total SO_2 emissions by source category for each geographic emission region.

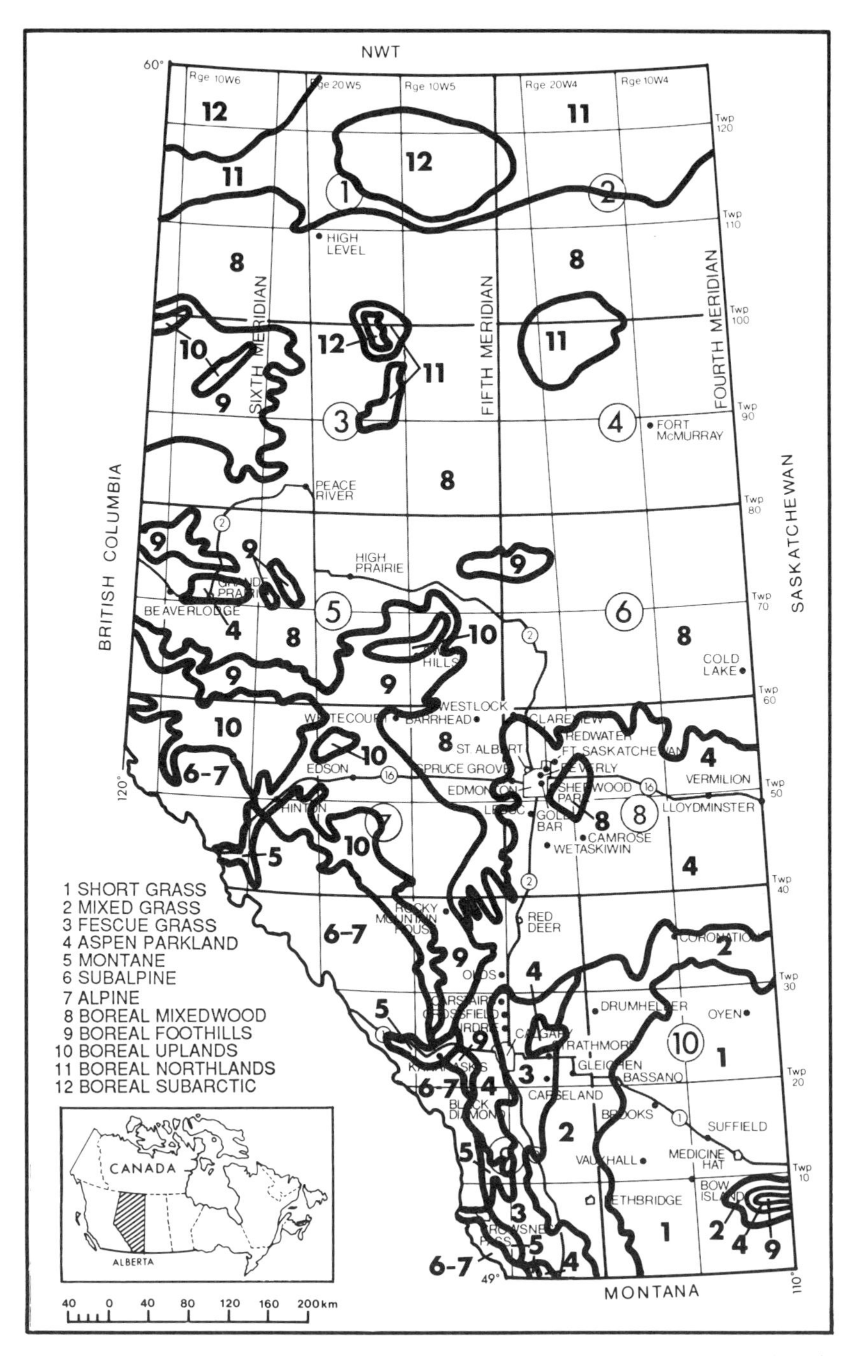

Figure 7.7 Map of Alberta showing ecoregions [Source: Strong and Leggat (1981)] with respect to major urban centres and major highways.

of Alberta with the geographic distribution of the 12 ecoregions within the Province.

A pie chart comparing combined SO_2 and NO_x emissions by ecoregion with relative sizes of the ecoregion is presented in Figure 7.8. Three ecoregions, 4, 8, and 9, accounted for approximately 76% of the emissions; these regions also comprised 63.9% of Alberta's land area. Ecoregions 3, 4, 5, 9, and 10 had relatively high emissions in comparison to their land areas, while ecoregions 1, 2, 6, 7, 8, 11, and 12 had comparatively low emissions (Table 7.2). The montane ecoregion (5) had by far the highest emissions per unit land area (19.56×10^{-3} t d^{-1} km^{-2}), and the sub-alpine (6), boreal northland (11), and boreal sub-arctic (12) ecoregions had the lowest at 0.11×10^{-3}, 0.14×10^{-3}, and 0.12×10^{-3} t d^{-1} km^{-2}, respectively.

Figure 7.9 presents histograms for each ecoregion showing the relative levels of NO_x and SO_2 emissions, and Table 7.2 presents a summary of emission levels and numbers of sources for each ecoregion. Ecoregion 4 contained the largest number of industrial sources (976 sources of NO_x and 172 sources of SO_2), urban centres (113) and length of major highway (669.3 km). Ecoregion 7 contained the fewest sources (4 sources of SO_2, no inventoried sources of NO_x, no urban centres, and no major highway segments).

NO_x emissions were higher than SO_2 emissions in ecoregions 1, 2, 4 and 6; SO_2 emissions were comparatively higher in the remaining eight ecoregions. Ecoregions with comparatively higher NO_x emissions were the short and mixed grass (1 and 2) ecoregions and aspen parkland (4) ecoregion where the influence of urban centres and highways was important. Ecoregions with higher SO_2 emissions than NO_x emissions were the fescue grass (3), boreal mixed-wood (8), boreal foothill (9), and boreal upland (10) ecoregions where sour gas processing was an important contributor to emission levels.

7.6 CONCLUDING REMARKS

Local or regional scale emissions modify the qualitative and quantitative characteristics of the incoming air relative to its pollutant composition. Consequently, the pollutant exposure dynamics and the total pollutant burden on the ecosystem components are modified. Emission inventories are not only important

Table 7.2 Summary of total provincial NO_x and SO_2 emissions by ecoregion.

Ecoregion			NO_x Emissions				SO_2 Emissions	
Name	Number	Area (km^2)	Total Emissions ($t\ d^{-1}$)	Number of Industrial Sources	Number of Urban Centres	Highway Segments (km)	Total Emissions ($t\ d^{-1}$)	Number of Sources
Short Grass	1	46,926	59.7	606	21	170.9	2.3	38
Mixed Grass	2	31,063	26.1	262	31	184.5	8.0	31
Fescue Grass	3	14,403	68.0	115	19	229.0	84.7	22
Aspen Parkland	4	73,268	206.8	976	113	669.3	195.0	175
Montane	5	3,538	28.7	16	1	252.7	40.5	6
Subalpine	6	23,133	2.6	11	2	0	0	2
Alpine	7	18,506	0	0	0	0	4.2	4
Boreal Mixed wood	8	285,611	212.0	770	47	111.1	554.4	134
Boreal Foothill	9	63,362	63.0	795	9	288.4	245.6	114
Boreal Upland	10	27,098	10.9	96	2	0	122.9	20
Boreal Northland	11	47,588	0.4	9	0	0	6.1	12
Boreal Subarctic	12	26,437	0	0	0	0	3.3	7

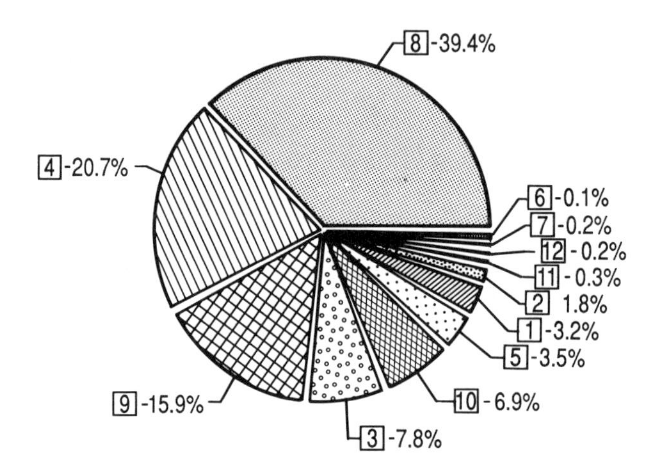

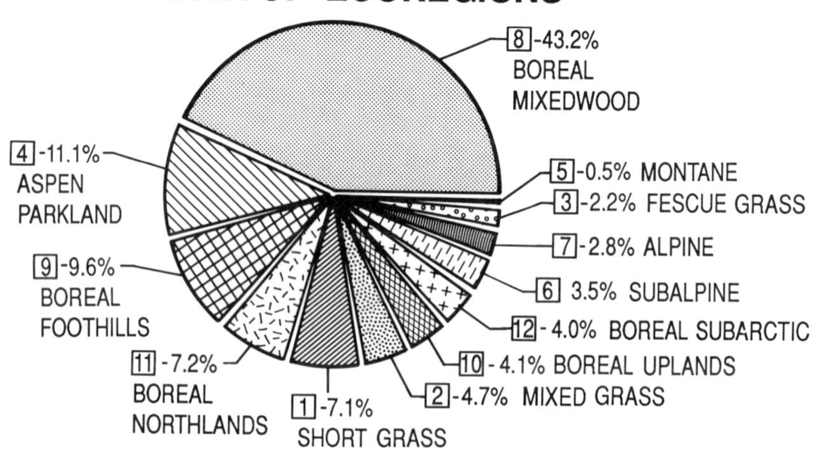

Figure 7.8 Comparison of combined NO_x and SO_2 emissions within each ecoregion as a percentage of total NO_x and SO_2 emissions within the percentage of the total area of Alberta occupied by each ecoregion.

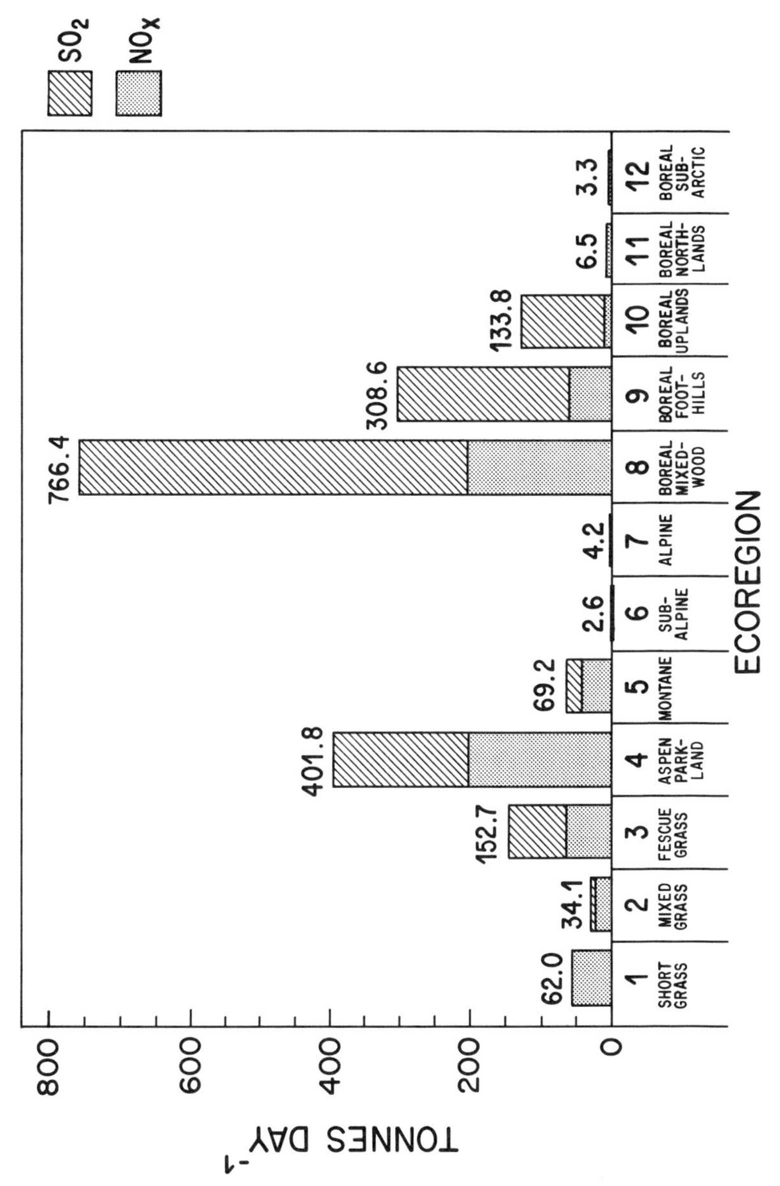

Figure 7.9 Total NO_x and total SO_2 emissions by ecoregion.

in understanding the aforementioned processes, but also in describing source-receptor relationships and in the application of mitigative strategies where warranted.

7.7 LITERATURE CITED

Alberta Environment, Pollution Control Division. 1984. Nitrogen Oxides Emissions for Alberta, 1978-1982. Edmonton, Alberta. 21 pp.

Alberta Environment, Pollution Control Division. 1985. Industrial Sulphur Emissions Inventory, 1979-83. Draft Copy. Edmonton, Alberta. 23 pp.

Alberta Municipal Affairs, Municipal Administrative Services Division. 1985. Official Population List. Edmonton, Alberta.

Alberta Transportation, Systems Planning Branch, Traffic Information Services. 1985. Traffic Volume Breakdown on Primary, Secondary and Other Roads. Edmonton, Alberta.

Energy Resources Conservation Board. 1985a. Alberta Electric Industry, Annual Statistics. Calgary, Alberta. ERCB Report ST 85-28. 33 pp.

Energy Resources Conservation Board. 1985b. Alberta Gas Plants to December 31. Calgary, Alberta. ERCB Report ST 85-13A.

Energy Resources Conservation Board. 1985c. Alberta Oil and Gas Industry, Annual Statistics. Calgary, Alberta. ERCB Report ST 85-17. 95 pp.

Energy Resources Conservation Board. 1985d. Alberta Oil Sands, Annual Statistics. Calgary, Alberta. ERCB Report ST 85-43. 27 pp.

Legge, A.H. and B.L. Baker. 1987. The Effects of Nitrogen Oxides on Native Vegetation Near Compressor Installations in Alberta. Part III: Compressor Installations in Alberta. Prepared for Alberta Environment, Research Management Division by the Kananaskis Centre for Environmental Research, The University of Calgary.

Edmonton, Alberta. Research Management Division RMD Report 87/43 (III). 197 pp.

Picard, D.J., D.G. Colley, and D.H. Boyd. 1987a. Overview of the Emission Data: Emission Inventory of Sulphur Oxides and Nitrogen Oxides in Alberta. Prep. for the Alberta Government-Industry Acid Deposition Research Program by Western Research. Calgary, Alberta. 87 pp.

Picard, D.J., D.G. Colley, and D.H. Boyd. 1987b. Design of the Emission Data: Emission Inventory of Sulphur Oxides and Nitrogen Oxides in Alberta. Prep. for the Alberta Government-Industry Acid Deposition Research Program by Western Research. Calgary, Alberta. 88 pp.

Picard, D.J., D.G. Colley, and D.H. Boyd. 1987c. Emission Data Base: Emission Inventory of Sulphur Oxides and Nitrogen Oxides in Alberta. Prep. for the Alberta Government-Industry Acid Deposition Research Program by Western Research. Calgary, Alberta. 335 pp.

Picard, D.J., D.G. Colley, and D.H. Boyd. 1987d. Results of the Emission Source Surveys: Emission Inventory of Sulphur Oxides and Nitrogen Oxides in Alberta. Prep. for the Alberta Government-Industry Acid Deposition Research Program by Western Research. Calgary, Alberta. 335 pp.

Strong, W.L. and K.R. Leggat. 1981. Ecoregions of Alberta. Alberta Resource Evaluation and Planning Division, Energy and Natural Resources. Edmonton, Alberta. 64 pp.

US-Canada Memorandum of Intent on Transboundary Air Pollution, Final Report, Work Group 3B, 1982.

CHAPTER 8

MODEL ESTIMATES OF PROVINCIAL SCALE ATMOSPHERIC SULPHUR DIOXIDE (SO_2) AND OXIDES OF NITROGEN (NO_x) IN ALBERTA

G. E. McVehil

8.1 INTRODUCTION

The deposition (dry and wet) of airborne species provides an important input of acidic and acidifying substances into terrestrial and aquatic ecosystems. To develop a meaningful assessment of the importance of these emissions, it is necessary to apply atmospheric models capable of predicting the deposition patterns resulting from the distribution of sources. For studying the acidic precipitation problem it is necessary that these models be of regional extent in order to treat the processes occurring during long-range transport of the important ambient species, i.e., SO_x and NO_x compounds (Carmichael and Peters 1984).

The objective of the work described in this chapter was to develop estimates of the average seasonal concentrations of major pollutants on a regional scale across Alberta. Such estimates are important to characterize and place in perspective, average Alberta pollutant levels, and also to provide a basis for regional assessment of environmental effects. Long-range transport modelling was applied to produce predictions of seasonal concentrations of SO_2, SO_4^{2-}, and NO_x. Other pollutants were not modelled because of the limitations in the availability of data on source emissions and pollutant chemistry.

8.2 THE APPROACH TO MODELLING

8.2.1 Regional Air Pollution Simulation Models

Models to simulate the distribution of pollutants in the atmosphere must include the consideration of three basic factors: the pollutant sources (source inventory); atmospheric transport and diffusion (dispersion model); and pollutant decay (transformation and deposition). All air quality models include the first two components; those involving purely local areas where pollutant travel time from the source is less than several hours often neglect the third. For regional and larger-scale models, all three components must be included.

In the following sections, the methods and data bases applied to each component in the present work are described. However, it is appropriate to first review the two basic types of regional scale models and to outline the approach taken here.

8.2.2 Existing Models

Modelling of pollutant transport over distances in excess of 50 km is generally referred to as long-range transport modelling. Long-range transport modelling may be performed on scales from hundreds to thousands of kilometres. The objectives vary from the assessment of the impacts of a single source on a specific receptor site to continental or global-scale effects of multiple sources.

A variety of approaches has been applied to long-range transport modelling. A general review of these methods was provided by Eliassen (1980). Another review, with reference to specific models for application in western Canada, was prepared by the Western Mesoscale Modelling Task Group (1985).

As described by Eliassen (1980), models for long-range transport fall into two types: models providing hourly or daily concentrations, and models of a statistical nature which give longer-term average concentrations.

The first type of model attempts to use observed meteorological conditions on an hour-by-hour basis to calculate the short-term transport and dispersion of pollutants. Such models can cover a wide range in complexity, depending upon the detail with which they treat atmospheric structure, land surface characteristics, and pollutant transformation and depletion. However, all such models have relatively large computing requirements because of the number of computations needed to treat spatially varying fields over many small time steps. In principle, it is possible to apply a short-term model over a long sequence of daily events to calculate seasonal or annual averages. However, in practice this procedure becomes impractical unless very extensive computer capabilities are available.

Statistical models use average values of input parameters such as wind conditions, mixing depths, and decay rates to provide time and space-averaged predictions of concentration or deposition. It can be argued that statistical models neglect important small-scale factors and that they cannot properly simulate nonlinear processes. However, the process of averaging can smooth out many of the random errors encountered in short-term models. Most importantly, statistical models have been shown to provide realistic estimates of long-term regional concentration and deposition values (Venkatram 1986). Statistical models offer a practical method for estimating long-term pollutant distributions where the statistics of pollutant emissions and regional scale meteorology can be defined.

8.2.3 Present Approach

It was reasonable to expect that average SO_2 and NO_x concentrations in Alberta can be estimated by considering sources of these pollutants in the Province, with some small additional contribution from the background air quality (transport into the Province from other sources). Prevailing winds in Alberta are from the northwest and west, and very few significant sources exist in these directions except at great distances. In fact, emissions of SO_2 and NO_x were considered to be small over large areas in all directions from the Province. Thus, it was assumed that sources outside Alberta do not require consideration in the model calculations and that constant background concentrations can be applied to account for outside sources.

A complete and detailed inventory of SO_2 and NO_x sources in Alberta was provided by Picard et al. (1987a,b,c,d) (Chapter 7). This inventory was used to generate the source definition for the present modelling. Since the inventory provided detailed location and emissions data for more than 4000 individual sources, it was desirable to utilize the spatial distribution information to provide as much resolution as possible in predicted concentrations.

Alberta consists of a large geographic area (over 500,000 km^2) with relatively sparse coverage of meteorological data. The combination of many sources, a large area, and limited meteorological data provided an argument against the practicality of short-term pollutant transport modelling. However, because of the described factors, and the need for long-term average concentrations, it was determined that a statistical model was most appropriate for the present objectives.

It was considered that the atmospheric dispersion model selected for use should be capable of treating a large number of point and area sources, and provide resolution comparable to the source description. Such a model should also utilize statistical, meteorological input in as great detail as possible for the Province. In satisfying these requirements, the Climatological Dispersion Model (CDM) (Busse and Zimmerman 1973) provided the necessary framework. The model was originally developed for application to urban areas on a scale of 25-50 km, but it could readily be modified for application on a regional scale encompassing rural conditions. The model utilizes a joint frequency distribution of meteorological parameters (wind speed, wind direction, and atmospheric stability) to define the range of meteorological conditions that occur over a season or year and their relative frequency of occurrence.

It can be concluded from the previous discussion of short-term pollutant transport versus statistical models that combinations of the two are possible and that, indeed, all models are partly statistical (hourly data are assumed to represent averages over a one-hour period, and smaller-scale processes are parameterized through statistical relationships). The present approach is fundamentally statistics oriented, but it does include the treatment of a wide range of meteorological and source conditions.

The model was applied on a basic rectangular grid of 20 km spacing, thus affording higher resolution than commonly found in statistical, long-range transport models (Fisher 1978; Venkatram et al. 1982). This resolution takes advantage of much of the detail available in the emission inventory. However, it should be recognized that details of concentration distributions near individual sources cannot be resolved; the results represent average concentrations smoothed on a scale of the order of 20 km. Further, though the model considers temporal variations in meteorology and chemistry, spatial variations are not considered except in the distribution of sources (statistics are assumed to be uniform across the Province).

8.3 REGIONAL EMISSION PARAMETERS

Complete emission inventories for SO_2 and NO_x sources in Alberta were prepared by Picard et al. (1987c). The inventories include 565 SO_2 sources with total emissions of 1267 t d^{-1} and 4025 NO_x sources with emissions totalling 678 t d^{-1}. The source inventory data included seasonal emission rates, stack parameters (height, diameter, exit velocity, and temperature), and source location expressed in terms of Section, Township, and Range within the Province.

All emissions from the inventory compiled by Picard et al. were included in the application of the statistical transport model. Source locations were converted from the Section-Township-Range designation to a rectangular coordinate system. Major sources (those with annual average emission rates greater than 1.0 t d^{-1}) were represented in the model as point sources at their true location. All smaller sources were combined within 20 x 20 km^2 grids and treated as area sources.

For point sources, actual stack parameters, as given in the inventory, were used. In the few cases where a particular parameter was missing, it was estimated based on typical values included elsewhere in the inventory for the same type of source. All area sources were assumed to have an emission height of 30 m and an initial vertical dispersion (σz) of 15 m. Separate emission rates were determined for the annual period, for the summer half-year (based on second and third quarter rates in the inventory), and for the winter half-year (average of first and fourth quarter rates). The emissions data were based primarily on 1984 information and should be considered to be representative of the 1984-87 time period.

The final modelling inventories consisted of three separate data files: one each for summer, winter, and annual time periods. Each file contained 659 area sources and 78 point sources. A summary of the seasonal emissions is provided in Table 8.1.

Table 8.1 Number of sources and total emissions (t d^{-1}) included in the provincial scale air quality model calculations for Alberta.

	Number of Sources	Total Emissions					
		Annual		Summer		Winter	
		SO_2	NO_x	SO_2	NO_x	SO_2	NO_x
Point Sources	78	1195.0	213.6	1167.4	207.9	1222.7	219.3
Area Sources	659	71.9	464.9	67.0	460.6	77.0	469.2
Total	737	1266.9	678.5	1234.4	668.5	1299.7	688.5

The seasonal variation in Alberta emissions was not large (Table 8.1). Both SO_2 and NO_x emissions were slightly higher during the winter half of the year, but the difference between summer and winter was less than 6% for both. A major portion of the SO_2 emissions originated from relatively large point sources. The 61 individual SO_2 sources with emissions greater than one tonne per day contributed 94% of all SO_2 emissions. Point source NO_x emissions from these 61 sources plus an additional 17 sources with NO_x emission rates greater than one tonne per day, contributed only 32% of all NO_x emissions. A significant portion of the NO_x emissions identified in the inventory was from motor vehicles and natural gas consumption in population centres. These emissions were all properly considered as distributed area sources. In addition, a large number of small individual NO_x point sources were combined into area sources to facilitate modelling.

8.4 THE DISPERSION MODEL

The CDM (Climatological Dispersion Model) was originally published by the U.S. Environmental Protection Agency (Busse and Zimmerman 1973). It is a statistical model designed to calculate seasonal or annual ground-level concentrations using average emission rates from point and area sources and a joint frequency distribution of wind direction, wind speed, and stability. The model was originally intended for studies of compliance with air quality standards in urban areas containing large numbers of pollution sources.

A modified version of the model was subsequently developed for application to rural areas by the Wyoming Department of Environmental Quality (Dailey 1982). This model, commonly referred to as the CDMW, differed from the original approach primarily through its inclusion of routines for calculating dispersion and plume rise under very stable conditions, not considered in the original urban version.

Several modifications were made to the CDM (beyond those in CDMW) for application in the present study. Expansion of the spatial scale of the model was easily achieved by increasing the distance increments used in the integration of area source contributions to each receptor. Within 10 km of a receptor, the integration increment was taken as 3750 m; for distances of 10-20 km the increment was set at 7500 m; beyond 20 km the increment was taken as 15 km. Preliminary tests of the model indicated that use of these radial integration increments, along with 1.125 degree angular increments, provide accurate and stable integrations of contributions from the 20 x 20 km area sources over the complete model grid (680 x 1200 km).

The model used a sector-averaged Gaussian dispersion equation to calculate contributions from both point and area sources. Plume centreline height was determined from the stack height and standard Briggs plume rise expressions. Wind speed was assumed to vary with height above the ground according to a stability-dependent power law. Vertical dispersion coefficients (plume spread) were taken from the standard Pasquill-Gifford (rural) curves. When the vertical plume dimension became comparable to the mixing depth, uniform mixing within the mixing depth was assumed.

The model required some modification in the treatment of mixing depths for application to long distances and pollutant travel times. In the original CDM, the mixing depth was varied according to stability class, with values of mixing depth ranging from average morning to average afternoon values. This treatment was not appropriate for long-range transport where dispersion over many hours of pollutant travel is considered. A simplified treatment was implemented in the model whereby an average seasonal or annual mixing depth was used uniformly for all stability classes.

The combined effects of plume rise (for point sources), dispersion, and mixing depth were considered in evaluating the applicability of the model to long-range transport. A number of trial calculations were made to assure that the model as implemented provided a reasonable, albeit simplified, simulation of atmospheric transport of pollutants.

For the range of source types, emission parameters, and mixing depths used in the present application, it was found that final plume height very rarely, if ever, exceeded the mixing depth. If the plume height was to exceed the mixing depth, the model would treat the pollutant as escaped from the surface boundary layer and would give no ground-level contribution at any distance from the source. Although in reality, on many occasions, sources with tall stacks will emit plumes above the surface mixing layer, the pollutant may well return to the surface layer at some distance downwind as a result of subsequent deepening of the mixed layer. The present application is therefore conservative in that virtually all the pollutant plume is retained within the mixing depth and there is no provision for escape of the pollutant from the boundary layer. However, this behaviour seems preferable to introducing a mechanism whereby a significant part of large point source emissions could escape but would never affect ground-level concentrations at any distance downwind.

As implemented, the model provides the following general representation of pollutant dispersion for elevated point source plumes.

a) Near-Source (0-10 km)
   - neutral and unstable - gradual pollutant dispersion from plume height to the ground, with the rate dependent on stability class.
   - stable - plume, in general, remains aloft because of slow vertical spreading.

b) Mid-Range (10-100 km)
   - neutral and unstable - after dispersion to ground level, plume becomes uniformly distributed through mixing depth.
   - stable - plume very slowly diffuses toward ground-level and may begin to affect ground concentrations at long distances.

c) Long-Range (100-1500 km)
   - neutral and unstable - uniform plume distribution in mixing layer.
   - stable - small ground-level contribution from elevated plume; may ultimately achieve fully mixed distribution at the longest distances.

All area sources were modelled as emissions at 30 m height with no plume rise. Thus, for the long distances considered in the model, area source emissions disperse essentially as ground-level releases, becoming fully mixed in the mixing layer at distances on the order of 10 km in unstable conditions and 200 km in neutral conditions, but remaining near the surface in stable conditions.

The treatment of dispersion and long-range transport as described follows from the structure of the CDM, the few modifications incorporated, and the magnitude of plume heights, mixing depths, etc. encountered in the application for Alberta. As in most long-range statistical models, a uniform distribution of pollutant in the mixed layer is achieved at long distances. However, closer to sources, this model provides some improved realism in terms of ground-level contributions that vary with stability and plume height.

No terrain or varying elevation effects were included in the model. Over most of Alberta this flat terrain assumption was not expected to impact model performance negatively. Of course, predicted concentrations in mountainous parts of western Alberta should be interpreted with caution.

The model was applied on a 20 km rectangular grid superimposed on a polyconic projection map of Alberta. Pollutant concentrations were calculated at 350 receptors distributed over the grid with a spacing of 20-40 km in southern Alberta and 60-80 km over northern Alberta.

8.5 METEOROLOGICAL DATA

8.5.1 Seasonal Meteorological Parameters

The application of the model required specification of annual and seasonal values of average ambient temperatures and mixing depth. Average surface temperatures on an annual basis, for summer (April 15-October 15), and winter (October 16-April 14), were estimated from Alberta climatological data. Annual average mixing depth, as well as average values for the summer and winter half of the years, were obtained from data provided by Leahey and Jamieson (1987). The Leahey and Jamieson analysis was performed specifically for the purpose of deriving input data for the application of statistical long-range transport models in western Canada. The values for parameters used for the modelling are shown in Table 8.2.

Table 8.2 Values of meteorological parameters used for seasonal and annual air quality modelling.

	Average Temperature (°C)	Mixing Depth (m)
Winter (Oct 16-Apr 14)	-3.0	750
Summer (Apr 15-Oct 15)	10.0	1400
Annual	3.0	960

8.5.2 Wind-Stability Joint Frequency Distributions

Data concerning the joint frequency distribution (JFD) for wind speed, wind direction, and stability, as determined by the Pasquill (1961) method from routine surface weather observations, were available for a number of locations in Alberta. However, the surface wind distributions at many of these stations were strongly influenced by local topography and the winds were not representative of regional pollutant transport.

Therefore, composite JFDs were developed for modelling, based on surface stability data and near-surface winds measured at upper air sounding stations.

Standard STability ARray (STAR) joint frequency data for the period 1974-83 were obtained for Calgary, Fort McMurray, Peace River and Vermilion. The average seasonal and annual distributions of stability classes for these four Alberta stations are shown in Table 8.3. These composite distributions were taken as representative of Alberta dispersion conditions.

Table 8.3 Mean distributions of Pasquill stability classes for Alberta (average of 1974-83 data from Calgary, Fort McMurray, Peace River, and Vermilion).

Stability Class	Frequency (%)		
	Winter	Summer	Annual
A	0.11	0.55	0.33
B	3.29	8.39	5.86
C	8.95	13.98	11.48
D	50.05	44.95	47.49
E	13.37	10.93	12.14
F	24.23	21.19	22.70

Following the work by Leahey and Jamieson (1987) to develop input data for statistical models, wind data for the mixing layer were obtained as a weighted average of 900 mb data reported at upper air observation stations. The 900 mb level occurs at an average height above the ground of about 250 m in Alberta; this height was considered to be appropriate for transport of surface and tall-stack emitted pollutants over most of the Province.

Leahey and Jamieson provided twice-a-day observations of 900 mb wind speed and direction obtained at Edmonton and Fort Smith for the 10-year period 1975-84. Weighted average wind distributions (2/3 Edmonton, 1/3 Fort Smith) were developed from these measurements and composite wind roses were tabulated (Table 8.4).

Table 8.4 Seasonal and annual composite 900 mb wind rose data for Alberta (obtained as a weighted average, 2/3 Edmonton, 1/3 Fort Smith of all 1975-84 measurements.) Frequencies in %.

Wind	Wind Speed (m s^{-1})			
Direction	<3.1	3.1-8.2	>8.2	Total
		a) Winter (Oct 16-Apr 14)		
N	1.56	3.91	1.43	6.90
NNE	1.07	1.24	0.28	2.59
NE	0.88	1.03	0.13	2.04
ENE	1.11	1.16	0.07	2.34
E	1.52	2.48	0.18	4.18
ESE	0.92	2.27	0.39	3.58
SE	0.92	3.20	1.26	5.38
SSE	1.13	3.13	1.45	5.71
S	1.58	3.65	2.15	7.38
SSW	1.22	1.88	0.87	3.97
SW	1.11	2.48	0.73	4.32
WSW	1.15	2.42	0.63	4.20
W	1.83	6.33	2.93	11.09
WNW	1.45	5.65	4.37	11.47
NW	1.32	6.81	6.64	14.77
NNW	1.20	5.14	3.75	10.09
Total	19.97	52.78	27.26	100.00
		b) Summer (April 15-Oct 15)		
N	1.82	4.40	1.64	7.86
NNE	1.19	2.00	0.39	3.58

Continued...

Table 8.4 (Concluded).

Wind Direction	Wind Speed ($m\ s^{-1}$) <3.1	3.1-8.2	>8.2	Total
NE	0.99	1.82	0.27	3.08
ENE	1.03	1.67	0.31	3.01
E	1.12	2.62	0.66	4.40
ESE	1.19	2.40	0.78	4.37
SE	1.11	3.34	1.03	5.48
SSE	1.10	3.36	1.62	6.08
S	1.97	4.53	1.57	8.07
SSW	1.29	2.33	0.55	4.17
SW	1.33	2.33	0.49	4.15
WSW	1.35	2.06	0.62	4.03
W	2.48	6.11	2.13	10.72
WNW	1.38	5.02	3.53	9.93
NW	1.64	6.58	4.28	12.50
NNW	1.49	4.92	2.21	8.62
Total	22.48	55.49	22.08	100.00
		c) Annual		
N	1.69	4.16	1.54	7.39
NNE	1.13	1.62	0.34	3.09
NE	0.93	1.42	0.20	2.55
ENE	1.07	1.41	0.19	2.67
E	1.32	2.55	0.42	4.29
ESE	1.06	2.34	0.58	3.98
SE	1.02	3.27	1.14	5.43
SSE	1.11	3.25	1.54	5.90
S	1.77	4.09	1.86	7.72
SSW	1.26	2.11	0.72	4.09
SW	1.22	2.40	0.61	4.23
WSW	1.25	2.24	0.63	4.12
W	2.15	6.22	2.52	10.89
WNW	1.42	5.33	3.95	10.70
NW	1.48	6.70	5.46	13.64
NNW	1.34	5.03	2.98	9.36
Total	21.21	54.12	24.67	100.00

The surface stability distributions were combined with the 900 mb wind distribution data to generate seasonal and annual JFDs as required for model input. Wind speed distributions from the surface and 900 mb data were related through the use of the power law:

$$u_2 = u_1 (z_2/z_1)^p$$

where u_1 and u_2 are wind speeds at heights z_1 (10 m), and z_2 (250 m), respectively, and P, the power law exponent, has values of 0.10, 0.15, 0.20, 0.25, 0.25, and 0.30 for stability classes A, B, C, D, E, and F, respectively.

Necessarily there is no unique combination of the surface and 900 mb data that provides a JFD perfectly consistent with both sets of distributions. However, distributions were readily constructed that maintained the overall distributions of the wind speed and wind direction of the 900 mb wind rose with the stability distribution of the surface data. The wind speed distributions of the surface data were maintained approximately for each stability class.

8.6 POLLUTANT TRANSFORMATION AND DECAY

The use of the transport and dispersion model previously described with an appropriate inventory of Alberta emissions will result in a prediction of average concentrations of the emitted pollutants at desired points in the grid. However, for a realistic simulation of SO_2 and NO_x concentrations at large distances from their sources, it was also necessary to account for chemical transformation during atmospheric pollutant transport and surface deposition due to wet and dry processes.

8.6.1 Loss of SO_2 and NO_x

The treatment of transformation and scavenging of SO_2 and NO_x follows the model of Venkatram et al. (1982). This model is similar to those also used by Rodhe and Grandell (1972), Fisher (1978), and Smith (1981). In the model, the pollutant is defined to exist in four states: a wet and a dry state corresponding to both the original pollutant (SO_2 or NO_x) and its transformation product (SO_4^{2-} or NO_3^-). The wet state exists when precipitation is occurring and the dry under other conditions. A system of simultaneous differential equations can then be written which represents the interactions or transformations between the four states. Terms in each of the equations represent chemical transformation, transition from a dry to a wet state and vice versa, dry deposition (in the dry state), and precipitation scavenging (in the wet state). With suitable initial conditions, the equations can be solved to provide a solution representing the evolution of total atmospheric pollutant mass as a function of time (Venkatram et al. 1982).

The solution for SO_2 or NO_x can be viewed as a sum of exponential terms; i.e.:

$$G = \sum_i A_i e^{-\alpha_i t}$$

where G is the fraction of pollutant (initially unity) remaining in the atmosphere at time t, and A_i and α_i are coefficients determined by the decay and transformation rates. The summation is over the four states, i=1-4. The exact sum can be closely approximated by one or two exponential terms. In the present model, these exponential terms were evaluated as a function of pollutant travel time (t=distance from the source/wind speed) and multiplied by the contribution from each source to each receptor site to account for pollutant loss. A similar sum of exponential terms can be derived to define the evolution of sulphate or nitrate mass, as described in the following section.

The constants A_i and α_i in the mathematical solution depend upon the transformation and loss parameters assumed. Since these parameters are taken as constants, only linear processes are represented in the model (decay and transformation parameters are independent of pollutant concentration). However, as pointed out by Venkatram (1986), reasonable values of the parameters have been shown to result in realistic predictions from statistical models, and results do not appear to be unduly sensitive to small changes in parameter values. The considerable success achieved with statistical models using this framework in varying geographic, meteorological, and pollution regimes (Venkatram and Pleim 1985; Venkatram 1986) offers some confidence for its application here.

The parameters needed for application of the decay model are:

k_d, k_w the conversion rates in dry and wet conditions,respectively, for transformation of sulphur dioxide to sulphate (or oxides of nitrogen to nitrate), expressed as a fraction converted per unit time.

τ_d, τ_w time scales for the average duration of dry and wet periods. The model assumes that the rates at which pollutant undergoes transition from dry to wet states and vice versa are given by $1/\tau_d$ and $1/\tau_w$.

λ_d rate at which pollutant is removed by dry deposition. The dry scavenging factor λ_d is usually taken as v_d/h, where v_d is an appropriate deposition velocity and h is the mixing depth.

λ_w wet scavenging rate, expressed as the fraction of pollutant removed per unit time in precipitation.

Table 8.6 lists the values of these parameters adopted for use in Alberta modelling. Table 8.5 shows the resulting decay rates for SO_2 and NO_x, expressed as the half-life for a simple exponential decay.

Table 8.5 Half-lives (hours) of SO_2 and NO_x from the statistical transformation and scavenging model.

	SO_2	NO_x
Summer	12.9	9.5
Winter	35.0	30.3
Annual	18.7	14.6

In general, parameter values were chosen following Venkatram's (1986) summary of past experience in testing statistical models. Venkatram provided a discussion of values that are "...generally consistent with our knowledge of the corresponding processes, (and) provide acceptable fits between model predictions and observations." However, where data exist or estimates could be obtained that appear to be more specific to Alberta, these data were incorporated into parameter selections.

Annual average chemical transformation rates are frequently quoted as being of the order of 1% per hour for SO_2, and 5-10% per hour for NO_x. After discussion with a number of researchers, the values selected for Alberta application were: SO_2, 1% per hour annual, 1.7% summer, and 0.3% winter; NO_x, 5% per hour annual, 7.5% summer, and 2.5% winter (personal communication; J.P. Lodge, Editor, Atmospheric Environment, Boulder, Colorado; J.H. Seinfeld, California Institute of Technology, Pasadena, California; and J.G. Calvert, National Center for Atmospheric Research, Boulder, Colorado to S. Krupa, ADRP). The substantial seasonal

Table 8.6 Pollutant scavenging parameters used in the model application ($second^{-1}$).

	SO_2			NO_x		
	Summer	Winter	Annual	Summer	Winter	Annual
Dry period chemical transformation rate	4.7×10^{-6}	8.3×10^{-7}	2.8×10^{-6}	2.1×10^{-5}	6.9×10^{-6}	1.4×10^{-5}
Wet period chemical transformation rate	4.7×10^{-6}	8.3×10^{-7}	2.8×10^{-6}	0	0	0
Dry deposition velocity (cm s^{-1})	1.1	0.2	0.5	0.2	0.04	0.1
Dry removal rate	8.0×10^{-6}	2.4×10^{-6}	5.2×10^{-6}	1.6×10^{-6}	0.5×10^{-6}	1.0×10^{-6}
Wet removal rate	4.0×10^{-5}	2.0×10^{-5}	3.0×10^{-5}	2.0×10^{-6}	0	1.0×10^{-6}
Dry to wet period transformation rate ($1/\tau_d$)	5.1×10^{-6}	6.9×10^{-6}	5.8×10^{-6}	5.1×10^{-6}	6.9×10^{-6}	5.8×10^{-6}
Wet to dry period transformation rate ($1/\tau_w$)	5.6×10^{-5}	4.3×10^{-5}	4.8×10^{-5}	5.6×10^{-5}	4.3×10^{-5}	4.8×10^{-5}

variations used, appeared to be in order for Alberta because of marked variation in solar radiation and ambient temperatures between summer and winter.

Mean lengths of wet and dry periods were determined for Alberta by Leahey and Jamieson (1987). Their annual values for most of Alberta are about 6 and 48 hours, respectively. Leahey and Jamieson, however, do not provide seasonal values. Summer precipitation tends to be more often of a convective or shower type as opposed to the large synoptic scale precipitation systems that dominate winter snowfall. On this basis, selected values of wet and dry time scales (hours) are

	τ_w	τ_d
Summer	5.0	55.0
Winter	6.5	40.0
Annual	5.75	47.5

Dry deposition rates were calculated from the deposition velocity and the seasonal mean mixing depths presented in Chapter 8.5. A typical mean deposition velocity for SO_2 was considered to be 1 cm s^{-1}. However, Venkatram notes that a somewhat smaller value is probably in order to account for reduced deposition during stable nocturnal conditions. For the same reason and also because of frozen and snow-covered surfaces, v_d is expected to be smaller in winter. Using summer, winter, and annual SO_2 deposition velocities of 1.1, 0.2, and 0.5 cm s^{-1}, respectively, dry deposition rates are 8 x 10^{-6} s^{-1} for summer, 2.4 x 10^{-6} s^{-1} for winter, and 5.2 x 10^{-6} s^{-1} as an annual average.

Dry deposition velocities for NO_x are less well established than for SO_2 but are considered to be much smaller. An annual average NO_x deposition velocity of 0.1 cm s^{-1} was used here, resulting in an annual deposition rate of 1.04 x 10^{-6} s^{-1}. The seasonal variation was taken as 0.5 x 10^{-6} s^{-1} in winter and 1.6 x 10^{-6} s^{-1} in summer.

The wet removal rate of SO_2 has usually been accepted in the range of 10^{-5} to 10^{-4} s^{-1}. Venkatram et al. (1982) found that a value of 3 x 10^{-5} s^{-1} provided reasonable agreement with observations. In Alberta, it is expected that summer scavenging is more efficient than winter scavenging due to generally higher precipitation rates and liquid versus frozen precipitation. Values of 2 x 10^{-5} for winter and 4 x 10^{-5} for summer were used here, with an annual average similar to that of Venkatram et al. (1982) (3 x 10^{-5} s^{-1}).

Wet removal of NO_x is expected to be much smaller than for SO_2 (Venkatram 1986). The annual value used here is 1.0 x 10^{-6} s^{-1}, ranging from zero in winter to twice the annual value in summer.

It can be noted from the order of magnitude of the various removal parameters that for NO_x the dominating removal mechanism is chemical transformation, with wet and dry scavenging playing relatively minor roles. For SO_2, wet removal is the most important process with both dry deposition and chemical transformation being significant but less efficient than wet scavenging.

8.6.2 Generation and Removal of Sulphate

The same statistical scavenging model produces a solution for the mass of sulphate (or nitrate) as a function of travel time from the SO_2 source (Venkatram et al. 1982). Additional scavenging parameters must be specified for the wet and dry removal of sulphate. The parameters used in the current application are shown in Table 8.7. They generally follow the recommendations of Venkatram et al. (1982) and Venkatram and Pleim (1985), with the same range of seasonal variation used for SO_2. For the annual values, a sulphate dry deposition velocity of 0.05 cm s^{-1} was assumed, and the wet removal rate was taken as 1.0 x 10^{-4} s^{-1}, a value extracted from Garland (1978).

Table 8.7 Sulphate scavenging parameters used in the model applications (fractions per second).

	Summer	Winter	Annual
Dry deposition velocity (cm s^{-1})	0.1	0.02	0.05
Dry removal rate (s^{-1})	8.0 x 10^{-7}	2.4 x 10^{-7}	5.2 x 10^{-7}
Wet removal rate (s^{-1})	1.25 x 10^{-4}	0.75 x 10^{-4}	1.0 x 10^{-4}

It is generally assumed that some small fraction of SO_2 emissions is actually in the form of sulphate; thus, values of 2-5% of the initial SO_2 mass are often used. Leahey and Jamieson (1987) point out that in Alberta only minimal sulphate emissions would be expected because a large fraction of Alberta emissions are from gas plants with negligible sulphate emissions. It was assumed here that 1% of Alberta SO_2 emissions are actually sulphate; i.e., initial sulphate mass is 1.5% of initial SO_2 mass.

The mathematical solution for the time-evolution of sulphate mass using the parameters defined previously can be closely approximated as the difference of two exponentials. The approximate solutions are shown in Table 8.8, with a tabulation of sulphate mass at various travel times for each case. The maximum sulphate mass, as a fraction of initial SO_2 mass, ranges from 29% in summer (at 28 hours travel time) to about 9% in winter (at greater than 50 hours).

Table 8.8 Evolution of total sulphate mass as a function of travel time from the source. Mass of sulphate expressed as fraction of the initial mass of sulphur dioxide. (Assumes that 1% of SO_2 inventory emitted is sulphate.)

Time (h)	Summer	Winter	Annual
0	.015	.015	.015
1	.038	.019	.028
2.8	.076	.025	.051
13.9	.226	.056	.152
27.8	.286	.079	.208
55.6	.250	.091	.208

Summer: $S = 0.678 \exp(-4.3 \times 10^{-6}t) - 0.663 \exp(-1.45 \times 10^{-5}t)$
Winter: $S = 2.095 \exp(-4.58 \times 10^{-6}t) - 2.080 \exp(-5.12 \times 10^{-6}t)$
Annual: $S = 0.744 \exp(-4.4 \times 10^{-6}t) - 0.730 \exp(-9.9 \times 10^{-6}t)$

Model predictions of Alberta sulphate concentration on a seasonal and annual basis were generated by applying the transformation rate equations of Table 8.8 to the dispersion model results for each source-receptor combination and each meteorological condition. The transformation equations were programmed into the model in place of the simple exponential decay terms used for SO_2 and NO_x. Because of the greater uncertainty surrounding appropriate values of transformation and scavenging parameters for nitrate, no calculations of the average nitrate concentration were attempted.

8.7 MODEL APPLICATION AND RESULTS

The Alberta regional statistical model was applied with the ADRP emission inventories, meteorological data, and pollutant loss parameters described in this report. Seasonal and annual average concentrations were computed for SO_2, SO_4, and NO_x. The computations are not specific to a given year or season. Since emission data are for the 1984-86 time frame and meteorological data represent an average over the period 1975-84, the results should be considered as being representative of average conditions during the past several years.

8.7.1 Background Concentrations

Since the model computations represent pollutant contributions from Alberta sources, a constant value of background concentration was added to all results to account for the presumably small transport into the Province from outside sources. To define background concentrations of sulphur dioxide and sulphate, data were extracted from the ADRP measurements at Fortress Mountain. The estimated background concentrations are 0.5 $\mu g\ m^{-3}$ for sulphur dioxide, and 0.4 $\mu g\ m^{-3}$ for sulphate.

For NO_x, monitoring data of adequate sensitivity to define an accurate background concentration for Alberta are not available. A background concentration of 0.2 $\mu g\ m^{-3}$ was adopted, based on data from remote areas as summarized by Logan (1983) and Logan et al. (1981).

8.7.2 Results

The results of the modelling efforts are shown as contour maps of average long-term concentration in Figures 8.1-8.9. All data include background concentrations; results are presented in units of micrograms per cubic metre ($\mu g\ m^{-3}$).

Figure 8.1 depicts annual average SO_2 concentrations. Typical values for Alberta ranged from $<1\ \mu g\ m^{-3}$ in the extreme north to 11 $\mu g\ m^{-3}$ near the oil sands plants at Fort McMurray. Aside from the oil sands area, maximum annual concentrations were in the range 5-9 $\mu g\ m^{-3}$ near Edmonton, Calgary, and areas in between. The distributio of the concentrations is seen to be determined

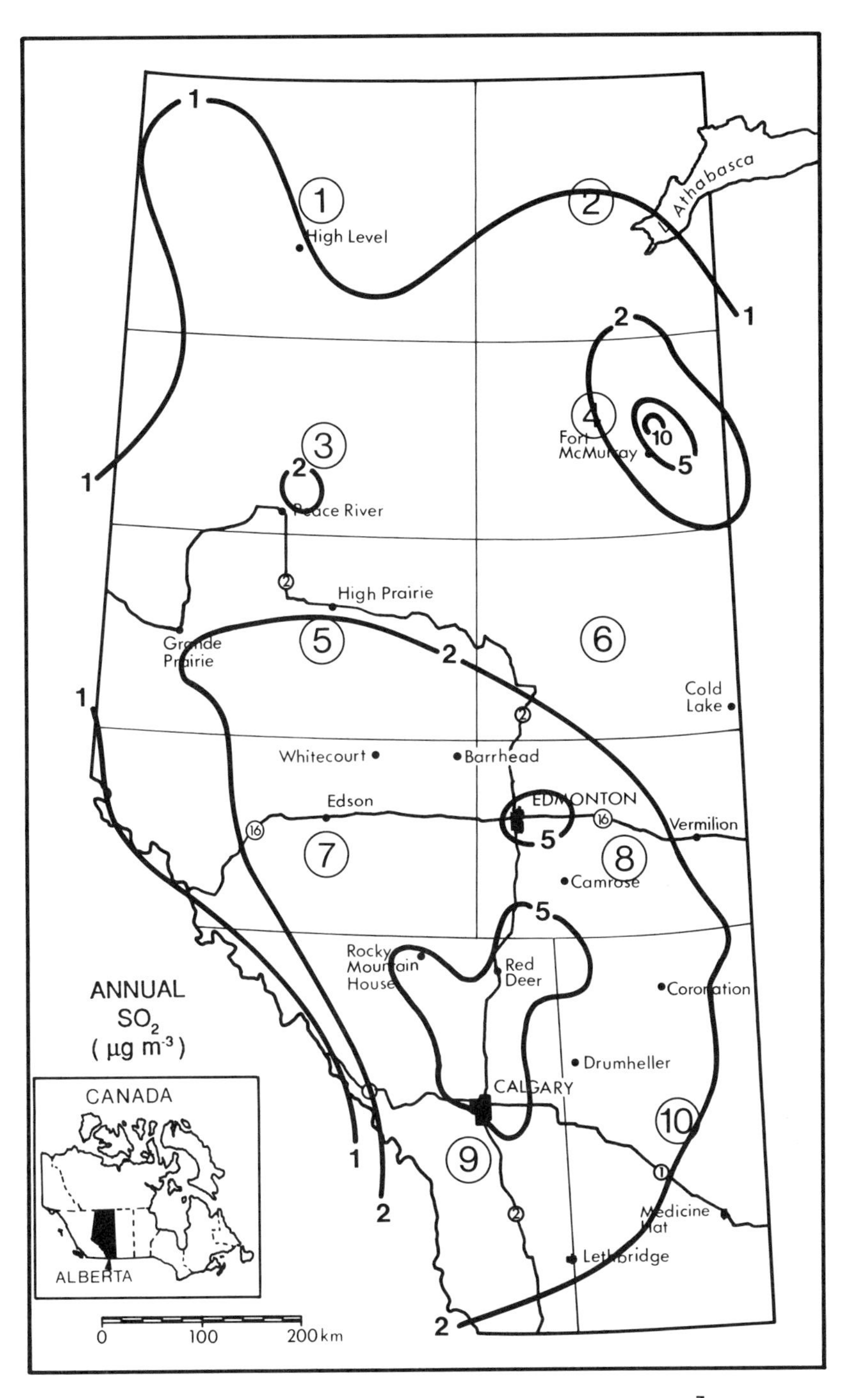

Figure 8.1 Average annual SO_2 concentration ($\mu g\ m^{-3}$).

primarily by the location of SO_2 sources, with some effect of prevailing meteorology evident in the northwest-southeast orientation of the contours.

The seasonal differences in average SO_2 concentration shown by Figures 8.2 and 8.3 were relatively small. Concentrations were somewhat higher in the winter, partly as a result of higher emissions and meteorological conditions but primarily due to slower scavenging and transformation processes. However, the distributions of the concentrations were not markedly different between the summer and the winter according to the results in Figures 8.2 and 8.3.

NO_x concentrations are shown in Figures 8.4 through 8.6. The distribution of NO_x sources in population centres and along major transportation routes was evident in the predicted NO_x distributions. Concentrations ranged from essentially the background in northern Alberta to over 20 $\mu g\ m^{-3}$ in Edmonton.

Seasonal NO_x concentrations showed similar patterns in the summer and the winter. However, mean winter concentrations averaged nearly twice the mean summer values. In the summer, there was virtually no NO_x contribution to northern Alberta, and typical concentrations over the southern half of the Province were in the range of 2-5 $\mu g\ m^{-3}$, except near Calgary and Edmonton. In the winter, concentrations greater than 5 $\mu g\ m^{-3}$ covered a large area of southern Alberta, and urban concentrations were indicated to exceed 30 $\mu g\ m^{-3}$ in limited areas surrounding Edmonton and Calgary.

Average sulphate concentrations are shown in Figures 8.7 through 8.9. Model-predicted sulphate distributions were smoother than those for SO_2 and NO_x because dispersion and transport of SO_2 emissions must take place during the production of sulphate. As seen in Figure 8.7, maximum annual sulphate concentrations of slightly over 1.0 $\mu g\ m^{-3}$ occurred downwind (east and southeast) from the primary SO_2 source areas.

The seasonal mean atmospheric sulphate concentration varied by nearly a factor of two. The reduced production of sulphate in the winter most likely caused the low concentrations, despite slightly higher SO_2 emissions, more restrictive atmospheric mixing, and decreased wet and dry deposition. The higher average sulphate concentrations in summer, up to 1.5 $\mu g\ m^{-3}$, imply nearly four times the total atmospheric sulphate burden of the winter, since the average summer mixing depth was twice that of the winter.

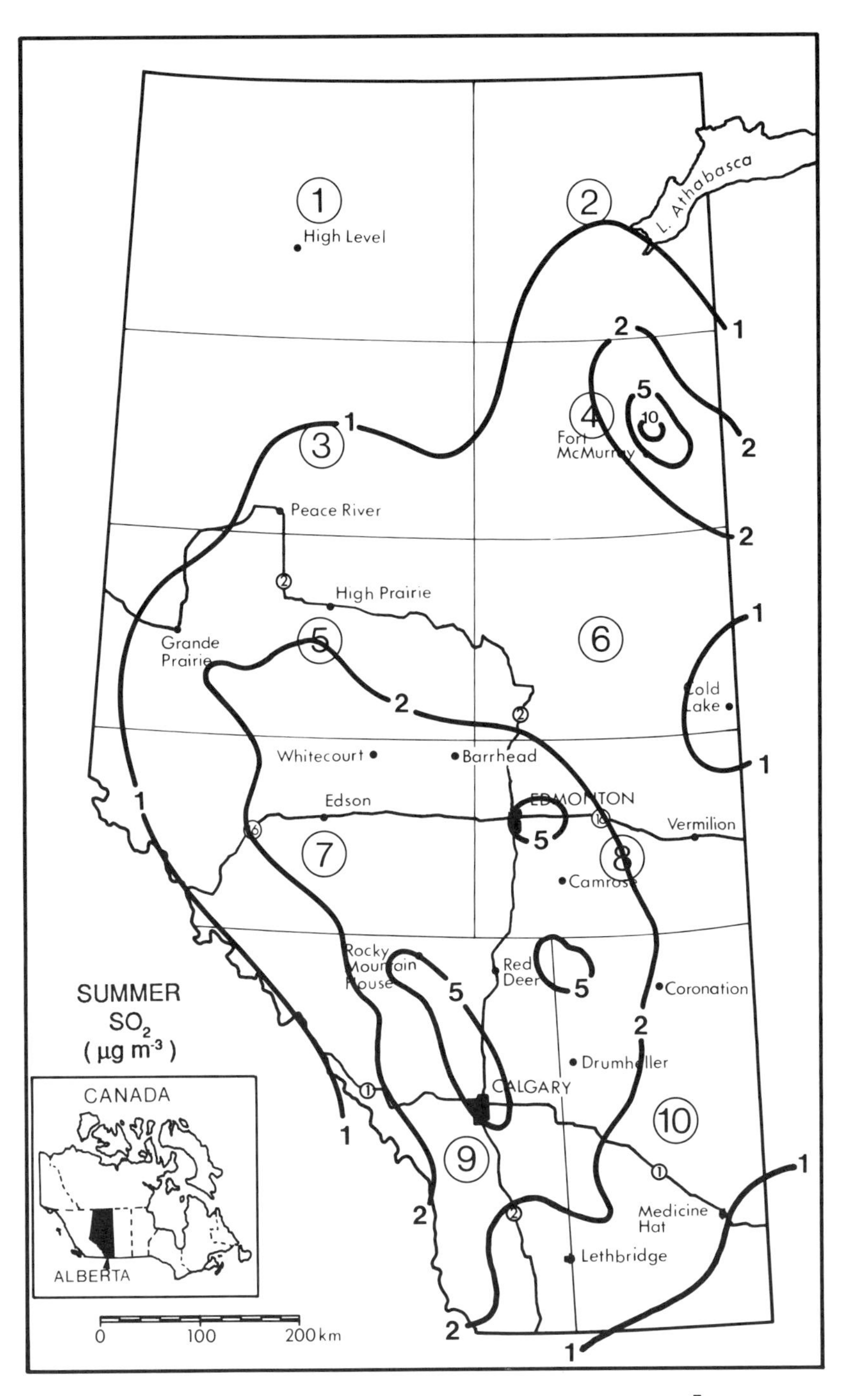

Figure 8.2 Average summer SO_2 concentration ($\mu g\ m^{-3}$).

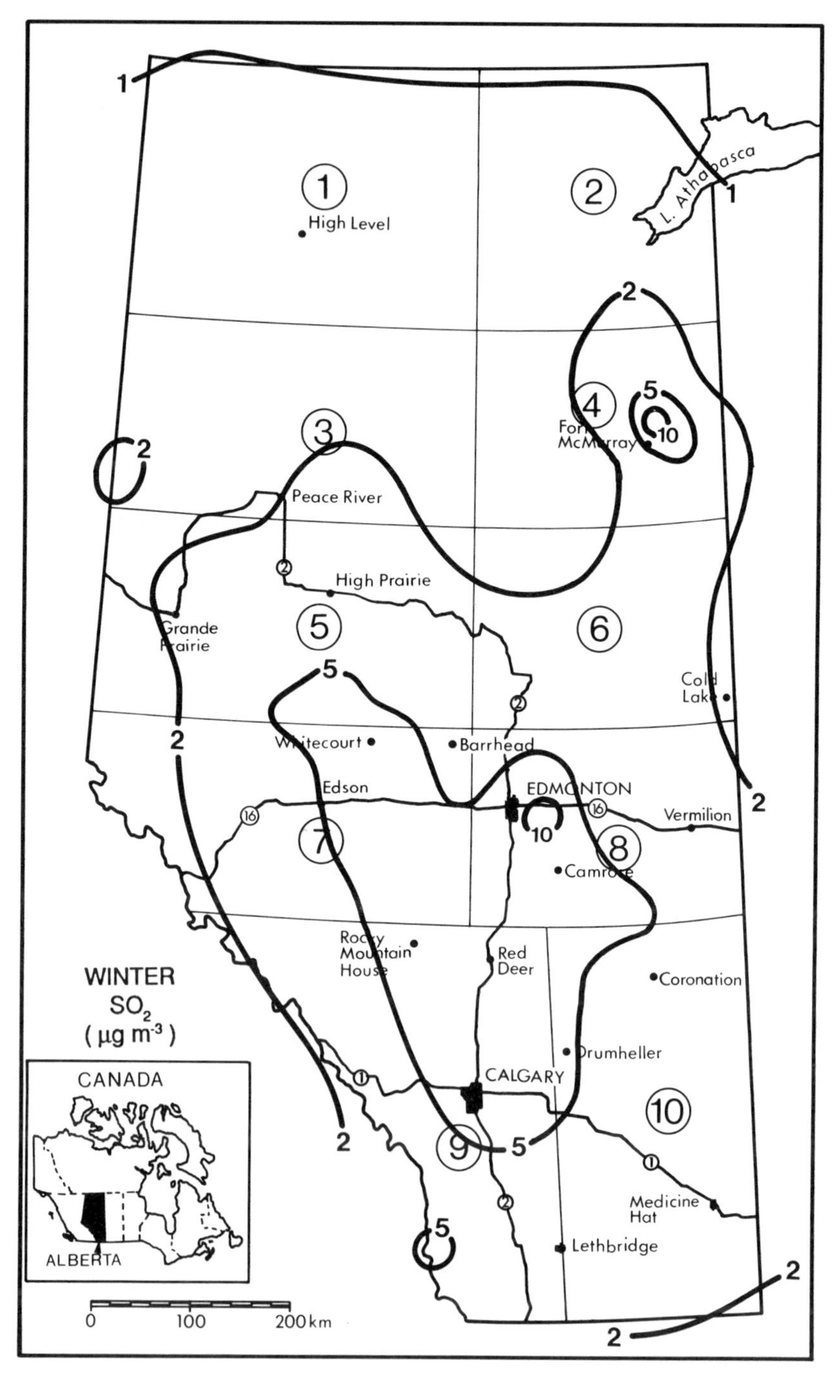

Figure 8.3 Average winter SO_2 concentration (μg m^{-3}).

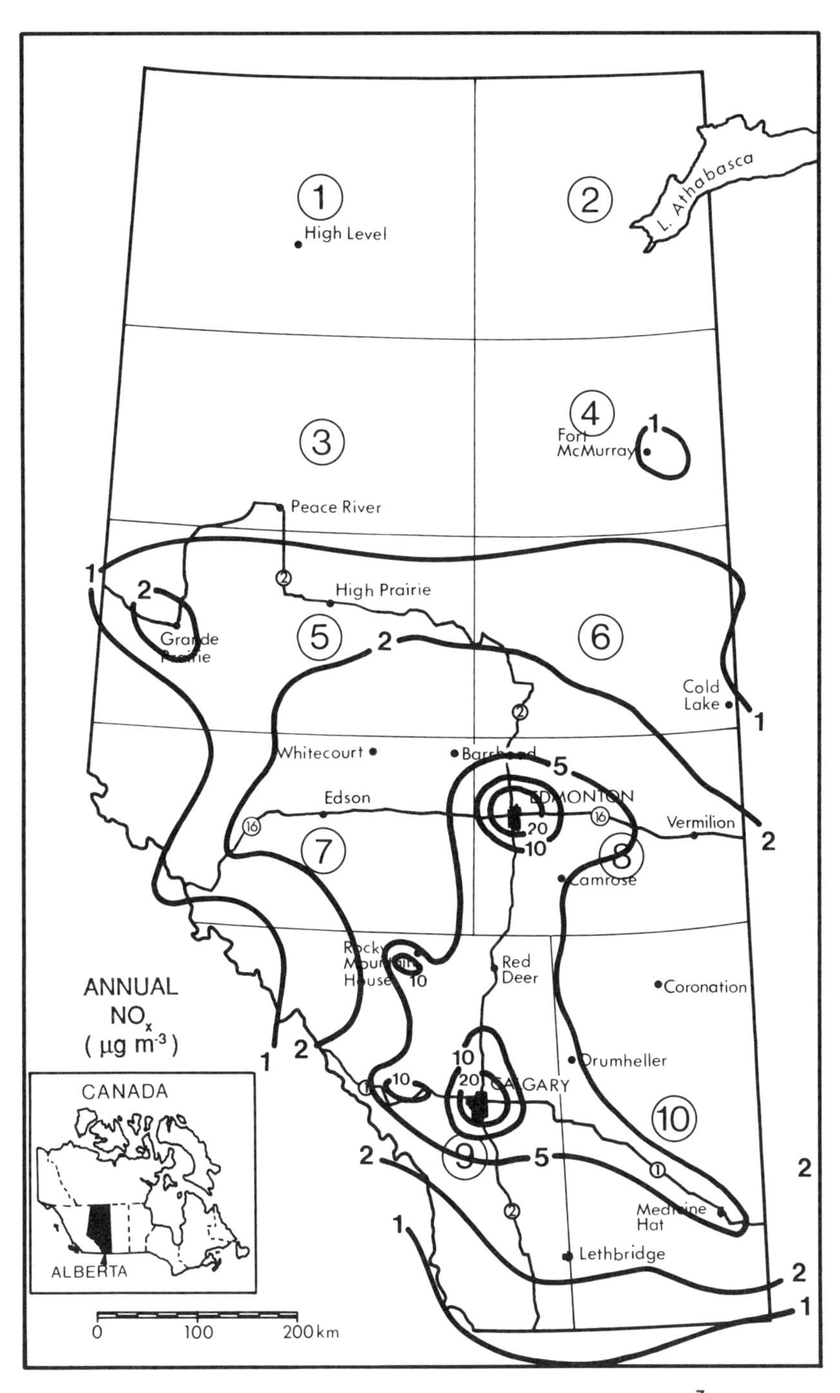

Figure 8.4 Average annual NO_x concentration (μg m^{-3}).

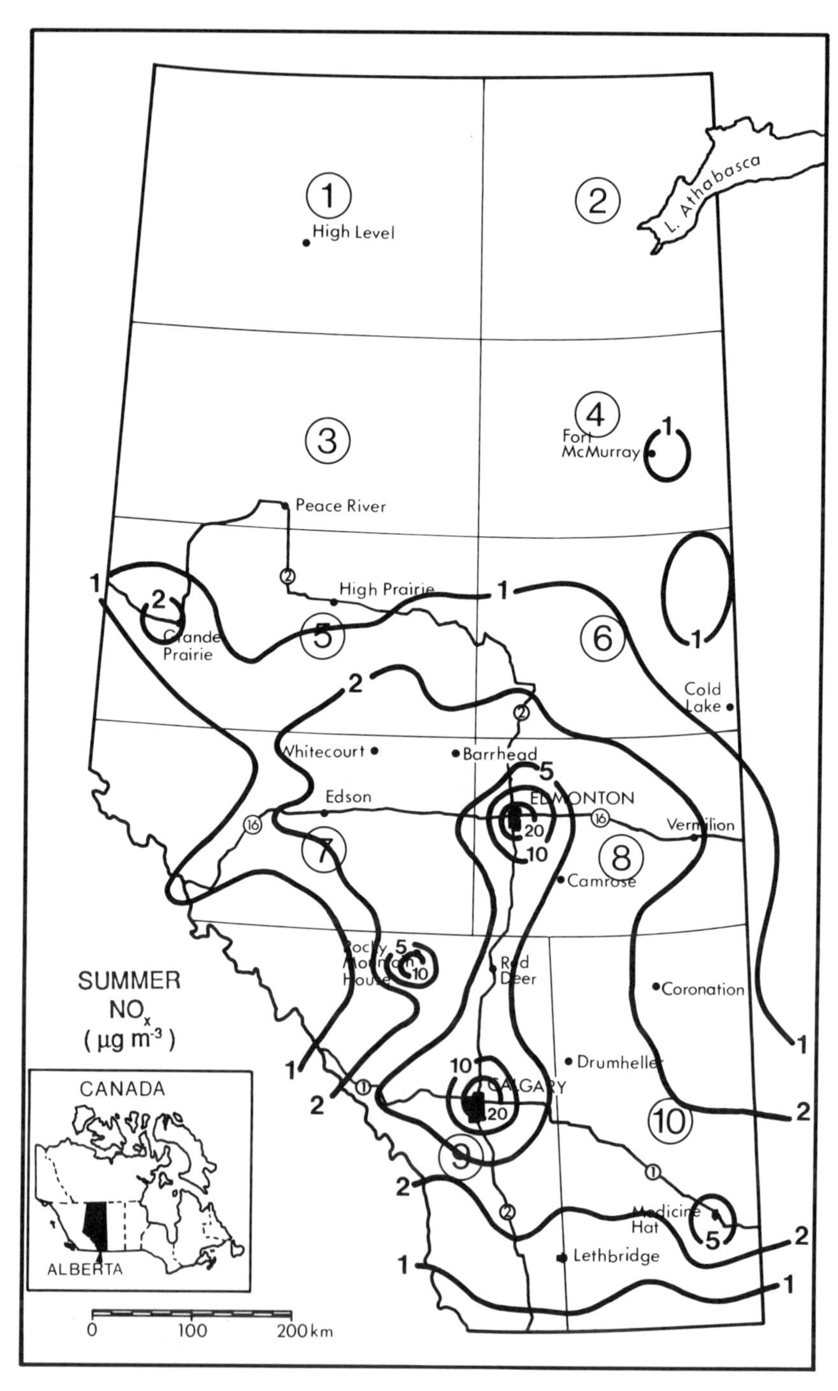

Figure 8.5 Average summer NO_x concentration ($\mu g\ m^{-3}$).

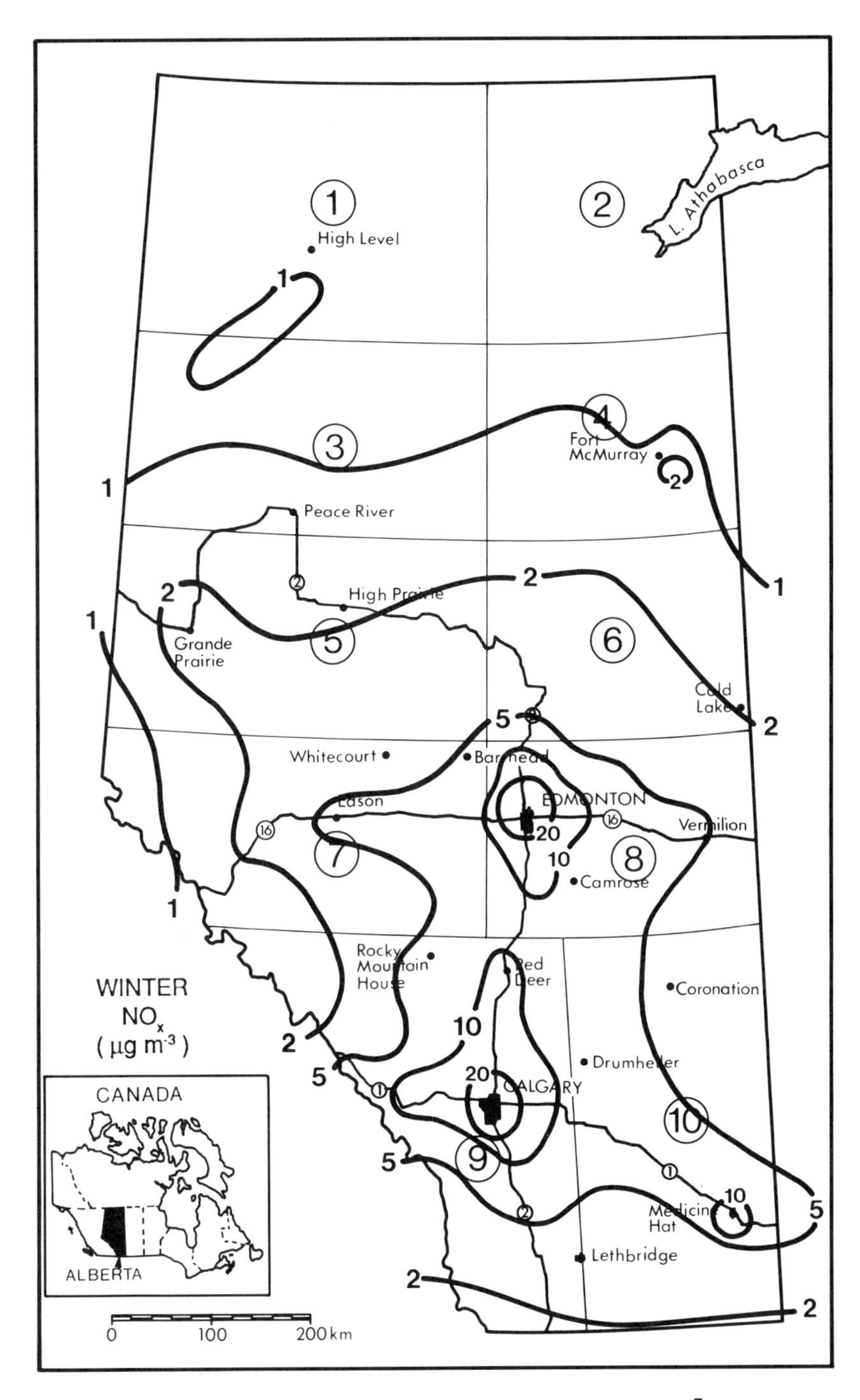

Figure 8.6 Average winter NO_x concentration ($\mu g\ m^{-3}$).

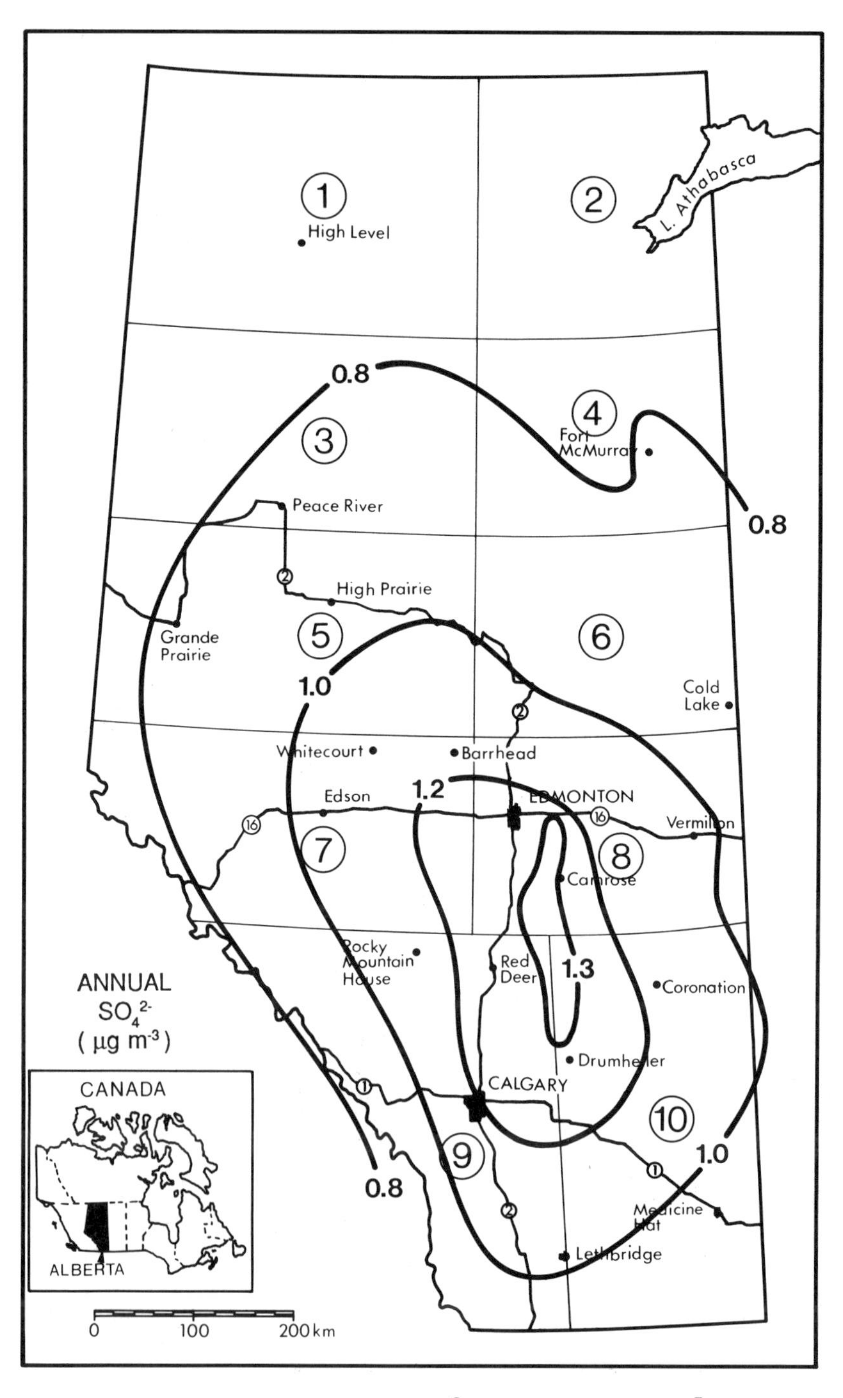

Figure 8.7 Average annual SO_4^{2-} concentration ($\mu g\ m^{-3}$).

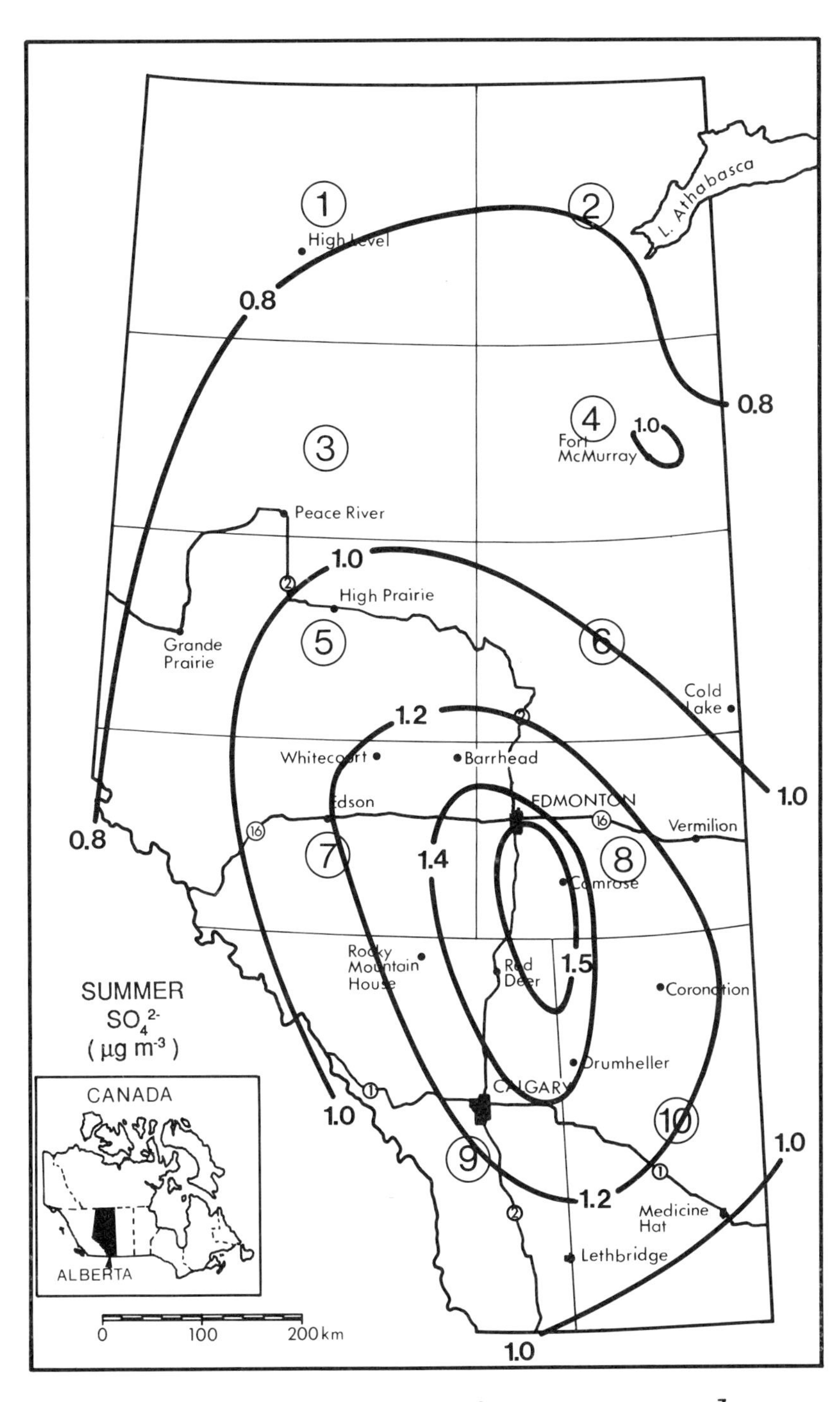

Figure 8.8 Average summer SO_4^{2-} concentration ($\mu g\ m^{-3}$).

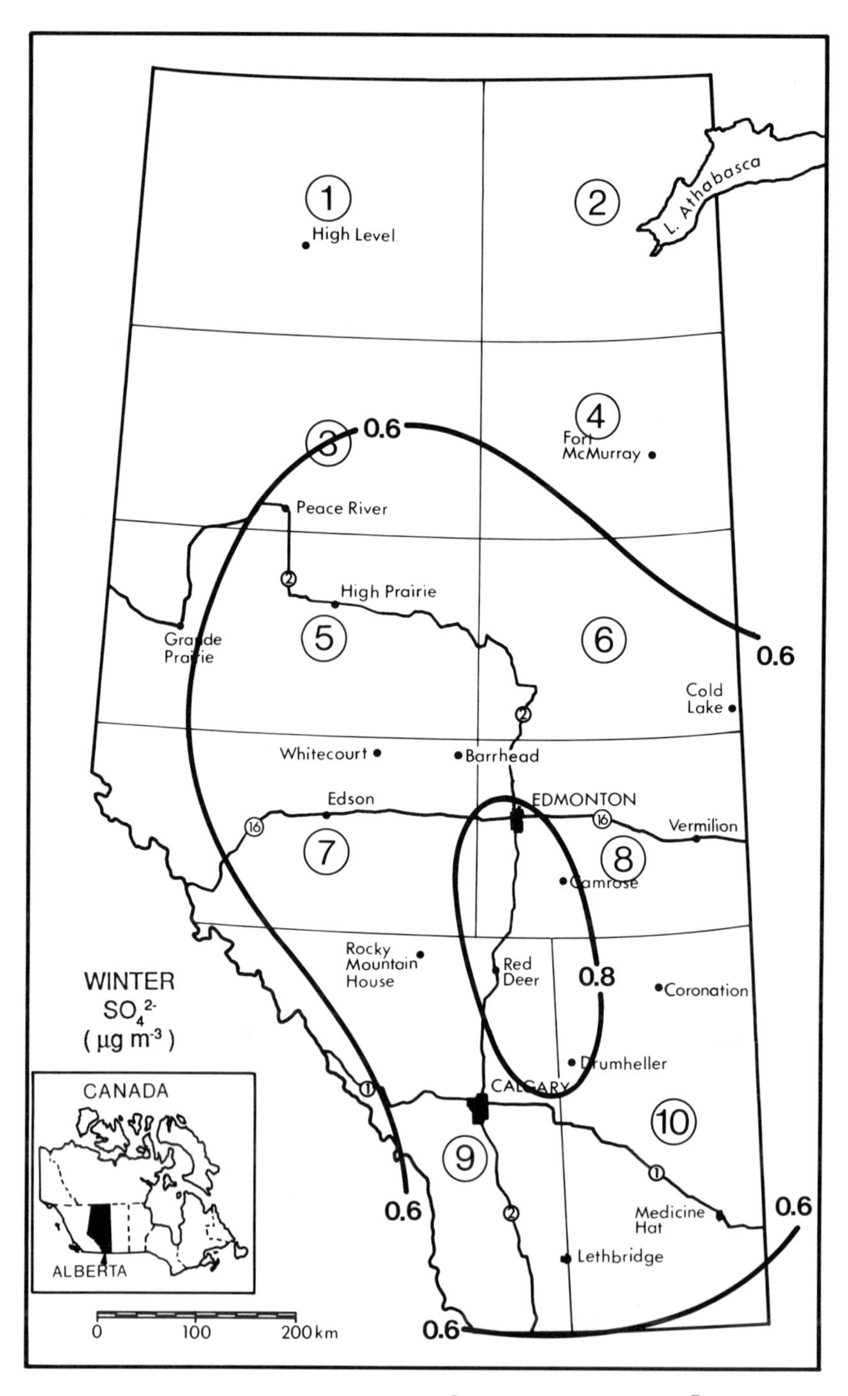

Figure 8.9 Average winter SO_4^{2-} concentration (μg m^{-3}).

The results indicated that patterns of mean pollution concentration across Alberta are dictated by source regions and meteorology, and on the average do not change markedly from summer to winter.

Average pollutant concentrations were relatively low in comparison with populated and industrialized areas of eastern Canada and the United States (Chapter 6). However, there were significant seasonal differences in the concentrations, resulting from seasonal changes in chemical transformation and scavenging rates. SO_2 concentrations were slightly higher in the winter, but sulphate concentrations were significantly higher in the summer owing to its more rapid production. For NO_x, winter concentrations were higher because of reduced transformation to nitrate and decreased wet and dry deposition.

Table 8.9 provides summary statistics on site specific deposition, and Table 8.10 provides data on regional scale pollutant deposition. The significance of these data is discussed in the following section. They are provided here for reference in review of the model results and to provide a summary of estimated Alberta regional deposition. The data in Tables 8.9 and 8.10 were not generated directly by transport modelling.

8.8 DISCUSSION

It is difficult to verify the accuracy of the model results presented in Chapter 8.7 because of the lack of accurate monitoring data in Alberta. Annual concentrations are so low that most values are below the detection limits of routine air quality monitoring. However, point comparisons for SO_2 can be made with integrated annular denuder data from the ADRP (Table 8.11).

ADRP integrated measurements and computations of annual average SO_2 concentrations at the Crossfield East and Crossfield West monitoring sites indicate values of 5.6 and 4.4 $\mu g\ m^{-3}$, respectively. Model results for the Crossfield area, derived from the data as shown in Figure 8.1, are on the order of 6.5 $\mu g\ m^{-3}$. Summer and winter model values are of the order 4.5 and 7.0 $\mu g\ m^{-3}$, respectively. Monitored long-term sulphate concentration at Crossfield is approximately 0.9-1.0 $\mu g\ m^{-3}$ which is close to the predicted value of 1.2 $\mu g\ m^{-3}$.

Some independent confirmation of existing SO_2 concentrations is available from sulphation data. D. Leahey (Western Research, Calgary, Alberta; personal communication) estimates annual SO_2

Table 8.9 Estimates of site specific ion deposition (kg ha^{-1} y^{-1}).

	Mean		
	Wet	Dry	Total
Fortress Mountain			
SO_4^{2-}	7.05	3.77	10.8
NO_3^-	3.47	3.86	7.33
Ca^{2+}	3.07	0.89	3.96
Mg^{2+}	0.38	0.03	0.41
Na^+	0.47	0.13	0.60
K^+	0.35	0.07	0.42
H^+	0.10	--	0.10
NH_4^+	1.90	1.52	3.42
Crossfield West			
SO_4^{2-}	10.0	14.6	24.6
NO_3^-	5.54	12.7	18.2
Ca^{2+}	2.40	3.28	5.68
Mg^{2+}	0.39	0.47	0.86
Na^+	0.47	0.03	0.50
K^+	0.37	0.66	1.03
H^+	0.069	--	0.069
NH_4^+	3.16	7.71	10.9
Crossfield East			
SO_4^{2-}	10.9	17.7	28.6
NO_3^-	5.54	12.7	18.2
Ca^{2+}	3.08	0.14	3.22
Mg^{2+}	0.57	0.56	1.13
Na^+	0.63	0.26	0.89
K^+	0.34	0.83	1.17
H^+	0.067	--	0.067
NH_4^+	3.07	9.74	12.8

* Based on measurements at the three sites; KAPS, dichotomous filters analyzed by XRF, precipitation analyzed by IC, AA, and colorimetry. Precipitation volumes at Crossfield East and West 457 mm y^{-1} and at Fortress Mountain 604 mm y^{-1}.

Table 8.10 Mean and "maximum" values to be used in modelling of acidic deposition-ecological effects relationships (kg ha^{-1} y^{-1}).

REGION 9*	Mean			Maximum		
	Wet	Dry	Total	Wet	Dry	Total
SO_4^{2-}	10.8	14.2	25	26.8	27.4	54
NO_3^-	5.6	5.7	11	15.1	20.9	36
Ca^{2+}	2.8	1.9	4.7	5.4	7.8	13
Mg^{2+}	0.45	0.29	0.74	0.98	1.05	2.0
Na^+	0.51	0.14	0.65	1.34	0.12	1.6
K^+	0.37	0.50	0.87	0.82	1.48	2.3
H^+	0.090	--	0.090	0.28	--	0.28
NH_4^+	2.4	4.1	6.5	9.6	17.6	27.2

Mean precipitation at Crossfield was 457 mm y^{-1} and at Fortress Mountain 604 mm y^{-1}. The estimated mean for Region 9 is 506 mm y^{-1}.
Dry to wet ratio of SO_4^{2-} at Crossfield was 1.48.
Estimated concentration of atmospheric particles (0 to 2.2 μm, 2.2 μm to 10 μm, and 10 μm) for Region 9 is 30 μg m^{-3}.
Concentration of NO_3^- in precipitation = 17 μmol L^{-1}.
Concentration of NH_4^+ in precipitation = 28 μmol L^{-1}.
Volume weighted pH = 4.79.

REGION 3*	Mean			Maximum		
	Wet	Dry	Total	Wet	Dry	Total
SO_4^{2-}	3.7	6.0	9.7	19	10	19
NO_3^-	2.1	0.7	2.8	5.2	1.4	6.6
Ca^{2+}	0.7	--	--	1.8	--	--
H^+	0.036	--	--	0.09	--	--

Precipitation at Beaverlodge in 1986 was 463 mm.
Estimated concentration of atmospheric particles for Region 3 is 20 μg m^{-3}.
Concentration of NO_3^- in precipitation = 7.3 μmol L^{-1}.
Concentration of NH_4^+ in precipitation = 8.7 μmol L^{-1}.
Concentration of HCO_3^- in precipitation = 0.66 μmol L^{-1}.
Volume weighted pH = 5.11.

Continued...

Table 8.10 (Continued).

REGION 7*	Mean			Maximum		
	Wet	Dry	Total	Wet	Dry	Total
SO_4^{2-}	10	12	22	25	18	43
NO_3^-	4.3	3.2	7.5	11.6	9.7	21
Ca^{2+}	5.3	--	--	14.3	--	--
H^+	0.005	--	--	0.014	--	--

Precipitation at Edson in 1984 was 547 mm.
Estimated concentration of atmospheric particles for Region 7 is 25 $\mu g\ m^{-3}$.
Concentration of NO_3^- in precipitation = 12.7 $\mu mol\ L^{-1}$.
Concentration of NH_4^+ in precipitation = 11.3 $\mu mol\ L^{-1}$.
Concentration of HCO_3^- in precipitation = 5.4 $\mu mol\ L^{-1}$.
Volume weighted pH = 6.02.

REGION 6*	Mean			Maximum		
	Wet	Dry	Total	Wet	Dry	Total
SO_4^{2-}	2.5	7.6	10	6	12	18
NO_3^-	1.0	2.7	3.7	2.6	5.5	8.1
Ca^{2+}	1.1	--	--	3.0	--	--
H^+	0.024	--	--	0.074	--	--

Precipitation at Cold Lake in 1986 was 417 mm.
Estimated concentration of atmospheric particles for Region 6 is 25 $\mu g\ m^{-3}$.
Concentration of NO_3^- in precipitation = 3.7 $\mu mol\ L^{-1}$.
Concentration of NH_4^+ in precipitation = 1.8 $\mu mol\ L^{-1}$.
Concentration of HCO_3^- in precipitation = 0.83 $\mu mol\ L^{-1}$.
Volume weighted pH = 5.21.

REGION 4*	Mean			Maximum		
	Wet	Dry	Total	Wet	Dry	Total
SO_4^{2-}	9.6	11	21	24	42	66
NO_3^-	1.8	0.7	2.5	4.5	2.8	7.2
Ca^{2+}	5.4	--	--	15	--	--
H^+	0.005	--	--	0.014	--	--

Continued...

Table 8.10 (Concluded).

REGION 4*	Mean			Maximum		
	Wet	Dry	Total	Wet	Dry	Total

Precipitation at Fort McMurray in 1984 was 516 mm.
Estimated concentration of atmospheric particles for Region 4 is 20 $\mu g\ m^{-3}$.
Concentration of NO_3^- in precipitation = 5.6 mol L^{-1}.
Concentration of NH_4^+ in precipitation = 1.2 mol L^{-1}.
Concentration of HCO_3^- in precipitation = 5.2 mol L^{-1}.
Volume weighted pH = 6.01.

* Refer to Chapter 6 of this report for other details.

Table 8.11 Comparison of two independent methods for estimating atmospheric concentrations of SO_x and NO_x ($\mu g\ m^{-3}$).

		Site Specific KAPS	Regional Model
Crossfield East			
	SO_2	4.9	6.5
	SO_4^{2-}	1.1	1.2
	NO_x	--	15.0
Crossfield West			
	SO_2	3.9	6.5
	SO_4^{2-}	1.0	1.2
	NO_x	--	15.0
Fortress Mountain			
	SO_2	0.9	2.5
	SO_4^{2-}	0.5	0.9
	NO_x	--	5.0

KAPS = Integrated annular denuder measurement.
Sampling duration = Fortress: 48 h; Crossfield (East and West): 24 h.

concentrations from sulphation data as 3.5-6.5 $\mu g\ m^{-3}$ in the Crossfield area, and 0.8 $\mu g\ m^{-3}$ at Fortress Mountain. These concentrations are comparable to the model estimates for the Crossfield area. The model appears to predict concentrations at Fortress Mountain which are higher than those actually observed.

This may be due to topographic effects which are not considered in the model.

Considerable SO_2 monitoring has been performed in the vicinity of the oil sands plants near Fort McMurray. Again, the frequent occurrences of SO_2 concentrations below the detection limit make annual averages derived from such monitored data unreliable. A variety of air quality modelling studies performed for the oil sands plants indicates maximum annual SO_2 concentrations in the order of 10-15 $\mu g\ m^{-3}$, in reasonable agreement with the present modelling results.

Data for NO_x are equally sparse. ADRP data from Crossfield, obtained by continuous monitoring and corrected for below-threshold concentrations by the stochastic probability-Weibull generator method (refer to Chapter 3), suggest annual average NO_x concentrations in the range of 6-20 $\mu g\ m^{-3}$. Routine monitoring data from Calgary (Alberta Department of Environment 1983) show annual average NO_x concentrations of about 45 $\mu g\ m^{-3}$.

The available data allow for much less than a comprehensive test of the model results. However, on the basis of the fragmentary measurements available, it can be concluded that the model predictions are reasonable. There is no evidence of systematic error in the predicted concentrations. It should be recognized in comparing predictions to monitoring data that monitors are frequently sited to measure the highest expected concentrations from a given source or group of sources. The model data are intended to represent spatially averaged concentrations and would not be expected to show maximum localized impacts.

Great emphasis should not be placed on the details of the contours in Figure 8.1 through 8.9. The contours were subjectively drawn, often through areas of relatively flat gradient, from the predicted grid point concentration values. They are intended to show general patterns of concentration only.

The sensitivity of the model results to variations in input values of transformation and deposition parameters can be judged from the seasonal variations shown. Actual annual values of the parameters are certainly within the range defined by summer and winter estimates used here. The results suggest that variations within this range result in differences in predicted concentrations of less than a factor of two. This conclusion is supported by the results of additional model runs not presented here and the findings of Venkatram (1986). In consideration of the uncertainty associated with the input model parameters and other potential model errors, it is expected that the results presented represent true average concentrations within a factor of two. The comparisons to ADRP monitoring data for Crossfield support this conclusion.

The apparent success of the modelling estimates would seem to confirm the initial assumption that pollutant concentrations in Alberta are determined almost completely by the magnitude of provincial sources. The model results are also consistent in indicating that contributions of Alberta sources in remote areas (i.e., extreme northern Alberta) are very small and the predicted total concentrations approach assumed background levels. The reader is cautioned that application of similar modelling methods in regional situations, where there is significant transport across boundaries into the region, will be more difficult.

Since predicted concentrations in the few areas near major sources where monitoring data exist seem to be reasonable, it can be expected that the regional variations shown in the model results are realistic. The spatial variations depend mostly on the source distributions and prevailing meteorology, both of which are rather well defined. However, actual distributions to be expected in any given year or season could differ appreciably from the mean results due to climatic anomalies, variation in specific source emissions, or other factors. Also, local areas where meteorology is strongly influenced by topographic effects are not simulated in the modelling.

The seasonal average concentrations and distributions should be considered more speculative than the annual results. Seasonal meteorological variations have been considered, but the seasonal changes in transformation and scavenging parameters are not well known. While the general seasonal trends shown are in agreement with our physical understanding of the processes, the magnitude of the seasonal differences cannot be confirmed.

8.9 A FUTURE SCENARIO

One of the major uses of long-range pollutant transport models has been the examination of the changes in impacts that would result from a potential change in location or magnitude of emissions. In this context it is important to estimate the change in regional pollutant concentrations in Alberta as implied by possible future changes in the emissions.

For the purposes of example calculations, it was assumed that six new major sources of SO_2 would be constructed in Alberta in the late 1990's. The assumed locations and emissions from these sources are listed in Table 8.12. The types and locations of sources were based on possible or suggested new developments in Alberta but are purely speculative at this time. They are intended to represent an example of a possible future scenario and are not a prediction of actual future developments.

Table 8.12 Hypothetical emission sources assumed for future scenario in Alberta.

Location	Number of Sources	SO_2 Emission ($t\ d^{-1}$)	NO_x Emission ($t\ d^{-1}$)
Peace River	1	14.0	0.0
Oil Sands (Fort McMurray)	2	64.6	13.2
Cold Lake	1	38.9	0.9
Caroline	1	42.7	0.0
Sundre	1	42.7	0.0

The emission and stack parameters for the hypothetical new sources were selected on the basis of similar existing sources. It was assumed that current technology for pollution control would be utilized but no major control improvements were considered.

The statistical long-range transport model was applied to calculate annual average SO_2, NO_x, and SO_4^{2-} concentrations as described in previous sections, including the six hypothetical new sources in the emissions inventory. All other Alberta emissions were included exactly as before. The resulting model predictions are shown in Figures 8.10 through 8.12.

The new sources add SO_2 emissions of 203 $t\ d^{-1}$ in Alberta. As shown in Figure 8.10 (compare with Figure 8.1), the effect of this addition to the average annual SO_2 concentrations in Alberta is quite small. As a provincial average, concentrations increase by approximately 0.2 $\mu g\ m^{-3}$. Greater increases, up to several micrograms per cubic metre, are indicated within 20 km of the new sources. The effect on annual sulphate concentration (Figures 8.12 and 8.7) is more apparent with a rather uniform increase in the concentration of 0.1 $\mu g\ m^{-3}$ (about 10%) over the Province.

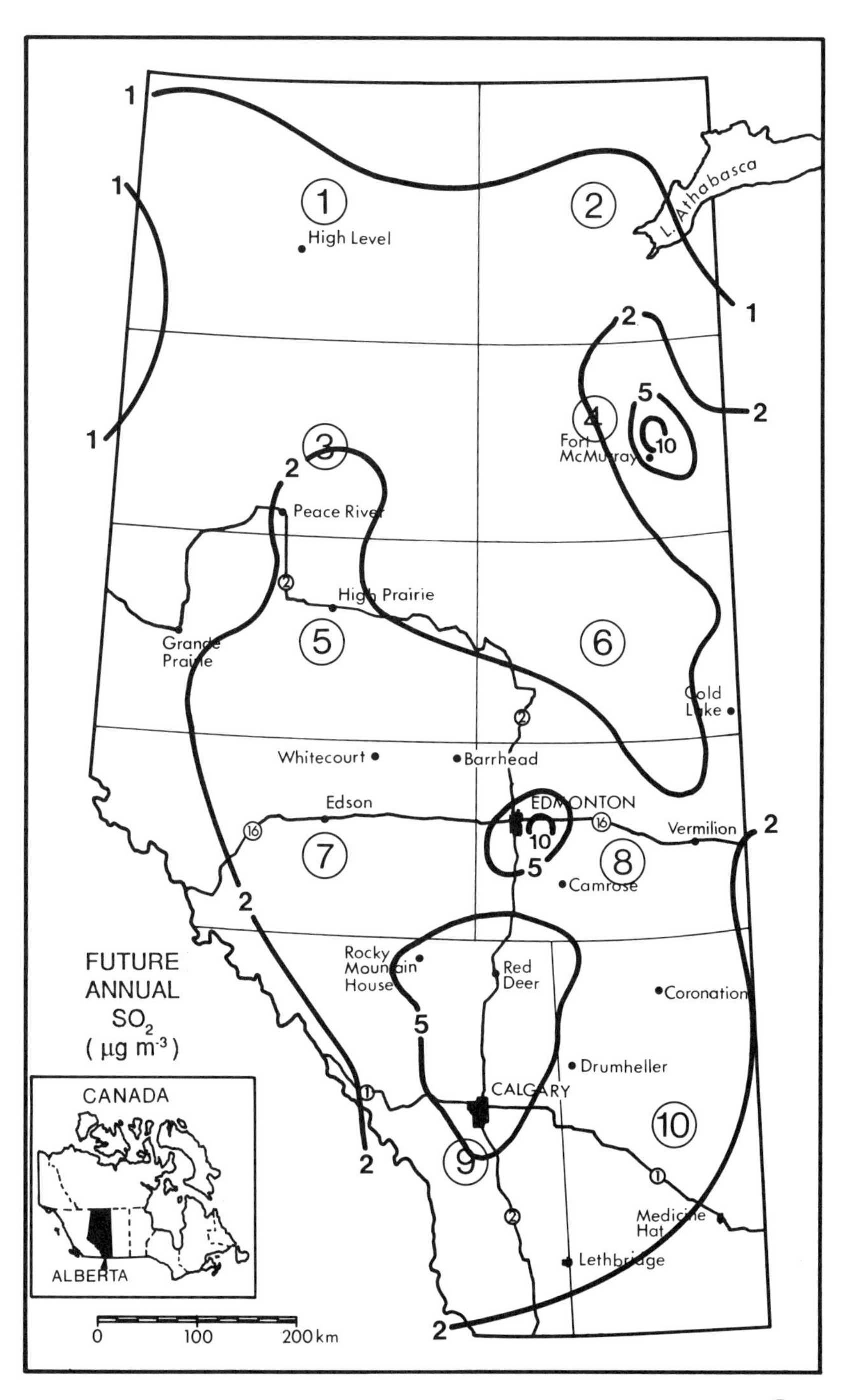

Figure 8.10 Average annual SO_2 concentration - future scenario ($\mu g\ m^{-3}$).

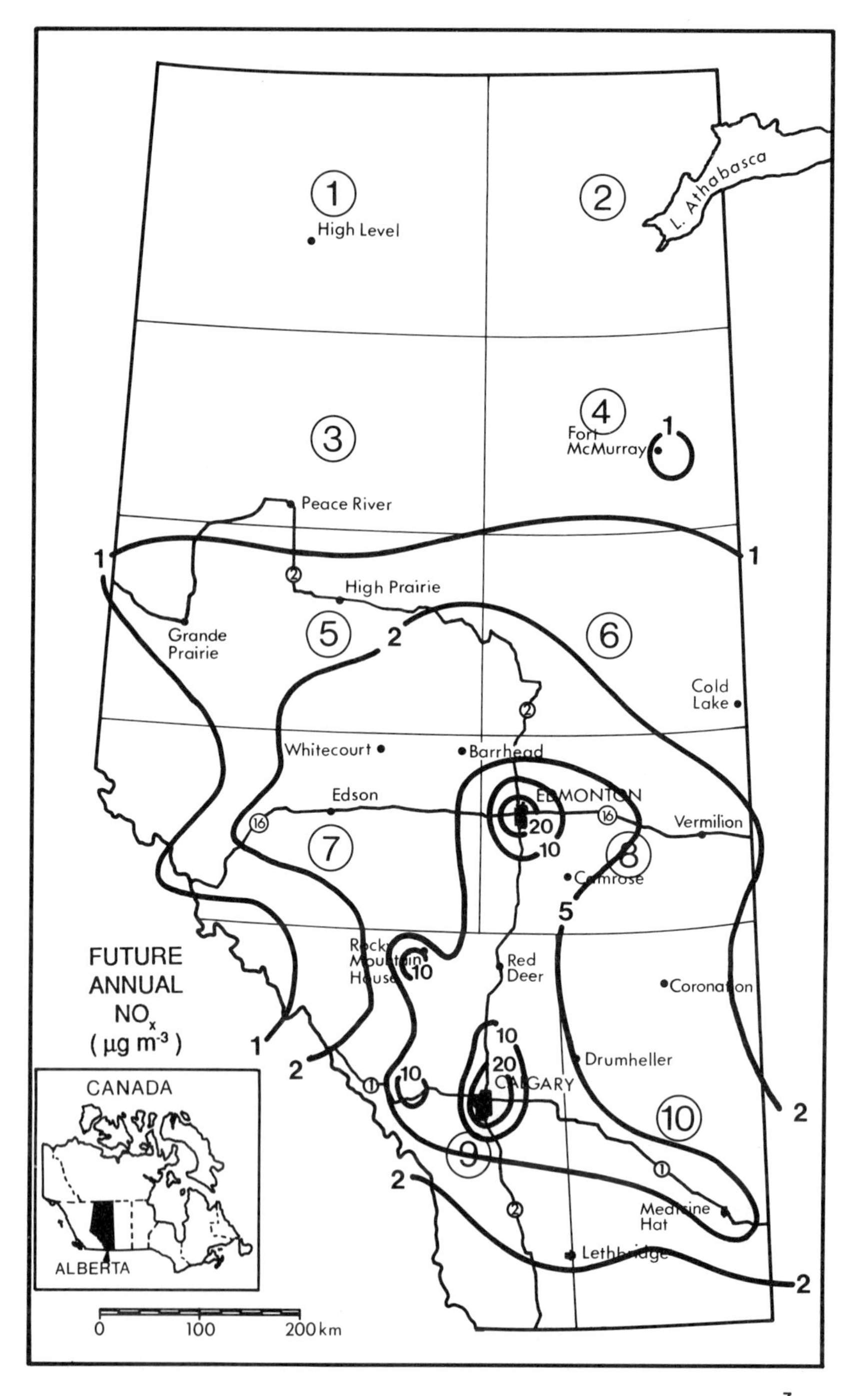

Figure 8.11 Average annual NO_x concentration - future scenario ($\mu g\ m^{-3}$).

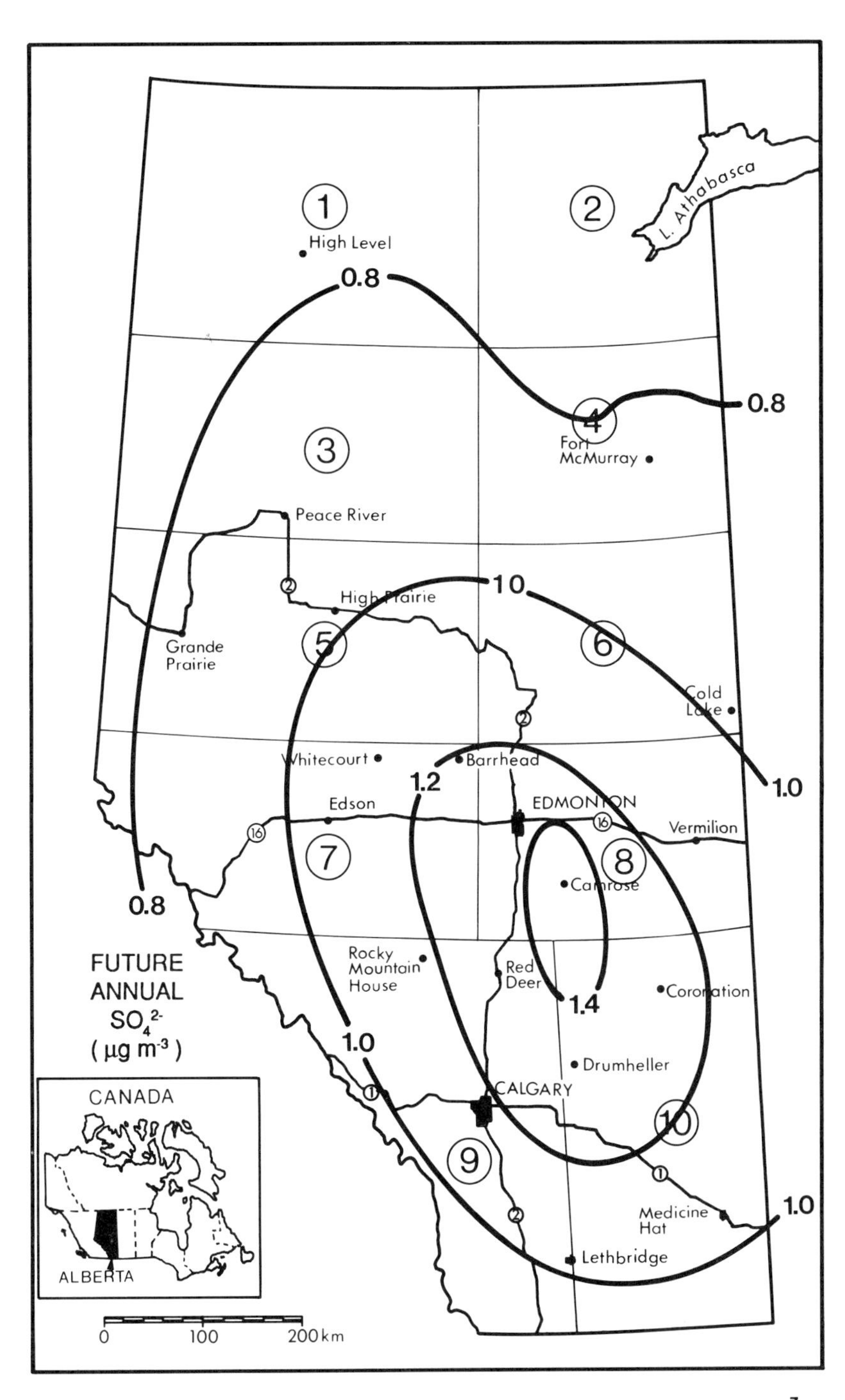

Figure 8.12 Average annual SO_4^{2-} concentration - future scenario (μg m^{-3}).

The increase in total provincial NO_x emissions represented by the future scenario is quite small, only 15 t d^{-1} or less than 2.5% of present emissions. As would be expected, predicted annual NO_x concentrations shown in Figure 8.11 are indistinguishable from those for the present emission inventory in Figure 8.4. The largest modelled increase was 0.4 $\mu g\ m^{-3}$ in the vicinity of Cold Lake.

8.10 CONCLUDING REMARKS

Seasonal and annual average concentrations of SO_2, SO_4^{2-}, and NO_x in Alberta have been estimated through the application of a regional air pollution simulation model. The model utilized average meteorological parameters for Alberta, a detailed inventory of source emissions, and pollutant transformation and scavenging parameters estimated from the best available information for Alberta.

Modelling results were in reasonable agreement with existing monitoring data and appeared to provide a realistic overview of average pollutant exposure in Alberta. They can be used, as in Chapter 9, to assist in the assessment of impacts on crops, soils, and surface waters.

8.11 LITERATURE CITED

Alberta Environment. 1983. Air monitoring report. City of Calgary. Air Quality Control Branch, Pollution Control Division, Environmental Protection Services. ISSN 0226-4005. Edmonton, Alberta.

Busse, A.D. and J.R. Zimmerman. 1973. User's Guide for the Climatological Dispersion Model. EPA-R4-73-024. U.S. Environmental Protection Agency. Research Triangle Park, North Carolina. 131 pp.

Carmichael, G.R. and L.K, Peters. 1984. Eulerian modeling of the transport of sulfur dioxide and sulfate. In: **Modeling of Total Acid Precipitation Impacts.** ed. J.L. Schnoor, Boston: Butterworth Publishers. pp. 25-51.

Dailey, B. 1982. Description of major program revisions for the CDMW version of CDM. Unpublished memo from the Wyoming Department of Environmental Quality, Air Quality Division. Cheyenne, Wyoming.

Eliassen, A. 1980. A review of long-range transport modeling. *Journal of Applied Meteorology* 19(3): 231-240.

Fisher, B.E.A. 1978. The calculation of long-term sulphur deposition in Europe. *Atmospheric Environment* 12: 489-501.

Garland, J.A. 1978. Dry and wet removal of sulphur from the atmosphere. *Atmospheric Environment* 12: 349-363.

Leahey, D.M. and A.L. Jamieson. 1987. Development of input data for three (3) statistical long range transport of air pollutant models. Report prepared under Contract KM643-6-0291-01-56 for Atmospheric Environment Service, Western Region by Western Research. Calgary, Alberta.

Logan, J.A. 1983. Nitrogen oxides in the troposphere: global and regional budgets. *Journal of Geophysical Research* 88(C15): 10785-10807.

Logan, J.A., M.J. Prather, S.C. Wofsy, and M.B. McElroy. 1981. Tropospheric chemistry: a global perspective. *Journal of Geophysical Research* 86(C8): 7210-7254.

Pasquill, F. 1961. The estimation of the dispersion of windborne material. *Meteorological Magazine* 90(1063): 33-49.

Picard, D.J., D.G. Colley, and D.H. Boyd. 1987a. Overview of the Emission Data: Emission Inventory of Sulphur Oxides and Nitrogen Oxides in Alberta. Prep. for the Alberta Government-Industry Acid Deposition Research Program by Western Research. Calgary, Alberta. 87 pp.

Picard, D.J., D.G. Colley, and D.H. Boyd. 1987b. Design of the Emission Data: Emission Inventory of Sulphur Oxides and Nitrogen Oxides in Alberta. Prep. for the Alberta Government-Industry Acid Deposition Research Program by Western Research. Calgary, Alberta. 88 pp.

Picard, D.J., D.G. Colley, and D.H. Boyd. 1987c. Emission Data Base: Emission Inventory of Sulphur Oxides and Nitrogen Oxides in Alberta. Prep. for the Alberta Government-Industry Acid Deposition Research Program by Western Research. Calgary, Alberta. 335 pp.

Picard, D.J., D.G. Colley, and D.H. Boyd. 1987d. Results of the Emission Source Surveys: Emission Inventory of Sulphur Oxides and Nitrogen Oxides in Alberta. Prep. for the Alberta Government-Industry Acid Deposition Research Program by Western Research. Calgary, Alberta. 335 pp.

Rodhe, H. and J. Grandell. 1972. On the removal time of aerosol particles from the atmosphere by precipitation scavenging. *Tellus* 24: 442-454.

Smith, F.B. 1981. The significance of wet and dry synoptic regimes on long-range transport of pollution and its deposition. *Atmospheric Environment* 15: 863-873.

Venkatram, A. 1986. Statistical long-range transport models. *Atmospheric Environment* 20(7): 1317-1324.

Venkatram, A. and J. Pleim. 1985. Analysis of observations relevant to long-range transport and deposition of pollutants. *Atmospheric Environment* 19(4): 659-668.

Venkatram, A., B.E. Ley, and S.Y. Wong. 1982. A statistical model to estimate long-term concentrations of pollutants associated with long-range transport. *Atmospheric Environment* 16(2): 249-258.

Western Mesoscale Modelling Task Group. 1985. A review of mesoscale modelling for application to Western Canada. Report prepared for Western Canada LRTAP Technical Committee. 29 pp.

CHAPTER 9
EFFECTS OF ACIDIC AND ACIDIFYING AIR POLLUTANTS ON SELECTED ENVIRONMENTAL COMPONENTS OF ALBERTA

9.1 INTRODUCTION

A. H. Legge

Air pollutants are emitted into the atmosphere by natural sources and through man's activities. In this context, the anthropogenic emissions of air pollutants and their present and potential impacts on the environment are of great concern.

The primary pollutants, sulphur dioxide (SO_2) and the oxides of nitrogen (NO_x),can be classified as acidifying pollutants since they become acids upon contact with moisture. These gases are also transformed in the atmosphere to their corresponding acid species, sulphuric (H_2SO_4) and nitric (HNO_3) acid. These air pollutants and all others can be deposited to the surfaces (such as vegetation, soils, and surface waters) by wet (acidic precipitation) and dry deposition (refer to Chapters 2 and 6). In addition to the conversion of NO_x to HNO_3, one of the two constituents of NO_x, nitrogen dioxide (NO_2),is also converted to ozone (O_3), peroxyacetyl nitrate (PAN), etc. Thus, ambient air is composed of a mixture of gases, vapours, and particulate matter. Depending on meteorological factors, chemical and physical processes in the atmosphere, air pollutants emitted and/or secondarily formed in a given location can be transported from a few up to thousands of kilometres.

Deposition processes limit the lifetime of pollutants in the atmosphere. These deposition processes are dynamic and stochastic in nature, varying significantly in time and space. Based on their sensitivity, environmental components vary in their response to a given pollutant or pollutant mixtures. These concepts are schematically represented in Figure 9.1.

It is beyond the scope of this introduction to provide a literature review of the effects of acidic and acidifying air pollutants on various environmental components. For details on this subject, the reader is referred to the previous literature reviews in the ADRP: Torn et al (1987), Mayo (1987), Turchenek et al. (1987), and Telang (1987). However, in the following part of this section, a brief discussion is provided on the concepts and issues relevant to the numerical analysis of the actual cause and effect relationships.

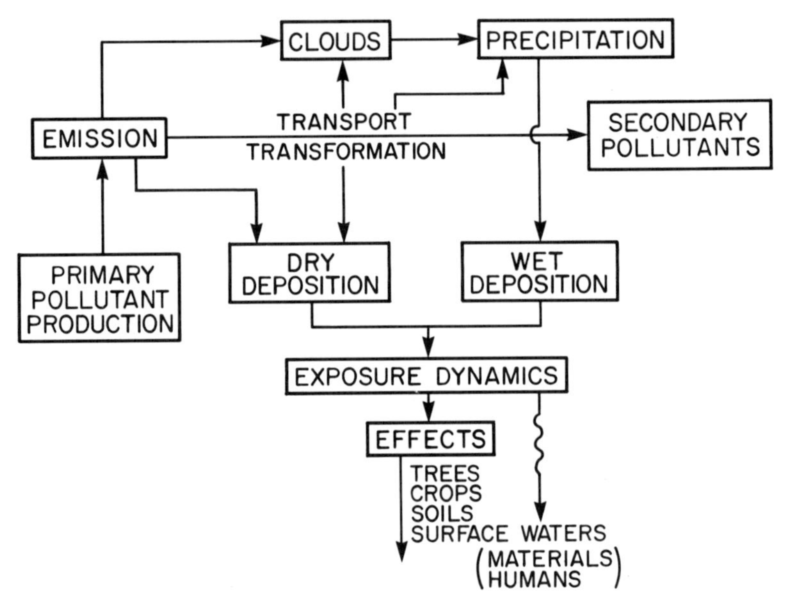

Figure 9.1 Flow diagram showing the relationships between atmospheric processes and environmental receptors.

The inherent relationship between air pollutant exposure and biological response is stochastic in nature. Models applied in explaining the numerical relationships between air pollutant deposition and changes in soil chemistry, surface waters, and in some cases, growth responses of crops and trees,utilize seasonal or annual mean values of single or multiple air pollutants (U.S. NAPAP 1987). While to observe changes in soil chemistry may require long-term inputs of air pollutants (wet and dry), such changes are known to occur rapidly following episodal inputs of precipitation ions into streams (Elzerman 1983). Similarly, crops and trees will respond rapidly to acute air pollutant exposures. However, of concern is their response to chronic exposures in terms of changes in growth, productivity, and quality. Chronic pollutant exposures consist of generally low to moderate concentrations for prolonged periods with intermittent occurrences of episodes or high concentrations. Krupa and Kickert (1987) and Lefohn and Runeckles (1987) reviewed the considerations relevant to pollutant exposure-vegetation response relationships. These authors concluded that long-term averages do not explain the importance of pollutant episodicity nor provide consistent relationships between the dynamic nature of the cause and the observed response.

Since many air pollutants exhibit non-normal frequency distributions of occurrence, computations of long-term means are considered to be inappropriate. Yet much of the effects research is based on long-term mean values. Unfortunately, cause-effect models which utilize distribution-free statistics (e.g., median and percentiles) have not been fully developed for wide application. Given this limitation, and since no new experimental studies on air pollutant exposure and vegetation response were involved in this phase (Critical Point I.) of ADRP, existing models had to be adapted for use in the evaluation of the Alberta situation. In the effects assessment, forest response evaluations were excluded; a literature review on this subject was completed, but no further work was mandated by the ADRP terms of agreement, since other agencies in Alberta were studying the subject and necessary data bases were not available during the course of this program.

9.1.1 Literature Cited

Elzerman, A.W. 1983. Effects of Acid Deposition (Rain) on a Piedmont Aquatic Ecosystem: Acid Inputs, Neutralization, and pH Changes. Final Technical Completion Report B-141-SC to Bureau of Reclamation, U.S. Department of the Interior, Washington, D.C. 79 pp.

Krupa, S. and R.N. Kickert. 1987. An analysis of numerical models of air pollutant exposure and vegetation response. *Environmental Pollution,* 44: 127-158.

Lefohn, A.S. and V.C. Runeckles. 1987. Establishing standards to protect vegetation - ozone exposure/dose considerations. *Atmospheric Environment,* 21: 561-568.

Mayo, J.M. 1987. The Effects of Acid Deposition on Forests. Prep. for the Acid Deposition Research Program by the Department of Biology, Emporia State University. ADRP-B-09-87. 80 pp.

Telang, S.A. 1987. Surface Water Acidification Literature Review. Prep. for the Acid Deposition Research Program by the Kananaskis Centre for Environmental Research, The University of Calgary, Calgary, Alberta. ADRP-B-01-87. 123 pp.

Torn, M.S., J.E. Degrange, and J.H. Shinn. 1987. The Effects of Acidic Deposition on Alberta Agriculture: A Review. Prep. for the Acid Deposition Research program by the Environmental Sciences Division, Lawrence Livermore National Laboratory. ADRP-B-08/87. 160 pp.

Turchenek, L.W., S.A. Abboud, C.J. Tomas, R.J. Fessenden, and N. Holowaychuk. 1987. Effects of Acid Deposition on Soils in Alberta. Prep. for the Acid Deposition Research Program by the Alberta Research Council, Edmonton. ADRP-B-05-87. 202 pp.

U.S. NAPAP (National Acid Precipitation Assessment Program). 1987. Volume IV: Effects of Acidic Deposition. Superintendent of Documents, Washington, D.C.: U.S. Government Printing Office.

9.2 MODEL-GENERATED AIR QUALITY STATISTICS FOR APPLICATION IN VEGETATION RESPONSE MODELS IN ALBERTA

G. E. McVehil and M. Nosal

9.2.1 Introduction

A number of statistical models have been developed to relate plant response to air pollutant exposure (Krupa and Kickert 1987). Some investigators have demonstrated that the response of vegetation is dependent upon the statistics of the time-varying air pollutant concentration, i.e., the frequency distribution, number and duration of exposures, and variability of short-term concentrations (Nosal 1984; Krupa 1987).

To test and apply vegetation response models in Alberta, air pollution statistics representative of various parts of the Province are required. At this time, air quality monitoring data of the requisite accuracy and time resolution are not available for most parts of Alberta. Therefore, there exists a need to develop appropriate air quality statistics. The objectives of the work reported here were to determine the applicability of model generated air quality statistics and to develop, by modelling, realistic and representative time series of hourly SO_2 concentrations that could be used to generate the statistics demanded by vegetation response models.

9.2.2 Generation of Air Pollution Statistics by Modelling

9.2.2.1 Model Selection

Numerous short-term air quality dispersion models have been developed for use in regulatory and environmental impact analyses. Most of the widely used models are fundamentally similar; they are of the Gaussian-plume type where the horizontal and vertical distribution of pollutant around the plume centerline is assumed to be Gaussian, with the standard deviations given by empirical functions of stability and distance from the source. Recent advances in dispersion modelling have generally involved measurements of atmospheric turbulence to better characterize dispersion rates; in these models site and time-specific data on boundary layer turbulence can be used to improve predictions, relative to models based on gross characterizations of atmospheric stability (Pasquill 1974).

For the present application, only routine meteorological data (hourly surface weather station reports) were available for most parts of Alberta. Thus, refined models requiring site-specific turbulence measurements were not feasible. Therefore, it was determined that a standard regulatory model was most appropriate. In addition to being well known and readily applied, such models have a substantial validation history which provides some basis for judging their expected performance.

The Industrial Source Complex (ISC) model in its short-term mode was selected for use in this study. The ISC model (U.S. EPA 1986) is a recommended guideline model of the United States Environmental Protection Agency and contains the standard plume rise and dispersion algorithms common to all EPA point source regulatory models. As used in this application, the model treats a number of individual point sources and provides one-hour average concentrations due to all sources at desired receptor locations. ISC is a steady-state Gaussian plume model that uses Briggs (1975) plume rise equations and Turner (1969) rural dispersion coefficients. For input meteorological data the model requires hourly surface weather data specifying stability class, wind speed and direction, temperature, and mixing height. Surface wind speeds are adjusted to the speed at stack height by a stability-dependent power law.

9.2.2.2 Applicability of Models

There have been a number of evaluations of the accuracy of Gaussian plume models. Several studies have specifically examined the characteristics of concentration frequency distributions generated by EPA point source models (Bowne 1987; Bowne et al. 1983; Liu and Moore 1984). It has generally been found that model predictions of the upper percentiles of the frequency distributions are realistic but performance is poor for lower percentiles (low concentration). A major reason for the failure to predict the occurrences of low concentrations is usually given as the neglect or inadequate treatment of background concentrations and effects of small and/or distant sources (Mills and Stern 1975).

It has been shown that the Gaussian plume regulatory models do a poor job at predicting the concentration at a particular time and receptor location, but they do much better at predicting the overall frequency and magnitude of the highest concentrations (Liu and Moore 1984). In other words, the models properly characterize peak concentrations and how often they will occur, but they are less successful at specifying the particular time and location of the high values. These results of model validation tests, however, are encouraging for the present application, since the interest is in the statistics of the higher concentrations rather than the precise time or location of the peaks. Furthermore, the problem of inaccurate prediction of low concentrations may be reduced in the present case since true background concentrations are quite low. Most model tests have focussed on a single major point source while ignoring much smaller or distant sources. In the present application, all sources within 50-100 km were included, so it was expected that improved simulation of low concentration effects from distant sources could be achieved.

One particular air quality model was designed specifically to generate concentration frequency distributions. This model, the Gaussian Frequency Distribution Model (FREDIS) (Davison et al. 1981) has been used in conjunction with a statistical tree growth response model (Nosal 1986). As its name implies, the model is a Gaussian plume formulation with some modifications to the values of plume spread and their dependence on local meteorological parameters. Nosal (1985) found that the model produced predicted frequency distributions that agreed well with observations, although accuracy was dependent on reliability of the input meteorological data. Given the same meteorological data base, it should be expected that the ISC model used in this application would provide results similar to FREDIS since they are similar in their treatment of plume dispersion.

9.2.2.3 Meteorological Data

Air quality modelling was performed for Alberta Emission Regions 7 and 9, as defined by Picard et al. (1987) in their emission inventory. Region 7 extends to the western border of Alberta from just west of Edmonton, and Region 9 includes Calgary and southwestern Alberta. Hourly surface weather data applicable to these regions were obtained for Edson (Region 7), and Calgary (Region 9).

The data consisted of hourly observations for the time period January 1, 1986 through September 4, 1987. Complete data files were compiled for the growing season (April 15-August 31) of both 1986 and 1987, for calendar year 1986, and for the available part of the calendar year 1987 (January 1 through September 4).

The input parameters needed for application of the air quality model were derived from the surface weather data by use of the EPA pre-processing computer program RAMMET (U.S. EPA 1977). The program takes hourly observed values of wind speed, wind direction, temperature, sky cover, and ceiling height, and prepares an hourly file containing wind speed, randomized wind direction (randomized within ten degree reporting sectors), temperature, and atmospheric stability, determined by the method of Pasquill (1974).

The pre-processor program also provides hourly values of mixing height. In the normal application of the program, twice-a-day (morning and afternoon) mixing height values are obtained from the nearest upper air sounding station as input to the program, and hourly values are then interpolated. Daily values of the mixing height were not available for a representative Canadian sounding station for the present application. Therefore, mean seasonal morning and afternoon mixing heights were estimated from climatological data summarized by Holzworth (1972) and Leahey and Jamieson (1987). These mixing heights, as listed below, were used in the pre-processing program to interpolate hourly values for each day.

Mean Seasonal Mixing Heights (m)

	Morning	Afternoon
Dec, Jan, Feb	400	500
Mar, Apr, May	500	1750
Jun, Jul, Aug	450	2500
Sep, Oct, Nov	425	1200

After application of the pre-processing program, the product was a complete data file for each weather station and each year containing hourly values of wind speed, wind direction, temperature, stability, and mixing height. These files contained all necessary meteorological data for execution of the air quality model program.

9.2.2.4 Model Application

In both of the emission regions, receptor locations were selected at which hourly SO_2 concentrations would be calculated by the model. Receptor sites were picked in an effort to define important pollution regimes representative of Alberta. To achieve this goal, three receptor locations were identified for each emission region. One location used was fairly close to a major SO_2 source, a second location was chosen somewhat further away, and a third location was picked to represent the influence by a number of regional sources but not dominated by any single large source. No attempt was made to pick the specific geographic point subject to the highest pollutant concentrations, since such points are unique and localized and not representative of the regional pollutant regimes.

Table 9.1 provides a listing of the situations which were modelled, the time periods represented, and the number of receptors modelled. Figures 9.2 and 9.3 show the specific geographic areas considered, the locations of the receptor points, and the SO_2 sources in each region that were included in the modelling.

Table 9.1 Cases for which air quality modelling was performed to generate estimates of hourly SO_2 concentrations.

Emmission Region	Vegetation Type	Meteorological Data Source	Time Periods	Number of SO_2 Sources	Number of Receptors
7	Alfalfa	Edson	15/04/86 to 30/09/86	12	3
			15/04/87 to 04/09/87	12	3
9	Alfalfa	Calgary	15/04/86 to 30/09/86	15	3
			15/04/87 to 04/09/87	15	3

The location, emission rate, and stack parameters for all SO_2 sources were obtained from the Alberta emission inventory prepared by Picard et al. (1987). Annual average SO_2 emission rates were used and were assumed to be constant for each source.

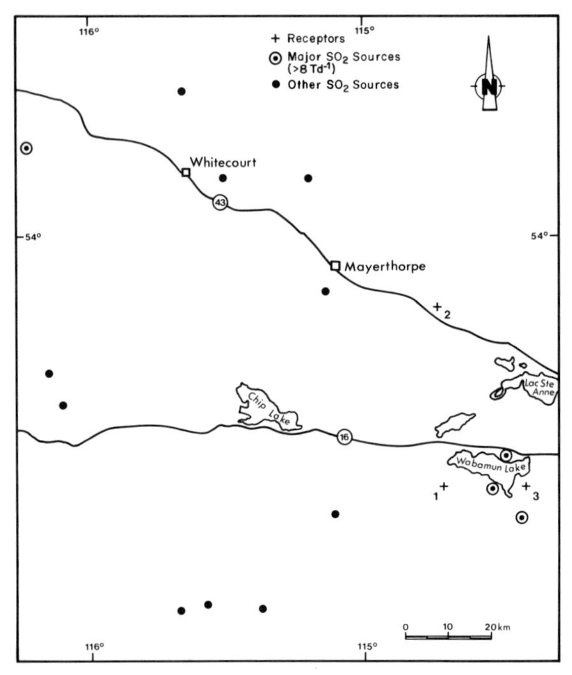

Figure 9.2 Study area modelled with SO_2 sources and receptors. Region 7: Alfalfa.

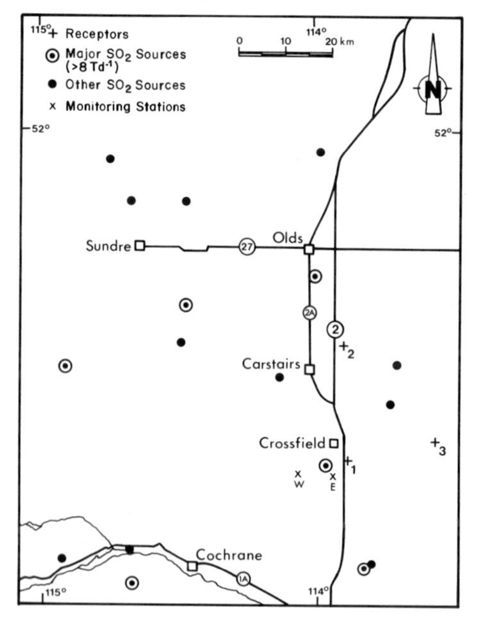

Figure 9.3 Study area modelled with SO_2 sources and receptors. Region 9: Alfalfa.

All significant SO_2 sources within 100 km of any receptor were included in the model calculations.

The ISC model was run for each region and time period (a total of 14 model runs) using the meteorological and source emission inputs as described previously. Model output consisted of predicted total hourly SO_2 concentrations at each receptor, due to the contributions from all sources that were affecting that receptor at that hour. A sample of the output files, showing predicted hourly SO_2 concentrations in $\mu g\ m^{-3}$ at each of the three receptor sites for a 48-hour period, is shown in Table 9.2.

Table 9.2 Example of predicted hourly SO_2 concentrations at three receptor sites.

Hour	Julian Day	Hourly SO_2 Concentrations ($\mu g\ m^{-3}$)		
		Site 1	Site 2	Site 3
1	110	.00	.00	12.93
2	110	18.30	.67	.00
3	110	.00	.00	.00
4	110	.00	.41	.00
5	110	.00	10.07	.00
6	110	.00	2.65	.00
7	110	.00	.00	51.88
8	110	.00	15.66	.13
9	110	15.43	17.18	.00
10	110	.00	1.38	.00
11	110	8.96	1.02	.00
12	110	17.03	.04	.00
13	110	.02	.00	.00
14	110	.00	.63	.00
15	110	.00	5.50	.00
16	110	.00	4.93	.00
17	110	.00	.25	5.24
18	110	12.67	1.84	2.37
19	110	2.65	14.64	7.90
20	110	.07	.02	18.07
21	110	2.94	4.87	36.21
22	110	.13	.17	32.36
23	110	1.28	2.93	44.95
24	110	.73	21.52	4.08
1	111	.01	3.76	14.75
2	111	7.25	16.50	1.15
3	111	1.19	.02	3.02

Continued...

Table 9.2 (Concluded).

Hour	Julian Day	Hourly SO_2 Concentrations ($\mu g\ m^{-3}$)		
		Site 1	Site 2	Site 3
4	111	7.67	7.16	.97
5	111	.00	12.08	.00
6	111	.00	.00	.00
7	111	.00	8.16	.00
8	111	16.56	4.86	64.10
9	111	.00	.01	13.97
10	111	13.56	4.87	13.72
11	111	9.32	2.61	4.74
12	111	6.82	1.87	1.43
13	111	.27	11.42	.00
14	111	.06	.01	1.13
15	111	1.00	6.71	2.81
16	111	6.07	1.03	1.16
17	111	.42	0.01	2.89
18	111	.45	.01	3.07
19	111	.01	.02	2.52
20	111	16.99	1.53	1.08
21	111	.00	.00	3.93
22	111	2.37	.61	.00
23	111	2.11	1.03	.00
24	111	.49	3.06	.00

9.2.2.5 SO_2 Concentration Statistics

Statistical vegetation response models developed by different investigators have used different statistical parameters of the SO_2 concentration history and different types of summary SO_2 data having varying averaging times. There is no consensus regarding the best parameters to describe pollutant exposure and, in fact, the optimum parameters probably depend on the crop species, the presence of other pollutants, and the study objective. Thus, it is not possible to derive a simple set of SO_2 concentration statistics that would be of general utility. However, several types of statistical parameters have been derived from the model-generated time series for use in a particular response model (Krupa 1987) that was applied in Alberta in the present case.

The problem of data averaging time is one that cannot be completely resolved at present. The alfalfa model of Krupa (1987) was based on continuous records of SO_2 concentrations. Model-generated data based on routine weather observations are limited to one-hour time resolution. Some statistical parameters can be adjusted for the averaging time, given a knowledge of the overall frequency distribution, but in general it is not possible to extrapolate parameters derived from one-hour samples to a shorter time resolution. Therefore it should be recognized that all statistics presented in this section are based on one-hour samples and are not necessarily directly comparable to statistics derived from other types of pollution records.

The Krupa (1987) alfalfa model utilized the following set of SO_2 parameters:

x_1 = number of episodes with concentrations exceeding a given threshold.

x_2 = time integral of SO_2 concentrations over the duration of each episode, cumulative through the time period analyzed.

$$= \int_{t_1}^{t_2} c(t)dt$$

x_3 = maximum value of $c(t)$

= peak concentration, $t_1 < t < t_2$

where $c(t)$ is the SO_2 concentration at time t. These parameters can be evaluated directly from the model-generated sequences of one-hour concentrations.

9.2.2.6 Results

For application of SO_2 statistics to alfalfa responses, as performed by Krupa (1987), it is necessary to define the time periods of alfalfa growth and harvest, and to divide each harvest period into three growth phases. Based on Alberta agricultural data for 1986 and 1987, harvests and growth phases were defined for Regions 7 and 9 as shown in Table 9.3. The resulting SO_2 concentration statistics from model-generated data for each receptor in each region are presented in Table 9.4. A concentration threshold of 6 ppb and a time unit of one hour were used in deriving the statistics.

Comparison to Krupa's (1987) SO_2 data is difficult since he used continuously monitored data rather than one-hour averages. However, it appears that the SO_2 exposures modelled here contain significantly higher concentrations than measured by Krupa in Minnesota. In Krupa's study, peak (instantaneous) SO_2 concentrations ranged from zero to 50 ppb, with an occasional peak of

Table 9.3 Time series of Alberta alfalfa harvests and growth phases.

Year	Harvest	Date		
	Region 7			
1986	Harvest 1	15 May	-	31 May
		01 Jun	-	16 Jun
		17 Jun	-	30 Jun
	Harvest 2	01 Jul	-	21 Jul
		22 Jul	-	11 Aug
		12 Aug	-	01 Sep
1987	Harvest 1	15 May	-	31 May
		01 Jun	-	16 Jun
		17 Jun	-	30 Jun
	Harvest 2	01 Jul	-	21 Jul
		22 Jul	-	11 Aug
		12 Aug	-	01 Sep
	Region 9			
1986	Single Harvest	15 Apr	-	15 May
		16 May	-	15 Jun
		16 Jun	-	16 Jul
1987	Harvest 1	20 Apr	-	13 May
		14 May	-	06 Jun
		07 Jun	-	30 Jun
	Harvest 2	01 Jul	-	19 Jul
		20 Jul	-	07 Aug
		08 Aug	-	25 Aug

around 100 ppb. The present results indicate one-hour average of concentrations of the same order of magnitude, implying considerably higher instantaneous peaks. Accordingly, the integrals determined here for alfalfa-growing areas of Regions 7 and 9 are significantly higher than the values reported by Krupa (1987).

9.2.2.7 Evaluation of Results

Statistics derived from the model-generated hourly concentrations appear, at least qualitatively, to be similar to the statistics used in the vegetation response models. However, direct comparison is hindered by differences in averaging times and types of

Table 9.4 SO_2 concentration statistics for the alfalfa model. Threshold = 6 ppb; all other concentrations also in ppb. Peak is the highest one-hour value.

	Site 1			Site 2			Site 3		
	No. of Episodes	Peak	Int.	No. of Episodes	Peak	Int.	No. of Episodes	Peak	Int.
Region 7									
1986									
Harvest 1									
Phase 1	13	23	157	5	15	67	11	23	190
2	16	16	207	3	11	40	22	35	373
3	10	21	214	4	11	56	4	15	50
Harvest 2									
Phase 1	3	13	36	6	15	67	17	28	286
2	9	22	105	7	11	64	25	24	327
3	10	17	156	5	11	49	15	19	148
1987									
Harvest 1									
Phase 1	5	18	102	1	8	8	10	21	148
2	15	14	158	2	8	15	7	30	163
3	8	17	134	6	11	54	7	23	115
Harvest 2									
Phase 1	11	19	148	1	8	8	18	20	224
2	11	20	167	3	11	26	15	26	309
3	16	15	142	9	14	94	12	15	141
Region 9									
1986									
Harvest 1									
Phase 1	49	67	1347	25	30	362	27	23	392
2	30	69	753	35	22	401	32	40	424
3	42	60	1092	27	42	434	34	23	410
1987									
Harvest 1									
Phase 1	42	73	910	19	26	199	25	46	413
2	30	73	1074	19	20	237	17	17	186
3	35	79	874	23	15	208	17	20	172
Harvest 2									
Phase 1	13	65	261	17	27	207	17	28	244
2	18	36	314	19	34	254	19	20	204
3	16	45	248	22	15	234	14	15	144

data used in the development of the response model. The most important question is to what extent the model-generated data represent actual Alberta SO_2 exposures. This question is addressed by comparison of the model results to existing SO_2 data.

Few data of sufficient accuracy and sensitivity exist for Alberta because of the low SO_2 concentrations usually observed. However, data collected by the ADRP in the Crossfield area are appropriate for use in verifying model predictions for that area. Statistics for average hourly SO_2 concentrations at the Crossfield West and East monitoring sites are provided in Chapter 4.

The locations of the model receptor points were previously shown in Figure 9.3. Note that the receptor site locations do not coincide with monitoring site locations. Receptor Site 1 might be expected to show effects most similar to Crossfield East since it is downwind and relatively close to the major SO_2 source at Crossfield. Receptor Sites 2 and 3 are more distant from the source (about 20 km) than either the Crossfield East or Crossfield West monitoring stations; they are in different directions and closer to other small sources. Because of the different locations, one cannot make hour-by-hour comparisons or expect direct correspondence between modelled and measured concentrations, but general comparisons are possible.

Table 9.5 shows data on measured and model-generated frequency distributions. Only the higher percentiles are shown since over 90% of both the observed and predicted SO_2 concentrations were below the monitor's minimum detection limit. The measured frequency distributions were obtained using the Weibull generator substitution (refer to Chapter 3 of this book). Recalling that Receptor Site 1 is most comparable to Crossfield East and Receptor Sites 2 and 3 to Crossfield West, the distributions of high SO_2 concentrations show reasonable agreement. That Crossfield West shows higher maximum SO_2 concentrations than Receptor Sites 2 and 3 is reasonable because of its proximity to the main SO_2 source.

The measured SO_2 concentrations at Crossfield were used to generate statistics used in Krupa's (1987) alfalfa response model. One-hour average SO_2 concentrations were processed to calculate peaks, numbers of episodes, and concentration integrals for the three phases of each of the 1986 alfalfa harvests. The results are compared with model-generated data for the three Crossfield area receptor points in Table 9.6.

Table 9.5 A comparison of model-generated and monitored SO_2 frequency distributions of one-hour concentrations (ppb).

	Percentiles			Maximum
	95	99	99.5	
Monitored Data				
Crossfield West				
1985-86	2.0	10.8	16.0	79
1986-87	1.7	8.6	11.9	84
Crossfield East				
1985-86	2.0	17.1	24.3	70
1986-87	4.9	22.8	31.5	92
Model-Generated Data				
Receptor 1				
1986	8.7	25.5	47.0	78
1987	8.3	27.2	41.0	79
Receptor 2				
1986	6.5	13.9	17.8	42
1987	5.9	12.0	14.6	34
Receptor 3				
1986	5.9	13.5	16.3	40
1987	5.4	12.1	16.8	46

Agreement between the two types of data within individual 15-day periods is not particularly good. However, this might be expected because of the different locations of the sites; impacts at any one site during a short period are strongly dependent on wind directions during that time. Considering the total period (again comparing Receptor Site 1 to Crossfield East and Receptor Sites 2 and 3 to Crossfield West) the overall correspondence is fairly good. Certainly the range of parameter values obtained from the model predictions is consistent with the observed range. There is a clear tendency for Receptor Sites 2 and 3 to experience more SO_2 episodes and higher concentration integrals than Crossfield West. Reference to Figure 9.3 will show that these receptor sites would be expected to receive more frequent SO_2 exposures, both from the major Crossfield source and from other sources, than the Crossfield West monitoring location (located 6.3 km west-southwest of the

Table 9.6 Comparison between model-generated and monitored SO_2 (ppb) statistics for the Crossfield area - 1986.

	15 Apr - 14 May			15 May - 14 Jun			15 Jun - 15 Jul		
	No. of Episodes	Peak	Int.	No. of Episodes	Peak	Int.	No. of Episodes	Peak	Int.
Monitored Data									
Crossfield West	9	49	217	15	43	234	5	17	74
Crossfield East	34	49	843	28	70	1028	32	43	682
Model-Generated Data									
Receptor Site 1	49	67	1347	30	69	753	42	60	1092
Receptor Site 2	25	30	362	35	22	401	27	42	434
Receptor Site 3	27	23	392	32	40	424	34	23	410

source). It is concluded that, considering the agreement that could be expected at dissimilar sites, the model-generated data appear to provide a reasonable representation of actual SO_2 exposure statistics.

9.2.3 Concluding Remarks

A conventional short-term air quality prediction model has been used to generate time series of hourly SO_2 concentrations for several pollutant exposure and crop growth regimes in Alberta. The model used weather data from local Alberta observation stations and a detailed inventory of SO_2 sources. Predicted concentration histories for three locations in each region were further processed to provide values of SO_2 exposure statistics used in a vegetation response model.

Predicted SO_2 concentrations and relevant statistics were compared with ADRP monitored SO_2 data from the Crossfield area and SO_2 data used previously in developing the vegetation response models. The comparisons show reasonable agreement in terms of peak SO_2 concentrations, long-term average concentrations, and frequency statistics for measurable concentrations. These results are in agreement with previous research which suggests that air quality models do a better job of predicting the frequency distribution and magnitude of high concentrations than in predicting time-and-point specific absolute concentrations. The results indicate that model-generated concentration statistics for SO_2 concentrations that are relevant to vegetation response studies are realistic and can provide a useful substitute for measured pollutant data.

Air pollutant data were generated to represent SO_2 exposure regimes for two different geographic areas of Alberta. These data were used in a crop response model (Chapter 9.3). Evaluation of the data indicates that they provide reasonable and representative examples of SO_2 concentrations that exist in those areas. They are not, of course, applicable to all areas. Neither the few localized areas of maximum impact adjacent to major sources, nor the many large areas remote from all sources, have been simulated. The intent was to characterize a range of Alberta agricultural areas which are subject to SO_2 exposures from typical sources. It is believed that the model results are representative of large areas of Alberta.

9.2.4 Literature Cited

Bowne, N.E. 1987. Observations and evaluations of plume models. Paper 87-73.2. Annual meetings of the Air Pollution Control Association. New York, New York, Air Pollution Control Association, Pittsburgh, Pennsylvania.

Bowne, N.E., R.J. Londergan, D.R. Murray, and H.S. Borenstein. 1983. Overview, results, and conclusions for the EPRI plume model validation and development project: plains site. EA-3074. The Electric Power Research Institute. Palo Alto, California.

Briggs, G.A. 1975. *Plume rise predictions.* Lectures on Air Pollution and Environmental Impact Analysis. American Meteorological Society. Boston, Massachusetts.

Davison, D.S., E.D. Leavitt, R.R. McKenna, R.C. Rudolf, and M.J.E. Davies. 1981. Airshed management system for the Alberta oil sands. Volume 1: A Gaussian frequency distribution model. Prepared for the Research Management Division, Alberta Environment, by INTERA Environmental Consultants Ltd. and Western Research and Development. AOSERP Report 119. Edmonton, Alberta. 149 pp.

Holzworth, G.C. 1972. Mixing heights, wind speeds and potential for urban air pollution throughout the contiguous United States. Publication No. AP-101. U.S. Environmental Protection Agency. Research Triangle Park, North Carolina.

Krupa, S.V. 1987. Responses of alfalfa to sulfur dioxide exposures from the emissions of the NSP-SHERCO coal-fired power plant units 1 and 2. Volume 1. Prepared for the Northern States Power Company, Minneapolis by the Department of Plant Pathology, University of Minnesota, St. Paul, Minnesota.

Krupa, S.V. and R.N. Kickert. 1987. An analysis of numerical models of air pollutant exposure and vegetation response. *Environmental Pollution*, 44: 127-158.

Leahey, D.M. and A.L. Jamieson. 1987. Development of input data for three (3) statistical long range transport of air pollutant models. Report prepared under Contract KM643-6-0291-01-56 for Atmospheric Environment Service, Western Region, by Western Research, Calgary, Alberta.

Liu, M.K. and G.E. Moore. 1984. On the evaluation of predictions from a Guassian plume model. *Journal of the Air Pollution Control Association*, 34: 1044-1050.

Mills, M.T. and R.W. Stern. 1975. Model validation and time-concentration analysis of three power plants. EPA-450/3-76-002. U.S. Environmental Protection Agency, Research Triangle Park, North Carolina.

Nosal, M. 1984. Atmosphere-biosphere interface: analytical design and a computerized regression model for lodgepole pine response to chronic atmospheric SO_2 exposure. RMD83-26, RMD83-27. Report prepared for the Research Management Division, Alberta Environment, Edmonton, Alberta.

Nosal, M. 1985. Atmosphere-biosphere interface: verification of the "FREDIS" model to determine the frequency distribution of SO_2 ground level concentrations. RMD85-26A. Report prepared for the Research Management Division, Alberta Environment, Edmonton, Alberta.

Nosal, M. 1986. Atmosphere-biosphere interface: application of a numerical model to assess growth response of jack pine to chronic SO2 exposure. RMD 83-26B. Report prepared for the Research Management Division, Alberta Environment, Edmonton, Alberta.

Pasquill, F. 1974. *Atmospheric Diffusion.* 2nd Edition. New York: John Wiley and Sons. 429 pp.

Picard, D.J., D.G. Colley, and D.H. Boyd. 1987. Emissions inventory of sulphur oxides and nitrogen oxides in Alberta. Four volumes. Prep. for the Acid Deposition Research Program by Western Research, Calgary, Alberta.

Turner, D.B. 1969. Workbook of Atmospheric Dispersion Estimates. PHS Publication No. 999-26. U.S. Environmental Protection Agency, Research Triangle Park, North Carolina. 84 pp.

U.S. EPA. Environmental Protection Agency. 1986. Industrial sources complex (ISC) dispersion model user's guide -- second edition. Volume 1. EPA-450-4-86-005a. Office of Air Quality Planning and Standards. Research Triangle Park, North Carolina.

U.S. EPA. Environmental Protection Agency. 1977. User's manual for single-source (CRSTER) model. EPA-450/2-77-013. Office of Air Quality Planning and Standards. Research Triangle Park, North Carolina.

9.3 REGIONAL SCALE EFFECTS OF SO_2 ON SOME AGRICULTURAL CROPS IN ALBERTA

R. N. Kickert

9.3.1 Regional Scale Effects of SO_2 on Alfalfa Productivity in Alberta

9.3.1.1 Introduction

Alfalfa (*Medicago sativa* L.) is one of the major crops of Alberta. Any possible negative impact on its production from industrial activity and air pollutants could have a significant economic impact. The general spatial distribution of Tame Hay production, as an indicator of alfalfa cultivation in Alberta, is shown in Figure 9.4. This spatial distribution is shown superimposed on a map of the ten provincial emission regions of sulphur and nitrogen oxides (Picard et al. 1987). The present assessment goal in Phase 1 of the ADRP is:

- At the present-time, are there observed or measurable adverse effects of sulphur dioxide, as an acidifying pollutant, on the alfalfa crop in Alberta?
- If not, where, when and under what conditions will such effects occur in Alberta?

It should be recognized that this initial assessment was performed using only the existing published literature on the effects of sulphur dioxide on alfalfa growth and productivity, coupled with the air quality data gathered in ADRP. This assessment was not based on any specific field or laboratory alfalfa experiments conducted by ADRP, since actual experimental research on crop responses were not included in the first phase, Critical Point I.

Initially the ADRP scientists decided that the evaluation and assessment of any environmental effects from air pollutants at the present time be focussed on Regions 7 and 9 (Figure 9.4) because of the present, relatively high emissions in those regions in comparison to others in the Province. In addition, it was decided to include Region 6 because of possible future expansion of industrial activities in that geographic area. From the figure, it may be seen that major portions of the Alberta tame hay production occur in the eastern portion of Emission Region 9, with Crossfield in the approximate centre; in the eastern portion of Emission Region 7, with the Barrhead area slightly to the north, but still within the major production area; and in the southern portion of Emission Region 6, with Cold Lake to the east within the band of moderate tame hay production (see arrows in Figure 9.4). The ADRP scientists also decided that such an assessment should be focussed on the area where an effect would most readily be found if it existed. For these reasons, the evaluation of ambient sulphur dioxide exposure regimes and their possible effects on Alberta alfalfa production was focussed on the Crossfield area in Emission Region 9, the Barrhead area in Emission Region 7, and for possible future provincial expansion of industry, the Cold Lake area of Emission Region 6. Since one of the objectives of ADRP was to examine possible regional effects of air pollutants, this report is not directed to any specific field site within these regions.

Because approximately half of the sulphur deposition in Alberta is thought to occur as dry deposition, and since pollutant exposure-crop response information for alfalfa growth under wet deposition is virtually non-existent, this assessment is confined to gaseous sulphur dioxide. In addition, according to Torn et al. (1987), evidence for sulphate deposition and hydrogen ion deposition by precipitation in intensively managed agricultural lands as posing a threat to crop growth is not available. Similarly, crop growth effects through the soil are also not expected because of cultivation practices, namely the application of fertilizers and liming. For these reasons, the present assessment was limited to the consideration of only the possible direct, above-ground effects on alfalfa from ambient SO_2 exposures in Alberta.

In Emission Regions 9 (Crossfield), 7 (Barrhead), and 6 (Cold Lake), estimated dates of emergence for a newly planted stand of alfalfa in 1986 were, respectively, April 27, May 27, and May 19. Corresponding dates for harvesting the crop in the Crossfield area were July 5 in 1986 (only one harvest reported), and June 24 and August 24, in 1987. Corresponding dates for harvesting the crop in the Barrhead area were June 30, and September 2, during both 1986 and 1987. Harvest dates for the alfalfa crop in the Cold Lake area were June 28, and August 24, 1986, and June 23, and August 24, in 1987. These data provide the lengths of the growing season, from emergence to final harvest for the Crossfield area of 119 days, for the Barrhead area, 100 days, and for the Cold Lake area, 98 days.

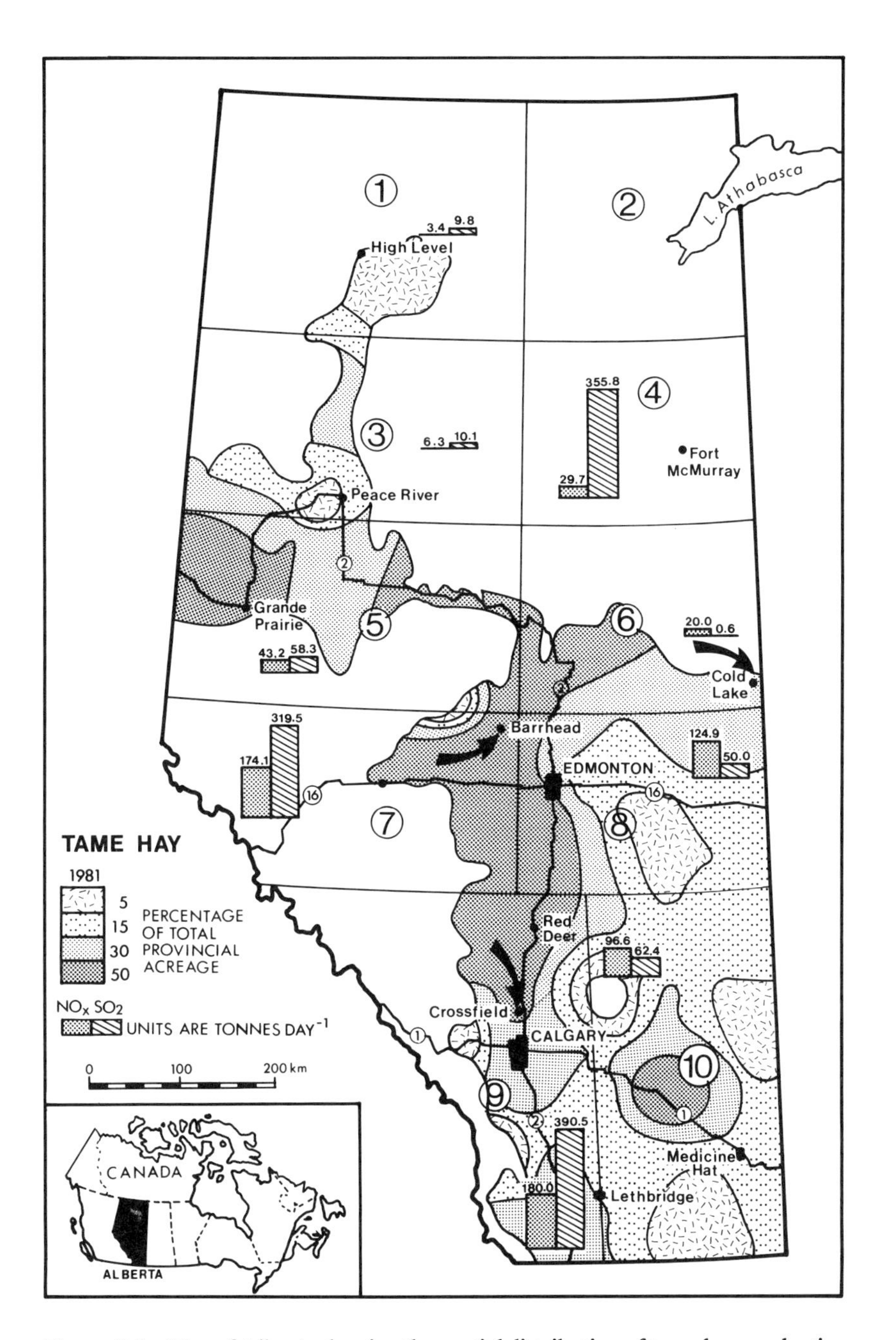

Figure 9.4 Map of Alberta showing the spatial distribution of tame hay production, and total NO_x and total SO_2 emissions as histograms in each of the 10 geographic emission regions.

9.3.1.2 Approach

The strategy used to determine whether, at the present time, there are observable adverse effects of SO_2 on alfalfa in Alberta was to first apply a computer simulation model of the daily growth (and regrowth after cutting) processes, under unpolluted conditions. This was for a pure, newly planted stand of alfalfa in all three areas: Crossfield, Barrhead, and Cold Lake, for 1986 and 1987. In this simulation, harvestable alfalfa biomass would respond to the daily amount of total solar radiation (direct as well as diffused sunlight flowing to, and absorbed by, the ground and the alfalfa crop), daily maximum and minimum air temperatures, and daily precipitation during each year. The model would predict what the expected alfalfa harvest biomass should be for a given location-dependent soil and seasonal weather pattern.

The ability to apply this model for these regions was dependent upon having information available for the conditions under which alfalfa is grown in these regions. The available information pertained to weather data and soil physical properties as discussed subsequently. On a regional scale, no information was available on differences in soil chemistry such as soil fertility and nutrient requirements, grower liming and tillage practices, and other cultivation practices such as pest management.

Following the application of the model, the results were compared with field reports of the actual alfalfa production based on the statistics compiled by various agricultural agency personnel. Because these field reports were obtained for the three areas on an "as-available" basis, there is no information on crop phenology available for existing alfalfa stands, nor for the annual average acreage of newly established versus older stands in Alberta. The varieties of alfalfa grown in Alberta include: Angus, Beaver, Rambler, and Algonquin (Bjorn Berg, personal communication). Additional varieties recommended (Goplen et al. 1980) include: Anik, Drylander, Kane, Rangelander, Roamer, Trek, Anchor, Chimo, Thor, Titan, and Vernal.

If the reported yield from the field was found to be far less than the yield reported by the model for the same location and year, it was interpreted as a condition under which some environmental processes other than weather alone (possibly air pollutants) were acting to depress the yields of alfalfa in the field.

9.3.1.3 Model Description

While a number of potentially useful biological process simulation, computer models and statistical models were described by Krupa and Kickert (1987) for examining vegetation response to air pollution stress, the alfalfa growth process model used in this assessment was ALSIM 1 (Level 2) as documented in Onstad and Fick (1981), and Fick (1981). Other related publications include Fick and Onstad (1981) and Fick (1984). The dynamic structure of this deterministic biomass-oriented alfalfa stand model is presented in Figures 9.5 and 9.6 from Fick (1981).

Eleven processes and five main structural components of an alfalfa stand are described by mathematical functions presented as sets of data coordinates accessed by a "look-up table" algorithm in the computer program (Fick 1981). The time-step is daily, and the model may be run for two complete growing seasons (each with multiple harvests) and the intervening over-wintering period. The integration method is the simple rectangular, Eulerian, because essentially finite difference equations are used to describe the state equations for the five main structural components: (1) materials available for top growth and storage; (2) leaves; (3) stems; (4) basal buds for regrowth; and (5) total nonstructural carbohydrates in taproots.

The ALSIM 1 model responds to data on daily solar radiation, daily temperature, and daily soil moisture available for growth. A part of the model (Figure 9.6) describes the daily changes in the soil water budget, because as soil water availability for crop growth diminishes over time, the growth rates of various components of the alfalfa crop are affected. Growth itself does not respond to soil chemistry in any way in the ALSIM 1 model. This might be considered to be one of its limitations.

ALSIM 1 has been used in various research efforts on alfalfa throughout the 1980's. Onstad et al. (1984) used it as a basis to study the potato leafhopper. Sawyer and Fick (1987) used it as a basis to explore the alfalfa blotch leafminer insect. Parsch (1987) recently published a validation of ALSIM 1 through its application in Michigan.

The FORTRAN computer code for ALSIM 1 (Level 2) was implemented on an 8086 personal computer, with 640 K random access memory (although less capacity would work), using IBM

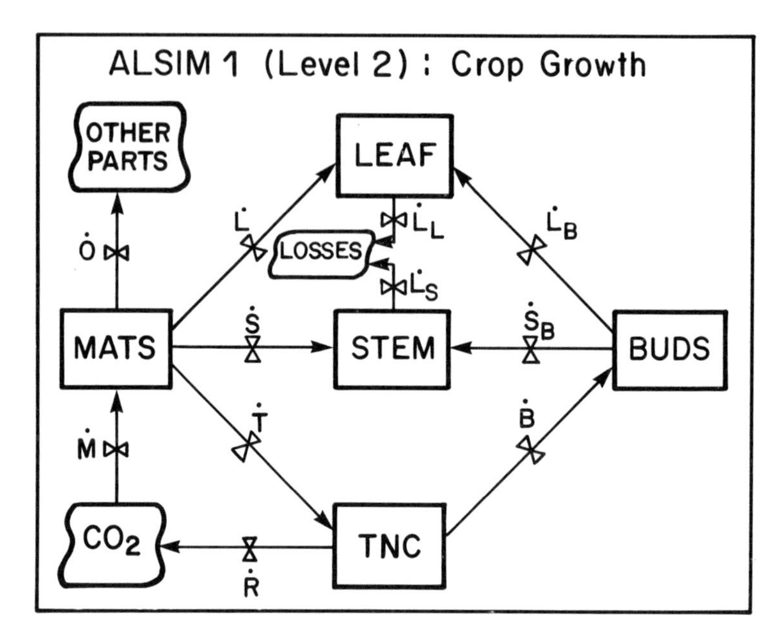

Figure 9.5 The relational diagram of the crop growth components of LEVEL 2 showing five main state variables and eleven processes to be simulated. The state variables represent parts of the stimulated alfalfa crop: MATS (materials available for top growth and storage), LEAF (leaves), STEM (stems), TNC (total nonstructural carbohydrates in the taproots), and BUDS (basal buds for regrowth). The processes are described by rate equations that simulate the transfer of material between the parts of the crop: M (crop growth rate), L (leaf growth rate), S (stem growth rate), T (TNC storage rate), R (TNC respiration rate), B (bud growth rate), S_B (growth rate of stems coming from buds), L_B (growth rate of leaves coming from buds), L_L (rate of leaf loss), L_S (rate of stem loss), and O (rate of other uses of MATS).

Professional FORTRAN and an 8087 math co-processor. The linked, executable program for ALSIM 1 requires only 56K magnetic disk space, while the input data file requires from 2 K (monthly average weather data version) to 30 K (actual daily weather data version) magnetic disk space. The output file produced requires 10 K disk space if two full growing seasons and the intervening over-wintering period are simulated, and if output is requested only every fifth day over a total of about 500 days

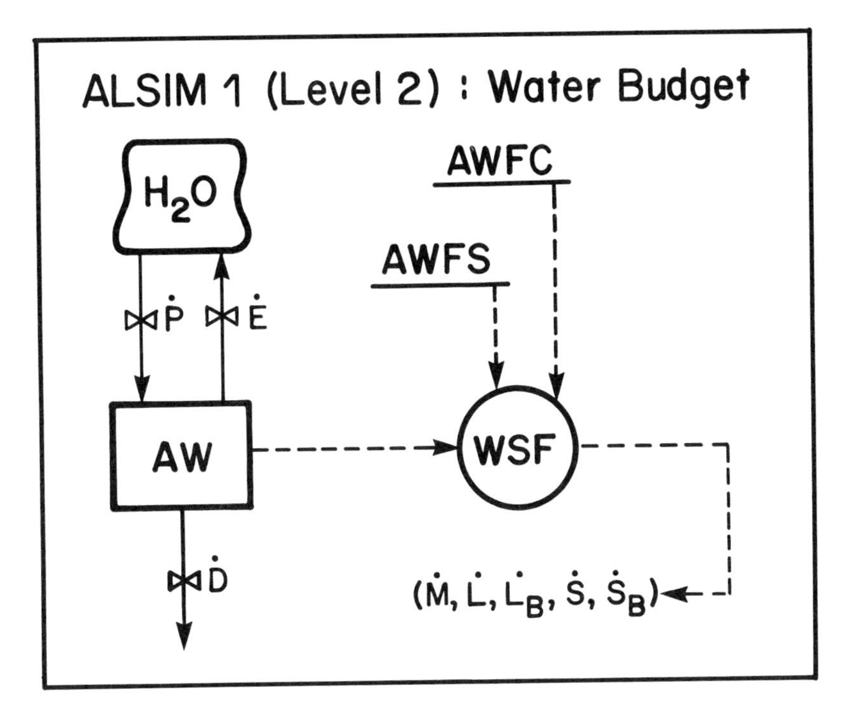

Figure 9.6 The relational diagram of the soil water budget component of LEVEL 2 showing the plant available water in the soil (AW) being increased by precipitation (P), and decreased by evapo-transpiration (E), and runoff and deep percolation (D). A water stress factor (WSF) is computed from AW and parameters for available water at field capacity (AWFC), and the available water fraction at which stress begins (AWFS). The growth rates M, L, L_B, S, and S_B (see Figure 9.10) are influenced by WSF.

simulated. A full 500-day simulation can be run in less than a minute. Verification runs were made using the 1979-80 data in Fick (1981) to compare results from this PC version with the results published in Fick (1981).

Input weather data files, for 1986 and 1987, for the Crossfield area, the Barrhead area, and the Cold Lake area were assembled. ALSIM 1 can be run to respond to average monthly weather, as Fick (1981) reported, or to daily weather data. Both of these approaches were used in this assessment to check on the difference between using monthly average values versus daily weather data. Furthermore, average monthly weather data can be either actual values for specific years, or for long-term normals (for comparing alfalfa growth response from long-term average values against that from specific years' monthly average values). Again, both types of weather data were used to drive alfalfa growth-regrowth simulations in this assessment for the purpose of comparing differences between resulting harvested yields. However, in all cases, the reported dates of emergence (for the first year of growth), and the cutting (harvesting) dates used were those reported previously for 1986 and 1987. This introduces some lack of realism when, for example, long-term normal monthly weather data are used since in reality it is likely that growers adjust their actual sowing dates for new stands (and consequently the emergence dates), and especially their harvesting dates in particular, for each year so those dates match the perceived optimum timing for carrying out the field practices by the grower.

The weather data in each of these files included average daily maximum and minimum air temperatures, average daily total solar radiation, and average daily precipitation, by month. For each of the three regions separate data files were assembled for: (a) actual 1986-87 monthly average values, (b) actual 1986-87 daily values, (c) long-term normal monthly average values, (d) long-term normal plus 1 standard deviation for monthly averages, and (e) long-term normal minus 1 standard deviation for monthly averages.

Each of these data files was used to provide the input data to drive a separate simulation using ALSIM 1 under the conditions observed for the Crossfield, Barrhead, and Cold Lake areas to determine simulated harvested yields of alfalfa without any mechanism (equation) for an air pollution effect in the model.

Actual 1986 monthly average values of daily solar radiation, and values of actual daily solar radiation for Crossfield were obtained from the ADRP data files for the Crossfield East air quality monitoring site. For Barrhead, data were obtained from the Edmonton-Stony Plain monitoring station as published by the Atmospheric Environment Service (1986a) for 1986, and January through March 1987, and from Ellersley for April 1987 through September 1987. The Cold Lake station does not monitor solar radiation, but Beaverlodge (on the far west side of the Province) was in the same isopleth band, that is, experienced about the same number of hours of sunshine per year. Therefore, Beaverlodge solar radiation data from Atmospheric Environment Service (1986a), were used for actual daily and monthly daily average solar radiation for the Cold Lake area. Long-term normals and standard deviations for monthly daily average solar radiation for Barrhead (Edmonton-Stony Plain) and Cold Lake (Beaverlodge) were obtained from the Atmospheric Environment Service (1982a). There was no comparable long-term record for solar radiation in the Crossfield area.

Monthly daily average solar radiation for 1986 and 1987 was relatively similar for Crossfield. Monthly daily average solar radiation at Barrhead was close to the long-term normal monthly values, although late summer 1987 was a little darker than normal. Cold Lake (Beaverlodge) monthly daily average solar radiation was generally normal for 1986 and 1987 although during both years, in the month of June, it was brighter than normal.

In terms of precipitation, Barrhead and Cold Lake were similar with respect to the isohyets, and Crossfield was wetter over the long term. Actual daily and monthly total precipitation data for 1986 and 1987 and long-term normal monthly totals were obtained from the Atmospheric Environment Service (1986b and 1982c). Data from Madden were used for Crossfield; data from Campsie were used for Barrhead; and data for Cold Lake were obtained from records at that location. Precipitation and air temperature data from Madden were used for the assessment in the Crossfield area. Figure 9.7 shows that the two locations are closely correlated for long-term normal precipitation, although Madden was somewhat drier during the wetter months.

Table 9.7 contains the monthly precipitation data used for all three areas. The same stations used for precipitation were also used to obtain the maximum and minimum air temperature data. The actual daily and monthly average maximum and minimum air temperature data for 1986 and 1987 and long-term normal monthly average values (Table 9.8) were obtained from the Atmospheric Environment Service (1986b and 1982b).

As a general summary, Tables 9.9 through 9.12 indicate the weather status for 1986 and 1987 for the three assessment areas. Salient features include: below normal solar radiation for Barrhead (Ellersley) in the summer of 1987 (Table 9.9); erratic summer precipitation and below normal precipitation for all three areas for

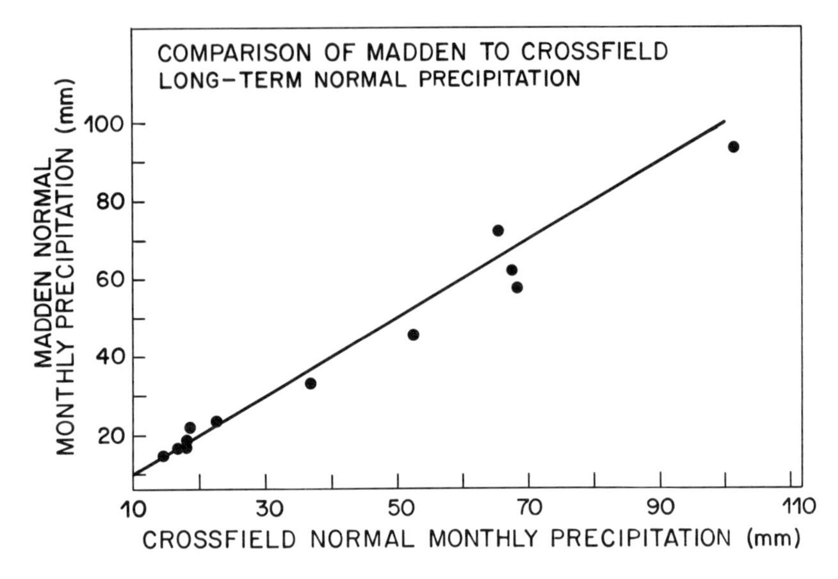

Figure 9.7 Comparison of long-term normal precipitation at Madden and at Crossfield.

the winter of 1986-87 (Table 9.10); above normal air temperatures for all three areas in the winter of 1986-87 (Tables 9.11 and 9.12).

Monthly average weather data might mask much of the day to day variability in the weather which can influence how the alfalfa crop would grow. For this reason, a version of ALSIM 1 was set up to respond to daily data on solar radiation, precipitation, and maximum and minimum air temperatures.

Sharratt et al. (1986) showed that over a 21-year period in Minnesota, spring yields of alfalfa (first harvests) for one-year-old stands were statistically related to precipitation during the preceding prehardening period in the fall ($R^2=0.40$). Such yields for a two-year-old stand were statistically more strongly related to the maximum air temperature during the preceding fall period, the minimum air temperature during the preceding winter, and the maximum air temperature during the spring growth period ($R^2=0.50$). Sharratt et al. (1987) reported similar changes in the importance among weather variables as the stand aged with respect to early summer harvests (the second cutting) and late summer harvests (where a third cutting is typical, such as in Minnesota). However, all these findings are related only statistically to the

Table 9.7 Data on monthly precipitation for Alberta Emission Regions 6, 7, and 9, used to execute different versions of the ALSIM alfalfa growth simulation model.

Month	Normal Monthly Total Precip. (mm)	Standard Deviation of Monthly Precip. (mm)	Monthly Total Less Std. Dev. (mm)	Monthly Total Plus Std. Dev. (mm)	Actual 1986 Monthly Total Precip. (mm)	Actual 1987 Monthly Total Precip. (mm)
Normal 1951-80 Campsie (19.3 km W of Barrhead)						
1	24.3	14.9	9.4	39.2		0.0
2	20.5	14.7	5.8	35.2		6.8
3	17.7	11.9	5.8	29.6	46.4	17.0
4	19.8	13.8	6.0	33.6	65.4	11.0
5	42.5	28.6	13.9	71.1	40.8	51.8
6	85.1	41.8	43.3	126.9	42.2	30.8
7	86.5	41.2	45.3	127.7	147.8	111.0
8	76.9	49.6	27.3	126.5	19.8	83.8
9	36.3	24.8	11.5	61.1	93.6	11.2
10	16.3	12.4	3.9	28.7	20.2	
11	17.4	13.0	4.4	30.4	26.2	
12	24.0	16.3	7.7	40.3	6.0	
Total:	467.3	283	184.3	750.3		
Normal 1951-80 Cold Lake A						
1	22.1	9.4	12.7	31.5		12.7
2	15.8	9.2	6.6	25.0		9.0
3	20.1	12.5	7.6	32.6	19.0	23.8
4	21.6	14.7	6.9	36.3	27.3	45.0
5	39.7	24.0	15.7	63.7	59.7	35.3
6	71.9	42.4	29.5	114.3	35.0	37.9
7	85.6	32.5	53.1	118.1	150.1	110.3
8	76.2	34.2	42.0	110.4	16.8	91.1
9	44.9	25.7	19.2	70.6	29.3	23.6
10	16.9	15.4	1.5	32.3	12.7	
11	20.3	12.8	7.5	33.1	28.6	
12	24.8	11.9	12.9	36.7	7.2	
Total:	459.9	244.7	215.2	704.6		

Continued...

Table 9.7 (Concluded).

Month	Normal Monthly Total Precip. (mm)	Standard Deviation of Monthly Precip. (mm)	Monthly Total Less Std. Dev. (mm)	Monthly Total Plus Std. Dev. (mm)	Actual 1986 Monthly Total Precip. (mm)	Actual 1987 Monthly Total Precip. (mm)
Normal 1951-80 Crossfield/Madden						
1	18.7	14.4	4.3	33.1		5.1
2	14.9	7.5	7.4	22.4		6.4
3	21.7	10.7	11.0	32.4	5.9	30.6
4	33.5	19.6	13.9	53.1	12.5	23.3
5	57.9	29.7	28.2	87.6	88.0	13.6
6	93.6	45.6	48.0	139.2	64.7	37.2
7	62.0	36.0	26.0	98.0	131.1	113.1
8	72.3	40.1	32.2	112.4	52.2	81.1
9	45.6	26.4	19.2	72.0	155.2	28.2
10	23.3	10.6	12.7	33.9	9.0	
11	16.9	12.5	4.4	29.4	12.5	
12	17.2	13.3	3.9	30.5	0.0	
Total:	477.6	266.4	211.2	744		

harvested yield, and were not presented by the authors as quantitative functions for the daily growth processes of photosynthesis, respiration, and/or carbohydrate partitioning among alfalfa plant organs. As a result, the findings of Sharratt et al. (1986, 1987) from a black box statistical model cannot be used to improve the ALSIM 1 process model, and it is questionable as to whether such improvement is needed because ALSIM 1 growth process variables respond to daily changes in these weather variables with the progress of the seasons.

ALSIM 1 is also designed in such a way that daily alfalfa growth in the model responds to inputs of soil moisture conditions. This requires setting numerical values for a few parameters for the soil physical properties. The "Total water available at field capacity" for the soils: Delacour, Dunvargan, and Antler in the Crossfield area, was computed as 149 mm for the top 45 cm of soil and then taken as a weighted average according to the spatial area of the region in each of these three alfalfa producing soils. The same approach was used for the Cooking Lake, Ponoka, and Peace Hills soils in the Barrhead area resulting in a regionally weighted average of 137 mm

Table 9.8 Data on monthly maximum and minimum air temperature for Alberta Emission Regions 6, 7, and 9, used to execute different versions of the ALSIM alfalfa growth simulation model.

Monthly Minimum Temperat.	Normal Mean Daily Maximum Temperat. (°C)	Normal Mean Daily Minimum Temperat. (°C)	Extreme Maximum Temperat. (°C)	Extreme Minimum Temperat. (°C)	1986 Mean Daily Maximum Temperat. (°C)	1986 Mean Daily Minimum Temperat. (°C)	1987 Mean Daily Maximum Temperat. (°C)	1987 Mean Daily Temperat. (°C)
Normal 1951-80 Campsie (19.3 km W of Barrhead)								
1	-11.2	-23.3	15.0	-51.7			-0.4	-14.0
2	-4.4	-18.0	25.5	-50.6			0.6	-11.0
3	0.2	-12.9	20.6	-43.9	5.5	-5.5	-0.5	-11.1
4	9.5	-3.9	29.4	-32.2	8.5	-2.5	14.5	-2.2
5	17.3	2.3	33.3	-12.2	18.3	2.6	19.0	1.9
6	20.5	6.4	37.8	-7.8	21.2	6.5	23.1	7.2
7	22.7	8.8	37.2	-2.8	20.5	9.2	20.5	9.2
8	21.3	7.4	33.9	-3.3	23.3	6.0	23.3	6.0
9	16.1	2.6	31.7	-17.2	13.3	1.6	13.3	1.6
10	10.9	-2.6	29.4	-26.7	13.2	-1.4		
11	-0.1	-11.0	22.2	-45.6	-3.4	-14.6		
12	-7.2	-18.9	15.0	-51.7	-0.1	-13.8		

Continued...

Table 9.8 (Continued).

Monthly Minimum Temperat.	Normal Mean Daily Maximum Temperat. (°C)	Normal Mean Daily Minimum Temperat. (°C)	Extreme Maximum Temperat. (°C)	Extreme Minimum Temperat. (°C)	1986 Mean Daily Maximum Temperat. (°C)	1986 Mean Daily Minimum Temperat. (°C)	1987 Mean Daily Maximum Temperat. (°C)	1987 Mean Daily Temperat. (°C)
Normal 1951-80 Cold Lake A								
1	-14.0	-23.9	10.6	-48.3			4.3	-15.3
2	-7.9	-19.3	11.7	-42.8			-2.4	-12.6
3	-1.7	-13.5	16.7	-41.1	3.3	-6.3	-1.3	-11.4
4	8.7	-2.9	29.4	-34.4	8.5	-2.7	12.7	-0.3
5	16.9	3.8	31.7	-7.8	17.6	5.5	18.7	4.8
6	20.6	8.4	35.6	-3.3	21.4	8.2	22.4	9.4
7	22.9	10.9	36.1	0.0	21.1	11.2	23.1	11.5
8	21.3	9.6	32.8	-0.5	23.0	8.4	18.9	7.2
9	15.4	4.2	32.8	-9.4	14.0	3.1	19.9	5.6
10	9.8	-0.9	27.4	-15.6	12.0	-1.3		
11	-1.9	-10.5	18.9	-36.7	-6.1	-14.8		
12	-9.6	-18.8	10.0	-42.8	-3.8	-14.6		

Continued...

Table 9.8 (Concluded).

Monthly Minimum Temperat.	Normal Mean Daily Maximum Temperat. (°C)	Normal Mean Daily Minimum Temperat. (°C)	Extreme Maximum Temperat. (°C)	Extreme Minimum Temperat. (°C)	1986 Mean Daily Maximum Temperat. (°C)	1986 Mean Daily Minimum Temperat. (°C)	1987 Mean Daily Maximum Temperat. (°C)	1987 Mean Daily Temperat. (°C)
Normal 1951-80 Crossfield/Madden								
	Crossfield				Madden			
1	-6.4	-17.6	12.2	-40.6			5.3	-7.4
2	-1.4	-12.4	15.6	-35.0			5.4	-7.7
3	2.1	-9.4	17.0	-34.4	8.5	-3.6	2.8	-7.8
4	9.2	-3.0	28.3	-22.2	10.2	-3.2	14.8	-0.1
5	15.9	2.4	30.0	-5.0	16.4	3.4	19.3	3.1
6	19.5	5.9	31.1	-1.1	21.1	7.8	23.3	7.8
7	22.5	8.3	34.4	0.0	19.2	7.0	21.6	8.3
8	21.7	7.6	33.9	-3.5	22.3	7.7	18.3	5.8
9	17.2	3.8	30.5	-10.6	10.8	2.0	20.9	5.8
10	11.9	-0.8	30.0	-19.4	14.7	1.5		
11	3.0	-8.4	20.6	-30.0	1.4	-11.5		
12	-2.0	-12.8	14.5	-39.5	4.5	-7.9		

Table 9.9 Comparison of solar radiation in 1986-87 to long-term normal monthly, daily average values in Alberta, Canada.

	1986			1987		
Location	Spring	Summer	Fall	Winter	Spring	Summer
Crossfield	Long-term normal not available					
Edmonton-1986 Stony Plain[1]	slightly below	normal	normal	normal		
Ellersley[2]					normal	below
Beaverlodge[3]	slightly below	above	near normal	normal	normal	slightly above

[1] Used for Barrhead area.
[2] Used for Barrhead.
[3] Used for Cold Lake, Alberta.

Table 9.10 Comparison of precipitation during 1986-87 to long-term normal monthly totals in Alberta, Canada.

	1986			1987		
Location	Spring	Summer	Fall	Winter	Spring	Summer
Crossfield/ Madden	erratic early below	very high mid-summer	very high	below	below	erratic low to high
Barrhead/ Campsie	above	erratic low to high	above normal	below	near low to high	erratic
Cold Lake	slightly above	erratic low to high	slightly below	below	above	erratic low to high

equivalent water depth in the top 45 cm of soil, and for the La Corey, Kehiwin, and Falun soils in the Cold Lake area, producing a regionally weighted average of 142 mm.

A similar approach was used to produce regionally weighted averages of the "Actual available soil water" in spring 1986. These values for Crossfield, Barrhead, and Cold Lake were 138, 124, and 129 mm, respectively.

Table 9.11 Comparison of 1986-87 values to long-term normal monthly mean, daily maximum air temperature in Alberta, Canada.

	1986			1987		
Location	Spring	Summer	Fall	Winter	Spring	Summer
Crossfield/ Madden	normal	normal	slightly below	above	above	normal
Barrhead/ Campsie	normal	normal	normal	above	slightly above	normal
Cold Lake	normal	normal	normal	above	slightly above	normal

Table 9.12 Comparison of 1986-87 values to long-term normal monthly mean, daily minimum air temperature in Alberta, Canada.

	1986			1987		
Location	Spring	Summer	Fall	Winter	Spring	Summer
Crossfield/ Madden	normal	normal	normal	above	near normal	normal
Barrhead/ Campsie	normal	normal	normal	above	normal	normal
Cold Lake	normal	normal	normal	above	slightly above	normal

The simulated soil water budget is partially driven by the amount of incoming solar radiation which is absorbed by bare soil between the crop in the stand. To compute this absorption, the soil albedo, which is the proportion of solar radiation reflected by the soil surface, is reported to vary from 5 to 15% (from moist to dry soil) for "dark soils" (Sellers 1965), such as for Black Chernozem soils in the Crossfield area, to between 10 to 20% for "soil, moist gray", and between 20 to 35% for "soil, dry clay or gray" (Sellers 1965), such as for gray luvisols in the Cold Lake and Barrhead areas. To prevent the model from becoming more complex, a single value of 0.20 (20%) was used for the soil albedo, and from Monteith (1959), and Chang (1968), albedo for the alfalfa crop was set at 0.26. At the beginning of the simulation, the soil albedo value controls the absorption of solar radiation. As the crop grows on a daily basis, the increase in the leaf area index causes a linear increase in the field albedo from that for the bare soil to a combination between soil and crop albedo up to a point at which the leaf area index is 4. At this time, solar radiation absorption is controlled by the alfalfa albedo because of canopy closure.

9.3.1.4 Sulphur Dioxide Exposure-Response Models

Sulphur dioxide exposure-plant response data for alfalfa growth and production were obtained from the literature review performed for this research program (Torn et al. 1987).

Estimated hourly sulphur dioxide concentrations for 1986 and 1987 alfalfa growth seasons were obtained from an air quality model (Chapter 9.2) applied for three different locations in the vicinities of each of the following locations:

- Barrhead (Emission Region 7)
- Crossfield (Emission Region 9).

The next step in the present analysis was to compare the estimated hourly SO_2 concentrations against the available knowledge based on SO_2-exposure-alfalfa-response relationships. If the estimated ambient concentrations were found to be in the range of concentrations known to cause negative effects on alfalfa growth and productivity, the computerized mathematical modelling approach for alfalfa growth simulation would then be modified to include an ambient SO_2 exposure-response function to determine the magnitude of the air pollutant effect on seasonal alfalfa yields.

A special, unique SO_2-exposure-alfalfa-response relationship was applied as found in Krupa (1987). In a central Minnesota Power Plant study of alfalfa growth under ambient sulphur dioxide pollution, Krupa found that the foliage production per unit stem had a strong statistical association to certain pollutant exposure statistics and certain ground-level meteorological statistics. While Krupa presented multivariate models for 1985, 1986, and both years combined in Minnesota, since the present assessment involved SO_2 data in Alberta for 1986 and 1987, only Krupa's 1986 statistical model was applied. The performance of this model in Minnesota showed that the SO_2 exposure and meteorological variables accounted for 97% of the variation in the biological response variable. An example of the Krupa 1986 model, applied to Barrhead, Location #2, during 1986, for harvest #1, is:

Leaf dry matter/100 Stems

= 111.369 -6.03 *EPIS1 + 1.26 *PEAK1 + -2.6 *TMAX1 + -0.97 *TMIN1 + -0.14 *PREC1 + 2.84 *$TMIN_2$ + -1.99 *TMAX3 + 0.27 *EPIS1**2 + -0.02 *PEAK1**2 + -0.0004 *INT1**2 + 0.001 *INT2**2 = 9.76 grams/ 100 stems

where, in order of their importance to the R^2 value of 0.97, EPIS1 = the number of SO_2 episodes in the first alfalfa growth phase (May 15 - May 31); PEAK1 = the maximum hourly concentration of SO_2 (ppb) during the first growth phase; TMAX1 = the weekly average maximum air temperature (°C) during the first growth phase; TMIN1 = the weekly average minimum air temperature (°C) during the first growth phase; PREC1 = the total precipitation (mm) during the first growth phase; $TMIN_2$ = the weekly average minimum air temperature during the second growth phase (June 1 - June 16); TMAX3 = the weekly average maximum air temperature (°C) during the third growth phase (June 17 - June 30); and INT1 and INT2 = the integrated total sum of SO_2 concentrations (ppb-h) over each of the respective alfalfa growth phases.

The model estimates of SO_2, provided in Table 9.4 show PEAKS, EPISodes, and INTegrals for hours with concentrations equal or exceeding 6 ppb. The data are for a total of three SO_2 sites at Crossfield and Barrhead, for three alfalfa growth phases per harvest during 1986 and 1987. The duration of the growth phases was determined by dividing the growth period prior to a harvest into three equal parts to reflect vegetative growth changes (Krupa, 1987). These SO_2 exposure statistics, together with the appropriate meteorological variables obtained from the corresponding weather data sets mentioned previously, were used to apply the Krupa 1987 multivariate model to the Crossfield area in Emission Region 9 and the Barrhead area in Emission Region 7.

9.3.1.5 Results and Discussion

Alfalfa Growth Simulation for 1986-87 Without Air Pollutants.

The response of the ALSIM 1 alfalfa growth model to various weather data sets is shown in Figures 9.8 through 9.22. Units used are "tons/acre" because these are the units used by Canadian growers and in field reports. Figure 9.8 shows the response to average monthly weather data during 1986 and 1987. The initial growth of the stand is shown from day 120 to around day 180 (the first and only harvest during 1986). Regrowth occurred rapidly, but with no second harvest, it simply ceased by around day 295 when the top growth was frost-killed. The first and second harvests are clearly traceable for the second year (1987) in Figure 9.8. During that year, downward loss of soil moisture was greater, and the number of days of water stress increased at a greater rate than in 1986.

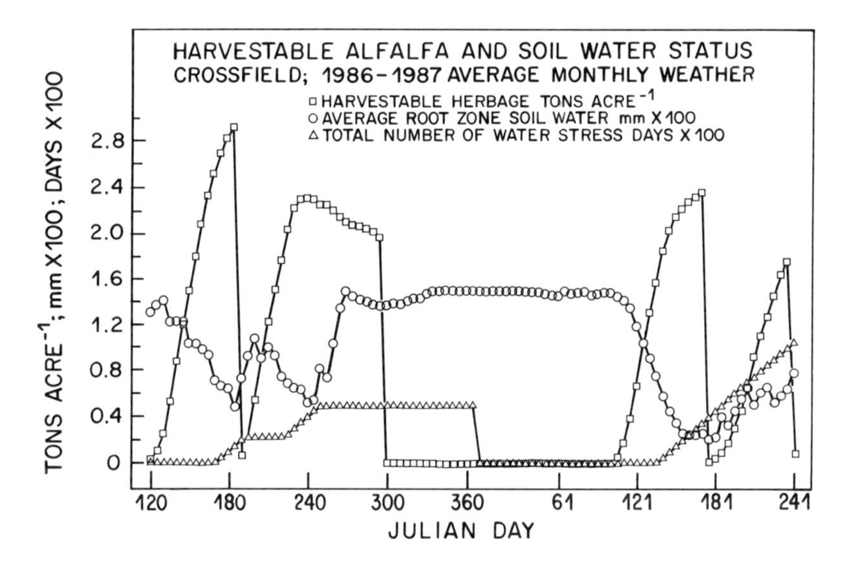

Figure 9.8 Harvestable alfalfa, available root zone water, and number of water stress days from ALSIM 1 simulation model every five days for Crossfield area, Emission Region 9, from estimated date of emergence for a new stand on April 27, 1986, through August 24, 1987, based on 1986-87 average monthly weather data.

The Barrhead area (Figure 9.9) shows a similar response except that there were two harvests in 1986, after which a smaller pre-frost shoot regrowth occurred before the first frost. Relative to the 1986-87 average monthly weather, Cold Lake shows an early soil moisture deficit, a number of days rapidly occurring with water stress, and less alfalfa shoot growth (Figure 9.10).

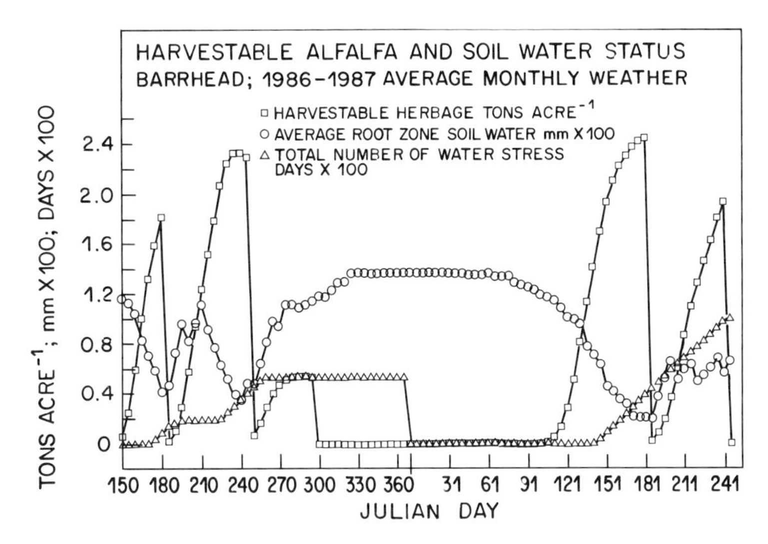

Figure 9.9 Harvestable alfalfa, available root zone water, and number of water stress days from ALSIM 1 simulation model every five days for Barrhead area, Emission Region 7, from estimated date of emergence for a new stand on May 27, 1986, through September 2, 1987, based on 1986-87 average monthly weather data.

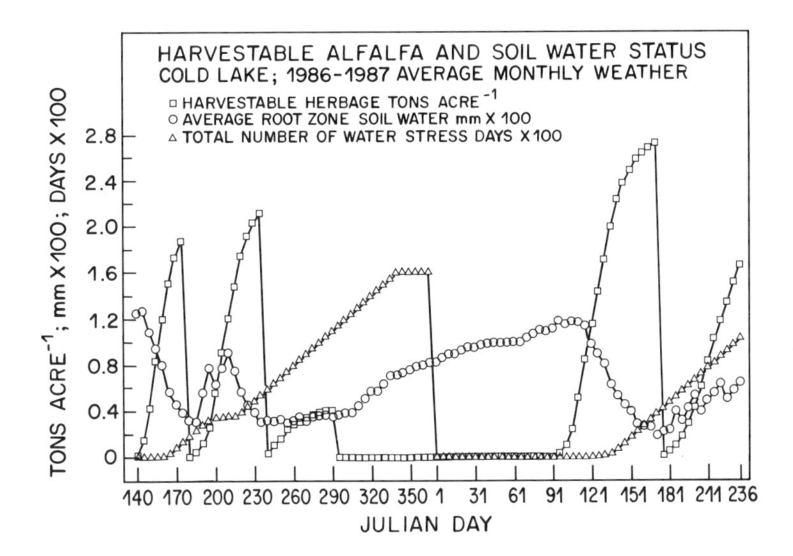

Figure 9.10 Harvestable alfalfa, available root zone water, and number of water stress days from ALSIM 1 simulation model every five days for Cold Lake area, Emission Region 6, from estimated date of emergence for a new stand on May 19, 1986, through August 24, 1987, based on 1986-87 average monthly weather data.

The response of the model to long-term normal average monthly weather is shown in Figures 9.11 through 9.13. A number of notable differences can be seen in the temporal dynamics of the alfalfa stand growth between 1986-87 and average monthly data versus the long-term normal weather.

Under a scenario that gives much brighter, warmer, and wetter weather (plus one standard deviation above normal), Barrhead shows the greatest change, especially in the second year, while crop growth in the Cold Lake area is not seen to change much (Figures 9.14 to 9.16). Alternatively, a scenario that gives much darker, colder, and drier weather, shows a drastic soil water deficit and crop water stress with a very diminished harvest in the Crossfield area (Figure 9.17), and a single-year crop of one harvest at Barrhead (Figure 9.18). Cold Lake alfalfa growth shows the least change, except that the second year can really support only one harvest (Figure 9.19).

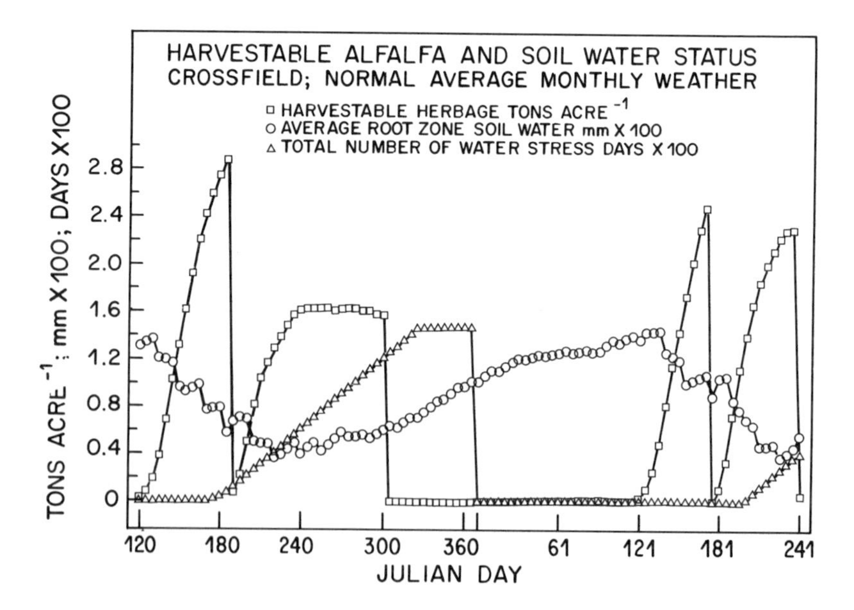

Figure 9.11 Harvestable alfalfa, available root zone water, and number of water stress days from ALSIM 1 simulation model every five days for Crossfield area, Emission Region 9, from estimated date of emergence for a new stand on April 27, 1986, through August 24, 1987, based on long-term normal average monthly weather data.

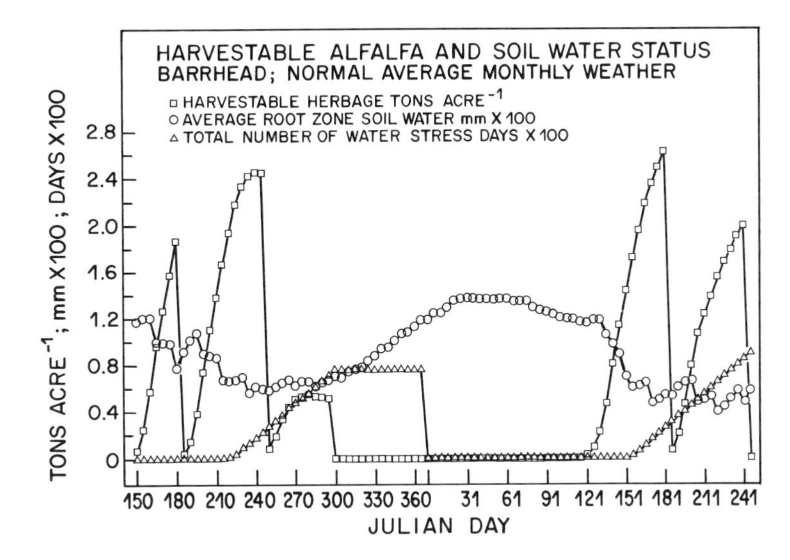

Figure 9.12 Harvestable alfalfa, available root zone water, and number of water stress days from ALSIM 1 simulation model every five days for Barrhead area, Emission Region 7, from estimated date of emergence for a new stand on May 27, 1986, through September 2, 1987, based on long-term normal average monthly weather data.

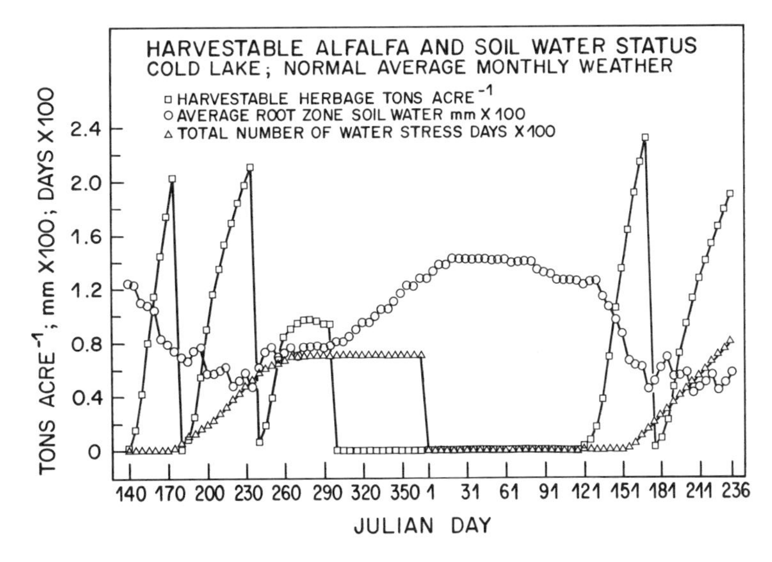

Figure 9.13 Harvestable alfalfa, available root zone water, and number of water stress days from ALSIM 1 simulation model every five days for Cold Lake area, Emission Region 6, from estimated date of emergence for a new stand on May 19, 1986, through August 24, 1987, based on long-term normal average monthly weather data.

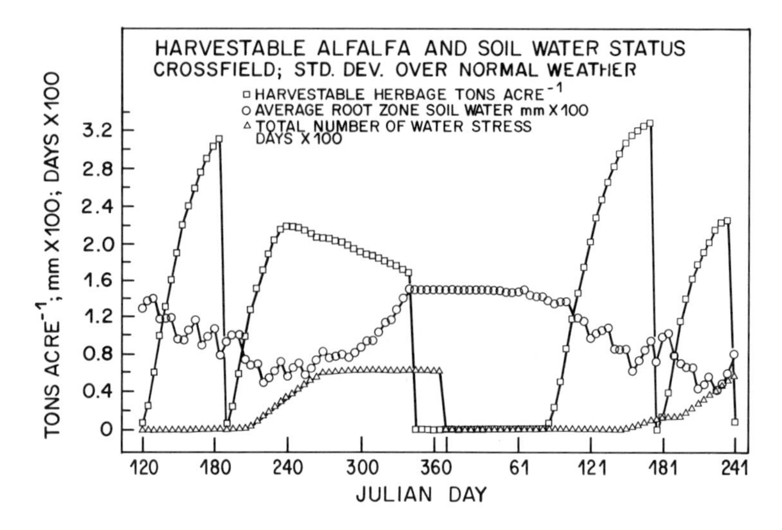

Figure 9.14 Harvestable alfalfa, available root zone water, and number of water stress days from ALSIM 1 simulation model every five days for Crossfield area, Emission Region 9, from estimated date of emergence for a new stand on April 27, 1986, through August 24, 1987, based on long-term normal plus one standard deviation average monthly weather data.

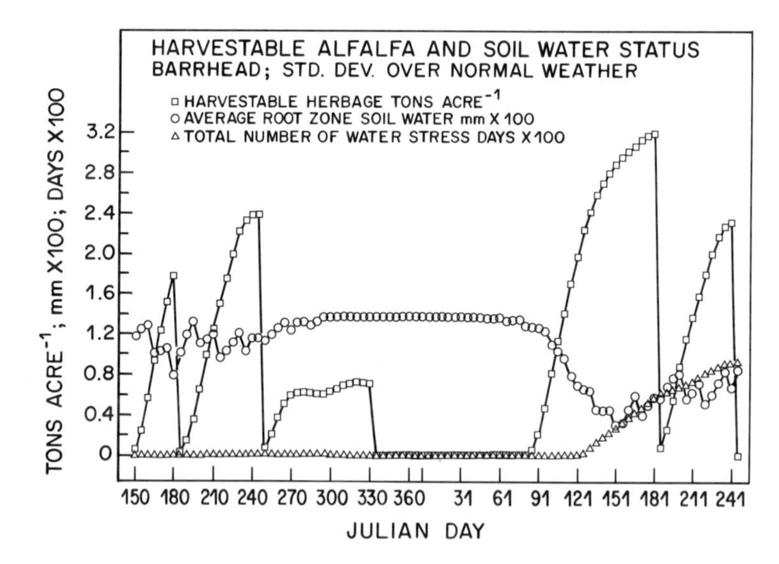

Figure 9.15 Harvestable alfalfa, available root zone water, and number of water stress days from ALSIM 1 simulation model every five days for Barrhead area, Emission Region 7, from estimated date of emergence for a new stand on May 27, 1986, through September 2, 1987, based on long-term normal plus one standard deviation average monthly weather data.

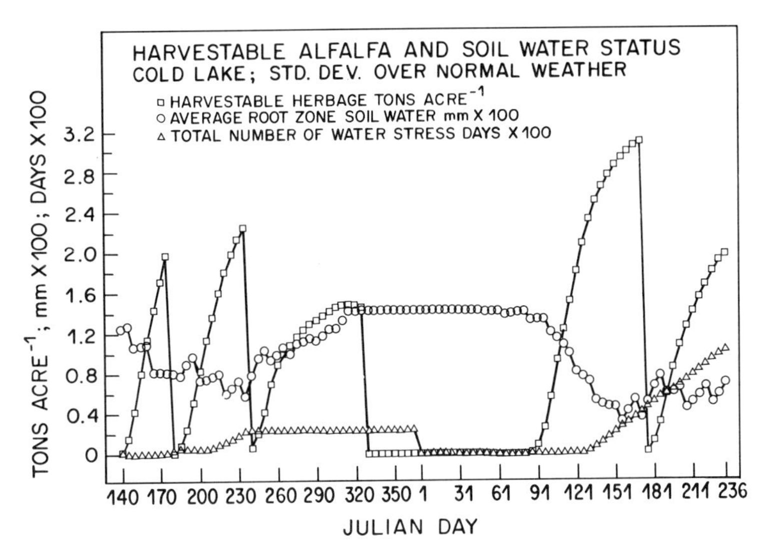

Figure 9.16 Harvestable alfalfa, available root zone water, and number of water stress days from ALSIM 1 simulation model every five days for Cold Lake area, Emission Region 6, from estimated date of emergence for a new stand on May 19, 1986, through August 24, 1987, based on long-term normal plus one standard deviation average monthly weather data.

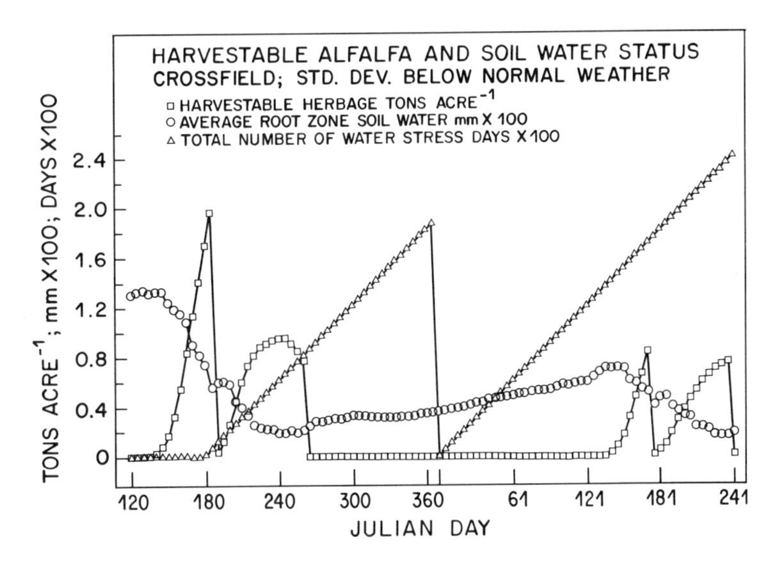

Figure 9.17 Harvestable alfalfa, available root zone water, and number of water stress days from ALSIM 1 simulation model every five days for Crossfield area, Emission Region 9, from estimated date of emergence for a new stand on April 27, 1986, through August 24, 1987, based on long-term normal less one standard deviation average monthly weather data.

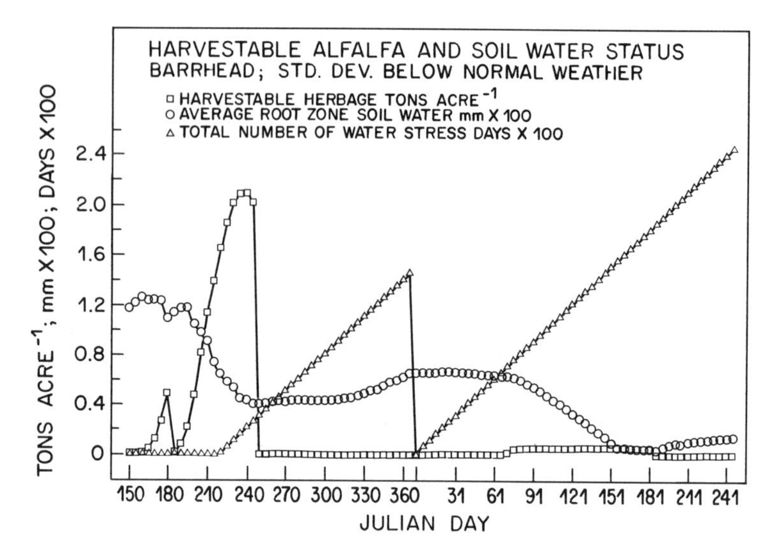

Figure 9.18 Harvestable alfalfa, available root zone water, and number of water stress days from ALSIM 1 simulation model every five days for Barrhead area, Emission Region 7, from estimated date of emergence for a new stand on May 27, 1986, through September 2, 1987, based on long-term normal less one standard deviation average monthly weather data.

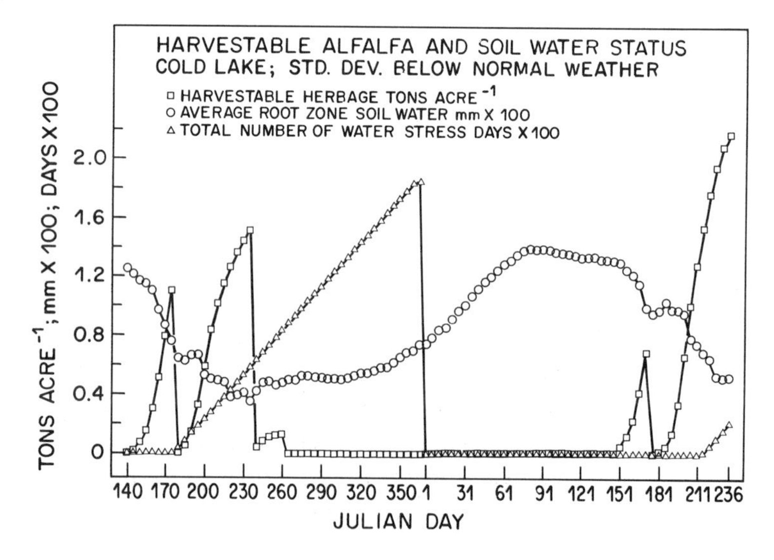

Figure 9.19 Harvestable alfalfa, available root zone water, and number of water stress days from ALSIM 1 simulation model every five days for Cold Lake area, Emission Region 6, from estimated date of emergence for a new stand on May 19, 1986, through August 24, 1987, based on long-term normal less one standard deviation average monthly weather data.

Figures 9.20 through 9.22 show the most realistic response, since actual 1986-87 daily weather data were used to drive the simulated system. Second season growth at Crossfield shows two springtime starts that were set back by late frosts. Cold Lake alfalfa growth shows the effect of a relatively warm, dry winter on poor growth in the second year (Figure 9.22).

All of the harvested alfalfa yields from the various weather-driven simulations are shown together in Table 9.13, along with the reported yields from the field in 1986 and 1987. It is important to realize that the field reports of observed alfalfa yields were obtained by contacting various field personnel as shown in the footnotes to Table 9.13. These individuals were not affiliated with the ADRP, nor do we know exactly how they arrived at the yield values they provided. Since no biological effects field research was included in the assessment part of the ADRP, this was the best available data on real yield for comparison to the model predictions.

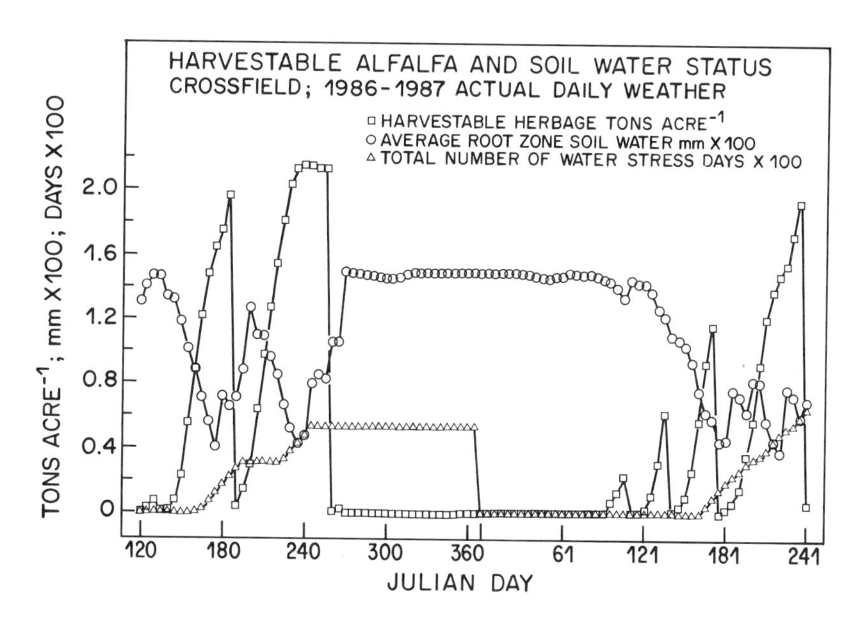

Figure 9.20 Harvestable alfalfa, available root zone water, and number of water stress days from ALSIM 1 simulation model every five days for Crossfield area, Emission Region 9, from estimated date of emergence for a new stand on April 27, 1986, through August 24, 1987, based on 1986-87 daily weather data.

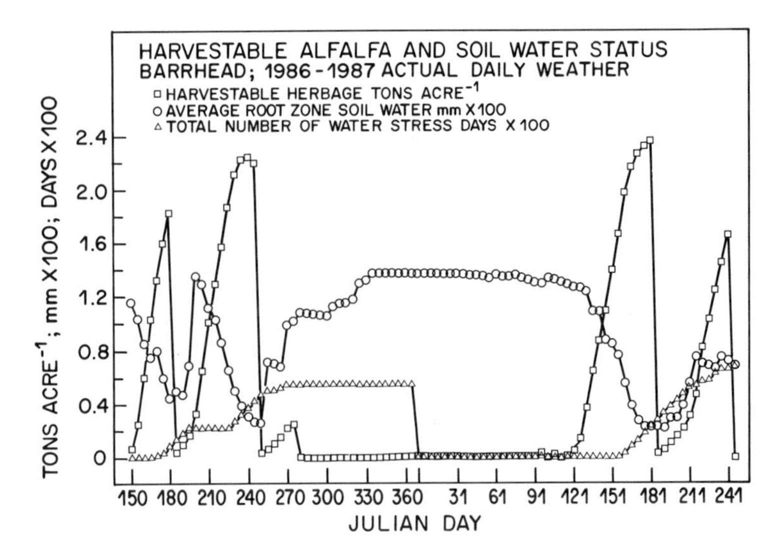

Figure 9.21 Harvestable alfalfa, available root zone water, and number of water stress days from ALSIM 1 simulation model every five days for Barrhead area, Emission Region 7, from estimated date of emergence for a new stand on May 27, 1986, through September 2, 1987, based on 1986-87 daily weather data.

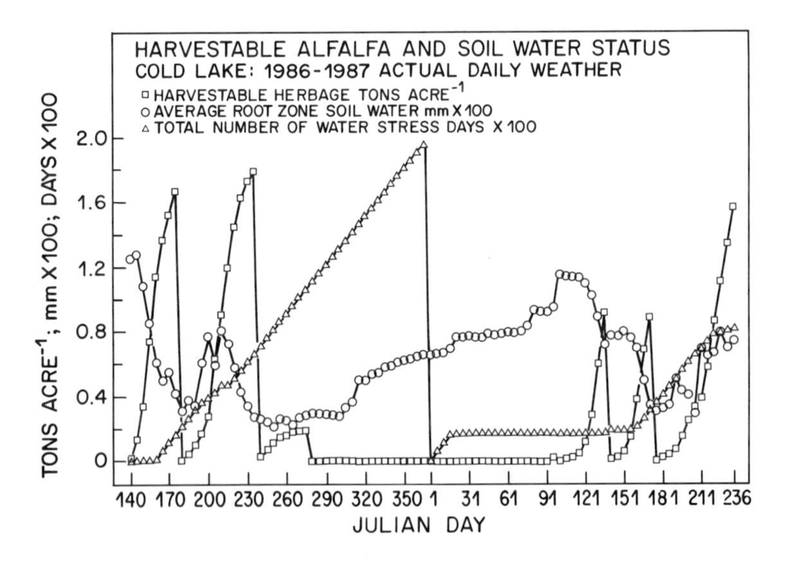

Figure 9.22 Harvestable alfalfa, available root zone water, and number of water stress days from ALSIM 1 simulation model every five days for Cold Lake area, Emission Region 6, from estimated date of emergence for a new stand on May 19, 1986, through August 24, 1987, based on 1986-87 daily weather data.

Table 9.13 Computer model simulated, and field reported, harvested yields (tons $acre^{-1}$) of alfalfa in Alberta, Canada, under various nonpollution weather regimes.

Location	Field Reported	Weather Patterns Used for Alfalfa Crop Growth Simulations				
		Actual[1] 1986-87 Monthly Average	Long-Term Normal Monthly Average	Actual 1986-87 Daily Data	Long-Term Normal+1 Std. Dev. Mon. Avg.	Long-term Normal-1 Std. Dev. Mon. Avg.
Crossfield						
Yr 1 Cut 1[2]	2.5[3]	2.94	2.91	2.00	3.12	1.99
Yr 2 Cut 1	0.5[4]	2.39	2.63	1.29	3.30	0.97
Yr 2 Cut 2	0.75-1.0[4]	1.76	2.31	1.94	2.26	0.75
Total	1.25-1.5[4]	4.15	4.94	3.23	5.56	1.72
Barrhead						
Yr 1 Cut 1	0.5	1.84	1.92	1.86	1.83	0.55
Yr 1 Cut 2	1.1[5]	2.30	2.44	2.20	2.39	2.02
Total	1.7[5] 2.0-2.6[6]	4.14	4.36	4.06	4.22	2.57

Continued...

Table 9.13 (Continued).

Location	Field Reported	Weather Patterns Used for Alfalfa Crop Growth Simulations				
		Actual[1] 1986-87 Monthly Average	Long-Term Normal Monthly Average	Actual 1986-87 Daily Data	Long-Term Normal+1 Std. Dev. Mon. Avg.	Long-Term Normal-1 Std. Dev. Mon. Avg.
Barrhead						
Yr 2 Cut 1	1.1[5]	2.45	2.62	2.37	3.20	0.05
Yr 2 Cut 2	0.9[5]	2.00	2.02	1.75	2.32	0.00
Total	2.0[5] 2.0-2.6[6]	4.45	4.64	4.12	5.52	0.05
Cold Lake						
Yr 1 Cut 1	-	1.95	2.25	1.77	2.18	1.32
Yr 1 Cut 2	1.76[7]	2.11	2.12	1.80	2.26	1.52
Total	2.0-2.4[8]	4.06	4.37	3.57	4.44	2.84
Yr 2 Cut 1	1.5-1.9[9]	2.75	2.39	0.96	3.13	0.88
Yr 2 Cut 2	1.4-1.7[10] 1.0[7]	1.66	1.89	1.57	1.96	2.18
Total	3.2-3.3[11]	4.41	4.28	2.53	5.09	3.06

Continued...

Table 9.13 (Concluded).

[1] Refers to solar radiation, air temperature, and precipitation data used.
[2] Only one cutting reported for 1986 in the Crossfield area.
[3] Larry Welsh, personal communication; for mixtures, not pure stands.
[4] Local crop specialist, personal communication; for mixtures, not pure.
[5] Fred Boise, personal communication.
[6] Alfalfa production plant operator for Legal and Barrhead areas, personal communication.
[7] Local grower producing very high crude protein, pure stand alfalfa for sale from second cutting; first cutting used as silage and not recorded; personal identity kept anonymous.
[8] Angus variety: 2.0; Beaver variety: 2.3; Rambler (creeping root) variety: 2.4; (Bjorn Berg, personal communication); pure stands.
[9] Algonquin (tap root) variety: 1.5; Rambler variety: 1.9; (Bjorn Berg, personal communication); pure stands.
[10] Rambler variety: 1.4; Algonquin variety: 1.7; (Bjorn Berg, personal communication); pure stands.
[11] Algonquin variety: 3.2; Rambler variety: 3.3; (Bjorn Berg, personal communication); pure stands.

Reports from Crossfield are for mixtures of alfalfa with other grasses. Therefore, they should be multiplied by 1.5 or 2.0 to compare against model simulated yields under "Actual 1986-87 Daily Data". With this modification, generally good agreement was observed between the model simulated yields (under no pollution) using actual daily 1986-87 weather data and the reported yields from the field at both Crossfield and Cold Lake. At Barrhead, however, reported yields from the field were only one-half the value that the ALSIM 1 alfalfa growth model simulated. Since ALSIM 1 responds to changes in sunshine, temperature, and water availability to the crop, this discrepancy was most likely caused by some other factor(s).

When individual harvest yield values and total year yield values for Crossfield and Cold Lake from the field reports of observers are plotted against the corresponding values produced by the ALSIM 1 model, the pattern is shown in Figure 9.23. This is a relatively good agreement. The R^2 value is 0.76, and the hypothesis that there is no difference between the observed (reported) values and the model-predicted values ($Y=X$) cannot be rejected at even the one percent probability level ($p=0.002$). Although the sample size is small, this is a better fit of ALSIM 1 to field values than any of the previously published results such as in Fick (1981), and Parsch (1987). Comparative results for the field test of ALSIM 1 for 4-cut systems in Michigan, recomputed by Kickert from data in Parsch (1987) are plotted in Figure 9.24.

Alfalfa Growth Conditions and Sulphur Dioxide Levels

Table 9.14 shows the SO_2 concentrations (ppb), over chronic or episodal portions of the cropping season, at which various alfalfa responses have been demonstrated. It should be noted that data indicating a positive, fertilization effect have not been emphasized here since this assessment is intended to evaluate whether any negative effects on alfalfa production might be occurring at the present levels of ambient sulphur dioxide on a regional scale within Alberta.

Most of the published information on alfalfa growth response under sulphur dioxide exposure refers to the average concentration of SO_2 over the growing season. Table 9.14 lists the various published exposure-response relationships. With respect to whole season average concentrations, it appears from this literature that

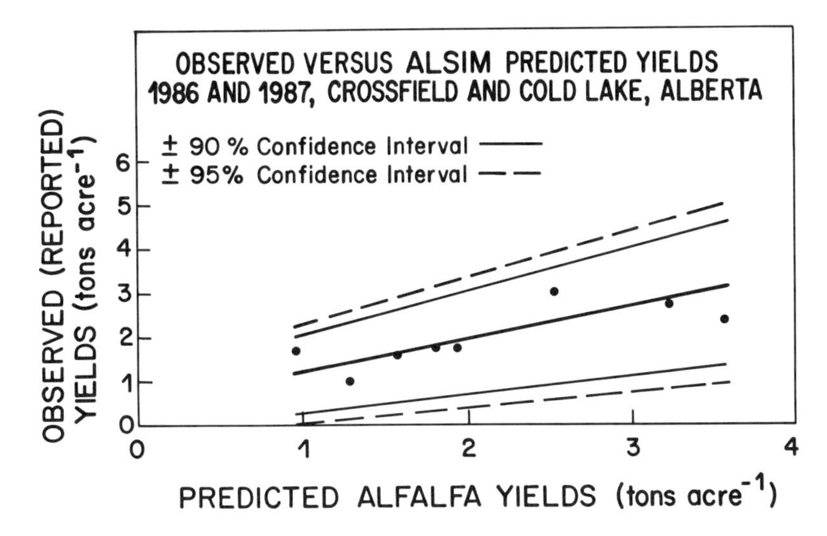

Figure 9.23 Reported alfalfa field yields compared to predicted yields from ALSIM 1 (Level 2) simulation model applied to the Crossfield and Cold Lake areas in Alberta for 1986 and 1987. Note that data for Barrhead, Alberta have not been included in this analysis. Data points include individual harvests and entire year harvest totals. Regression equation is Y = 0.71 X + 0.43: R^2 = 0.76.

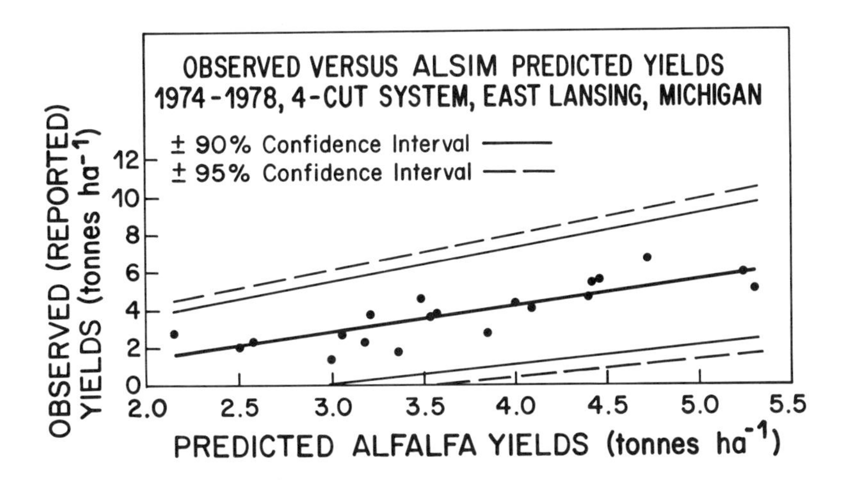

Figure 9.24 Observed versus predicted alfalfa yields from ALSIM 1 (Level 2) simulation model for 4-cut systems in East Lansing, Michigan. Regression equation is Y = 1.42X - 1.53; R^2 = 0.68. Data from Parsch (1987).

Table 9.14 SO_2 concentrations (in increasing order) over the whole, or portions, of the field growing season at which effects on alfalfa growth or productivity have been documented.

Effect/ Response	Average SO_2 Conc. (ppb)	Effect Parameter	Publication
threshold	7.6	Alfalfa yield reduction	p. 76, Torn et al. (1987), [Guderian (1977)]
-4.4%	10	Alfalfa yield reduction	p. 79, Torn et al. (1987), [Guderian (1977)]
0.0%	15	Alfalfa yield change	p. 78, Torn et al. (1987), [Godzik and Krupa (1982)]
-2.6%	20	Alfalfa yield reduction	p. 79, Torn et al. (1987), [Guderian (1977)]
-0.8%	29	Alfalfa yield change	p. 78, Torn et al. (1987), [Godzik and Krupa (1982)]
+13%	< 30 median/27d	Alfalfa yield change	Male et al. (1983)
threshold	32.7	Alfalfa yield reduction	p. 76, Torn et al. (1987), [Godzik and Krupa (1982)]
-1.8%	36	Alfalfa yield change	p. 78, Torn et al. (1987), [Godzik and Krupa (1982)]
+0.4%	38	Alfalfa yield change	p. 78, Torn et al. (1987), [Godzik and Krupa (1982)]
+9%	> =30 & <50 median/27d	Alfalfa yield change	Male et al. (1983)
-1.4%	40	Alfalfa yield change	p. 78, Torn et al. (1987), [Godzik and Krupa (1982)]
-12.0%	47	Alfalfa yield change	p. 78, Torn et al. (1987), [Godzik and Krupa (1982)]
-6.9%	51	Alfalfa yield reduction	p. 79, Torn et al. (1987), [Guderian (1977)]
-14.3%	58	Alfalfa yield change	p. 78, Torn et al. (1987), [Godzik and Krupa (1982)]
-18.0%	60	Alfalfa yield change	p. 78, Torn et al. (1987), [Godzik and Krupa (1982)]
-23.7%	62	Alfalfa yield change	p. 78, Torn et al. (1987), [Godzik and Krupa (1982)]

Continued...

Table 9.14 (Concluded).

Effect/ Response	Average SO_2 Conc. (ppb)	Effect Parameter	Publication
-4%	> =50 & <80 median/27d	Alfalfa yield change	Male et al. (1983)
-21.7%	68	Alfalfa yield change	p. 78, Torn et al. (1987), [Godzik and Krupa (1982)]
-30.0%	79	Alfalfa yield change	p. 78, Torn et al. (1987), [Godzik and Krupa (1982)]
-30.3%	82	Alfalfa yield change	p. 78, Torn et al. (1987), [Godzik and Krupa (1982)]
-19.0%	83	Alfalfa yield reduction	p. 79, Torn et al. (1987), [Guderian (1977)]
-23%	> =80 & <120 median/27d	Alfalfa yield change	Male et al. (1983)
-19.0%	83	Alfalfa yield reduction	p. 57, Torn et al. (1987), [Guderian and Stratmann (1968)]
-36.2%	141	Alfalfa yield reduction	p. 79, Torn et al. (1987), [Guderian (1977)]
-49%	> =120 & <170 median/27d	Alfalfa yield change	Male et al. (1983)
-76%	> =170 median/27d	Alfalfa yield change	Male et al. (1983)
variable[1]		Alfalfa leaf dry wt per 100 stems in harvest period	Krupa (1987)
-1.00%	241	Threshold for leaf destruct. over entire season (1008 hrs)	Stevens & Hazleton (1976)

[1]0 to 46 episodes, with 10 to 128 ppb peak concentration, and 0 to 459 ppb integrated exposure over first, second, and third growth phases within a harvest period.

negative effects on biomass production only begin somewhere within the range of 10 to 40 ppb. In fact, Torn et al. (1987), referring to SO_2 exposures and agricultural crops, state "To avoid deleterious effects on growth and yield to agricultural crops, average concentrations of ambient SO_2 should not exceed 0.01 ppm (10 ppb) and hourly average values should not exceed 0.06 ppm (60 ppb)."

Recently, several investigators have pointed out the problems of lack of realism in using "average" values of air pollutant concentrations computed over extended time periods in examining biological responses and possible effects. The statistic called the "average" is only a valid measure of the central tendency of a statistical distribution, such as from a time series, if that distribution is normally (Gaussian) distributed. When the latter is not the case, then the "average" is not a realistic statistic of the central tendency. Modelled ambient SO_2 concentrations used in this assessment for Alberta were not normally distributed. In addition, as an analogy, even though the average flow of water in a stream over several months might be several thousands of cubic metres of water per hour, it only requires one hour of zero-flow to kill many fish. The reality of this effect would never be detectable in the average stream flow over several months.

Hence, other statistical measures describing the air pollutant stress in vegetation effects assessment, such as the median of the concentration distribution, the peak (maximum) value, the total, integral concentration to which receptor organisms are exposed over some exposure time period, and the number of times they are exposed, should all be examined.

Two other approaches listed in Table 9.14, which attempt to avoid the problems associated with using a whole season average value to characterize the SO_2 exposure,were considered:

- the use of the 27-day median concentration for daylight hours from 0500 through 2000, as used by Male et al. (1983);
- the use of number of air pollutant episodes, peak (maximum) concentration, and the total integral of all concentrations, for each of three growth phases of the alfalfa crop for each harvest, as used by Krupa (1987).

For the Crossfield and Barrhead areas, Table 9.15 presents results of sulphur dioxide exposure characteristics during the respective alfalfa shoot growth seasons. The highest hourly seasonal average for both areas, both years, and all locations during the alfalfa growth seasons is 1.91 ppb, far below the levels shown in

Table 9.15 Sulphur dioxide exposure statistics for 1986-87 alfalfa growing season, Alberta, Canada (units are ppb).

Area	Location	Growing Season	Hourly Average	24-hour Median	5am-8pm Median Peak	Max. hourly Peak	Integrated Exposure (ppb-h)
Crossfield 1986	1	Day 117-186[1]	1.91	0.09	5.78 13.37[2]	69.39	3204
	2	Day 117-186	1.24	0.18	1.98	30.35	2083
	3	Day 117-186	1.12	0.00	2.06	39.91	1882
Crossfield 1987	1	Day 105-236[1]	1.79	0.05	12.89	79.21	5807
	2	Day 105-236	1.23	0.17	2.43	34.07	3907
	3	Day 105-236	1.11	0.00	1.28	45.86	3519
Barrhead 1986	1	Day 147-245[1]	0.75	0.06	2.97	22.77	1785
	2	Day 147-245	0.44	0.02	0.73	15.11	1043
	3	Day 147-245	0.87	0.04	0.92	35.42	2078
Barrhead 1987	1	Day 147-245	0.78	0.06	3.16	19.99	1861
	2	Day 147-245	0.42	0.06	1.24	19.99	1001
	3	Day 147-245	0.77	0.01	3.07	30.13	1823

[1] Estimated date of emergence to reported date of final harvest.
[2] On day 198, after the single 1986 harvest.

Table 9.14 to be capable of producing negative effects on alfalfa yield.

Table 9.16 shows the SO_2 exposure-alfalfa response data for 27-day medians from temporally fluctuating exposures of sulphur dioxide (Male et al. 1983). Figure 9.25 shows a graph of this relationship, and as a comparison, the curve for alfalfa yield response to seasonal average values originally published by Warteresiewicz (1979), and cited by Godzik and Krupa (1982). The median-curve from Male et al. shows a positive fertilization phase within the range of 0 to about 62 ppb, after which negative effects occur, accelerating beyond a 27-day median of 100 ppb.

Table 9.15 shows that the peak value for all model generated daily median hourly SO_2 concentrations between 0500 through 2000 was only about 12.89 ppb (Crossfield, Location #1, 1987). Because of this, the median approach as used by Male et al. (1983) would only indicate a fertilization effect under the modelled SO_2 exposures during 1986-87 in Alberta Emission Regions 7 and 9.

The Krupa (1987) multivariate approach mentioned previously has many risks when extrapolating to other locations and other years than those for which the equation was derived. For this reason, a comparison of the range of data between the original Minnesota 1986 data set used to derive this statistical model, and the corresponding ranges for Alberta locations and years are shown in Table 9.17. In applying this model to 1986-87 Alberta SO_2 and meteorological data, there were time periods for which the Alberta data occurred outside the range of the respective input variable used in Minnesota. For example, for specific alfalfa growth phases at Crossfield, the number of exposure episodes, the total integral of the concentrations, the maximum and minimum air temperature, and precipitation (although only by 2 mm; 88 versus 86) were outside the range. Similar exceedances for the Barrhead area were observed only for the minimum air temperature (0.7 versus 2.1°C) and for precipitation (134 versus 86 mm).

Bearing in mind these risks in applying the Krupa 1986 Minnesota statistical model to Alberta 1986-87 conditions and using the modelled SO_2 exposure statistics (refer to Chapter 9.2), the results are shown in Table 9.18. To obtain an estimate of alfalfa growth "without" air pollution, and yet remain within the ranges for the data (Table 9.17) used by Krupa to define the model, minimal values consisting of the following were used:

Table 9.16 Effect of stochastic SO_2 exposure on alfalfa growth. (Male et al. 1983).

27 Day Median Stochastic Concent. (ppb)	Per Pot Mean Alfalfa Yield (g)	27-day Fract. Change	Per Pot Mean Alfalfa Yield/day (g day^{-1})	(Optimal Growth Conditions) Total Time Period 27 days: Equiv. Field Yield (kg m^{-2})	Equiv. Field Yield (lbs ac^{-1})	Equiv. Field Yield (tons ac^{-1})	Net Change (tons ac^{-1})
0	23	0.00	0.852	0.6284	5586	2.79	0.00
15	26	0.13	0.963	0.7104	6315	3.16	0.36
40	25	0.09	0.926	0.6831	6072	3.04	0.24
65	22.1	-0.04	0.819	0.6038	5367	2.68	-0.11
100	17.6	-0.23	0.652	0.4809	4274	2.14	-0.11
145	11.7	-0.49	0.433	0.3197	2842	1.42	-1.37
190	5.5	-0.76	0.204	0.1503	1336	0.67	-2.13

27 days: June 14, 1977 to July 12, 1977.
Pot area: $(10.8)^2$ (3.1416) = 366 cm^2.

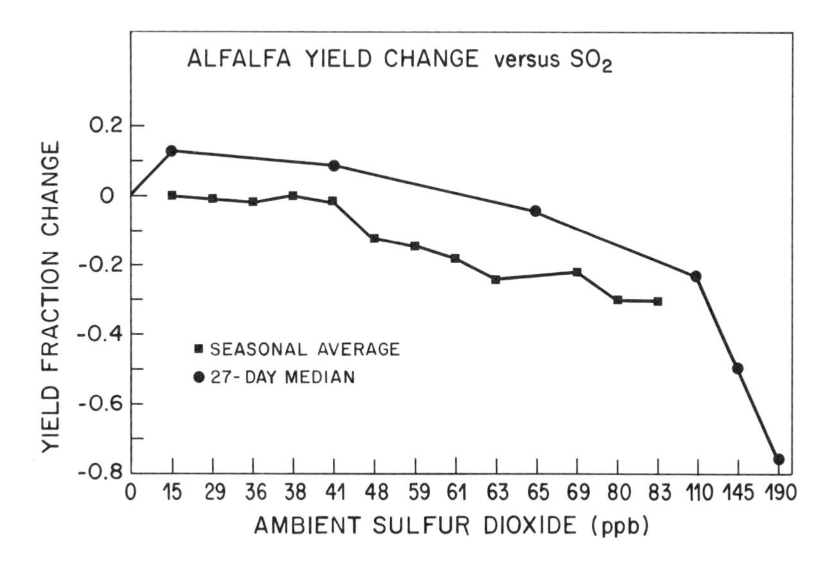

Figure 9.25 Sulphur dioxide exposure-alfalfa yield response curves based on (a) seasonal average concentration from Warteresiewicz (1979) reported in Godzik and Krupa (1982), and (b) 16-hour daytime median concentration over 27 days (from Male et al. 1983). Note: horizontal axis is variable, not a constant increment.

Table 9.17 Ranges of data used to apply the Krupa multivariate sulphur dioxide exposure-response model for alfalfa to Crossfield and Barrhead areas in Alberta.

	EPIS (Number)	PEAK (ppb)	INT (ppb-h)	Tmax (C)	Tmin (C)	Prec (mm)
Krupa (1987) for 1986 in Minnesota	0-46	10-128	0-459	15.5-30.9	2.1-15.7	3-86
Crossfield/ Madden 1986-87	13-49	14-79	144-1347	11.1-23	-0.8-9.0	8-88
Barrhead/ Campsie 1986-87	1-25	8-35	8-373	18.4-23.2	0.7-10.9	15-134

Table 9.18 Comparison of alfalfa leaf dry weight in grams (g) per 100 stems without and with modelled ambient sulphur dioxide pollution using the Krupa multivariate exposure-response model (Krupa 1987).

				Alfalfa Leaf dry wt g 100 stems^{-1}		
Area	Location	Year	Harvest	Without	With SO_2	Change
Crossfield	1	1987	2	40.39	71.48	+
Crossfield	2	1987	1	22.66	60.23	+
Crossfield	2	1987	2	40.39	77.83	+
Crossfield	3	1987	1	22.66	17.77	-
Crossfield	3	1987	2	40.39	48.41	+
Barrhead	1	1986	1	23.86	37.59	+
Barrhead	1	1986	2	16.19	19.13	+
Barrhead	1	1987	1	35.42	44.12	+
Barrhead	1	1987	2	37.54	34.82	-
Barrhead	2	1986	1	23.86	9.76	-
Barrhead	2	1987	1	35.42	33.76	-
Barrhead	2	1987	2	37.54	36.33	-
Barrhead	3	1986	1	23.86	128.39	+
Barrhead	3	1987	1	35.42	32.67	-
Barrhead	3	1987	2	37.54	104.19	+

a) 1 episode for growth phase 1,
b) 10 ppb-h as the integral exposure during growth phases 1 and 2,
c) 10 ppb as the peak exposure during growth phase 1.

All the meteorological data for a given area, location, year, harvest, and growth phase were the same for the "without" and "with" calculations with respect to SO_2, and these were taken from the actual data records.

Results for some of the locations, years, and harvests, where the input exposure data were far outside the ranges used by Krupa to derive the model obviously caused very unrealistic results and, therefore, are not listed in Table 9.18. An example is Crossfield 1986.

The results in Table 9.18 cannot be treated in a quantitative fashion as if they were on an interval or ratio scale in terms of measurement scale concepts. They should be viewed only as ordinal scale indicators. This means that absolute differences or percentages for "alfalfa leaf dry wt. per 100 stems" - "without" and "with SO_2" should not be attempted. Information presented in Table 9.18 suggests that within the two Alberta Emission Regions 7 and 9, during some Alberta growth periods, SO_2 can have a fertilization effect on growth (on a per stem basis), and during other periods can have a negative effect. However, of the 15 cases listed, six location-year-harvest combinations are indicated to result in a negative effect, and the other nine combinations in a positive effect, so there is no consistent indication of a constant area-wide impact on alfalfa growth.

Since model-generated SO_2 concentrations were generally higher in the Crossfield area than at Barrhead, it is of interest to examine results for the two areas separately. In Region 9 (Crossfield), four of the five cases in Table 9.18 indicate a positive effect on growth. Such an effect is consistent with Krupa's original results for Minnesota, where he found the measured SO_2 exposures to be related to a modest increase in per-stem growth. It can be seen in Table 9.4 that the air quality model-generated exposure statistics for the Crossfield area were indeed in the same general range as those observed by Krupa in Minnesota.

For Barrhead, half of the ten model applications resulted in an indication of positive effect, and half showed a negative effect. It should be recalled (Table 9.13) that in the Barrhead area, reported field yields were far below those simulated by the ALSIM 1 model for pollution free conditions. This set of results suggests that alfalfa growth in the two regions of Alberta discussed in this book, especially in Emission Region 7, could be fluctuating back and forth from season to season between a positive fertilization response versus a negative response in terms of "leaf dry weight per 100 stems."

In summary, application of the Krupa alfalfa response-model in Alberta might not provide any conclusive evidence of continuous systematic negative impact on alfalfa growth due to SO_2 exposure. However, there are several factors that preclude a firm conclusion that no adverse SO_2 effects are occurring.

Actual short-term SO_2 peak concentrations at Crossfield (and possibly also in the Barrhead area) are occasionally higher than any concentrations used in development of the Krupa model. These high short-term (a few minute) concentrations could not be simulated by the model-generated one-hour statistics (Chapter 9.2.2.6). The maximum 10-minute SO_2 concentration observed at Crossfield by ADRP was 168 ppb (Table 4.18), significantly higher than the observed Minnesota peak of 128 ppb and the maximum model-generated one-hour peak of 79 ppb. Because these higher short-term concentrations are beyond the range of the Krupa response model, no estimate of their possible effects can be made.

The possible combined effects of SO_2 and ozone or other pollutants have not been evaluated because of the unavailability of exposure data and applicable response models.

The Krupa model applied here did suggest possible negative effects for some combinations of SO_2 exposure, climate, and growth period variables.

Definitive conclusions on potential adverse pollution effects on alfalfa,therefore, will require development of response models using exposure and growth data specific to Alberta pollution and crop conditions.

9.3.1.6 Concluding Remarks

There is some evidence, although only circumstantial, that in the Barrhead area of Emission Region 7 during certain alfalfa growth periods and harvests, ambient sulphur dioxide time series exposures, together with meteorological variables at the ground level, might be associated with negative effects on alfalfa biomass production.

Alfalfa yields were found to be lower in the Barrhead area than would be expected under conditions without air pollution. Further laboratory and experimental field research would be required to verify the conclusions derived from modelling. Attention should also be directed to the nutritional quality of the crop.

9.3.1.7 Literature Cited

Atmospheric Environment Service. Canada. 1982a. Canadian Climate Normals. 1951-1980. Vol. 1. Solar Radiation. Downsview, Ontario, Canada.

Atmospheric Environment Service. Canada. 1982b. Canadian Climate Normals. 1951-1980. Vol. 2. Temperature. Downsview, Ontario, Canada.

Atmospheric Environment Service. Canada. 1982c. Canadian Climate Normals. 1951-1980. Vol. 3. Precipitation. Downsview, Ontario, Canada.

Atmospheric Environment Service. Canada. 1986a. Solar Radiation Summary. Downsview, Ontario, Canada.

Atmospheric Environment Service. Canada. 1986b. Monthly Record. Downsview, Ontario, Canada.

Chang, J.H. 1968. Climate and Agriculture: an Ecological Survey. Aldine Publ. Co., Chicago, Ill. 304 pp.

Fick, G. 1981. ALSIM 1 (Level 2) - User's Manual. Mimeo 81-35. Department of Agronomy, Cornell University, Ithaca, New York. 41 pp.

Fick, G.W. 1984. Simple simulation models for yield prediction applied to alfalfa in the Northeast. *Agronomy Journal,* 76: 235-239.

Fick, G.W., and D. Onstad. 1981. Simple computer simulation models for forage-management applications. pp. 483-485. In: **Proc. XIV International Grassland Congress.**

Godzik, S., and S.V. Krupa. 1982. Effects of sulphur dioxide on the growth and yield of agricultural and horticultural crops. In: **Effects of Gaseous Air Pollution in Agriculture and Horticulture.**, eds., M.H. Unsworth and D.P. Ormrod, London: Butterworth Scientific. pp. 247-265

Goplen, B.P., H. Baenziger, L.D. Bailey, A.T.H. Gross, M.R. Hanna, R. Michaud, K.W. Richards, and J. Waddington. 1980. Growing and managing alfalfa in Canada. Publication No. 1705, Agriculture Canada, Ottawa. 49 pp.

Guderian, R. 1977. *Air Pollution - Phytotoxicity of Acidic Gases and Its Significance in Air Pollution Control.* New York: Springer-Verlag. 127 pp.

Guderian, R., and H. Stratmann. 1968. Freilandversuche zur Ermittlung von Schwefeldioxidwirkungen auf die Vegetation. III. Teil Grenzwerte schadlicher SO_2^- Immissionen für Obstund Forstkulturen sowie für landwirtschaftliche und gartnerische Pflanzenarten. Forschungsberichte des Landes Nordrhein Westfalen. 114 pp.

Krupa, S.V. 1987. Responses of alfalfa to sulfur dioxide exposures from the emissions of the NSP-SHERCO coal-fired power plant units 1 and 2. Vol. 1. Report submitted to Northern States Power Company, Minneapolis, Minnesota. 236 pp.

Krupa, S., and R.N. Kickert. 1987. An Analysis of Numerical Models of Air Pollutant Exposure and Vegetation Response. Prep. for the Acid Deposition Research Program and Kananaskis Centre for Environmental Research. University of Calgary, Calgary, Alberta, Canada. ADRP-B-10-87. 113 pp.

Male, L., E. Preston, and G. Neely. 1983. Yield response curves of crops exposed to SO_2 time series. *Atmospheric Environment*, 17(8): 1589-1593.

Monteith, J.L. 1959. The reflection of short-wave radiation by vegetation. *Quarterly Journal of the Royal Meteorological Society*, 85: 386-392.

Onstad, D., C.A. Shoemaker, and B.C. Hansen. 1984. Management of potato leafhopper, *Empoasca fabae (Homoptera: Cicadellidae)*, on alfalfa with the aid of systems analysis. *Environmental Entomology*, 13: 1046-1058.

Onstad, D. and G.W. Fick. 1981. ALSIM 1 (Level 2) in FORTRAN IV. Agronomy Mimeo 81-38. Department of Agronomy, Cornell University, NY. 10 pp.

Parsch, L.D. 1987. Validation of ALSIM 1 (Level 2) under Michigan conditions. *Agricultural Systems*, 25: 145-157.

Picard, D.J., D.G. Colley, and D.H. Boyd. 1987. Overview of the Emission Data: Emission Inventory of Sulphur Oxides and Nitrogen Oxides in Alberta. Prep for the Acid Deposition Research Program by Western Research, Division of Bow Valley Resource Services Ltd., Calgary, Alberta, Canada. ADRP-A-1-87. 87 pp.

Sawyer, A.J., and G.W. Fick. 1987. Potential for injury to alfalfa by alfalfa blotch leafminer (*Diptera: Agromyzidae*): simulations with a plant model. *Environmental Entomology*, 16: 575-585.

Sellers, W.D. 1965. *Physical Climatology.* University of Chicago Press, Chicago, Ill. 272 pp.

Sharratt, B.S., D.G. Baker, and C.C. Sheaffer. 1986. Climatic effect on alfalfa dry matter production. Part I. Spring harvest. *Agricultural and Forest Meteorology* 37: 123-131.

Sharratt, B.S., D.G. Baker, and C.C. Sheaffer. 1987. Climatic effect on alfalfa dry matter production. Part II. Summer harvests. *Agricultural and Forest Meteorology* 39: 121-129.

Stevens, T.H., and T.W. Hazelton. 1976. Sulfur dioxide pollution and crop damage in the Four Corners region: a simulation analysis. New Mexico Agricultural Experiment Station Bull. 647. Las Cruces, New Mexico.

Torn, M.S., J.E. Degrange, and J.H. Shinn. 1987. The Effects of Acidic Deposition on Alberta Agriculture: A Review. Prep. for Acid Deposition Research Program by Environmental Sciences Division, Lawrence Livermore National Laboratory, Livermore, California. ADRP-B-08-87. 160 pp.

Warteresiewicz, M. 1979. *Archiv Ochrony Srodowiska*, 7: 95-166.

9.3.2 Regional Scale Effects of SO_2 on Wheat Productivity in Alberta

9.3.2.1 Introduction

Wheat is one of the important crops of Alberta and possible negative impact on its production from air pollutants could have a significant economic impact. The general spatial distribution of wheat production in Alberta is shown in Figure 9.26. This spatial distribution is shown overlayed on a map of ten provincial emission regions of sulphur oxides and nitrogen oxides emissions as reported in Picard et al. (1987). The present assessment consists of:

- At this time, are there observed or measurable adverse effects of sulphur dioxide, as an acidifying pollutant, on wheat crop in Alberta?
- If not, where, when and under what conditions will such effects occur in Alberta?

It should be recognized that this initial assessment was performed using only the existing published literature on the effects of sulphur dioxide on wheat growth and productivity in the context of air quality data gathered in ADRP. This assessment is not based on any field or laboratory research experiments performed on wheat within ADRP since experimental crop research was not planned within Critical Point I.

For the first phase of this assessment, scientists in the ADRP decided that the evaluation and assessment of any effects from air pollutants at the present time be focussed on Regions 7 and 9 in Figure 9.26 because of the present, relatively high emissions in those regions. From the Figure, it may be seen that the westernmost part of the southeastern Alberta wheat region lies within the eastern portion of Emission Region 9. It was further agreed by the ADRP scientists that such an assessment should be focussed on an area where an effect would most readily be found if it exists. For these reasons, this evaluation of ambient sulphur dioxide concentrations and their possible effects on Alberta wheat production was focussed on the Crossfield area (see arrow in Figure 9.26). Since one of the objectives of ADRP was to examine the possible regional scale effects of air pollutants, this report is not focussed on any one field site at, or near, Crossfield.

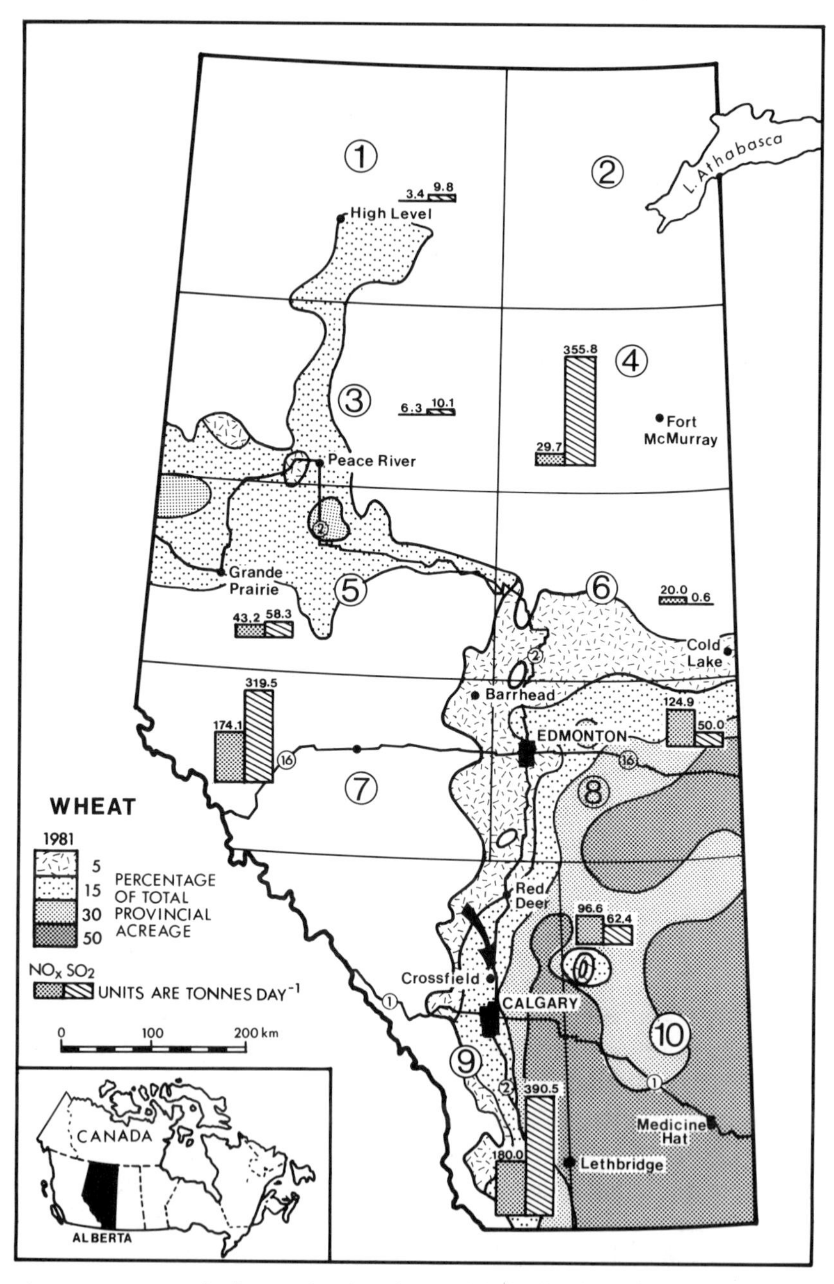

Figure 9.26 Map of Alberta showing the spatial distribution of wheat production and total NO_x and total SO_2 emissions as histograms in each of the 10 geographic emission regions.

Because approximately half of the sulphur deposition in Alberta is thought to occur as dry deposition (refer to Chapter 6), and since exposure-response information for wheat growth and productivity under wet deposition in natural field conditions is virtually non-existent, this assessment was confined to sulphur dioxide. In addition, Torn et al. (1987) discussed the evidence for sulphur and hydrogen ion deposition in intensively farmed lands as not posing a threat to crop growth via soil impacts due to the greater importance of cultivation practices (fertilizer use and liming). For this reason, the present assessment was limited to considering only the possible direct above-ground effects of sulphur dioxide and the growth and yield of spring wheat in Alberta.

For the study area, estimated date of 90% spring wheat seedling emergence for 1986 and 1987 was June 16. Corresponding dates for swathing the crop were September 23, 1986 and September 15, 1987. These dates provide growing season lengths of 100 days in 1986, and 92 days in 1987.

As a frame of reference, Dumanski (Agriculture Canada, personal communication), using an FAO Spring Wheat Model (Stewart 1981; Stewart et al. 1984), showed that spring wheat production on a Black Chernozemic soil, such as the Crossfield area, will have a maximum genetically-determined potential yield of 6.2 tonnes ha^{-1} (92.2 bu $acre^{-1}$). This is simply a result of genetic potential yield from long-term climatological solar radiation and thermal regimes at a given location. Other growth-affecting factors, such as (1) shortage of water availability, (2) soil nutrient availability/shortage, (3) amount of deviation from long-term solar radiation and thermal regimes for a specific year (the actual year's weather), (4) pests (insects and diseases), (5) suboptimal management and cultivation practices, and (6) the presence of any toxic substances, including air pollutants at toxic levels, that act to reduce the potential yield down to some actual yield realized for a given growing season and plot of land. With reference to the aforementioned upper limit, Dumanski indicates that the Crossfield area, when considering soil and moisture availability (climate) constraints, is anticipated to yield 4.6 tonnes ha^{-1} (68.4 bu $acre^{-1}$). This is a 25.8% reduction due to suboptimal soil and moisture. Table 9.19 shows a comparison of these upper limits on wheat production in the Crossfield area of Emission Region 9, and the actual yields reported to Alberta Agriculture for 1979-83, 1986 and 1987. The actual yields for 1986 and 1987 are only 48% to 56% (33/68.4;

Table 9.19 Potential and actual yields of spring wheat in sulphur Emission Region 9, Crossfield area, Alberta, Canada (long tonnes).

Maximum Genetic Potential Yield		Emission Region 9 Black Chernozem[1] Anticipated Yield		1979 - 1983 Crop Reporting District Yield		Actual Yield 1986 1987 Park Cultivar		Actual Yield 1986 1987 Neepawa Cultivar	
tonnes ha^{-1}	bu $acre^{-1}$	tonnes ha^{-1}	bu $acre^{-1}$	tonnes ha^{-1}	bu $acre^{-1}$	tonnes ha^{-1}	bu $acre^{-1}$	tonnes ha^{-1}	bu $acre^{-1}$
6.2[2]	92.2	4.6[2]	68.4	2.4[2]	35.7	2.56	38[3]	2.89-3.03	43-45[3]
						2.22-2.35	33-35[3]	2.56-2.69	38-40[3]
						Range of Decreases from 1986 to 1987			
						7.9%-	13.2%	11.1% -	11.6%

[1] Soil map unit A3/115.
[2] Source: Letter from Julian Dumanski, Land Resource Research Centre, Agriculture Canada, Ottawa, January 25, 1988.
[3] Source: Data from Larry Welsh, Alberta Agriculture, contacted October 2, 1987; Alberta Agricultural Statistics and Estimated Yields.

38/68.4) of what should be obtainable when comparing the Park cultivar, and 56% to 66% (38/68.4; 45/68.4) when comparing the Neepawa cultivar. This represents a loss in yield for 1986 and 1987 ranging between 34 to 52%, compared to the long-term anticipated yield from the FAO Spring Wheat Model. Yields can vary by at least 8 to 13% between two successive years for the Park cultivar, and a little over 11% for Neepawa, because of the "other growth-affecting factors" mentioned previously. The question which remains to be addressed is the role of SO_2 in these reduced yields.

9.3.2.2 Methods and Procedures

Sulphur dioxide exposure-response data for wheat growth and production were obtained from the literature review performed for this research program by Torn et al. (1987). Table 9.20 displays the SO_2 concentrations (ppb), over chronic or episodal portions of the cropping season, at which various wheat responses of a negative nature have been found. It should be noted that data indicating a positive, fertilization effect have not been used here since this assessment is intended to evaluate whether any negative effects on wheat production might be occurring at the present levels of regional scale ambient sulphur dioxide exposures within Alberta.

Estimated sulphur dioxide concentrations on an hourly basis over both the 1986 and 1987 wheat growing seasons were obtained (refer to Chapter 9.2). These were generated from model output for three different locations in the vicinities of each of the following:

- Barrhead (Emission Region 7)
- Crossfield (Emission Region 9).

The first step in this analysis was to compare estimated hourly SO_2 concentrations against the available knowledge base on SO_2 exposure-wheat response relationships. If these estimated ambient concentrations were found to be in the range of possible negative effects on wheat growth and production, then a computerized mathematical modelling method for wheat growth, including an ambient SO_2 exposure-response function, would be used to determine the magnitude of the effect on seasonal yields.

9.3.2.3 Results

During 1986, at Location 1 near Crossfield, the highest hourly estimated concentration representing both years, both regions, and

Table 9.20 Sixteen SO_2 concentrations (in increasing order) over the field growing season at which effects on wheat growth or production have been documented.

Effect/ Response	Average SO_2 Conc. (ppb)	Effect Parameter	Publication
threshold	4.5	Winter wheat yield reduction	page 76, Torn et al. (1987), [Guderian (1977)]
threshold	7.7	Spring wheat yield reduction	page 76, Torn et al. (1987), [Guderian (1977)]
- 1%	10	Spring wheat yield reduction	page 79, Torn et al. (1987), [Guderian (1977)]
- 1.4%	20	Spring wheat yield reduction	page 79, Torn et al. (1987), [Guderian (1977)]
-11.7%	51	Spring wheat yield reduction	page 79, Torn et al. (1987), [Guderian (1977)]
-26.6%	83	Spring wheat yield reduction	page 79, Torn et al. (1987), [Guderian (1977)]
-16%	100	Winter wheat seed wt. reduct.	page 71, Torn et al. (1987), [Heagle et al. (1979)]
-33%	130	Winter wheat seed wt. reduct.	page 71, Torn et al. (1987), [Heagle et al. (1979)]
-36%	141	Spring wheat yield reduct.	page 79, Torn et al. (1987), [Guderian (1977)]
-30%	200	Yield reduction	page 71, Torn et al. (1987), [Shannon and Mulchi (1974)]
-1.00%	241	Threshold for leaf destruct. over entire season (1008 h)	Stevens & Hazleton (1976)
-26.60%	440	Spring wheat yield effect	page 57, Torn et al. (1987), [Guderian and Stratmann (1968)]
-30.30%	600	Thatcher(cv.)Hard Red Spring wheat dry wt. loss with 100 hour exposure	page 60, Torn et al. (1987), [Laurence (1979)]
-20.00%	1315	Wheat yield decrease	page 58, Torn et al. (1987), [Maly (1974)]
-4.00%	1470	Percent foliar injury	page 59, Torn et al. (1987), [Noggle and Jones (1982)]
-6.00%	1470	Percent reduction seed wt.	page 59, Torn et al. (1987), [Noggle and Jones (1982)]

all three locations in each region, was projected to be about 78.1 ppb (204.6 mg m^{-3}). The other two locations at Crossfield showed whole season highest hourly estimated SO_2 concentrations of 41.9 ppb and 24.4 ppb. Similarly for 1987, at the three locations at Crossfield the highest hourly estimated SO_2 concentrations were 73.3, 34.1, and 30.7 ppb. All other projected ambient concentrations were usually far below these levels. Whole season estimated median hourly concentrations were 0.04 ppb, 0.15 ppb, 0.00 ppb, 0.00 ppb, 0.11 ppb, and 0.00 ppb. These are justified statistical indices for representing the data because the data did not exhibit a normal Gaussian frequency distribution. However, for the sake of comparison to data in Table 9.20, the whole season mean hourly estimated concentrations for 1987 were 1.41 ppb, 1.23 ppb, and 1.03 ppb, while for 1986, they were 1.65 ppb, 1.28 ppb, and 1.05 ppb. This being the case, it is impossible that in these data sets, means or medians for periods of time longer than one hour (exposure episodes, specific phases of plant phenology, or whole-season) could have values even approaching, much less exceeding, the maximum estimated hourly value (78.1 ppb). It is also highly improbable that values would be within most of the range of SO_2 concentrations (Table 9.20) or 20 ppb seasonal average (Table 9.20) which could be considered as the threshold to be exceeded to justify further analysis for possible above-ground, direct negative effects on the yield of spring wheat. Most often the modelled values were much smaller than this. Hence, there does not appear to be any justification for more intensive and detailed analysis for possible above-ground, direct effects on spring wheat growth in the eastern end of Emission Region 9.

9.3.2.4 Closing Remarks

The information on wheat response to SO_2 exposure, as shown in Table 9.20, is defined from several-week, if not whole-season, averages. This is unfortunate because recent research has shown "averages" to be associated with the low side of plant effects. But, this situation does reflect the current state of knowledge (or lack of it) with regard to this crop and SO_2. To our knowledge, no data exist from field experiments which have been used in the evaluation of wheat response to the three SO_2 exposure statistics: peak concentration, total integral of the exposures, and number of SO_2

episodes, either for the entire growing season, or by stages of development over the season.

Furthermore, this base of knowledge only applies to above-ground exposure to ambient sulphur dioxide. It provides no information on possible below-ground, indirect effects from soil sulphur loading, and subsequent chemical changes, on root growth processes and their consequences on yield over the long term. Torn et al. (1987) indicate that agricultural fertilization practices have a far greater effect on plant-soil relations than sulphur deposition from air pollution. With the estimated ambient SO_2 concentrations, the range of growing season deposition of sulphur could be around 0.39 to 1.22 kg ha^{-1} for Region 9 (Table 9.21).

Table 9.21 Calculated growing season sulphur deposition from ambient sulphur dioxide in Emission Region 9, Crossfield area, Alberta, Canada.

Year	Location #1	Location #2	Location #3
1986	3192 ppb-h 8363 $\mu g\ m^{-3}$-h 2.11 kg SO_2 ha^{-1} 1.05 kg S ha^{-1}	1197 ppb-h 3136 $\mu g\ m^{-3}$-h 0.79 kg SO_2 ha^{-1} 0.39 kg S ha^{-1}	2423 ppb-h 6348 $\mu g\ m^{-3}$-h 1.6 kg SO_2 ha^{-1} 0.8 kg S ha^{-1}
1987	3681 ppb-h 9644 $\mu g\ m^{-3}$-h 2.43 kg SO_2 ha^{-1} 1.22 kg S ha^{-1}	1339 ppb-h 3508 $\mu g\ m^{-3}$-h 0.88 kg SO_2 ha^{-1} 0.44 kg S ha^{-1}	1363 ppb-h 3571 $\mu g\ m^{-3}$-h 0.90 kg SO_2 ha^{-1} 0.45 kg S ha^{-1}

Beaton and Soper (1986) showed that 70% (11 million ha) of the area with Black Chernozemic soils in the Canadian Prairie Provinces are neither sulphur deficient nor potentially deficient. McLelland (1985), reporting the results of a survey conducted in 1983-84 with 43 above-average producers of soft white wheat in Alberta, states that "Very few top producers are applying supplemental sulphur." In addition, the actual nitrogen compounds used as fertilizers were not identified as to their total chemical composition, including sulphur constituents, but "Irrigation water contains about 30 lb of sulphur for every 12 acre inches applied". The marginal wheat lands in Region 9 are generally not irrigated. But, for example in the more southern portions of Alberta where irrigation is practiced, he further reports that the total water applied as irrigation was about 75 cm-95 cm hectare. This means that through irrigation water, around 30 lb acre^{-1} (34 kg ha^{-1}) of sulphur is applied to the soil in these areas during every year of irrigation.

9.3.2.5 Literature Cited

Beaton, J.D. and R.J. Soper. 1986. Plant response to sulfur in Western Canada. In: **Sulfur in Agriculture.** Number 27, ed., M.A. Tabatabai, American Society of Agronomy, Crop Science Society of America, and Soil Science Society of America, Madison, Wisconsin. pp. 375-403

Guderian, R. 1977. *Air Pollution - Phytotoxicity of Acidic Gasses and Its Significance in Air Pollution Control.* Springer-Verlag, New York. 127 pp.

Guderian, R., and H. Stratmann. 1968. Freilandversuche zur Ermittlung von Schwefeldioxidwirkungen auf die Vegetation. III. Teil Grenzwerte schadlicher SO_2 Immissionen fur Obstund Forstkulturen sowie fur landwirtschaftliche und gartnerische Pflanzenarten. For-schungsberichte des Landes Nordrhein Westfal-en. 114 pp.

Heagle, A.S., S. Spencer, and M.B. Letchworth. 1979. Yield responses of winter wheat to chronic doses of ozone. *Canadian Journal of Botany*, 57: 1999-2005.

Laurence, J.A. 1979. Response of maize and wheat to sulfur dioxide. *Plant Disease Reporter*, 63: 468-471.

Maly, V. 1974. *Scientia Agriculturae Bohemoslovacae*, 6(XXIII): 147-159.

McLelland, M.B. 1985. Soft white wheat production. AGDEX 112/20-2. Field Crops Branch, Alberta Agriculture, Lacombe, Alberta. 14 pp.

Noggle, J.C., and H.C. Jones. 1982. Effects of air pollutants on wheat and grains. pp. 79-90. In: **Effects of Air Pollutants on Farm Commodities**, eds., J.S. Jacobson and A.A. Millen. The Izaak Walton League, Arlington.

Picard, D.J., D.G. Colley, and D.H. Boyd. 1987. Overview of the Emission Data: Emission Inventory of Sulphur Oxides and Nitrogen Oxides in Alberta. Prep for the Acid Deposition Research Program by Western Research, Division of Bow Valley Resource Services Ltd., Calgary, Alberta, Canada. ADRP-A-1-87. 87 pp.

Shannon, J.G. and C.L. Mulchi. 1974. Ozone damage to wheat varieties at anthesis. *Crop Science*, 14: 335-337.

Stevens, T.H. and T.W. Hazelton. 1976. Sulfur dioxide pollution and crop damage in the Four Corners region: a simulation analysis. New Mexico Agricultural Experiment Station Bull. 647. Las Cruces, New Mexico.

Stewart, R.B. 1981. Modeling methodology for assessing crop production potentials in Canada. Contribution 1983-12E. Agriculture Canada, Research Branch, Land Resource Research Center, Ottawa, Ontario. 29 pp.

Stewart, R.B., J. Dumanski, and D.F. Acton. 1984. Production potential for spring wheat in the Canadian Prairie Provinces - an estimate. *Agricultural Ecosystems and Environment*, 11: 1-14.

Torn, M.S., J.E. Degrange, and J.H. Shinn. 1987. The Effects of Acidic Deposition on Alberta Agriculture: A Review. Prep. for Acid Deposition Research Program by Environmental Sciences Division, Lawrence Livermore National Laboratory, Livermore, California. ADRP-B-08-87. 160 pp.

9.4 ASSESSMENT OF PRESENT AND POTENTIAL EFFECTS OF ACIDIC AND ACIDIFYING AIR POLLUTANTS ON ALBERTA SOILS

S. A. Abboud and L. W. Turchenek

9.4.1 Introduction

A qualitative assessment of the potential impacts of acidic and acid forming substances on Alberta soils has been provided by Turchenek et al. (1987). This effort involved review, analysis, and interpretation of appropriate world literature in the context of Alberta (refer to Chapter 2 for a summary).

Since the preparation of the report by Turchenek et al. (1987) the ADRP completed the collection of air quality data (both wet and dry deposition) at three locations in Alberta over a two year period. In addition, based on these data and through modelling, regional scale estimates of atmospheric deposition have been computed (refer to Chapters 6 and 8). These efforts formed the basis for a more thorough evaluation of the impacts of acidic and acid forming air pollutants on Alberta soils, as described in the following sections. This evaluation was achieved by computer simulations of soil responses (pH, base saturation, and soluble aluminum levels) to present and possible future acid loadings.

9.4.2 Model Description

9.4.2.1 Background

The application of process models in assessing the impacts of acidic deposition on soils has been previously reviewed by Turchenek et al. (1987). Models which were developed specifically for acidic deposition effects simulations include those described by Arp (1983), Reuss (1980), Gherini et al. (1985), and Bloom and Grigal (1985). The model of Bloom and Grigal (1985) was used for an initial assessment of responses of Alberta soils to acidic deposition (Turchenek et al. 1987). Other models have been described in the literature since then. Kauppi et al. (1986) have described a model which calculates soil pH as a function of hydrogen ion input and the buffer mechanisms of the soil. The model has been incorporated into the RAINS (Regional Acidification Information and Simulation) model system of the International Institute for Applied Systems Analysis (Laxenberg, Austria) for analyzing the acid deposition problem in Europe. Levine and Ciolkosz (1988) described a model used to determine the sensitivity of Pennsylvania soils to acid deposition. The model is similar to that of Bloom and Grigal (1985) but incorporates additional sub-routines to account for more of the soil processes influencing acidity.

These various models have different capabilities and limitations with regard to assumptions made, data required, ease of execution, and other considerations. The Bloom and Grigal (1985) model was used for assessment of acid deposition impacts on soils in Alberta because of its use of readily available input data. Changes in soil reaction, base status, and solution Al^{3+}, over specified periods of time, can be predicted from the model. Bloom and Grigal argued that other numerical models developed to predict long-term effects of acidic deposition on soils, such as those indicated previously, require input of detailed data which is usually not available. The model is suitable for applications on a regional basis, and is, therefore, appropriate for broad application by planning or regulatory agencies. Several assumptions are used in the model, and land management effects are not considered.

The Bloom and Grigal model was applied for a more thorough analysis of changes in soil system components as a consequence of input of acidic and acid forming substances. The model was modified in some ways since its initial application to Alberta soils as indicated in the following sections.

9.4.2.2 Description of the Modified Bloom and Grigal Model

9.4.2.2.1 General Description. The model is described by Bloom and Grigal (1985) as semiempirical, and was developed for estimation of the long-term effects of acidic deposition on non-agricultural soils. Agricultural soils were excluded because of the much greater effects of management practices, such as fertilization and liming, as compared with acidic deposition. It could possibly be applied to soils in which such practices are minimal as in rangelands. The model simulates changes in exchangeable bases, soil pH, and soil solution Al^{3+} with time. The assumptions used in the model are conservative, and the rates of change that are calculated are maximal rates (Bloom and Grigal 1985). The main features of the model are presented herein, but the original paper should be referred to for a full description.

9.4.2.2.2 Assumptions. The assumptions used in the model, condensed from Bloom and Grigal (1985) are as follows:

a. Sulphate is not adsorbed by the soil and is treated as a mobile anion.
b. The depth of soil affected by acidic inputs is 25 cm. This is the depth in forest soils within which most plant roots important for forest tree nutrition are located.
c. The current measured properties of the soils reflect a steady state condition. The pH and base status in an undisturbed soil represent an integration of rates of soil weathering, biocycling, additions of organic acids, and leaching of bases from the surface soil. Addition of acidic deposition disturbs the steady state. This is the main reason the model is not applicable to soils under cultivation, under regenerating forest stands, or under other disturbed situations.
d. The acidic deposition reacts completely with the top 25 cm of the soil.
e. The effective acidity in rain and snow equals H^+ plus NH_4^+ minus NO_3^-. This equation was modified in this project by multiplying the NO_3^- by a factor of 0.7 to account for the less than equal replacement of nitrate by bicarbonate during vegetational uptake. This effect is compounded if NO_3^- is formed by nitrification of the NH_4^+; the NH_4^+ is, therefore, multiplied by a factor of 1.15 to account for this. Derivation of these correction factors can be found in Coote et al. (1982).
f. The volume of water flowing downward from the top 25 cm soil layer is equal to precipitation minus evapotranspiration. This also assumes that plants obtain all their water from the top 25 cm of soil.
g. Soil solution pH can be calculated from the equation,

$$pH + 1/2 \log [Ca^{2+} + Mg^{2+}] = \text{constant}.$$

In Minnesota, $[Ca^{2+} + Mg^{2+}]$ is estimated to be 2×10^{-4} mmol L^{-1} in forest soils. The soil solution pH is, therefore, 0.8 units greater than soil pH measured in 0.01 mmol L^{-1}

$CaCl_2$. Examination of data from 117 samples indicated that a similar relationship exists for soils of Alberta.

h. The partial pressure of CO_2 (PCO_2) in soil air is 0.005 atmospheres (0.0005 MPa).
i. The chemical activity of Al^{3+} in the soil solution of surface soils can be calculated from the equation,

$$\log (Al^{3+}) = 2.46 - 1.66\ pH.$$

j. An assumption in the original Bloom and Grigal model states that the rate of mineral weathering does not increase with a decrease in pH. This assumption does not apply in the present study because a weathering function was included in the model as discussed in Section 9.4.3.2.6.
k. The pH of soil solutions is related to the base saturation by a linear function in which,

$$pH = a(BS) + b$$

where BS is the base saturation of the soil. This is a modification of the relationship between pH and BS (Henderson-Hasselbach) originally used by Bloom and Grigal (1985). The validity of this equation relating pH and base saturation has been questioned, for example, by Magdoff and Bartlett (1986). Kauppi et al. (1986) used a power function, and Levine and Ciolkosz (1988) used a linear function to relate these two soil attributes. Both of these functions were examined using data for soils in Alberta. The Henderson-Hasselbach equation gives unrealistically low pH values at low BS levels, while the power function overestimated pH at low BS values. The linear function appeared to give the most reasonable value of the soil pH at low BS values. The linear relationship also had a slightly higher R^2 value than the others (Abboud and Turchenek, 1988) and was therefore applied in the model.

9.4.2.2.3 Model Calculations. Inputs into the model are as follows:

1. annual precipitation (cm);
2. annual evapotranspiration (cm);
3. acidity of wetfall (pH);

4. NH_4^+ in wetfall (μmol L^{-1});
5. NO_3^- in wetfall (μmol L^{-1});
6. dryfall acidity (kmol (H^+ equivalents) $ha^{-1}y^{-1}$);
7. soil pH;
8. sum of bases in soil (kmol(+)ha^{-1});
9. CEC (kmol(+)ha^{-1});
10. PCO_2 (atmospheres);
11. increment of time between reported periods (year); and
12. length of simulation period (year).

The annual loss of bases is calculated by subtraction of acid leached from the top 25 cm of soil, of bases released by weathering, and of decrease in bicarbonate weathering due to decrease in soil solution pH from the effective acidity in wetfall and dryfall. A new sum of bases is calculated each year by subtracting the annual loss of bases. The new pH is calculated from the equation relating soil pH and base saturation. Al^{3+} in solution is calculated from the equation in the previously stated assumption (i) (Bloom and Grigal 1985). Details of the computations are presented Section 9.4.3.3.

9.4.2.2.4 Capabilities and Limitations. Bloom and Grigal (1985) found that results of simulations for Minnesota soils were comparable in pH, base saturation, and soluble Al^{3+} values to actual data for acidified soils in the eastern United States and in Germany. The computations were rapidly carried out on a microcomputer using a program written in BASIC. Due to the assumptions made, simulations result in overestimation of changes in pH, base saturation, and solution Al^{3+}. This can be either an advantage or a disadvantage, depending on how the results are to be used. Graphic display of change in pH, base saturation, or Al^{3+} concentration over time enables quick comparison of different soils.

Bloom and Grigal (1985) indicated that the model is especially well suited for calculation of the responses of soils that are buffered by cation exchange. These are soils in the pH(H_2O) range of 4.0 to 5.0, and >5.0 for many soils. At pH(H_2O) values of 4.0-2.8 soils are buffered by dissolution of Al^{3+}, and the magnitude of changes in pH and base saturation with acidic deposition will be much less than for cation-exchange buffered soils.

Limitations of the model are related to the assumptions which had to be made to keep it simple and to enable its use of easily obtainable data inputs. Other models, indicated in previous sections, have the same limitations. For example, use of annual precipitation and evapotranspiration data does not take into consideration factors such as intensity of individual precipitation events or snowmelt effects. Bloom and Grigal (1985) indicated, however, that where surface runoff data are available, values for surface runoff can be subtracted from precipitation. A modification of the model which uses monthly data inputs has also been prepared (P. Bloom and D. Grigal, University of Minnesota, personal communication 1987).

Other limitations are discussed by Bloom and Grigal (1985). They indicated that more work is required to determine if the pH-Al^{3+} relationship in assumption (i) is applicable to a wide range of soils. They also indicated that effects of management of forest soils need to be taken into account.

The model of Bloom and Grigal was developed for mineral soils, and the organic portion of these soils is assumed to make no contribution to buffering and to remain unaffected by acidic inputs. This assumption is invalid, but a model which includes buffering by organic materials is not yet available. Due to this and other limitations, the model results must be considered as overestimates of responses. The model is also not applicable to organic soils, and alternative means of assessing impacts on these soils need to be developed. The basic structure of the model may be applicable, but the pH - base saturation - [Ca^{2+} + Mg^{2+}] - Al^{3+} relationship for calculations may be very different for organic soils than for mineral soils.

9.4.3 Materials and Methods

9.4.3.1 Soil Selection

Soils occurring within Regions 3, 4, 6, 7, and 9 of the geographic emission regions described by Picard et al. (1987) were selected for simulations (Figure 9.27). Soils were also selected from the alpine and subalpine areas,collectively referred to as the Mountain Region. Descriptions of soil profiles from these regions were obtained from various reports of the Alberta Soil Survey. The soils selected for study were predominantly those interpreted by Holowaychuk and Fessenden (1987) as being highly and moderately sensitive to acid deposition. In addition, some of the soil types used in the assessment of vegetation and aquatic effects within the

ADRP were also selected for study (Sections 9.3 and 9.5). Some properties of these soils are indicated in Tables 9.24 to 9.28. More complete descriptions are provided in Abboud and Turchenek (1988).

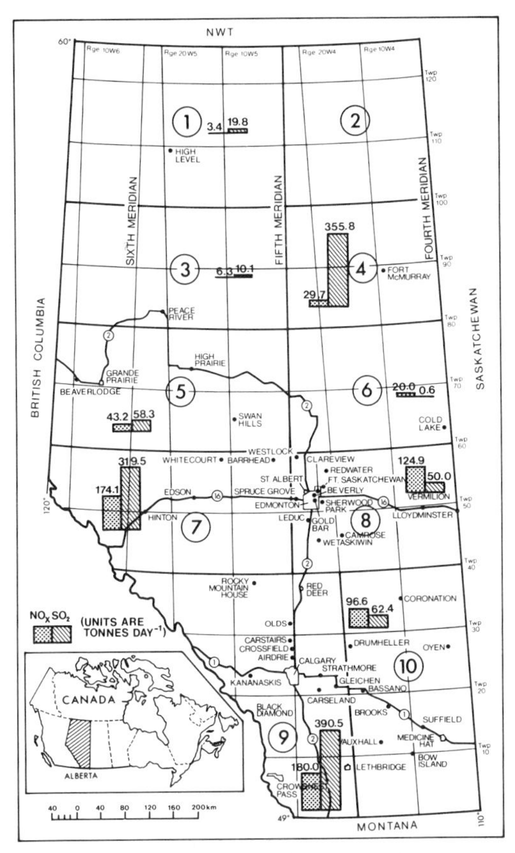

Figure 9.27 Map of Alberta showing total NO_x and total SO_2 emissions as histograms in each of the 10 geographic emission regions.

9.4.3.2 Data Inputs

9.4.3.2.1 Soil Data

A. pH, Cation Exchange Capacity, and Exchangeable Bases. The soil pH values reported in earlier soil survey reports were generally obtained by the soil paste method whereas pH determined in $CaCl_2$ is used in the model. The difference between pH(H_2O) and pH($CaCl_2$) is about 0.5 units for Alberta soils. Therefore, 0.5 was subtracted from pH(H_2O)

values for model inputs. Otherwise, pH($CaCl_2$) values as reported for sampled pedons were used.

Exchangeable bases in soils were obtained by the ammonium acetate extraction method. Cation exchange capacity, determined from the sum of bases plus $BaCl_2$ triethanolamine titratable acidity, was used by Bloom and Grigal (1985). Since data for titratable acidity are not generally available for Alberta soils, total CEC values obtained by the buffered (pH 7) ammonium acetate method were used in the simulations and in prior determinations of the pH - base saturation relationship referred to in assumption (k), Section 9.4.2.2.2.

Little data for partial pressure of CO_2 in the soil atmosphere is available, and since PCO_2 is quite variable, the value of 0.005 atmospheres used by Bloom and Grigal (1985) was also used in this project. A sensitivity analysis showed that the pH difference is minor if PCO_2 is ≤ 0.005 atmosphere, but the pH after 50-100 years can be ≤ 0.6 units compared to the aforementioned value if PCO_2 is >0.005 atmospheres.

Activity coefficients of Al^{3+} and $Al(OH)^{2+}$ were 0.82 and 0.92, respectively, as used by Bloom and Grigal (1985).

The input data for soil pH, cation exchange capacity, and sum of bases were weighted mean values for the top 25 cm of air-dried mineral soil. Litter layers, or L, F, and H horizons, were excluded from the calculations. The thickness of the soil horizons and the bulk density were considered in computing the means. As bulk density data were not provided in the soil survey reports, mean values determined for different kinds of soil horizons in Alberta as reported by Tajek et al. (1989) were applied to the selected pedons.

The model is sensitive to soil pH and base saturation which are related parameters as discussed in Section 9.2.2.2. Variation in pH($CaCl_2$) by about 0.5 units equates to about a 20% difference in base saturation. Thus, a relatively small difference in pH translates to a large difference in base saturation and ultimately in the base status and buffering capability of a soil. The base saturation multiplied by the cation exchange capacity determines the total base status, hence the buffering capacity, of a soil. Sensitivity analysis showed that the model is sensitive to CEC at low values.

However, it is more sensitive to base saturation because the proportion of the CEC available to neutralize acidity is governed by this parameter. It is apparent that accurate soil data are required to maximize confidence in model results.

The weighted mean values for each of the soils used in the computer simulations are given in Tables 9.23 to 9.28 in Section 9.4.3.2.3.

B. Equation constants in the pH-base saturation relationship. The values of a and b in the equation relating pH and base saturation were assumed to be similar for soils of the same order but to differ between orders. Data for pH and base saturation were collected from a number of pedons sampled during recent soil surveys. The regression of pH on BS was obtained, and the values for a and b, and their correlations, were determined as indicated in Table 9.22.

The correlation for Luvisolic soils was relatively high as compared with those for Chernozemic and Brunisolic soils. The correlation coefficients were slightly higher or about the same as those obtained with the Henderson-Hasselbach equation previously used with the same data set reported by Turchenek et al. (1987). Use of the linear function also eliminated another problem with the Henderson-Hasselbach equation. The pH becomes very sensitive to BS as BS reaches low values in the range of about 0 to 0.10, and calculated values of pH become unrealistic. A power function was used by Kauppi et al. (1986) in the soils component of the Regional Acidification Information and Simulation (RAINS) model. It would appear that more investigation of the pH - base saturation relationship is required to determine which function is most appropriate for soils in Alberta. This would require sampling and analysis of a wide variety of soils. Based on available information, the previously described linear function appears to be the most suitable at this time.

9.4.3.2.2 Precipitation Chemistry Data. The chemistry of wet deposition in Region 9 is based on monitoring results from the Crossfield air quality monitoring stations established by the ADRP. Data for the Mountain Region is based on the ADRP air quality monitoring station at Fortress Mountain. Data for each of the

Table 9.22 Values for a and b for different soil orders calculated from the equation pH = a(BS) = b.

Soil Order	N	a	b	R^2
Chernozemic	56	2.25	3.52	0.63***
Luvisolic	78	2.32	3.35	0.73***
Brunisolic	28	3.18	3.14	0.55***

*** Significant at the 0.001 level

other regions have been derived from CANSAP and Alberta Environment precipitation chemistry stations as described in Chapter 6. The wet chemistry data required for soil simulations are pH, and NH_4^+ and NO_3^- concentrations. For all regions, data for present and future scenarios were derived from values for two years, from late 1985 to late 1987. The actual values used for each soil and region are indicated in Tables 9.23 to 9.28 in Section 9.4.3.2.7.

9.4.3.2.3 Dry Deposition Inputs. The effective acidity (EA) in precipitation, according to Bloom and Grigal, is calculated as

$$EA = H^+ + NH_4^+ - NO_3^-.$$

The modified calculation, according to Coote et al. (1982) is

$$EA = H^+ + 1.15\ NH_4^+ - 0.7\ NO_3^-.$$

The model required dryfall acidity as a ratio of the dryfall to the wetfall acidic inputs which is usually calculated as the ratio of dryfall S to wetfall S. The calculation of the total acidity consisted of the summation of wetfall acidity plus wetfall acidity multiplied by the dry to wet ratio. When the dry:wet ratio is used in the Bloom-Grigal model, it is implied that all wetfall acidity is due to or correlates strongly with SO_4 in the precipitation. This is generally accepted in eastern North America but has been refuted in the west (e.g., Summers, 1986). Hence, it is inappropriate to use the dry:wet ratio for calculating total acidity in applying the model to Alberta soils. Therefore, the acidity calculation was modified to:

$$EA = [\text{wetfall } H^+ + 1.15NH_4^+ - 0.7NO_3^-] + [\text{dryfall aerosol-S} + SO_2\text{-S (expressed as equivalents of } H^+)]$$

Data for aerosol SO_4 and SO_2, expressed as total SO_4, used in this study were those described in Chapters 6 and 8 of this report. Those values were converted to equivalent acidity by assuming that each molecule of SO_4 and SO_2 was associated with or would generate two H^+ ions.

Dry acidity and total acidity were calculated for each region. Two scenarios were evaluated for the soils with "mean" and "maximum" values presented in Chapter 6 for the "present" and "future estimated" loadings,respectively. The total annual input of acidity for each scenario was held constant over the simulation period of 300 years. The data required for model execution are indicated in Tables 9.23 to 9.28.

9.4.3.2.4 Climate Data. Annual precipitation data for late 1985 to late 1987 were compiled from the records of Atmospheric Environment Service, Environment Canada. For annual evapotranspiration, values for average areal evapotranspiration determined by Bothe and Ames (1984) were used. Stations selected to represent each of the regions evaluated in this project were as follows: Peace River for Region 3; Fort McMurray for Region 4; Athabasca climate station for precipitation and Cold Lake area for evapotranspiration for Region 6; Edson for Region 7; Madden (Crossfield) for precipitation and Calgary for evapotranspiration for Region 9; and, Fortress Mountain for precipitation and Banff for evapotranspiration for the Mountain Region. No effort was made to estimate surface runoff for subtraction from total precipitation in any of the soil simulations.

9.4.3.2.5 Time. The model can be executed for any specified length of time, and simulation results can be reported for any specified increment of time within the total simulation period. Predictive soil effects data are of greatest interest and need for the immediate and near future. Therefore, it would be desirable to predict soil responses for the period during which pollutant emissions can be forecasted. It would also be of interest, from a soil development point of view, to determine soil responses to acidic deposition over very long periods of time since changes in soils occur slowly. Three hundred years was arbitrarily selected for the simulation period.

Table 9.23 Input data for soil acidification simulations using the Bloom and Grigal model - Region 3.

Soil Subgroup - Canada	Eluviated Dystric Brunisol	Orthic Gray Luvisol	Orthic Gray Luvisol	Solonetzic Gray Luvisol
Soil Subgroup - U.S.A.	Spodic Cryopsamments	Typic Cryoboralfs	Dystric Cryochrepts	Natriborafs
Soil Series	Heart	Demmitt	Boundary	Alcan
Parent Material	Eolian	Morainal	Shale	Marainal
Particle Size Class	Sandy	Fine-loamy	Fine-clayey	Fine-clayey
Annual Precipitation (cm)	463	463	463	463
Annual Evapotranspiration (cm)	392	392	392	392
pH of Precipitation	5.11, 4.62[1]	5.11, 4.62	5.11, 4.62	5.11, 4.62
NH_4^+ of Precipitation (μmol L^{-1})	8.7, 25.0	8.7, 25.0	8.7, 25.0	8.7, 25.0
NO_3^- of Precipitation (μmol L^{-1})	7.3, 20.0	7.3, 20.0	7.3, 20.0	7.3, 20.0
Dryfall- Equiv H^+ (kmol $ha^{-1}y^{-1}$)	0.13, 0.21	0.13, 0.21	0.13, 0.21	0.13, 0.21
pH ($CaCl_2$) of Soil	5.3	5.3	4.1	4.4
Total Bases in Soil (kmol(+)ha^{-1})	117	288	99	268
CEC of Soil (kmol(+)ha^{-1})	174	346	825	470
Base Saturation (%)	67	83	12	57

Continued...

Table 9.23 (Concluded).

a in pH = a(BS) + b	2.25	2.32	2.32	2.32
b in pH = a(BS) + b	3.52	3.35	3.35	3.35
Total H^+ Input (kmol $ha^{-1}y^{-1}$)	0.19, 0.39	0.19, 0.39	0.19, 0.39	0.19, 0.39
Total Equiv S Input (kg $ha^{-1}y^{-1}$)	3.0, 6.2	3.0, 6.2	3.0, 6.2	3.0, 6.2

[1] Numbers separated by commas indicate present (mean) and future (maximum) loadings.

Table 9.24 Input data for soil acidification simulations using the Bloom and Grigal model - Region 4.

Soil Subgroup - Canada	Eluviated Dystric Brunisol	Eluviated Dystric Brunisol	Orthic Gray Luvisol	Orthic Gray Luvisol
Soil Subgroup - U.S.A.	Spodic Cryopsamments	Spodic Cryopsamments	Typic Cryoboralfs	Typic Cryoboralfs
Soil Series	Firebag	Mildred	Kinosis	Legend
Parent Material	Glaciofluvial	Glaciofluvial	Morainal	Morainal
Particle Size Class	Sandy	Sandy	Fine-loamy	Fine-loamy
Annual Precipitation (cm)	516	516	516	516
Annual Evapotranspiration (cm)	301	301	301	301
pH of Precipitation	6.01, 5.52[1]	6.01, 5.52	6.01, 5.52	6.01, 5.52
NH_4^+ of Precipitation (μmol L^{-1})	1.2, 3.5	1.2, 3.5	1.2, 3.5	1.2, 3.5
NO_3^- of Precipitation (μmol L^{-1})	5.6, 15.0	5.6, 15.0	5.6, 15.0	5.6, 15.0
Dryfall- Equiv H+ (kmol $ha^{-1}y^{-1}$)	0.23, 0.88	0.23, 0.88	0.23, 0.88	0.23, 0.88
pH ($CaCl_2$) of Soil	4.2	4.8	4.0	4.1
Total Bases in Soil (kmol(+)ha^{-1})	10	66	82	228
CEC of Soil (kmol(+)ha^{-1})	32	120	256	445
Base Saturation (%)	30	55	32	51

Continued...

Table 9.24 (Concluded).

a in pH = a(BS) + b	2.25	2.25	2.32	2.32
b in pH = a(BS) + b	3.52	3.52	3.35	3.35
Total H^+ Input (kmol $ha^{-1}y^{-1}$)	0.22, 0.86	0.22, 0.86	0.22, 0.86	0.22, 0.86
Total Equiv S Input (kg $ha^{-1}y^{-1}$)	3.5, 13.8	3.5, 13.8	3.5, 13.8	3.5, 13.8

[1] Numbers separated by commas indicate present (mean) and future (maximum) loadings.

Table 9.25 Input data for soil acidification simulations using the Bloom and Grigal model - Region 6.

Soil Subgroup - Canada	Eluviated Dystric Brunisol	Eluviated Eutric Brunisol	Eluviated Eutric Brunisol	Orthic Gray Luvisol	Orthic Gray Luvisol	Orthic Dark Gray Chernozemic
Soil Subgroup - U.S.A.	Spodic Cryopsamments	Spodic Cryopsamments	Typic Cryochrepts	Typic Cryoboralfs	Typic Cryoboralfs	Boralfic Cryoborolls
Soil Series	Nestow	Nicot	Codesa	Tawatinaw	La Corey	Falun
Parent Material	Eolian	Eolian	Morainal	Morainal	Morainal	Morainal
Particle Size Class	Sandy	Sandy	Coarse-loamy	Coarse-loamy	Fine-loamy	Fine-loamy
Annual Precipitation (cm)	417	417	417	417	417	417
Annual Evapotranspiration (cm)	346	346	346	346	346	346
pH of Precipitation	5.21, 4.72[1]	5.21, 4.72	5.21, 4.72	5.21, 4.72	5.21, 4.72	5.21, 472
NH_4^+ of Precipitation (μmol L^{-1})	1.8, 5.3	1.8, 5.3	1.8, 5.3	1.8, 5.3	1.8, 5.3	1.8, 5.3
NO_3^- of Precipitation (μmol L^{-1})	3.7, 10.0	3.7, 10.0	3.7, 10.0	3.7, 10.0	3.7, 10.0	3.7, 10.0
Dryfall-Equiv H^+ (kmol $ha^{-1}y^{-1}$)	0.16, 0.25	0.16, 0.25	0.16, 0.25	0.16, 0.25	0.16, 0.25	0.16, 0.25
pH ($CaCl_2$) of Soil	4.2	5.8	5.9	5.8	5.9	5.4
Total Bases in Soil (kmol(+)ha^{-1})	17	94	364	402	150	207
CEC of Soil (kmol(+)ha^{-1})	34	126	419	473	182	277
Base Saturation (%)	50	75	87	85	82	75

Continued...

Table 9.25 (Concluded).

a in pH = a(BS) + b	2.25	2.25	2.32	2.32	2.32	3.18
b in pH = a(BS) + b	3.52	3.52	3.35	3.35	3.35	3.14
Total H^+ Input (kmol $ha^{-1}y^{-1}$)	0.18, 0.33	0.18, 0.33	0.18, 0.33	0.18, 0.33	0.18, 0.33	0.18, 0.33
Total Equiv S Input (kg $ha^{-1}y^{-1}$)	2.9, 5.3	2.9, 5.3	2.9, 5.3	2.9, 5.3	2.9, 5.3	2.9, 5.3

[1] Numbers separated by commas indicate present (mean) and future (maximum) loadings.

Table 9.26 Input data for soil acidification simulations using the Bloom and Grigal model - Region 7.

Soil Subgroup - Canada	Orthic Gray Luvisol	Orthic Gray Luvisol	Orthic Gray Luvisol	Podzolic Gray Luvisol	Eluviated Eutric Brunisol	Orthic Black Chernozemic	Eluviated Black Chernozemic
Soil Subgroup - U.S.A.	Typic Cryoboralfs	Typic Cryoboralfs	Typic Cryoboralfs	Typic Cryorthods	Typic Cryochrepts	Cumulic Haploborolls	Boralfic Argiborolls
Soil Series	Culp	Cooking Lake	Hubalta	Nosehill	Berland	Red Willow	Ponoka
Parent Material	Glaciofluvial	Morainal	Morainal	Morainal	Morainal	Glaciofluvial	Glaciolarustria
Particle Size Class	Sandy	Fine-loamy	Fine-loamy	Fine-loamy	Coarse-loamy	Sandy	Fine-loamy
Annual Precipitation (cm)	547	547	547	547	547	547	547
Annual Evapotranspiration (cm)	383	383	383	383	383	383	383
pH of Precipitation	6.02, 5.53[1]	6.02, 5.53	6.02, 5.53	6.02, 5.53	6.02, 5.53	6.02, 5.53	6.02, 5.53
NH_4^+ of Precipitation (μmol L^{-1})	11.3, 33.0	11.3, 33.0	11.3, 33.0	11.3, 33.0	11.3, 33.0	11.3, 33.0	11.3, 33.0
NO_3^- of Precipitation (μmol L^{-1})	12.7, 34.0	12.7, 34.0	12.7, 34.0	12.7, 34.0	12.7, 34.0	12.7, 34.0	12.7, 34.0
Dryfall- Equiv H^+(kmol $ha^{-1}y^{-1}$)	0.25, 0.38	0.25, 0.38	0.25, 0.38	0.25, 0.38	0.25, 0.38	0.25, 0.38	0.25, 0.38
pH ($CaCl_2$) of Soil	5.8	6.2	5.6	4.4	5.7	5.6	5.7
Total Bases in Soil (kmol(+)ha^{-1})	195	423	408	397	127	239	541
CEC of Soil (kmol(+)ha^{-1})	222	466	445	242	162	331	682
Base Saturation (%)	88	91	92	61	78	72	79

Continued...

Table 9.26 (Concluded).

a in pH = a(BS) + b	2.32	2.32	2.32	2.32	2.25	3.18	3.18
b in pH = a(BS) + b	3.35	3.35	3.35	3.35	3.52	3.14	3.14
Total H^+ Input (kmol $ha^{-1}y^{-1}$)	0.28, 0.47	0.28, 0.47	0.28, 0.47	0.28, 0.47	0.28, 0.47	0.28, 0.47	0.28, 0.47
Total Equiv S Input (kg $ha^{-1}y^{-1}$)	4.5, 7.5	4.5, 7.5	4.5, 7.5	4.5, 7.5	4.5, 7.5	4.5, 7.5	4.5, 7.5

[1] Numbers separated by commas indicate present (mean) and future (maximum) loadings.

Table 9.27 Input data for soil acidification simulations using the Bloom and Grigal model - Region 9.

Soil Subgroup - Canada	Orthic Black Chernozemic	Orthic Black Chernozemic	Orthic Black Chernozemic	Orthic Gray Luvisol	Dark Gray Luvisol
Soil Subgroup - U.S.A.	Typic Cryoborolls	Typic Cryoborolls	Typic Cryoborolls	Typic Cryoboralfs	Mollic Cryoboralfs
Soil Series	Antler	Delacour	Dunvargan	Spruce Ridge	Elbow
Parent Material	Morainal	Morainal	Morainal	Morainal	Glaciolacustrine
Particle Size Class	Fine-loamy	Fine-loamy	Fine-loamy	Fine-loamy	Fine-clayey
Annual Precipitation (cm)	506	506	506	506	506
Annual Evapotranspiration (cm)	412	412	412	412	412
pH of Precipitation	4.79, 4.30[1]	4.79, 4.30	4.79, 4.30	4.79, 4.30	4.79, 4.30
NH_4^+ of Precipitation (μmol L^{-1})	28, 82	28, 82	28, 82	28, 82	28, 82
NO_3^- of Precipitation (μmol L^{-1})	17, 46	17, 46	17, 46	17, 46	17, 46
Dryfall- Equiv H^+ (kmol $ha^{-1}y^{-1}$)	0.30, 0.57	0.30, 0.57	0.30, 0.57	0.30, 0.57	0.30, 0.57
pH ($CaCl_2$) of Soil	5.8	5.7	5.9	4.4	5.0
Total Bases in Soil (kmol(+)ha^{-1})	970	338	827	379	898
CEC of Soil (kmol(+)ha^{-1})	1042	485	827	605	1341
Base Saturation (%)	93	70	100	63	67

Continued...

Table 9.27 (Concluded).

a in pH = a(BS) + b	3.18	3.18	3.18	2.32	2.32
b in pH = a(BS) + b	3.14	3.14	3.14	3.35	3.35
Total H^+ Input (kmol $ha^{-1}y^{-1}$)	0.84, 1.14	0.84, 1.14	0.84, 1.14	0.84, 1.14	0.84, 1.14
Total Equiv S Input (kg $ha^{-1}y^{-1}$)	7.7, 18.2	7.7, 18.2	7.7, 18.2	7.7, 18.2	7.7, 18.2

[1] Numbers separated by commas indicate present (mean) and future (maximum) loadings.

Table 9.28 Input data for soil acidification simulations using the Bloom and Grigal model - Mountain Region.

Soil Subgroup - Canada	Orthic Dystric Brunisol	Eluviated Dystric Brunisol
Soil Subgroup - U.S.A.	Dystric Cryochrepts	Dystric Cryochrepts
Soil Series	Egypt	Topaz
Parent Material	Morainal	Morainal
Particle Size Class	Coarse-loamy	Sandy
Annual Precipitation (cm)	604	604
Annual Evapotranspiration (cm)	300	300
pH of Precipitation	4.78, 4.29[1]	4.78, 4.29
NH_4^+ of Precipitation (μmol L^{-1})	17, 50	17, 50
NO_3^- of Precipitation (μmol L^{-1})	13, 35	13, 35
Dryfall- Equiv H^+(kmol $ha^{-1}y^{-1}$)	0.08, 0.16	0.08, 0.16
pH ($CaCl_2$) of Soil	4.0	4.6
Total Bases in Soil (kmol(+)ha^{-1})	58	23
CEC of Soil (kmol(+)ha^{-1})	349	75
Base Saturation (%)	17	31
a in pH = a(BS) + b	2.25	2.25
b in pH = a(BS) + b	3.52	3.52
Total H^+ Input (kmol $ha^{-1}y^{-1}$)	0.24, 0.67	0.24, 0.67
Total Equiv S Input (kg $ha^{-1}y^{-1}$)	3.8, 10.7	3.8, 10.7

This time frame would not obscure the data for interpretation of short term effects, yet would provide a longer term view of soil changes. Ten years was selected as the increment of time between reported values in the simulations.

9.4.3.2.6 Effect of Weathering. The model can be applied either with or without weathering. If the rate of weathering is known at any pH, the rate at another pH can be calculated from the equation

$$r = r_o 10^{-0.5(pH-pH_o)}$$

where r_o and pH_o are the initial conditions. The constant 0.5 is assumed for soils but has values between 0.5 and 1 for different minerals (Bloom and Grigal 1985). Bloom and Grigal (1985) used the value of 0.07 kmol(+)ha^{-1} y^{-1} at a pH_o of 5.0. This was based on calculations from weathering experiments using lysimeters in Minnesota and Wisconsin. Using these initial values, example weathering rates are as follows (in kmol(+)$ha^{-1}y^{-1}$): 0.007 at pH 7; 0.022 at pH 6; 0.07 at pH 5; 0.22 at pH 4; and 0.70 at pH 3.

9.4.3.2.7 Summary of Data Inputs. The starting parameters for soils used in simulations are given in Tables 9.23 to 9.28. The taxonomy and some general descriptive features of the soils are indicated along with input data described previously. The total acid input, calculated from wet and dryfall acidity is also provided in Tables 9.23 to 9.28 in terms of H^+ input and in terms of S equivalent inputs. The S equivalents are provided to facilitate comparison with deposition data reported elsewhere in the literature.

9.4.3.3 Computations

The loss of bases is calculated on an annual basis by

$$S = I - A - C - W$$

where S is the sum of bases lost, I is the effective acidity in the precipitation plus dryfall, A is the acid leached out of the top 25 cm of soil, C is the decrease in bicarbonate weathering due to the decrease in soil solution pH, and W is the base contribution due to weathering. The value for A is given by

$$A = (3[Al^{3+}] + 2[AlOH^{2+}] + [H^+])(\text{Precip} - \text{ET})(10^2)$$

where Precip is the rainfall plus snowfall and ET is the evapotranspiration. The pH-dependent value for $[Al^{3+}]$ is computed using the equation in assumption (i), Section 9.4.2.2.2 and the pH at the beginning of the computation year. The concentration of $AlOH^{2+}$ is calculated using pH, $[Al^{3+}]$ and the stability constant for $AlOH^{2+}$. The decrease in bicarbonate weathering, C, is given by

$$C = ([HCO_3^-]_o - [HCO_3^-])(\text{Precip} - \text{ET})10^2$$

where $[HCO_3^-]_o$ is the initial concentration of bicarbonate and $[HCO_3^-]$ is the bicarbonate concentration at the beginning of the year of computation. Bicarbonate concentrations are given by

$$[HCO_3^-] = \text{inverse log } (\log PCO_2 + \text{pHCaCl}_2 + 0.8 - 7.81)$$

where 0.8 is included to calculate soil solution pH from soil $pH(CaCl_2)$.

At the end of each year a new sum of bases is calculated by subtraction from the sum of bases for the previous year. A new value for pH is calculated using the equation described in assumption (k) in Section 9.4.2.2.2.

9.4.3.4 Model Execution and Data Outputs

The model was modified and compiled using the Turbo Basic language and the simulations were run on a COMPAQ 386 desktop computer. Computations were made for changes in soil properties on an annual basis, but data were reported for 10 year intervals as indicated previously. Output data for each time interval included (1) year, (2) pH of soil, (3) acid input, (4) acid output, (5) protonation, (6) change in pH, (7) base saturation, (8) sum of bases, (9) bases lost, and (10) Al^{3+}. The output of major interest consists of the changed values of soil pH, base saturation, and Al^{3+}. These data were transferred to the data management and statistics package RS1 (BBN) on a VAX 780 mini computer, and tables and plots of pH, BS and Al^{3+} versus time were then generated.

9.4.4 Results

9.4.4.1 Soil Responses to Acid Loadings

Results of the computer simulations were tabulated in terms of values of pH, base saturation, and solution Al^{3+} attained after 20, 50, and 100 years of the simulation (Table 9.29). This was carried out to facilitate examination of the results and comparison of soils in the short term. Twenty years was selected to identify soil impacts which might require immediate remedial or protective action. Fifty years was selected as a period within which effects might be predicted with some accuracy and within which some management and planning activities could be carried out. One hundred years was chosen simply to provide some medium term indications of effects. The results have also been presented graphically by Abboud and Turchenek (1988). Examples of the model output for two of the soils are shown in Figures 9.28 and 9.29.

All soils selected for simulations in Region 3 changed only slightly in pH, base saturation, and Al^{3+} concentration over 100 years. There was little difference between the mean and maximum

Table 9.29 Changes in pH, base saturation, and solution Al content resulting from continued long-term loadings of acidity at estimated present (mean) and future (maximum) levels.

Soil	Acid Loading $kmol(H^+)ha^{-1}y^{-1}$	pH				Base Saturation				Al(μmol L^{-1})		
		0 yr	20 yr	50 yr	100 yr	0 yr	20 yr	50 yr	100 yr	20 yr	50 yr	100 yr
Region 3												
Heart	0.19 [1]	5.3	5.3	5.2	5.1	0.67	0.65	0.62	0.58	0.4	0.4	0.5
	0.39	5.3	5.2	5.1	4.9	0.67	0.63	0.57	0.48	0.4	0.5	0.7
Demmitt	0.19	5.3	5.3	5.2	5.2	0.83	0.82	0.81	0.78	0.3	0.4	0.4
	0.39	5.3	5.3	5.2	5.1	0.83	0.81	0.78	0.73	0.4	0.4	0.5
Alcan	0.19	4.4	4.4	4.4	4.3	0.57	0.56	0.55	0.53	2.1	2.2	2.5
	0.39	4.4	4.4	4.3	4.2	0.57	0.56	0.53	0.50	2.2	2.5	3.1
Boundary	0.19	4.1	4.1	4.1	4.1	0.12	0.12	0.11	0.10	4.5	4.7	5.0
	0.39	4.1	4.1	4.1	4.0	0.12	0.12	0.10	0.08	4.6	5.0	5.8
Region 4												
Firebag	0.22	4.2	4.0	3.7	3.6	0.31	0.21	0.11	0.06	6.3	13.0	19.6
	0.86	4.2	3.5	3.5	3.5	0.31	0.01	0.01	0.01	23.9	23.9	23.9
Mildred	0.22	4.8	4.7	4.6	4.5	0.55	0.52	0.47	0.40	0.9	1.2	1.7
	0.86	4.8	4.5	4.1	3.6	0.55	0.41	0.22	0.01	1.5	4.9	21.3

Continued...

Table 9.29 (Continued).

Soil	Acid Loading $kmol(H^+)ha^{-1}y^{-1}$	pH				Base Saturation				Al($\mu mol\ L^{-1}$)		
		0 yr	20 yr	50 yr	100 yr	0 yr	20 yr	50 yr	100 yr	20 yr	50 yr	100 yr
Kinosis	0.22	4.0	4.0	3.9	3.9	0.32	0.31	0.29	0.27	6.4	7.2	8.7
	0.86	4.0	3.9	3.7	3.4	0.32	0.26	0.18	0.06	8.9	17.0	42.1
Legend	0.22	4.1	4.1	4.1	4.0	0.51	0.51	0.49	0.48	4.6	5.0	5.6
	0.86	4.1	4.0	3.9	3.7	0.51	0.48	0.42	0.34	5.6	8.1	15.0
						Region 6						
Nestow	0.18	4.2	4.0	3.7	3.3	0.50	0.40	0.27	0.12	6.2	16.3	52.7
	0.33	4.2	3.8	3.3	3.1	0.50	0.32	0.10	0.01	10.6	59.3	124.9
Nicot	0.18	5.8	5.7	5.7	5.6	0.75	0.72	0.68	0.63	0.2	0.2	0.2
	0.33	5.8	5.7	5.6	5.3	0.75	0.70	0.63	0.54	0.2	0.2	0.3
Codesa	0.18	5.9	5.9	5.9	5.8	0.87	0.86	0.85	0.83	0.1	0.1	0.1
	0.33	5.9	5.9	5.8	5.7	0.87	0.85	0.84	0.80	0.1	0.1	0.2
Falun	0.18	5.8	5.8	5.7	5.7	0.85	0.84	0.83	0.82	0.2	0.2	0.2
	0.33	5.8	5.8	5.7	5.6	0.85	0.84	0.82	0.79	0.2	0.2	0.2

Continued...

Table 9.29 (Continued).

Soil	Acid Loading $kmol(H^+)ha^{-1}y^{-1}$	pH				Base Saturation				Al($\mu mol\ L^{-1}$)		
		0 yr	20 yr	50 yr	100 yr	0 yr	20 yr	50 yr	100 yr	20 yr	50 yr	100 yr
Tawatinaw	0.18	5.9	5.9	5.8	5.7	0.82	0.81	0.78	0.75	0.1	0.1	0.2
	0.33	5.9	5.8	5.7	5.6	0.82	0.79	0.75	0.68	0.1	0.2	0.2
La Corey	0.18	5.4	5.4	5.3	5.3	0.75	0.73	0.72	0.69	0.3	0.3	0.3
	0.33	5.4	5.4	5.3	5.1	0.75	0.72	0.69	0.64	0.3	0.3	0.4
						Region 7						
Culp	0.28	5.8	5.8	5.7	5.6	0.88	0.86	0.83	0.80	0.2	0.2	0.2
	0.47	5.8	5.7	5.6	5.5	0.88	0.84	0.80	0.74	0.2	0.2	0.7
Cooking Lake	0.28	6.2	6.2	6.2	6.1	0.91	0.90	0.89	0.88	0.1	0.1	0.1
	0.47	6.2	6.2	6.1	6.1	0.91	0.89	0.87	0.85	0.1	0.1	0.1
Hubalta	0.28	5.6	5.6	5.5	5.5	0.92	0.91	0.89	0.86	0.2	0.2	0.2
	0.47	5.6	5.6	5.5	5.4	0.92	0.90	0.87	0.83	0.2	0.2	0.3
Berland	0.28	5.7	5.6	5.6	5.5	0.78	0.75	0.72	0.68	0.2	0.2	0.2
	0.47	5.7	5.6	5.4	5.3	0.78	0.73	0.67	0.59	0.2	0.2	0.3
Red Willow	0.28	5.6	5.6	5.5	5.4	0.72	0.71	0.69	0.66	0.2	0.2	0.3
	0.47	5.6	5.5	5.4	5.3	0.72	0.70	0.66	0.61	0.2	0.3	0.3

Continued...

Table 9.29 (Continued).

Soil	Acid Loading $kmol(H^+)ha^{-1}y^{-1}$	pH				Base Saturation				Al($\mu mol\ L^{-1}$)		
		0 yr	20 yr	50 yr	100 yr	0 yr	20 yr	50 yr	100 yr	20 yr	50 yr	100 yr
Ponoka	0.28	5.7	5.7	5.6	5.6	0.79	0.79	0.78	0.76	0.2	0.2	0.2
	0.47	5.7	5.7	5.6	5.5	0.79	0.78	0.76	0.74	0.2	0.2	0.2
Nosehill	0.28	4.4	4.4	4.3	4.3	0.61	0.60	0.58	0.55	2.1	2.4	2.9
	0.47	4.4	4.4	4.3	4.1	0.61	0.59	0.55	0.50	2.2	2.8	3.9
Region 9												
Spruce Ridge	0.48	4.4	4.4	4.3	4.2	0.63	0.61	0.59	0.55	2.2	2.5	3.1
	1.14	4.4	4.3	4.2	4.0	0.63	0.59	0.53	0.44	2.4	3.4	6.3
Elbow	0.48	5.0	5.0	5.0	4.9	0.67	0.66	0.65	0.63	0.6	0.6	0.6
	1.14	5.0	5.0	4.9	4.8	0.67	0.65	0.63	0.59	0.6	0.7	0.8
Antler	0.48	5.8	5.8	5.7	5.7	0.93	0.92	0.91	0.89	0.2	0.2	0.2
	1.14	5.8	5.7	5.6	5.5	0.93	0.91	0.88	0.83	0.2	0.2	0.2
Delacour	0.48	5.7	5.6	5.6	5.4	0.70	0.68	0.65	0.61	0.2	0.2	0.3
	1.14	5.7	5.6	5.4	5.0	0.70	0.65	0.59	0.49	0.2	0.3	0.5
Dunvargan	0.48	5.9	5.9	5.8	5.7	0.99	0.98	0.96	0.94	0.1	0.1	0.2
	1.14	5.9	5.8	5.7	5.5	0.99	0.97	0.93	0.87	0.1	0.2	0.2

Continued...

Table 9.29 (Concluded).

Soil	Acid Loading $kmol(H^+)ha^{-1}y^{-1}$	pH				Base Saturation				Al($\mu mol\ L^{-1}$)		
		0 yr	20 yr	50 yr	100 yr	0 yr	20 yr	50 yr	100 yr	20 yr	50 yr	100 yr
						Mountain Region						
Egypt	0.24	4.0	4.0	4.0	3.9	0.17	0.16	0.15	0.13	6.2	6.8	7.7
	0.67	4.0	3.9	3.8	3.7	0.17	0.13	0.09	0.02	8.4	10.1	15.0
Topaz	0.24	4.6	4.5	4.3	4.1	0.31	0.25	0.18	0.08	1.6	2.5	4.6
	0.67	4.6	4.2	3.9	3.9	0.31	0.14	0.01	0.01	2.9	6.9	6.9

[1] Top number is the present or "mean" loading value; bottom number is the estimated future or "maximum" loading value.

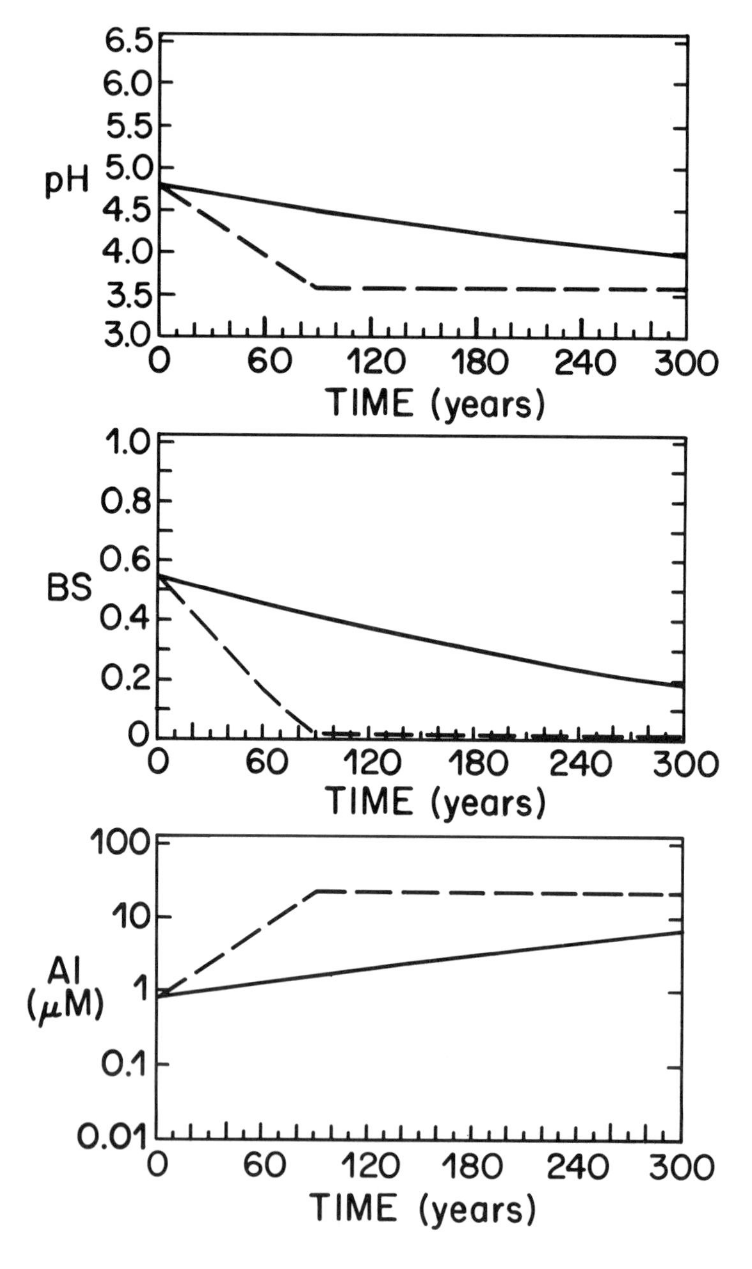

Figure 9.28 Example of model predictions for response of pH, base saturation, and solution Al concentration to acid loadings in the Mildred series, a soil with potentially high sensitivity to acid deposition (—— present, or mean loading; ------ future or maximum loading).

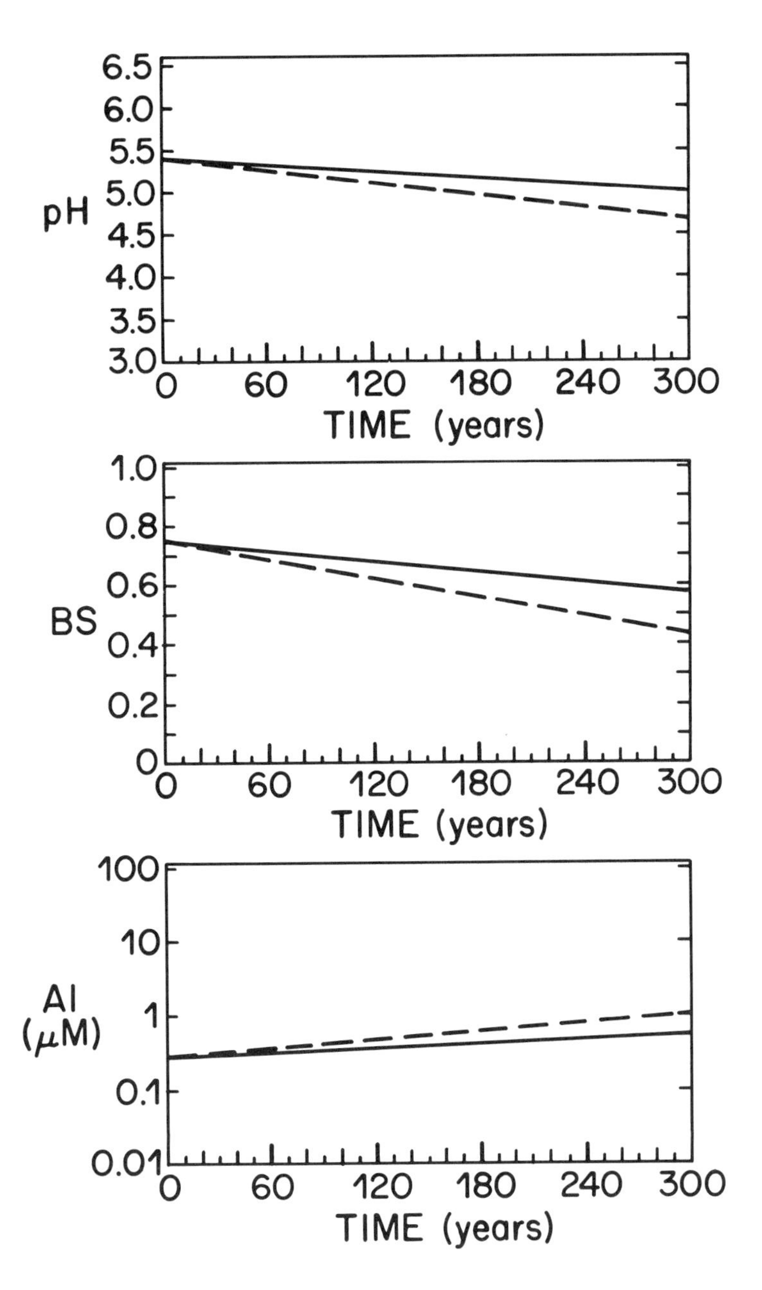

Figure 9.29 Example of model predictions for response of pH, base saturation, and solution Al concentration to acid loadings in the La Corey series, a soil with potentially moderate sensitivity to acid deposition (——— present, or mean loading; ------ future or maximum loading).

loadings of acidity and this resulted in minor differences in the soil responses.

Soil chemical properties changed markedly in Region 4. Changes in pH ranged from 0.4 units in the Legend soil to 1.2 units in the Mildred soil at the maximum levels of acid input. Most of these soils are already strongly acidic, and the acid deposition increases the acidity even more. The magnitude of the pH drop in the Mildred soil is perhaps of greatest significance in these comparisons because the results show that even in soils which are already strongly acidic, the acid inputs reduced the pH to considerably lower levels.

In Region 6, the Nestow soil pH dropped by 0.9 units at the mean input level and by 1.1 units at the maximum input level. This soil is an Eluviated Dystric Brunisol developed on sandy materials, similar to the Firebag and Mildred soils in Region 4. The base saturation dropped to 1% in these soils after 100 years of the simulation. The Firebag BS also dropped to a very low level at 6%. These soils also attained the highest Al^{3+} levels of all soils used in the simulations.

In Regions 7 and 9, and in the Mountain Region, soils did not change greatly within 100 years in any of the soil properties. Exceptions are the Delacour and Topaz soils, each of which decreased by 0.7 pH units at the maximum levels of acid input.

9.4.4.2 Soil Responses in Relation to Threshold Values

The number of years required to reduce soil pH to threshold levels was estimated for each soil from the simulation results. The results were tabulated for responses at both mean and maximum levels of acid inputs and are presented in Table 9.30. The concept of threshold or critical values of soil properties, and observations and discussion with regard to the results presented in Table 9.31 are presented in Section 9.4.5.2 below.

9.4.5 Discussion

9.4.5.1 General Impacts of Acid Deposition on Soils

Predictions for changes in pH, base saturation, and solution Al^{3+} levels in Alberta soils using the Bloom-Grigal simulation model have previously been reported by Turchenek et al. (1987). In these

Table 9.30 Model predictions of time (years) required for soils to reach threshold pH levels at estimated current and future loadings of acidic deposition.

Soil	Acid Loading $kmol(H^+)ha^{-1}y^{-1}$	Initial pH	Threshold pH 5.0	Threshold pH 4.0	Threshold pH 3.5
		Region 3			
Heart	0.19[1]	5.3	145	>300	>300
	0.39	5.3	65	>300	>300
Demmitt	0.19	5.3	280	>300	>300
	0.39	5.3	130	>300	>300
Alcan	0.19	4.4	-	>300	>300
	0.39	4.4	-	175	>300
Boundary	0.19	4.1	-	230	>300
	0.39	4.1	-	100	>300
		Region 4			
Firebag	0.22	4.2	-	15	160
	0.86	4.2	-	5	20
Mildred	0.22	4.8	-	280	>300
	0.86	4.8	-	50	90
Kinosis	0.22	4.0	-	0	>300
	0.86	4.0	-	0	80
Legend	0.22	4.1	-	110	>300
	0.86	4.1	-	20	160
		Region 6			
Nestow	0.18	4.2	-	15	70
	0.33	4.2	-	10	35
Nicot	0.18	5.8	>300	>300	>300
	0.33	5.8	195	>300	>300
Codesa	0.18	5.9	>300	>300	>300
	0.33	5.9	>300	>300	>300
Falun	0.18	5.8	>300	>300	>300
	0.33	5.8	>300	>300	>300
Tawatinaw	0.18	5.9	>300	>300	>300
	0.33	5.9	>300	>300	>300
La Corey	0.18	5.4	295	>300	>300
	0.33	5.4	155	>300	>300

Continued...

Table 9.30 (Concluded).

Soil	Acid Loading $kmol(H^+)ha^{-1}y^{-1}$ [1]	Initial pH	Threshold pH		
			5.0	4.0	3.5
		Region 7			
Culp	0.28	5.8	>300	>300	>300
	0.47	5.8	>300	>300	>300
Cooking Lake	0.28	6.2	>300	>300	>300
	0.47	6.2	>300	>300	>300
Hubalta	0.28	5.6	>300	>300	>300
	0.47	5.6	>300	>300	>300
Berland	0.28	5.7	>300	>300	>300
	0.47	5.7	210	>300	>300
Red Willow	0.28	5.6	>300	>300	>300
	0.47	5.6	200	>300	>300
Ponoka	0.28	5.7	>300	>300	>300
	0.47	5.7	>300	>300	>300
Nosehill	0.28	4.4	-	280	>300
	0.47	4.4	-	155	>300
		Region 9			
Spruce Ridge	0.48	4.4	-	220	>300
	1.14	4.4	-	95	215
Elbow	0.48	5.0	0	>300	>300
	1.14	5.0	0	>300	>300
Antler	0.48	5.8	>300	>300	>300
	1.14	5.8	270	>300	>300
Delacour	0.48	5.7	295	300	>300
	1.14	5.7	105	275	>300
Dunvargan	0.48	5.9	>300	>300	>300
	1.14	5.9	255	>300	>300
		Mountain Region			
Egypt	0.24	4.0	-	0	>300
	0.67	4.0	-	0	110
Topaz	0.24	4.6	-	125	150[2]
	0.67	4.6	-	35	140

[1] Top number: present or "mean" loading; bottom number: future estimated or "maximum" loading.

[2] Minimum pH reached in this soil was 3.9.

Table 9.31 Criteria for relating risk of forest damage to some chemical characteristics of soils.

Soil Property Risk	Increasing Risk	High Risk	Very High
$pH(H_2O)$	≥ 4.2	4.0 to 4.2	< 4.0
Base Saturation	≥ 0.05	< 0.05	0.0
Al^{3+} ($\mu mol\ L^{-1}$)	< 25	25 to 40	> 40
Ca/Al Ratio	> 0.4	0.1 to 0.4	< 0.1

Source: Ulrich et al. (1984).

studies it was concluded that sandy soils with low cation exchange capacity are the most vulnerable to change in the above properties, and that adverse effects would occur soonest in those soils which already have relatively low pH and base saturation levels. The simulation of soil responses carried out in this project corroborates the findings of the previous reports. Changes in soil properties were greatest in coarse-textured soils having low CEC and exchangeable base contents. Significant reductions in pH and base saturation occurred in coarse-textured soils within 100 years in the simulations. An increase of solution Al^{3+} concentrations to potentially toxic levels occurred in Brunisols and Luvisols with initially low pH values. Many Luvisolic soils can be significantly affected due to the sandy nature and low organic matter content of their Ahe and Ae horizons. Therefore, they have low CEC and exchangeable base levels, and hence, low buffering capacities; they approach sandy Brunisolic soils in terms of magnitude of response to given acid loadings. The cation exchange capacities and exchangeable base cation levels in coarse-textured Chernozemic soils are most likely a consequence of their higher organic matter contents. Soils such as the Red Willow series did not respond to as great an extent as did the similarly textured Luvisolic (Culp) and Brunisolic (Firebag, Mildred, Nestow) soils.

The model predictions are probably overestimations of actual conditions; that is, the calculated soil responses are worse than can be expected because of some of the assumptions made. An assumption made is that exchange of H^+ for base cations is complete. Also, the neutralization by basic substances in the polluting medium itself (i.e., bases contributed by atmospheric dust deposition), nutrient cycling, and buffering by organic matter in litter layers were not considered.

9.4.5.2 Soil Responses in Relation to Threshold Values

The concept of a critical or threshold value of a soil property has been used in evaluations of acidic deposition impacts on soils. A critical or threshold value can be defined as a quantitative value of a soil property above or below which a system would be regarded as being adversely affected. Such adverse effects are not considered in terms of the soil itself, but generally in terms of the life it supports or influences. The concept enables evaluations of soil responses in terms of a standard as opposed to making comparisons based on relative differences. There is some difficulty, however, in determining threshold values because of different requirements and tolerances among organisms in relation to soil properties. Nilsson (1986) discussed the maximum loadings of acidic and acid-forming substances to which soils could be exposed. He suggested that the loadings should be no higher than those at which weathering would totally offset acid inputs such that pH and base saturation percentage of the soil would not be affected. Thus, a threshold value other than the current value of pH and base saturation was not considered to be permissible.

Posch et al. (1985) have described the concept of "critical pH" in their evaluations of the effects of acidic deposition in Europe. Critical pH (H_2O) refers to an increased risk for adverse effects on forests due to changes in soil chemistry. The value of 4.2 was suggested for the critical pH because buffering changes at this pH from the cation exchange range to the aluminum range (Ulrich 1981, 1983). Adverse effects on vegetation could occur at this pH due to increased dissolved Al^{3+} concentrations in the soil solution. Aluminum toxicity depends on many factors, however, and there is no certainty that there is no risk above the critical pH, nor that adverse effects will definitely occur below it. Some criteria for relating the risk of forest damage by soil acidity to some chemical characteristics of soils have been prepared by Ulrich et al. (1984) (Table 9.31). These criteria were used by the International Institute for Applied Systems Analysis for evaluation of risks based on predicted emission and effects scenarios simulated by the RAINS model.

Levine and Ciolkosz (1988) applied a simulation model to determine the sensitivity of Pennsylvania soils to acidic deposition. Soils were classified into various sensitivity classes based on the amount of time required for the soil to reach any of the following threshold values: pH ($CaCl_2$) of 4.0 or less in both A and B horizons, or soil solution Al^{3+} concentration of 1.0 mg L^{-1} in the A horizon, or Al^{3+} concentration of 0.1 mg L^{-1} in the B horizon. This pH value was chosen because of the high acidity and associated conditions such as low base saturation and microbial activity, high rates of mineral weathering and leaching of heavy metals, and presence of significant levels of exchangeable H^+ and Al^{3+}. The level of 1.0 mg L^{-1} Al^{3+} in the A horizon was chosen to represent a level toxic to some plant species, and the 0.1 mg L^{-1} in the B horizon was chosen to represent a level potentially toxic in aquatic systems. This latter value was selected under the assumption that soil water would move into the groundwater and other water bodies with little additional change.

The sensitivity classes of Levine and Ciolkosz (1988) consisted of the following: (1) very sensitive soils, or those which will reach a threshold value between 0 and 30 years; (2) sensitive soils, or those which will reach a threshold value between 30 and 60 years; (3) slightly sensitive soils, or those which will reach a threshold value between 60 and 90 years; and (4) nonsensitive soils, or those which will not reach a threshold value within 90 years.

The aforementioned studies are relevant to the objectives of the current project because the critical or threshold values for soil properties may be applicable to evaluation of Alberta soils. Nilsson (1986) and Posch et al. (1985) established critical levels in soil properties in order to arrive at critical loads. The concept of critical load (i.e., the highest loading that will not cause changes leading to long-term harmful effects on the most sensitive ecological systems) has been discussed by Environmental and Social Systems Analysts Ltd. and Concord Scientific Corporation (1987). Nilsson (1986) argued that the critical or threshold levels for soil pH and base saturation should be the current ones. Posch et al. (1985), based on Ulrich et al. (1984), seem to suggest that some change in the soil is permissible as long as properties remain above certain levels. Levine and Ciolkosz (1988) used criteria similar to those of Ulrich et al. (1984). A discrepancy exists, however, since a pH of 4.2 reported by Ulrich et al. is based on its determination in H_2O, while that of Levine and Ciolkosz (1988) is based on determination in 0.01 M $CaCl_2$. The threshold values of Levine and Ciolkosz (1988) are also of interest because of the introduction of a time factor.

Simulation results were presented in terms of three different threshold values of soil pH in Table 9.31, Section 9.4.4.2.

The highest value (i.e., 5.0) was selected as a pH level below which soil conditions adverse to crop production would occur. The pH calculated in the soil acidification model would be obtained by measurement of a soil sample in $CaCl_2$ solution. The equivalent pH determined in water would be 0.5 to 0.8 units higher. Discussions of plant growth in relation to soil acidity are generally in terms of pH determined in water. Soils in the range of pH(H_2O) 5.6-6.0 are sufficiently acidic to cause serious losses in yields of most crops in Alberta (Penney et al. 1977; Hoyt et al. 1981). The lower limit of this range, equal to pH($CaCl_2$) of about 5.0, was selected as the threshold value for the purposes of this study.

It was stated previously that the model is applicable only to soils in a steady state condition; that is, it cannot take into account other acidifying or acid neutralizing influences in soils under agriculture, regenerating forest stands, and other situations. Thus, the threshold pH ($CaCl_2$) level of 5.0 is applicable only in situations such as identification of soils in which problems could occur should they be converted from native forest to crop production.

Loss of crop productivity in acidic soils is due to reduced availability of some plant nutrients, metal elemental toxicities, and restriction of nitrification and nitrogen fixation by microorganisms (Hoyt et al. 1981). *Rhizobium* species in particular are sensitive to soil pH (H_2O) below roughly 6.0 (Visser et al. 1987).

The second threshold level, pH($CaCl_2$) 4.0, was selected solely on the basis of prior use of that value by Levine and Ciolkosz (1988). From the discussion on agricultural soils, however, it is apparent that this pH cannot be applied in relation to crop growth and that it is more relevant only to evaluations of forested soils.

A low threshold pH($CaCl_2$) of 3.5 was also selected for discussion of simulation results. Many forested soils in Alberta already have surface soil pH values of about 4.0. For example, Turchenek (1982) reported pH values of less than 4.0 for numerous profiles in permanent sample plots in the Athabasca oil sands area in northeastern Alberta. The median pH($CaCl_2$) in the 0 to 5 cm layer in sandy Eluviated Dystric Brunisols was 4.4. The median pH at 15 to 20 cm was 4.7. Ranges in pH were 3.5 to 5.5 at 0 to 5 cm, and 3.6 to 5.6 at 15 to 20 cm. Of 109 horizons sampled, 15 had pH of 4.0 or less in the 0 to 5 cm layer, and 3 had pH of 4.0 or less in the 15 to 20 cm layer. Thus, many of the sandy Eluviated Dystric

Brunisols in the oil sands area are already at threshold pH values based on the criteria of Levine and Ciolkosz (1988). Solution Al^{3+} and base saturation levels were not reported for these soils.

Vegetation at all of the soil sample sites in the oil sands study consisted of jack pine-lichen communities. Functioning of these communities in soils with pH as low as 3.5-3.6 suggests that the critical (threshold) pH ($CaCl_2$) of 4.0 may be high. Indeed, the pH(H_2O) 4.2 level suggested by Ulrich et al. (1984) may be more appropriate. This pH value is equivalent to about pH 3.4 to 3.7 for determinations in 0.01 M $CaCl_2$.

It is desirable to provide evaluations based on properties other than, or in addition to, soil pH. Soil pH varies both spatially and temporally, and measurement and monitoring in relation to loadings are difficult. Therefore, use of multiple criteria such as those of Ulrich et al. (1984) and Levine and Ciolkosz (1988) is preferred. Nevertheless, only threshold values for pH were considered in this project for reasons outlined below.

The two other properties simulated by the acidification model are base saturation and solution Al^{3+} content. Aluminum levels are calculated directly from pH values. Thus, each of the threshold pH values is accompanied by a specific Al^{3+} level as follows: 0.5 μM at pH 5.0; 6 μM at pH 4.0; and 30 μM at pH 3.5. Different threshold levels for Al^{3+} which have been suggested are 10 μM (Bloom and Grigal 1985), 25 μM (Ulrich et al. 1984), and 40 μM (Levine and Ciolkosz 1988). Plant tolerances to Al^{3+} are highly variable, and it is not possible to consider only one threshold Al^{3+} level with either crop or forest species (Foy 1984; Morrison 1984). It would appear, however, that at pH 4.0, the Al^{3+} levels from the simulations would be sufficiently low for most species, and even at pH 3.5 the Al^{3+} level could be tolerable. However, the dependence of solution Al^{3+} on pH becomes very sensitive at pH 3.5. At pH 3.4 the Al^{3+} is about 40 μmol or 10 μmol higher than at pH 3.5. A pH($CaCl_2$) of 4.0 for forest soils thus seems to be an appropriate threshold level, notwithstanding the fact that some soils already are more acidic than this value.

The other soil property given in simulations output is base saturation. Base saturation does not exactly correspond to pH as might be expected from the linear relationship between the two properties. The regression of pH on BS is based on a large number of samples with considerable scatter about the regression line. Base saturation in the model is calculated from the slope of the regression line. The intercept differs for each soil modelled, depending on whether it originally lies above or below the line. The minimum pH attained is dependent on the intercept. Thus, pH differs at the same BS in different soils, and vice versa.

For soils in which the pH - BS relationship conforms exactly or close to the linear relationship, the BS ranges from 0.58 to 0.66 among the soil orders at pH ($CaCl_2$) 5.0, from 0.21 to 0.28 at pH 4.0, and from 0.00 to 0.11 at pH 3.5. Ulrich et al. (1984) indicated that a BS of 0.05 might be a threshold value for forest soils. Thus, the BS in Alberta soils may reach values below this at pH 3.5. Once again, it would appear that pH 4.0 is a more suitable threshold value than pH 3.5, while the BS at pH 5.0 indicates a system that is still dominated by base cations.

To summarize, different pH threshold values have been selected for evaluation of acidic deposition effects on soils. A pH($CaCl_2$) of 5.0 was determined to be a level at which most agricultural crops would not be adversely affected. A pH($CaCl_2$) of 4.0 appears to be a level above which forest ecosystems will remain healthy. A pH($CaCl_2$) of about 3.5 occurs in some forest soils and could be regarded as a minimum permissible level, and adverse effects could be expected below this pH.

9.4.5.3 Potential Soil Responses in the Emission Regions

Summaries of the major features of each region and general conclusions based on the previous discussions are presented in this section.

9.4.5.3.1 Region 3

a. Most soils in Region 3 are under forest cover, although extensive areas have been cleared for farming. Many of these have pH values below 5.0 as exemplified by the Boundary and Alcan series selected for this study. Farming practices on these soil types must include lime application to increase surface soil pH to levels suitable for crop growth.
b. Both of the estimated present and future loadings are relatively low at 0.19 and 0.39 kmol(H^+)$ha^{-1}y^{-1}$; therefore, present and future scenarios do not differ greatly for each soil.

c. The Heart soil typifies sandy soils with low buffering capacity in Region 3. Reduction of pH to a threshold level of 5.0 in soils such as the Heart could occur in 150 to 200 years at present acid loading rates and in 50 to 100 years at future loading rates.
d. The Demmitt soil, with pH of 5.0 to 5.5, is representative of many Luvisolic soils of Region 3. It is similar to the Braeburn soil which occurs over an extensive area in the Peace River region. Both current and future levels of acidic deposition would have minor effects on these soils in the short term. Reduction of pH to the threshold level of 5.0 would require over 100 years at the higher input rate.
e. Some Gray Luvisols already have pH values near the forest soil threshold level of pH 4.0, but they also have high buffering capacities. Even with an initial pH of 4.1 in one of the soils, 100 years or more was required to lower it to the forest soil threshold value of pH 4.0.
f. It is estimated that soils similar to the Heart account for 1.4% or about 9300 km^2, of the area of Alberta. Of this area only about 900 km^2 occur in Region 3. Similar areas of these soils occur in the adjacent Regions 1 and 5. In total, soils occupying an area between 2500 and 3000 km^2 in these three regions could be affected by acid inputs in the same way as the Heart soil investigated in this study.
g. As a general conclusion, changes in most soils in Region 3 at both present and estimated future loading rates are expected to be minor. Soil pH reductions of about 0.5 units would require about 100 years of acidic inputs at the projected rates.

9.4.5.3.2 Region 4

a. All soils in Region 4 are under forest and have initial pH levels less than 5.0
b. There is a bigger difference between present and future acid loading rates in this region than in the others, with the exception of Region 9; the rates are 0.22 and 0.86 $kmol(H^+)ha^{-1}y^{-1}$.
c. The pH, base saturation, and solution Al^{3+} concentration were significantly altered at the estimated future loading rate in all soils selected for study.

d. The Firebag soil is characteristic of sandy Dystric Brunisols having very low buffering capacities and low initial pH values. Even at the present loading rate, the pH in these soils could drop within 20 years to the lower threshold pH of 3.5 accompanied by significant base loss and by Al^{3+} rising to potentially toxic concentrations.
e. Soils such as the Mildred series have slightly higher buffering capacities and higher initial pH values than the Firebag soil. Acidic deposition effects at the present loading rate on these soils are minor, but acidification to threshold levels of pH 4.0 and 3.5 could occur within 100 years.
f. There are about 21,000 km^2 of sandy Dystric Brunisols in the Province. About two-thirds of this area (about 15,000 km^2) occurs in the northeast, within Region 4 and in the adjacent Region 2. A large proportion of the soils in this area are similar to the Firebag soil which was highly susceptible to acidic deposition impacts. Others are similar to the Mildred soil which had higher pH and buffering capacity but which nevertheless is considered to be highly sensitive.
g. Soils such as the Legend and Kinosis series have low pH values and are more strongly buffered than the Firebag and Mildred soils. Such soils will not have severe pH reductions, but their susceptibility to base loss and Al^{3+} solubilization is high, particularly at the estimated future loading rate. The modelling results indicated that Al^{3+} concentration could rise to potentially toxic levels within 50 years at the future loading rate. There are approximately 10,000 km^2 of these soils within Region 4.

9.4.5.3.3 Region 6

a. Many of the soils in Region 6 are used for agriculture. The pH values of soils selected for study are mainly above pH 5.0.
b. Both present and future acid loading rates (0.18 and 0.33 $kmol(H^+)ha^{-1}y^{-1}$) are relatively low and are similar to those in Region 3.
c. Dystric Brunisols such as the Nestow series are already very strongly acidic (pH 4.2); therefore, base loss and Al^{3+} solubilization are of concern. Reduction of pH to 3.5 and severe base loss could occur within 50 years at the future

loading rate, and within 100 years at the present loading rate. Potentially toxic Al^{3+} levels could be reached within 20 and 50 years at future and present loading rates, respectively. About 6000 km^2 of these soils occur within Region 6, including some areas at the western edge of the region which extend into Region 5.

d. Sandy soils with relatively high pH values such as the Nicot series (Eutric Brunisol) occupy another 5000 km^2 within Region 6 and adjacent parts of Region 5. Such soils may drop about 0.5 pH units in 100 years, and the accompanying base loss and Al^{3+} solubilization would be minor at both present and future loading rates.
e. Most of the Luvisolic soils in Region 6 have moderate buffering capacities and the changes in these soils will be minor at both current and future estimated loading rates. Some of these soils have very sandy surface horizons and would suffer some pH reduction similar to soils in (d) above. These soils account for about 20,000 km^2 in Region 6.

9.4.5.3.4 Region 7

a. Region 7 consists of forested soils in its western portions and some agricultural soils in the eastern parts.
b. The acid loading rates in Region 7, at 0.28 and 0.47 $kmol(H^+)ha^{-1}y^{-1}$) are intermediate to the high rates of Regions 4 and 9, and the relatively low rates of Regions 3 and 6.
c. Changes in pH, base saturation, and solution Al^{3+} in Chernozemic, Brunisolic, and Luvisolic soils occurring in this region were generally minor with both present and estimated future acid loading scenarios. The Culp soil, a sandy Gray Luvisol, was the most sensitive of the soils, decreasing from pH 5.8 to 5.0 within 60 years at the high acid input rate.
d. Soils such as the Nosehill series have initially low pH values. Effects were minor at both loadings, but acid loading rates which are higher than that of the future scenario could result in extensive base loss and solubilization Al^{3+} to toxic levels.
e. Soils with very low buffering capacities may occur in small pockets throughout Region 7. Where these occur, their responses to acid loadings would be similar to those of Region 6.

9.4.5.3.5 Region 9

a. Soils in Region 9 are used in a variety of ways including forestry, agriculture, and rangeland.
b. The highest acid loading rates, 0.48 and 1.14 kmol(H^+) $ha^{-1}y^{-1}$, occur in Region 9.
c. Most soils selected for study in this region were well buffered. The changes in soil chemistry were consequently minor, even at the future loading rate.
d. Some Chernozemic soils with moderately high buffering capacity, such as the Delacour series, could decrease in pH to the agricultural threshold of pH 5.0 in about 100 years. Liming would be required to neutralize acid inputs and maintain this pH.
e. Luvisolic and Brunisolic soils with moderate buffering capacity are extensive in the foothills portion of Region 9. The Luvisolic soil selected for study (Spruce Ridge) had an initially low pH (4.4) which decreased to pH 4.0 in 90 to 100 years. In soils with lower buffering capacities than the Spruce Ridge, a greater loss of bases and solubilization of Al^{3+} would be expected.
f. Sandy soils with very low buffering capacities may occur in small pockets in Region 9. Where these occur, their responses to acid loadings would be similar to those of Region 4 which has acid loadings approaching those of Region 9.

9.4.5.3.6 Mountain Region

a. Acid loading rates in the Mountain Region (0.24 and 0.67 kmol(H^+)$ha^{-1}y^{-1}$) are intermediate when compared to the other regions.
b. The Egypt soil typifies mountain soils of high buffering capacity. However, the initial low pH (4.0) and acid inputs could result in base depletion and increase in Al^{3+} concentration. These changes could be severe enough to adversely affect growth of vegetation within 50 years at the future estimated loading rate. Soils such as this appear to be of spotty occurrence throughout the Mountain Region.
c. Soils such a the Topaz series, having very low buffering capacity and initial pH greater than 4.5, could reach the

threshold value within 50 years at the future loading rate. Bases could diminish to very low levels, and Al^{3+} could increase to near threshold levels within 50 years as well. These very sensitive soils occupy about 2500 km^2 in the Mountain Region.

9.4.5.4 Implications Regarding Soil-Plant Relationships

9.4.5.4.1 Microbial Activity and Organic Matter Dynamics. The potential effects of acidic deposition on soils were previously reviewed by Turchenek et al. (1987). The effects of acidic deposition on organic matter turnover were summarized as follows:

a. decreased rate of C mineralization due to acidification and/or associated trace metal toxicity;
b. decreased CO_2 flux from the land to the atmosphere; and,
c. as a consequence of the above, increased retention of organic matter.

Another postulated effect related to organic matter was the possible breakdown of organo-mineral linkages due to disruption of bonding mechanisms such as cation bridge linkages. A consequence of decreased organic matter decomposition is the non-availability of plant nutrients which become tied up in the organic soil layers.

There is little information available to relate organic matter impacts to specific acid loadings or other changes in soil chemical properties. The information in this area has been summarized in a literature analysis by Visser et al. (1987). Controlled field and laboratory studies have shown that litter decomposition is retarded only if plant residues are treated with extremely acidic (pH 2.0) simulated rain or fumigated with very high concentrations (530 ppb) of SO_2. With uncontrolled inputs of acid-forming substances in the field, reductions in forest floor pH from 4.8 to 3.3, 3.5, or 3.9 have resulted in retardation of organic matter turnover. Similarly, laboratory and field experiments have shown that reduction of naturally acidic forest soils to pH 3.0 or less would have an inhibitory effect on soil respiration. Microbial biomass has also been shown to be reduced at this soil pH. It would appear from these studies that the lowest level of soil acidity at which adverse effects with regard to microbial activity and organic matter dynamics would occur is at about pH 4.0. It was shown in previous sections that many forest soils in Alberta are already at these pH levels. All those soils with pH values below this level could be expected to experience reduced microbial activity and/or altered community composition, decreased soil respiration, and increased organic matter accumulation.

9.4.5.4.2 Plant Nutrition. Availability of nutrients to plants is affected by changes in soil pH. While nitrogen, sulphur, phosphorus, molybdenum, boron, calcium, magnesium, and potassium are all reduced in their availability with decrease in pH (H_2O) below about 5.0 to 6.0, iron, manganese, zinc, copper, and cobalt increase in availability with reduced pH. Some other impacts of acidic deposition on these nutrients have been reviewed by Turchenek et al. (1987). Similarly some of the influences of acidic deposition on agricultural crops have been reviewed by Torn et al. (1987). Crop growth can be adversely affected at soil pH(H_2O) levels of less than 6.0. Alfalfa is an example of a crop that is sensitive at these pH levels. A soil pH (H_2O) of 5.0 or less is considered to be strongly acidic, and the growth of a number of crops such as alfalfa, barley, canola, and wheat can be seriously reduced. Soil acidification due to acidic deposition, however, is obscured by other influences on soil chemistry in farmed soils. Torn et al. (1987) indicated that current agricultural practices have a much greater effect on soil acidity than does acidic deposition, with atmospheric acid inputs estimated to be 1 to 2 orders of magnitude smaller than the acidic or alkaline inputs from common agricultural practices such as nitrogen fertilization and liming. Sound management practices include measures such as liming to maintain soil pH at suitable levels for crop growth which, coincidentally, mitigate the effects of atmospheric acidic deposition as well.

9.4.5.4.3 Toxic Elements. Toxicity of Al^{3+} and Mn^{2+} is probably the most limiting factor to crop growth caused by soil acidity (Foy 1984). Liming of soils is required to neutralize all or part of the toxicity of Al^{3+} and Mn^{2+} to improve crop yield. Thus, the solubilization of Al^{3+} and Mn^{2+} by acid deposition clearly is not an issue where sound management practices are followed to prevent toxic soil conditions resulting from natural soil development or from fertilization practices.

Mobilization of Al^{3+} by anthropogenic acidity to levels toxic to forest trees is, however, a concern. Little information is available on Al^{3+} toxicity to forest species. Some of the information has been reviewed by Morrison (1984) and in an ADRP report by Mayo

(1987). The increase of soil Al^{3+} concentrations to potentially toxic levels is discussed in terms of threshold values in Section 9.4.5.4.

9.4.6 Concluding Remarks

The results of simulations using the modified Bloom and Grigal (1985) model corroborated results of previous studies in which coarse-textured soils within various soil orders were found to be the most readily affected by acidic inputs. It is predicted that changes due to acid stress will be greatest in sandy soils of the Dystric Brunisol taxonomic group. Significant reductions in pH and base saturation occurred in these soils within 100 years in the simulations. The largest extent of these soils is in the Athabasca oil sands region. Relatively small areas also occur sporadically throughout the rest of the forested parts of the Province. The model results show that some of these soils may be acidified even at the current acid loading rates.

Many Luvisols and Eutric Brunisols can be significantly affected by acidic inputs due to the sandy nature and low organic matter contents of their surface horizons. These horizons have low CEC and exchangeable base levels, hence low buffering capacities, and they approach sandy Dystric Brunisols in terms of magnitude of response to given acid loadings. Many Brunisols and Luvisols have naturally low pH values. Increases in solution Al^{3+} to potentially toxic concentrations were indicated by the simulations of these soils.

Chernozemic soils examined in this study were well buffered. Even coarse-textured Chernozemics had relatively high cation exchange capacities and exchangeable base cation levels as a consequence of relatively high organic matter contents in their surface horizons.

Other soil types which occur in Alberta include those of the Solonetzic, Regosolic, Gleysolic, Podzolic, Organic, and Cryosolic orders. These were not considered in this study because their areal extent is small, because they are recognized as having strong buffering capabilities, or because the model is not applicable to them.

In interpreting the modelling results, it is again stressed that not all acid neutralizing factors were taken into consideration in the acidification simulations and, therefore, the responses found are more severe than may actually occur in ecosystems. Natural soil variability must also be considered in analyzing the results.

Changes in pH, for example, may fall within the temporal and spatial variability of acidity in soils. It is especially important that further modelling efforts include buffering due to organic matter in the litter layers of forest soils. However, the model results are considered to be reliable in the case of sandy Brunisolic soils since these have very little organic matter contributing to the acid neutralizing capacity.

9.4.7 Literature Cited

Abboud, S.A. and L.W. Turchenek. 1988. Assessment of present and future effects of acidic and acid forming substances on Alberta soils. Prep. for the Acid Deposition Research Program by Alberta Research Council, Edmonton. Open File Report, Kananaskis Centre for Environmental Research, University of Calgary. 114 pp.

Arp, P.A. 1983. Modeling the effects of acid precipitation on soil leachates. *Ecological Modelling*, 19: 107-117.

Bloom, P.R. and D.F. Grigal. 1985. Modeling soil response to acid deposition in nonsulfate adsorbing soils. *Journal of Environmental Quality*, 14: 489-495.

Bothe, R.A. and J.R. Ames. 1984. Evapotranspiration in Alberta 1912 to 1982. Hydrology Branch, Technical Services Division, Water Resources Management Services, Alberta Environment, Edmonton.

Canada Soil Survey Committee. 1978. The Canadian system of soil classification. Canada Department of Agriculture Publication 1646. Ottawa: Supply and Services Canada. 164 pp.

Coote, D.R., D. Siminovitch, S. Singh, and C. Wang. 1982. The Significance of Acid Rain to Agriculture in Eastern Canada. Land Resource Research Institute Contribution No. 119. Ottawa: Agriculture Canada. 26 pp.

Environmental and Social Systems Analysts Ltd., and Concord Scientific Corporation. 1987. Interim Target Loadings for Acidic Deposition in Western Canada: A Synthesis of Existing Information. Technical Committee for the Long-Range Transport of Atmospheric Pollutants in Western and Northern Canada. Victoria, British Columbia. 214 pp.

Foy, C.D. 1984. Physiological effects of hydrogen, aluminum, and manganese toxicities in acid soil. In: **Soil Acidity and Liming, Number 12 in the series Agronomy**, ed., F. Adams, American Society of Agronomy, Madison, Wisconsin. pp. 57-97

Gherini, S.A., L. Mok, R.J.M. Hudson, G.F. Davis, C.W. Chen, and R.A. Goldstein. 1985. The ILWAS model: formulation and application. *Water, Air, and Soil Pollution*, 26: 425-459.

Holowaychuk, N. and R.J. Fessenden, 1987. Soil Sensitivity to Acid Deposition and the Potential of Soil and Geology to Reduce the Acidity of Acidic Inputs. Earth Sciences Report 87-1. Alberta Research Council, Edmonton, Canada. 38 pp.

Hoyt, P.B., M. Nyborg, and H. Ukrainetz. 1981. Degradation by acidification. Proceedings of the 18th Annual Soil Science Workshop. 1981 February 24-25; Edmonton, Alberta; pp. 41-71.

Kauppi, P., J. Kamari, M. Posch, and L. Kauppi. 1986. Acidification of forest soils: Model development and application for analyzing impacts of acidic deposition in Europe. *Ecological Modelling*, 33: 231-253.

Kjearsgaard, A.A. 1972. Reconnaissance soil survey of the Tawatinaw map sheet (83-I). Alberta Soil Survey Report Number 29. Edmonton: Alberta Institute of Pedology. 103 pp.

Levine, E.R. and E.J. Ciolkosz. 1988. Computer simulation of soil sensitivity to acid rain. *Journal of the Soil Science Society of America*, 52: 209-215.

Magdoff, F.R. and R.J. Bartlett. 1986. Letters to the editor. Comments on "modeling soil response to acidic deposition in non-sulfate adsorbing soils." *Journal of Environmental Quality*, 15: 199.

Mayo, J.M. 1987. The Effects of Acid Deposition on Forests. Prep. for the Acid Deposition Research Program by the Department of Biology, Emporia State University, Kansas. ADRP-B-09-87. 80 pp.

Morrison, I.K. 1984. A review of literature on acid deposition effects in forest ecosystems. *Forestry Abstracts*, 45: 483-506.

Nilsson, S.I. 1986. Critical deposition limits for forest soils. In: **Critical Loads for Sulphur and Nitrogen,** Report from a Nordic Working Group, J. Nilsson, ed. Copenhagen: Nordic Council. 151 pp.

Penney, D.C., M. Nyborg, P.B. Hoyt, W.A. Rice, B. Siemans, and D.H. Laverty. 1977. An assessment of the soil acidity problem in Alberta and northeastern British Columbia. *Canadian Journal of Soil Science*, 57: 157-164.

Picard, D.J., D.G. Colley, and D.H. Boyd. 1987. Overview of the Emission Data: Emission Inventory of Sulphur Oxides and Nitrogen Oxides in Alberta. Prep. for the Acid Deposition Research Program by Western Research, Division of Bow Valley Resource Services Ltd., Calgary, Alberta, Canada. ADRP-A-1-87. 87 pp.

Posch, M., L. Kauppi, and J. Kamari. 1985. Sensitivity analysis of a regional scale acidification model. Collaborative Paper-85-45, International Institute for Applied Systems Analysis. Laxenberg, Austria. 33 pp.

Reuss, J.O. 1980. Simulation of soil nutrient losses resulting from rainfall acidity. *Ecological Modelling*, 11: 15-38.

Summers, P.W. 1986. The role of acid and base substance in wet and dry deposition in western Canada. In: **Acid Forming Emissions in Alberta and Their Ecological Effects, Proceedings of the Second Symposium/Workshop**, eds., H.S. Sandhu, A.H. Legge, J.I. Pringle, and S. Vance. Prep. by Research Management Division, Alberta Department of the Environment and Kananaskis Centre for Environmental Research, University of Calgary. 1986 May 12-15. Calgary, Alberta. pp. 85-116

Tajek, J., W.W. Pettapiece, and K.E. Toogood. 1989. Water Supplying Capacity of Alberta Soils. Alberta Institute of Pedology Report Publication M89-2. Edmonton, Alberta. 20 pp. plus maps.

Torn, M.S., J.E. Degrange, and J.H. Shinn. 1987. The Effects of Acid Deposition on Alberta Agriculture: A Review. Prep. for the Acid Deposition Research Program by the Environmental Sciences Division, Lawrence Livermore National Laboratory, California. ADRP-B-08-87. 106 pp.

Turchenek, L.W. 1982. Soils of Permanent Sample Plots in the Athabasca Oil Sands Area. Prep. for Research Management Division, Alberta Environment by Alberta Research Council. RMD Report OF-36. Edmonton, Alberta. 64 pp.

Turchenek, L.W., S.A. Abboud, C.J. Tomas, R.J. Fessenden, and N. Holowaychuk. 1987. Effects of Acid Deposition on Soils in Alberta. Prep. for the Acid Deposition Research Program by the Alberta Research Council, Edmonton, ADRP-B-05-87. 202 pp.

Ulrich, B. 1981. Theoretische Betrachtungen des Ionenkreislaufs in Waldokosystemen. *Z. Pflanzenernahr. Bodenkunde,* 144: 647-659.

Ulrich, B. 1983. Soil Acidity and Its Relation to Acid Deposition. In: **Effects of Accumulation of Air Pollutants in Forest Ecosystems**, eds., B. Ulrich and J. Pankrath. Proc. Workshop. Gottingen, FRG, May 16-19, 1982. D. Reidel Publ. Co., Dordrecht, pp. 127-146.

Ulrich, B., K.J. Miewes, N. Konig, and P.K. Khanna. 1984. Untersuchungsverfahren und Kriterion zur Bewertung der Versauerung und ihrer Folgen in Waldboden. *Forst. u. Holzwirt,* 39: 278-286.

Visser, S., R.M. Danielson, and J.F. Parr. 1987. Effects of Acid-Forming Emissions on Soil Microorganisms and Microbially-Mediated Processes. Prep. for the Acid Deposition Research Program by the Kananaskis Centre for Environmental Research, The University of Calgary, and the U.S. Department of Agriculture, Beltsville, Maryland. ADRP-B-02-87. 86 pp.

9.5 ASSESSMENT OF THE SENSITIVITY OF ALBERTA SURFACE WATERS TO ACIDIFICATION

R. A. Crowther and E. C. Krug

9.5.1 Introduction

A study of the effects of acidic deposition on the surface water resources of Alberta was undertaken to evaluate the risk associated with current regional deposition and future worst-case atmospheric loading of acid-forming emissions. For these purposes the Province was divided into ten emission regions (Figure 9.30). Division of these emission regions was according to the emission inventory work completed by Picard et al. (1987). The effects of acidic deposition on the rivers and lakes of five major regions were examined; these were Regions 3, 4, 6, 7, and 9. Regions 4, 7, and 9 presently exhibit relatively high emissions of acid-forming compounds; Regions 3 and 6 are expected to undergo industrial growth in the future which may cause their emission levels to increase. Regions 7 and 9 were further sub-divided along the foothills to account for the vast physiographic and depositional differences between the mountains and the plains in those regions.
Sub-divisions of these regions were designated as Sub-regions 7b and 9b.

There are many uncertainties in predicting aquatic acidification by acidic deposition. Where either the data base or the knowledge is inadequate, risk assessment analyses provide worst-case scenarios. The rationale for the development of such scenarios is that, when in doubt, it is better to examine the potential for impacts, however improbable they may be, than not to consider the potential for impacts at all. The problems associated with this type of approach are that predictions developed on the basis of worst-case scenarios are often accepted and presented as being reasonable and actual without presentation of some important limitations (Krug et al. 1985). In the present study, worst-case scenarios were developed as a screening mechanism to identify where and under what conditions surface water acidification is possible, or impossible, within the context of the regions of Alberta that were examined.

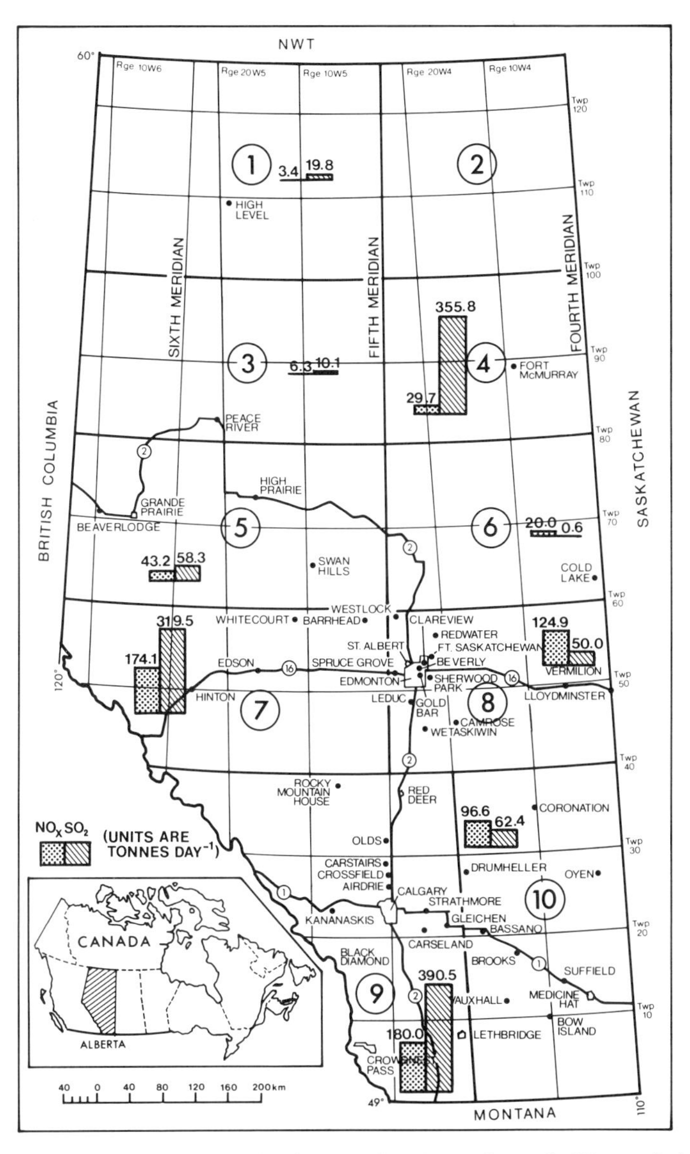

Figure 9.30 Map of Alberta showing total NO_x and total SO_2 emissions as histograms in each of the 10 geographic emission regions.

The analytical techniques used during this study, and the assumptions associated with them, are presented below. In cases where the potential for surface water acidification was detected, the consequences of accepting the analytical assumptions are stated and discussed.

9.5.2 Modelling Aquatic Acidification/Sensitivity of Waters

It can be considered that there are three general classes of models used in defining acidification, or sensitivity to surface water acidification. The simplest "models" assign "sensitivity classes" to ranges of surface water alkalinity. Examples of this class of model are Erickson (1987) and Erickson and Trew (1987). The other two classes of models are more fastidious in their requirements for surface water chemistry. Both classes of models treat the acidification of surface waters by sulphuric acid as a titration, i.e., acid titration of basic solutions. These models assume that surface waters are the result of carbonic acid weathering of mineral bases and the alkalinity produced is titrated by the deposition of H_2SO_4. Generally, these models assume that little or no deposited HNO_3^- gets into surface waters, therefore, this parameter is usually ignored.

The simplest class of titration models considers the titration to be literally like that of a basic solution in an unreactive beaker. In other words, deposited acid is 100% effective in reducing the alkalinity of a receiving water.

The most sophisticated class of titration models considers watershed/acidic deposition interactions. Examples of such models are the ILWAS (Gherini et al. 1984; Goldstein et al. 1984), MAGIC (Cosby et al. 1985), and Enhanced Trickle Down (Schnoor et al. 1984) models. All these models treat the watershed as a reactive funnel where most of the incoming water usually does not fall directly into a beaker but rather onto a funnel which contains a filter capable of neutralizing some undefined portion of the incoming acid. The funnel and filter used in these models are hypothesized to represent the drainage basin and the important processes that occur within it. Most of the effort in these types of models is expended by quantifying the portion of deposited acid that is neutralized by the watershed. The data requirements of these models are extensive, requiring input data for watershed topography, geology, hydrogeology, mineralogy, soils, and biology. Within the Alberta context, these data resources, in general, are simply not available except for extremely localized areas. Therefore, for a regional assessment, the simpler titration models are the only appropriate tools available to estimate potential effects of acidic deposition.

The interaction/titration-acidification models generally consider surface waters to be less susceptible to acidification than do the literal titration models because ameliorating effects of the watershed are considered to have neutralized some of the deposited acid. However, Krug (1987) reported that the watershed interaction models still tend to overestimate potential acidification as a result of deposition. Krug showed that the models do not predict the existence of naturally acidic, clear, and coloured surface waters in unpolluted areas; nor do the models agree with the paleoecological data. For example, Krug (1987) showed that of 19 lakes examined in the northeastern United States which currently have values lower than is possible on the basis of carbonate chemistry alone (pH<5.5), 16 of the 19 lakes examined were found to have had pH values of <5.5 prior to industrialization. Of the 16 lakes, 14 are currently clear-waters. The models do not predict that pH <5.5 clear-waters could have existed in pre-industrial times (Krug 1987).

A good example of the type of error generated by these interaction models may be seen in the predictions made for Woods Lake (one of the three ILWAS lakes, Gherini et al. 1984). Of the three ILWAS lakes, Woods Lake was the only one with a pH lower than 5.5. According to the Interim Assessment Report (U.S. NAPAP 1987), ILWAS predicted that the pH of Woods Lake would rise from pH 5 to about pH 6 if sulphate deposition could be cut in half (Malanchuk and Turner 1987). In the same report, paleoecological records were cited that indicated that the present pH of Woods Lake was very similar to its pre-industrial pH. Similar findings were also reported by the National Academy of Science (Charles and Norton 1986). The ILWAS model is the most mechanistic and is considered the best available watershed interaction/titration model. It is so complex that it usually requires a super-computer to run. However, as indicated, in spite of the complexities the model predictions may not be accurate.

As previously mentioned, within Alberta, comprehensive watershed data of sufficient quality and quantity are not available for use in mechanistic acidic deposition (watershed interaction) models. This, in fact, is usually the case where regional water surveys represent the data base. For these reasons, simple and less data-intensive titration models are usually used to predict trends toward, or the potential for, regional acidification. Simple titration models have been used to predict the regional sensitivity of surface waters to acidification in Scandinavia (Henriksen 1979), in the

Canadian Maritimes (Thompson 1982), and in the United States (Sullivan et al. 1988).

The Henriksen (1979) model assumes that unpolluted waters are principally of the calcium bicarbonate type with constant Ca^{2+}/Mg^{2+} ratios. However, alumino-silicate rock, which is common in Alberta, may contain proportionately more Na^{+} + K^{+} than Ca^{2+} + Mg^{2+} which, on weathering, produces Na^{+}/K^{+} waters. The Thompson (1982) model takes this into account (Kramer and Tessier 1982; Haines and Akielaszak 1983) and is considered more appropriate for use in Alberta.

Therefore, the potential sensitivity of Alberta's surface waters to acidic deposition was modelled using the Thompson (1982) cation denudation model. It was recognized that significant physio-chemical differences exist between the Maritimes (where the model was developed for use) and Alberta as well as in the types of input data available. These differences required that the model be modified for use in Alberta prior to assigning any level of confidence to the output and/or predictions generated.

9.5.3 Materials and Methods

9.5.3.1 The Thompson Model

The Thompson (1982) cation denudation model assumes that non-marine, or excess sulphate in the rivers of Maritime Canada, are the result of acidic deposition. Pre-acidification alkalinity is assumed to equal the sum of non-marine Ca^{2+} + Mg^{2+} + Na^{+} + K^{+}. Non-marine sulphate is defined by the Thompson model as the loss of alkalinity. Loss of alkalinity is related to pH (an ecologically significant parameter) by the equation

$$pH = pK + pPCO_2 - pHCO_3^-$$

where K equals the equilibrium constant for the dissociation of carbonic acid to bicarbonate, p represents negative logarithm, and P the partial pressure of carbon dioxide in atmospheres. The value for pK varies only slightly over the temperature range normally found in Canadian waters (Garrels and Christ 1965). Therefore, if the carbon dioxide pressure is known, there is a direct, calculable relationship between pH and HCO_3^-. In the atmosphere, $pPCO_2$ is about 3.5. This equation, plus the relationship between non-marine anions and cations indicated by the following equation, form the basis of the Thompson (1982) model:

$$\text{Sum of cations} - SO_4^{2-} = HCO_3^-$$

These equations have been shown to be valid only for carbonate solutions with pH under 7.5 (Kramer and Tessier 1982).

The following assumptions also apply to the model:

a. All non-marine sulphate in precipitation is anthropogenic in origin and is in the form of sulphuric acid (H_2SO_4);
b. All non-marine sulphate in surface waters is derived from atmospheric deposition;
c. All H^+ associated with deposited sulphate passes through watersheds into the surface waters unaltered (e.g., acidic deposition does not result in increased weathering of minerals or the depletion of exchangeable bases from the soil, or F = 0.0);
d. All watershed processes are at steady state from pre- through post-deposition; and
e. Water chemistry is perfectly described by the carbonic acid weathering of minerals.

Point (c) also assumes that biological and other watershed processes do not alter the concentration of incoming sulphate in any way.

Comprehensive regional wet and dry pollutant deposition data were obtained by ADRP. These data were used throughout the modelling process and are shown in Tables 9.32A & B. Additional data that pertain to regional precipitation and evapotranspiration rates were obtained from Alberta Department of Environment, and Atmospheric Environment Service of Environment Canada. In combination, these data were used to directly estimate the influence of atmospheric deposition on surface water chemistry. Using these data sources, direct estimates of the atmospheric contribution to lakewater sulphate were derived. These estimates were then compared with measured values (Tables 9.32A & B) to test the validity of model assumptions regarding sulphur chemistry.

9.5.3.2 Sources of Water Quality Data

Water quality data were gathered from a number of sources, including the following:

Table 9.32A Mean and maximum wet and dry deposition of air pollutants by region in kg ha^{-1} y^{-1} (refer to Chapter 6).

	Mean			Maximum		
	Wet	Dry	Total	Wet	Dry	Total
REGION 3						
SO_4^{2-}	3.7	6.0 *	9.7	19	10	19
NO_3^-	2.1	0.9 **	3.0	5.7	1.4	7.1
Ca^{2+}	0.7	--	--	1.8	--	--
H^+	0.036	--	--	0.09	--	--
REGION 4						
SO_4^{2-}	9.6	11	21	24	42	66
NO_3^-	1.8	0.9	2.7	4.9	2.9	7.8
Ca^{2+}	5.4	--	--	15	--	--
H^+	0.005	--	--	0.014	--	--
REGION 6						
SO_4^{2-}	2.5	7.6	10	6	12	18
NO_3^-	1.0	2.7	3.7	2.6	5.7	8.3
Ca^{2+}	1.1	--	--	3.0	--	--
H^+	0.024	--	--	0.074	--	--
REGION 7						
SO_4^{2-}	10	12	22	25	18	43
NO_3^-	4.3	3.3	7.6	11.6	10.0	22
Ca^{2+}	5.3	--	--	14.3	--	--
H^+	0.005	--	--	0.014	--	--
REGION 9						
SO_4^{2-}	8.2	14.2 *	22.4	26.8	27.4	54
NO_3^-	4.9	8.1 **	13.0	15.9	23	39
Ca^{2+}	2.8	1.9	4.7	5.4	7.8	13
Mg^{2+}	0.45	0.29	0.74	0.98	1.05	2.0
Na^+	0.51	0.14	0.65	1.34	0.12	1.6
K^+	0.37	0.50	0.87	0.82	1.48	2.3
H^+	0.09	--	0.09	0.28	--	0.28
Cl^-	0.76	0.01	0.77	1.34	0.12	1.5
NH_4^+	2.4	4.1	6.5	9.6	17.6	27.2

* Includes SO_2 and fine and coarse particle SO_4^{2-}.

** Includes NO_x, HNO_2, HNO_3, and fine and coarse particle NO_3^-.

Table 9.32B Deposition of air pollutants at the three ADRP monitoring sites ($kg\ ha^{-1}y^{-1}$).

	SO_4^{2-}	NO_3^-	Ca^{2+}	Mg^{2+}	Na^+	K^+	H^+	NH_4^+
Fortress Mountain								
Wet deposition	7.1	4.6	3.1	0.38	0.47	0.35	0.10	1.9
Deposition of gases	4.3	5.1	--	--	--	--	--	1.2
Fine particle deposition	0.13	0.04	0.00	0.00	0.00	0.01	--	0.06
Coarse particle deposition	0.51	0.48	0.89	0.03	0.13	0.28	--	0.26
Total deposition	12.0	10.2	4.0	0.41	0.61	0.64	0.10	3.4
Crossfield West								
Wet deposition	10.8	5.5	2.4	0.39	0.86	0.37	0.069	3.2
Deposition of gases	14.6	15.9	--	--	--	--	--	7.2
Fine particle deposition	0.28	0.19	0.02	0.00	0.00	0.01	--	0.09
Coarse particle deposition	1.29	0.87	3.3	0.46	0.14	0.65	--	0.41
Total deposition	26.2	22.5	5.7	0.85	1.0	1.0	0.069	10.9
Crossfield East								
Wet deposition	10.9	5.6	3.1	0.57	0.63	0.34	0.067	3.1
Deposition of gases	18.4	18.4	--	--	--	--	--	9.4
Fine particle deposition	0.30	0.22	0.03	0.00	0.00	0.01	--	0.11
Coarse particle deposition	1.39	1.03	4.19	0.55	0.25	0.82	--	0.49
Total deposition	31.0	25.2	7.3	1.13	0.88	1.17	0.067	13.1

- Alberta Department of Environment (D.Trew NAQUADAT files (Demayo 1970))
- Alberta Fish and Wildlife, St. Paul (H. Norris, pers. comm.)
- ESSO Resources Canada Limited (Cold Lake Regional Water Quality Network, A. Kennedy, pers. comm.)
- Shell Canada Resources Limited (Waterton Gas Plant Water Quality Network, N. Graham, pers. comm.)
- Alberta Department of Environment (A. Soziak, pers. comm.) Southern and Central Regions

In total, the water chemistry of 484 lakes was evaluated. This data base was broken down among regions as follows:

Region 3 - 17 lakes
Region 4 - 44 lakes
Region 6 - 19 lakes
Region 7 - 140 lakes
Sub-region 7b - 126 lakes
Region 9 - 142 lakes
Sub-region 9b - 104 lakes

Sub-regions 7b and 9b are composed of the Cordilleran portion of each of the main Regions 7 and 9. The lakes shown as belonging to Sub-regions 7b and 9b were included in the total for each of the main regions.

The remaining 122 lakes, for which data were collected, were within Emission Regions 1, 2, 5, 8, and 10, regions not chosen for study. Consequently these data were not included in this analysis.

In addition to lakes, the water chemistry of several rivers and streams was also evaluated. These streams were as follows:

Region 3	Smoky River
Region 4	Firebag, Steepbank, Muskeg, McKay, Hangingstone, and Ells River
Region 6	Jackfish Creek, Marie Creek, Sinclair Creek, and Sand River
Region 7	McLeod, Pembina, and North Saskatchewan (above Edmonton) River
Region 9	Three Hills Creek, Red Deer River (above Dickson Dam), Willow Creek, North Drywood Creek, and Kneehills Creek

In each case, for both lakes and rivers, the following data were archived into a permanent computerized data base:

Name
Legal site description, longitude and latitude
Date of sampling
Mean depth of lake
Surface area of lake
pH
Alkalinity
Calcium concentration
Magnesium concentration
Sodium concentration
Potassium concentration
Chloride concentration
Sulphate concentration
Total organic carbon concentration
Total ammonium concentration
Nitrate-nitrite concentration (or TKN)

These data were sorted by field and record number using the standard data manipulation system: Lotus-Symphony (IBM version 3.3). Following the original sort, data were then sorted by region to form concentration groupings with respect to total alkalinity and sulphate content.

A number of lakes from Regions 1, 2, and 5 were also evaluated during the process of modelling. The results of these analyses form part of the archival system but do not form part of the results section of this report although they may be discussed in a general way.

Water Quality Testing Protocols: Often when these types of regional assessments are conducted, the water chemistry used is somewhat suspect because of differences in the analytical techniques used over either the years or between different laboratories. In an attempt to clarify the validity of the present data base, the analytical techniques employed were identified and were recorded by parameter and test period.

Testing procedures that have a bearing on the modelling results are: calcium, magnesium, potassium, sodium, and sulphate. The changes in the analytical techniques of calcium, magnesium, and sulphate noted above between the years 1961 and 1980 and their consequences were examined. In all cases, cited NAQUADAT codes were derived either from source documentation (where this was available) or from discussions with the original investigators.

The techniques used were then matched to the NAQUADAT dictionary codes supplied by Alberta Department of Environment.

The original test for calcium (20101L) was based on an EDTA titration using Calver II indicator for the detection of the end point. If the sample was turbid prior to titration, it was first passed through a 0.45 μm filter. The filtrate was then treated with sodium hydroxide solution and Calver II and finally titrated with standard EDTA solution (disodium dihydrogen ethylendiamine tetracetate). Interference occurred only if heavy metal concentrations of greater than 0.5 mg L^{-1} were found in the test solution. Using this analytical technique, the detection limit for calcium was 0.5 mg L^{-1}. The test was used from 1961 through 1973 throughout Alberta by all agencies. In reviewing the data it was found that in a number of instances calcium concentrations during these years were reported as being equivalent to the detection limit. Since the accuracy of these values was uncertain, a value equivalent to 1/2 the detection limit was used.

For the remainder of the period in the record, calcium concentrations were determined using atomic absorption spectroscopy (A.A.S.) with a detection limit of 2 μg L^{-1}. If the samples were turbid, the sample was first filtered using a 0.45 μm filter, and lithium chloride was added to the filtrate prior to analysis. The A.A.S. procedure followed was that of Traversy (1971). This technique has been in use from 1974 up to the present time by all government and industrial agencies from which data were obtained.

Sulphate measurement methodology also appears to have changed during the periods 1961-73 and 1974-present. The basic technique was the same: a $BaCl_2$ colorimetric titration with a detection limit of 1.0 mg L^{-1}. However, the major difference appears to be in the use of alcohol and thorin in the earlier measurements and methylthymol blue in the latter period as an indicator. On the basis of this, no real differences in analytical results were thought to be present in the data. However, review of the actual data indicated that a less accurate, modified field method with a minimum detection limit of 5 mg L^{-1} was used for a substantial number of lakes. Accordingly, the more conservative value of 2.5 mg L^{-1} (one-half of the minimum detection limit) was used for sulphate in these lakes.

The third parameter, for which changes in analytical methodology were found over the period of the data collection, was magnesium. Initially, magnesium concentrations were calculated, not measured . Calculations were based on the determinations of total hardness and calcium concentration as follows:

$$[Mg^{2+}] = ([Tot.Hard] * (0.01998) - [Ca^{2+}] * (0.0499) * (12.16))$$

where; [ ] denotes concentration. This technique was used between 1974 and 1978.

Dissolved magnesium concentrations were determined between 1974-78 using A.A.S. following the procedure outlined in Traversy (1971). In addition, in the data sets examined, dissolved magnesium concentrations were not calculated during 1961-73 or in 1980.

As shown in the analytical protocols outlined in Table 9.33, it appears that both calculated and A.A.S. measured values were used to report Mg^{2+} concentrations in surface waters during the period for which the data were available. A comparison of the two techniques showed remarkable agreement. The detection limit for Mg^{2+} by the A.A.S. technique is 0.1 mg L^{-1}.

Table 9.33 Analytical techniques used for measurement of Alberta Surface water chemistry by time period.

	NAQUADAT CODES				
Parameter/Year	1961-73	1974-76	1978	1980 - Present	
pH (laboratory)	10301L	10301L	10301L	10301L	
Total alkalinity	10101L	10101L	10101L	10101L	
Calcium dissolved	20101L	20103L	20103L	20103L	*
Magnesium dissolved	12102L	12101L	12101L	12102L	*
Sodium dissolved	11103L	11103L	11103L	11103L	
Potassium dissolved	19103L	13103L	19103L	19103L	
Chloride dissolved	17203L	17206L	17206L	17206L	*
Sulphate dissolved	16303L	16306L	16306L	16306L	
Nitrate-Nitrite Nitrogen	07105L	07106L	07110L	07110D	***
Ammomium	07551L	missing	07506P	07506P	*

L = a laboratory analysis without any on-site preservation.
P = a laboratory analysis with on-site preservation of sample.
D = Samples filtered in the field and analyzed in the laboratory.
* = a change in sampling technique.
*** = several changes in analytical technique over the years.

On the basis of this review, it would appear that the available data are compatible among and between years with the exception of the noted problems with early calcium determinations. However, the corrections applied to these values make even these data quite usable.

All water quality data used in the analysis were for the open water season, and for lakes they represented mid-lake samples collected from 1.0 m depth.

9.5.3.3 Data Analysis

The water chemistry data were sorted by region, alkalinity, and sulphate concentration using the sort merge function of "Statistics II." Unweighted linear regression analyses of the natural logarithms of sulphate concentration versus concentration of the sum of total cations (Na^+ + Ca^{2+} + K^+ + Mg^{2+}) were performed for all regions and lakes (Figure 9.31). In addition, the relationship of atmospheric sulphate contribution to lake sulphate concentration was plotted by region against lake number (Figures 9.32-9.37).

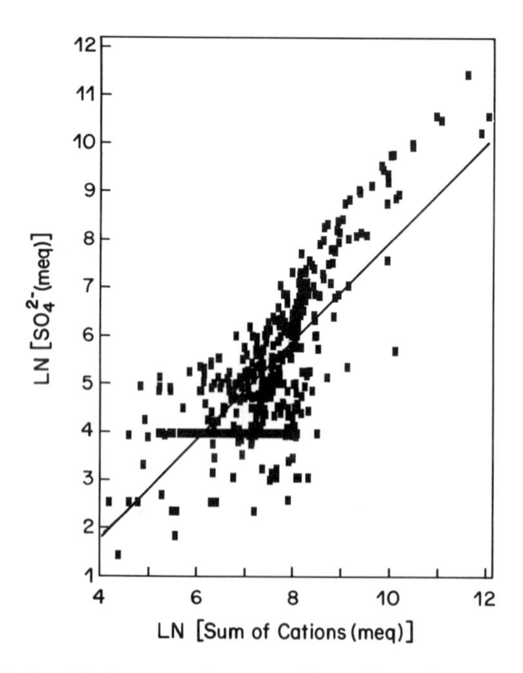

Figure 9.31 Relationship between the natural logarithm of sulphate concentration and the natural logarithm of total cation concentration (Na^+ + Ca^{2+} + K^+ + Mg^{2+}) for 484 lakes.

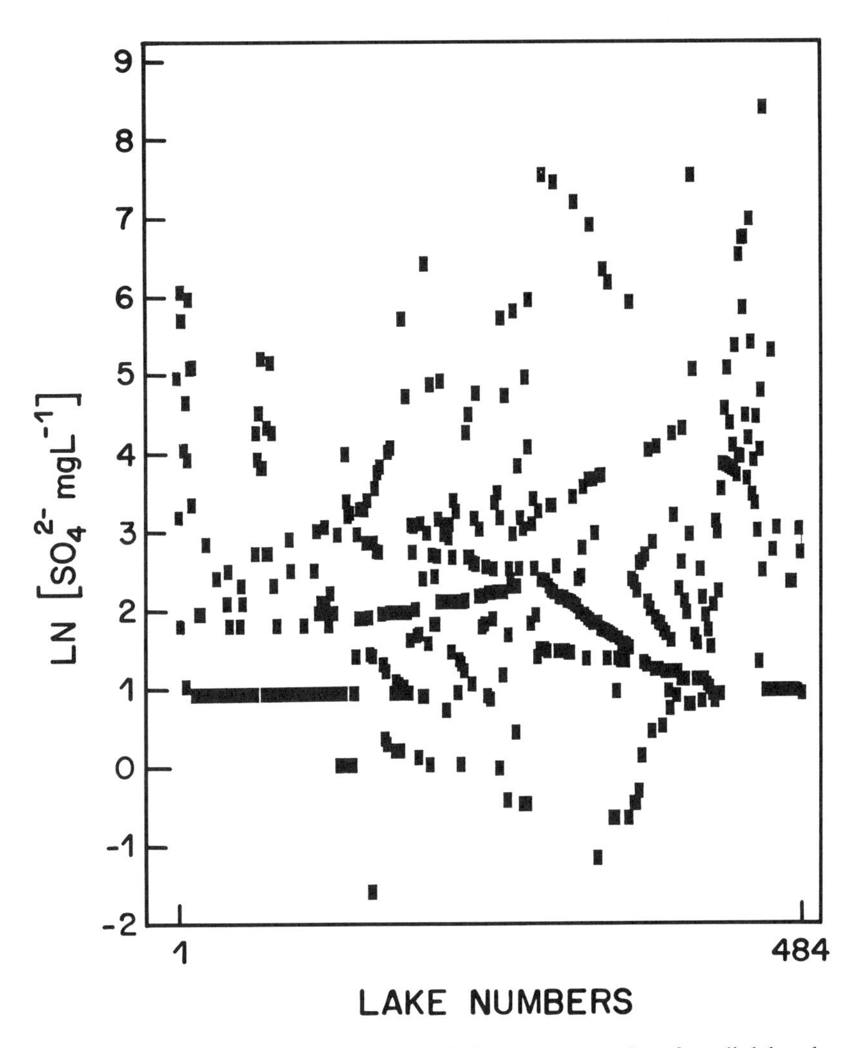

Figure 9.32 Natural logarithm of lake sulphate concentration for all lakes by number.

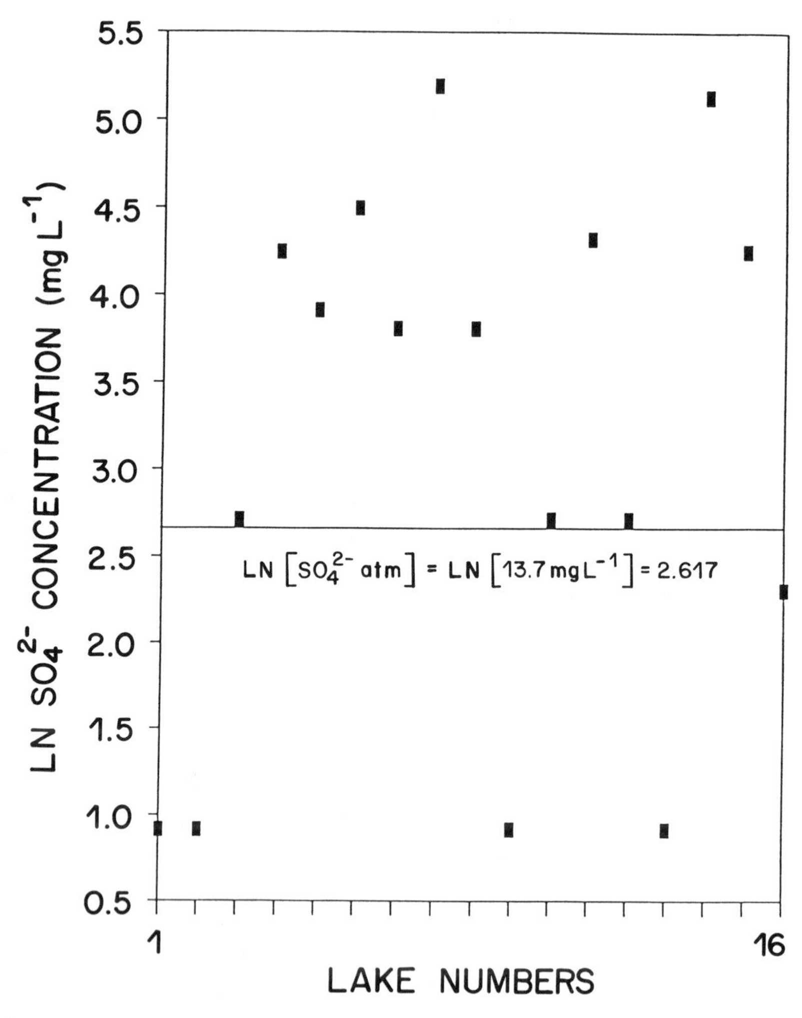

Figure 9.33 Natural logarithm of lake sulphate concentration for all Region 3 lakes by number showing the relationship to estimated contribution of atmospheric sulphate to lake sulphate concentration.

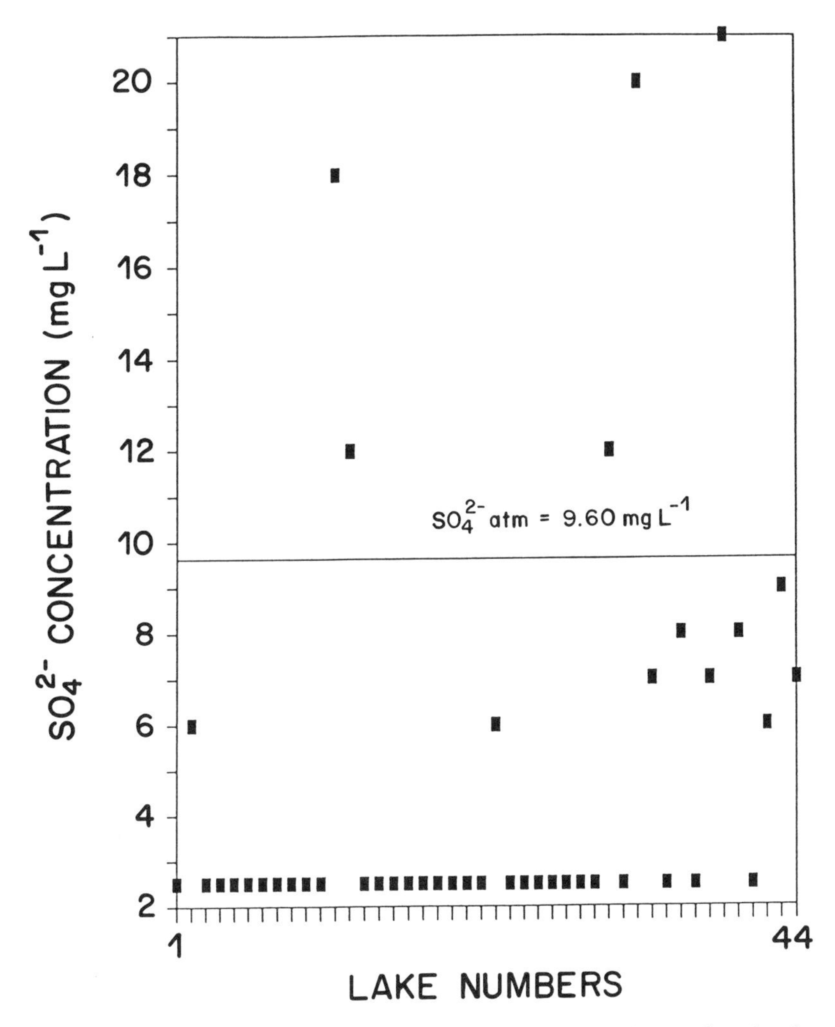

Figure 9.34 Lake sulphate concentration for all Region 4 lakes by number showing the relationship to estimated contribution of atmospheric sulphate to lake sulphate concentration.

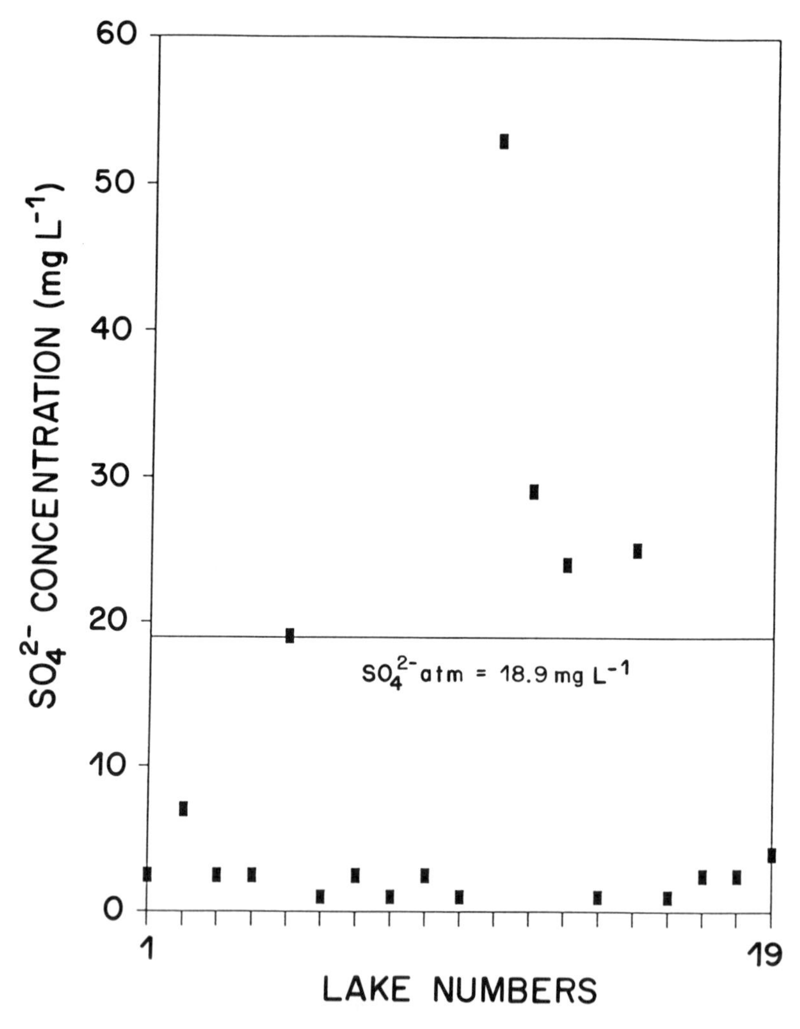

Figure 9.35 Lake sulphate concentration for all Region 6 lakes by number showing the relationship to estimated contribution of atmospheric sulphate to lake sulphate concentration.

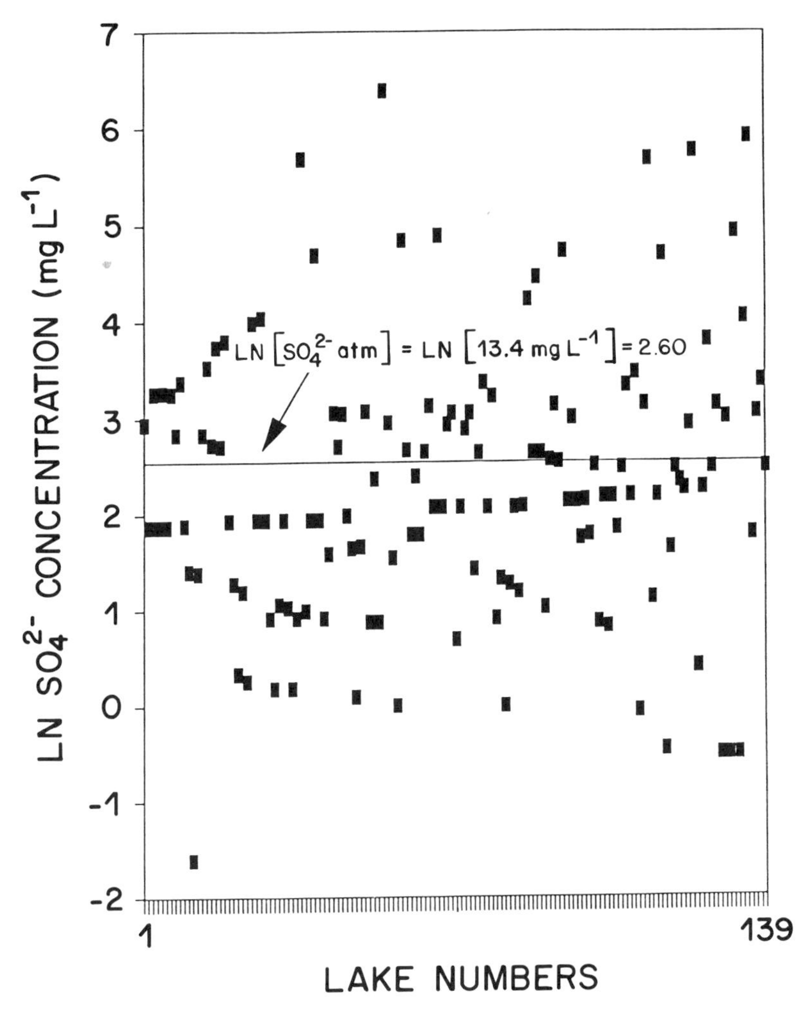

Figure 9.36 Natural logarithm of lake sulphate concentration for all Region 7 lakes by number showing the relationship to estimated contribution of atmospheric sulphate to lake sulphate concentration.

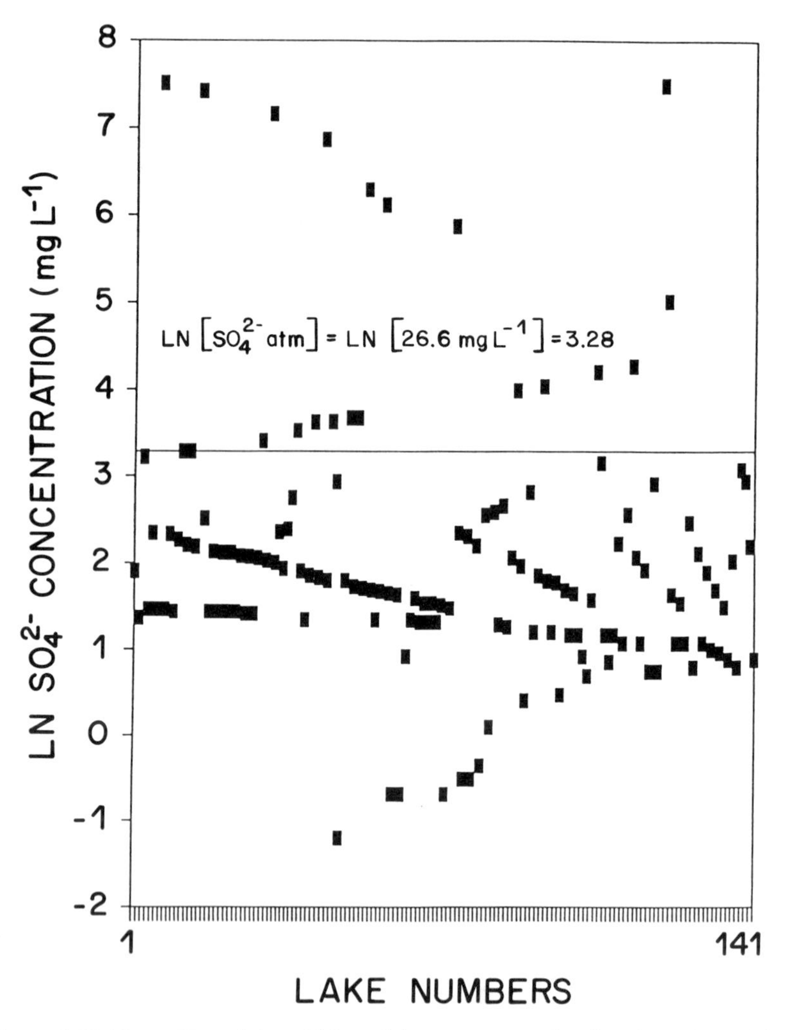

Figure 9.37 Natural logarithm of lake sulphate concentration for all Region 9 lakes by number showing the relationship to estimated contribution of atmospheric sulphate to lake sulphate concentration.

During the modelling routine the following values of evapotranspiration, precipitation, and the calculated ET concentration factors were used to determine potential alkalinity loss (Table 9.34).

Table 9.34 Regional mean annual values for evapotranspiration.

Region	Evapotranspiration (ET)	Precipitation	ET Concentration Factor *
3	392 mm	463 mm	6.52
4	301 mm	416 mm	2.40
6	364 mm	417 mm	7.87
7	383 mm	547 mm	3.335
9	412 mm	457 mm	5.38
Mountains 7b, 9b	300 mm	604 mm	1.99

Evapotranspiration values were obtained from R. Bothe, Hydrology Branch, Alberta Department of Environment, pers. comm.

9.5.4 Results

9.5.4.1 Modelling the Lakes

The relationship between the concentration of SO_4^{2-} and total cations for all surveyed lakes is shown in Figure 9.29. The results of evapotranspiration values were obtained from R. Bothe, Hydrology Branch, Alberta Department of Environment, pers. comm.

$$\text{ET concentration} = \frac{\text{Precipitation}}{\text{Precipitation - ET}}$$

This analysis produced a highly significant and positive regression (P=0.0000) with the following predictive line:

$$Y = -2.34 + 1.03X$$
$$R^2 = 0.57$$

In general, this analysis indicates that there is a positive correlation between concentration of sulphate and total cations in all lakes in all emission regions (1 through 10).

Lake sulphate concentrations for all regions, as indicated by number, is shown in Figure 9.32. This figure indicates that sulphate concentration is highly variable both between and among regions. A comparison between the estimated mean annual atmospheric loading of SO_4^{2-} within each study region to observed ranges in sulphate concentration within lake waters is shown in Table 9.35. These data indicate that in lakes, concentrations of sulphate often exceed that estimated to be available from atmospheric deposition. Equally as often, lake concentrations are well below the estimated regional atmospheric loading value (Figures 9.33 to 9.37). The deposition value used in the present case is highly conservative since it assumes that all deposited sulphate passes through watersheds into lakes. In Region 3, median and mean, and in Regions 7, and 9 mean lake sulphate concentrations exceed estimated atmospheric loading by considerable amounts (Table 9.33). In Regions 4 and 6, estimated atmospheric sulphate deposition exceeds mean lake sulphate by a factor of approximately two and median values by even greater margins. In all regions, however, maximum lake sulphate concentrations greatly exceed estimated atmospheric deposition.

Table 9.35 Estimated contribution of annual total atmospheric sulphur deposition to concentration of sulphate in surface waters for ADRP emission regions, sub-regions, and Crossfield East versus measured values in mg L^{-1}.

Location	Atmospheric SO_4^{2-}	Lakewater SO_4^{2-} Mean	Median	Max.	Min.
Region 3	13.7	53.6	45.0	180.0	2.5
Region 4	9.6	5.0	2.5	21.0	2.5
Region 6	18.9	9.7	2.5	53.0	1.0
Region 7	13.4	31.0	9.0	594.0	0.2
(Jasper, 7b)*	3.6	19.4	8.5	294.0	0.2
Region 9	26.6	73.0	5.6	2130.9	0.3
(Banff, 9b)*	3.6	19.6	5.1	1282.0	0.3
(Crossfield East)	68.3	-	-	-	-

* Sub-regions 7b and 9b represent the mountainous portions of the main Emission Regions 7 and 9, respectively.

Deposition values for Banff and Jasper were supplied by Atmospheric Environment Service of Canada and Alberta Department of Environment.

The relationship of lake sulphate concentration to estimated atmospheric contribution for Region 3 is shown in Figure 9.33. This figure illustrates that only 5 out of 17 regional lakes have sulphate concentrations below that available from the atmosphere. In spite of this, the mean pH of Region 3 lakes is 7.47 (with a median of 7.50) and the maximum pH found in the region was 8.10. This suggests that sulphate deposition has not markedly acidified these waters.

The relationship of lake sulphate to estimated atmospheric contribution for Region 4 is shown in Figure 9.34. Unlike the conditions in Region 3, the majority of lakes in Region 4 exhibit sulphate concentrations well below estimated atmospheric contribution. In fact, the majority of lakes (30 of 44) exhibited extremely low sulphate concentrations. Mean values for pH in Region 4 were found to be 7.37 (with a median of 7.30). Based on these data it does not appear that sulphate deposition has appreciably acidified lakes in this region.

Region 6 lake sulphate concentrations exhibited a trend similar to that observed in Region 4 (Figure 9.35) with respect to their relationship to estimated atmospheric sulphate deposition.

Lakes in Region 7 appeared to have sulphate concentrations that were distributed almost equally above and below the estimated atmospheric sulphate contribution (Figure 9.36). Likewise, atmospheric deposition appears to have had a negligible influence on the acidity of Region 7 lakes.

In Region 9, the majority of lakes exhibited sulphate concentrations that were below estimated regional atmospheric contribution (Figure 9.37). However, a number of lakes also exhibited very high sulphate concentrations with the maximum recorded value being 2130.9 mg L^{-1}. It is unlikely that such high readings are attributable to atmospheric loadings, and this will be discussed in detail in the following sections.

Regional lake pH, alkalinity, and sulphate concentrations (mean, median, minimum, and maximum) are shown in Table 9.36. Using these data "original alkalinity" was calculated as follows:

$$\text{Present Alkalinity} + 1.04\,(SO_4^{2-}) = \text{Original Alkalinity (as mg } L^{-1}\ CaCO_3)$$

This calculation assumes that all lake sulphate represents deposition of H_2SO_4, all of which has decreased the original alkalinity of lakes.

Table 9.36 Regional and sub-regional estimated original (pre-acidification) alkalinity, current alkalinity, current lake sulphate concentrations, and current pH values (all values in mg L^{-1} except pH, alkalinity mg L^{-1} as $CaCO_3$).

Emission Region	pH	SO_4^{2-} mg L^{-1}	Alkalinity Current mg L^{-1} as $CaCO_3$	Alkalinity Original mg L^{-1} as $CaCO_3$
Sub-region 7b				
mean	7.93	19.40	83.8	103.9
median	8.10	8.50	74.0	82.8
min.	6.10	0.20	0.2	0.4
max.	8.80	294.00	309.0	614.8
Sub-region 9b				
mean	7.90	19.60	119.2	139.6
median	7.90	5.10	55.0	60.3
min.	6.80	0.30	6.8	6.8
max.	10.20	1282.00	5451.0	6794.3
Region 3				
mean	7.47	53.56	104.3	160.0
median	7.50	61.25	105.0	168.7
min.	6.40	2.51	15.0	17.6
max.	8.10	180.00	179.0	355.8
Region 4				
mean	7.37	5.10	59.2	64.5
median	7.30	2.98	51.5	54.6
min.	5.90	2.5*	6.0	8.6
max.	9.20	20.0	140.0	158.7
Region 6				
mean	8.10	9.61	128.0	138.0
median	8.20	15.39	135.0	151.0
min.	7.50	2.5*	50.0	52.6
max.	8.80	29.04	200.0	230.2
Region 7				
mean	7.97	30.96	86.4	118.6
median	8.10	15.29	80.8	96.7
min.	6.10	0.20	0.2	0.4
max.	9.26	426.98	724.1	309.0

Continued...

Table 9.36 (Concluded).

Emission Region	pH	SO_4^{2-} mg L^{-1}	Alkalinity Current mg L^{-1} as $CaCO_3$	Alkalinity Original mg L^{-1} as $CaCO_3$
Region 9				
mean	7.96	72.98	184.8	260.7
median	7.90	5.58	65.5	71.3
min.	6.80	0.96	6.8	7.5
max.	10.20	2130.9	6300.0	8188.6

Original Alkalinity = Alkalinity + 1.04 [SO_4^{2-}] as mg L^{-1} as $CaCO_3$.

* Value of SO_4^{2-} based on 1/2 Analytical Detection Limit.
Sub-regions 7b and 9b represent the mountainous portions of the major Emission Regions 7 and 9, respectively.

The pH data (Table 9.36) indicate that slightly acidic lakes (pH less than 7.0) occur in all emission regions except Region 6. Considerable amounts of alkalinity are available in lakes in all regions except in some instances in Region 7 where the minimum available alkalinity was found to be 0.2 mg.L^{-1} (measured as $CaCO_3$).

The total estimated atmospheric sulphur deposition as H_2SO_4 (converted to alkalinity) is shown in Table 9.37 for the emission regions studied. The partitioning of deposited sulphur as wet and dry deposition is also shown in Table 9.37 (columns 2 and 3) reported as alkalinity.

The data in columns 1, 2, and 3 of Table 9.37 assume that all sulphate deposition is in the form of H_2SO_4, as is assumed by the Thompson model for precipitation in Maritime Canada. Alberta has significant amounts of alkaline dust to neutralize atmospheric acidity. In wet deposition there is sufficient basic material to act as a net contributor of alkalinity to watersheds (Table 9.37). Accordingly, estimated mean annual total and net acidic deposition is overestimated.

Column 5 in Table 9.37 contains a more realistic estimation of potential acidification by current rates of atmospheric deposition: total dry deposition of sulphate is considered to be H_2SO_4 minus estimated mean annual net deposition of alkalinity.

Even these values of dry acidic deposition are overestimates because in Alberta there is dry deposition of alkaline, neutralizing substances which is not taken into account. For example, mean wet

Table 9.37 Estimated loss of alkalinity of surface waters (mg L^{-1} as $CaCO_3$) due to atmospheric sulphur deposition using various assumptions. (The acid/base chemistry of deposition is 100% effective in reducing alkalinity; no watershed modification of precipitation chemistry by soils, biology, geology, or aquatic sediments).

		(1)	(2)	(3)	(4)	(5)	(6) Potential Additional Alkalinity Loss With Crossfield Deposition	
Location	Precipitation pH	Total S (H_2SO_4)	Wet Only (H_2SO_4)	Dry Only (H_2SO_4)	HCO_3^- Wet Deposition	Potential Alkalinity Loss at Present	Additive Loss	Total Loss
Region 3	5.11	14.2	5.4	8.8	0.5	8.3	30.8	39.1
Region 4	6.01	9.9	4.6	5.3	0.8	4.5	4.9	9.4
Region 6	5.21	19.6	4.9	14.7	0.1	14.6	40.2	54.8
Region 7	6.02	14.0	6.4	7.6	0.9	6.7	5.5	12.2
(Jasper 7b)	5.07	3.8	2.5	1.3	-	1.3	6.6	7.9
Region 9	5.08	27.7	11.9	15.7	0.2	15.5	6.3	21.8
(Banff, 9b)	5.07	3.8	2.5	1.3	-	1.3	5.8	7.1
(Crossfield East)	5.10	70.7	27.0	43.9	17.0	26.9	0	26.9

* Precipitation pH = mean annual volume.
1 = H_2SO_4 as converted alkalinity (1.04 x SO_4^{2-}) mg L^{-1} as $CaCO_3$.
1 = Column 2 + Column 3.
5 = Column 3 - Column 4.

deposition values obtained in Region 9 with samples subject to contamination by dust were found to have mean Ca^{2+} concentration values 3 times greater than mean values obtained with samples not subject to contamination by dust (Chapter 6). Also, for dust contaminated wet deposition samples, mean values of H^+ deposition were 13 times less than for uncontaminated samples (Chapter 6). However, because we still cannot derive good estimates of the degree of dry deposited alkaline substances which actually participates in neutralizing acid on a regional basis, we used the worst case condition, i.e., that there are no reactive bases in dry deposition.

The sixth column of Table 9.37 represents the total potential increase in alkalinity loss if the worst case scenario (Crossfield East deposition rates) were superimposed on each region. The increase in sulphur deposition represented by this worst case, therefore, represents a direct increase in H_2SO_4 deposition.

This scenario assumes that increased atmospheric acidity will not increase rates of reaction and the overall amount of atmospheric acidity neutralized by alkaline substances in the air. The scenario also assumes that anthropogenic activities will not increase the flux of dust to the atmosphere.

However, rate, extent, and nature of acid/base reactions are strongly influenced by the amount and concentration of acid available to react with a fixed amount of base. Chemical principles predict that increased atmospheric acidity results in more complete reaction of acid with base in the atmosphere.

Anthropogenic activities also influence the amount of crustal aerosols and concomitantly, the amount of bases in the atmosphere that are available to react with atmospheric acidity. It was determined that "the contribution to fine and coarse particles by soil and road dust were about two fold greater at Crossfield East as compared with West" (Chapter 5). Crossfield East is downwind of the gas plant and Crossfield West is upwind of the gas plant. Concentrations of atmospheric particulate matter were clearly higher at both Crossfield East and West when the gas plant plume was influencing the sampling sites (Chapter 5).

Cluster analysis showed that wet deposition of base cations (Ca^{2+}, Mg^{2+}) and acid anions (SO_4^{2-}, NO_3^-) at Crossfield East and West were related, while the deposition of H^+ at these sites showed virtually no positive relationship with any of the measured variables and negative correlations with most variables (Chapter 4).

In summary, the assumption that increased sulphur deposition resulting from the superimposition of a local Crossfield equivalent point source, is all H_2SO_4 represents a worst-case deposition scenario. Some portion of the hypothesized increase in sulphur deposition will be basic sulphate. However, because we do not know the extent to which the hypothesized increase in sulphur deposition is neutralized by increased reaction of acid with base we used the worst-case condition that the hypothesized increase in sulphur deposition is all H_2SO_4.

The results of these analyses show that, at the present estimated levels of atmospheric acidic deposition, the highest potential for surface water acidification exists in Regions 9 and 6 (Table 9.37, Column 5). Annual deposition rates of sulphuric acid in these regions have the potential to titrate 15.5 and 14.6 $mg\,L^{-1}$ of alkalinity, respectively (Table 9.37, Column 5). However, when the deposition rates associated with a local point source (such as those estimated near the Crossfield East air quality monitoring station), Regions 6 and 3 exhibited the highest potential for acidification followed by Region 9 (Table 9.37, Column 6). Increases in potential alkalinity loss increased by approximately five-fold in Region 3, two and one-half times in Region 6, and by half in Region 9 in comparison with present conditions. Significant increases in the potential for loss of alkalinity were also detected in both mountain Sub-regions 7b and 9b (Table 9.37, Column 6).

The percentage of lakes in each region where all mean annual alkalinity would be consumed under both present and point source deposition scenarios is shown in Table 9.38. In general, these analyses indicate that in all emission regions some lakes could be potentially sensitive to acidification under the assumptions of the present model and superimposition of a local point source. The highest potential for acidification under these conditions is in Region 7 with up to 20.7% of its lakes becoming susceptible. However, when the mountain lakes are removed from this grouping and examined independently, under worst case conditions 9.5% appear to have the potential for losing all alkalinity. These differences were caused by the lower estimated deposition for the mountains in comparison with the plains and by the lower evapotranspiration factor which resulted in lower concentration of acid and concomitant lower estimated loss of alkalinity. It was for precisely this reason that the main region was broken down into sub-regions. By lumping the two physiographic regions together into one, the potential for acidification is overestimated. Under present loading conditions Regions 7 and 9 are the most susceptible to acidification, with 7.2 and 2.9% of their lakes showing the potential to lose all alkalinity. As noted previously, for the major part this was caused by the inclusion of the mountain lakes in the calculation of sensitivity for these regions. However, these mountain lakes appear to be somewhat susceptible to acidification even using the low deposition loading values generated from Fortress Mountain.

Table 9.38 Percent of lakes examined by region which have the potential to lose all mean annual alkalinity by superimposition of (A) deposition equivalent to Crossfield East (Table 9.36, Column 6, Total Loss) or (B) under present deposition levels (Table 9.36, Column 5).

	Percent of Lakes Examined With the Potential to Lose All Alkalinity	
Region	A	B
3	11.8	0.0
4	13.6	0.0
6	5.0	0.0
7 (whole region)	20.7	7.2
7b (mountains only)*	9.5	1.6
9 (whole region)	5.6	2.9
9b (mountains only)*	3.8	1.0

* Sub-regions 7b and 9b represent the mountainous portions of the whole Regions 7 and 9, respectively, and were derived on the basis of deposition values for Fortress Mountain and evapotranspiration values for Jasper.

Superimposing a local point source, the potential sensitivity of lakes within Regions 3, 4, and 6 increased substantially from no sensitivity under current loading conditions to 11.8, 13.6, and 5.0% of all lakes in each respective region having the potential to lose all alkalinity. Similarly, with a local point source, the potential sensitivity of lakes to lose all alkalinity increased considerably for all regions.

9.5.4.2 Modelling the Rivers

The mean annual alkalinity, pH, sulphate concentration, original alkalinity, and potential alkalinity losses under present and worst case scenarios for all regional streams and rivers examined are shown in Table 9.39. These data indicate that on a regional basis none of the streams or rivers examined have any potential for acidification. This does not infer that local point sources or event-related acid effects could not be produced within the regions studied or even in the streams examined. On the basis of these results, no further modelling was attempted on the rivers examined.

9.5.5 Discussion

The use of the atmospheric deposition data (Table 9.39) and the Thompson (1982) model with the assumptions stated, have allowed the determination of the sensitivity of surface waters to acidification in Alberta. This was done by determining the potential loss of alkalinity under two acid loading scenarios: 1) current and 2) estimated worst case loadings if additional deposition equivalent to that estimated for Crossfield East occurred in each emission region being studied. The analyses indicated that except for the western Cordilleran areas of the Province, regional and sub-regional estimates of potential alkalinity loss were quite large for the local point source situation.

The observed water chemistry data support the conclusion drawn from atmospheric chemistry and deposition data that model predictions of acidification are worst case overestimates. For example areas where the estimated loss of alkalinity (pre-acidification alkalinity minus current alkalinity) was found to be lowest, also contained waters with the lowest natural alkalinities. This strongly suggests that sulphate, principally related to the products of mineral weathering (base cations), was the principal alkalinity-producing process of most watersheds. This hypothesis is supported by the observed significant and positive correlation of sulphate to the sum of the base cations (Ca^{2+}, Mg^{2+}, K^{+}, and Na^{+}). The concentrations of SO_4^{2-} in surface waters were also shown to be highly variable being either significantly less than, greater than, or approximately equal to regional and sub-regional estimates of atmospheric sulphate. It is likely that at least some of this variability is due to differences between local and regional/sub-regional estimates for sulphate deposition and evapotranspiration. However, analysis showed that SO_4^{2-} concentrations in surface waters varied by several orders of magnitude on both regional/sub-regional basis as well as on a very local and proximate basis. Thus it appears that sources as well as sinks of sulphate have a prominent influence on concen-

Table 9.39 Relationship between mean annual pH, SO_4^{2-}, alkalinity, original alkalinity and potential atmospheric deposition of acid forming sulphates (converted to alkalinity equivalents) for all regional rivers. (All values in mg L^{-1} except pH in units; alkalinity mg L^{-1} as $CaCO_3$.)

Emission Regions	pH	SO_4^{2-}	Present Alkalinity ($CaCO_3$)	Original Alkalinity	Present Potential Alkalinity Loss*	Future Possible Alkalinity Loss**	Assessment of Sensitivity
3	8.01	37.57	130.8	169.87	8.3	39.1	Not Sensitive
4	7.50	11.42	210.6	222.48	4.5	9.4	Not Sensitive
6	7.83	2.46	209.6	212.16	14.6	40.2	Not Sensitive
7	7.53	7.00	119.4	126.68	6.7	5.5	Not Sensitive
9	8.17	214.30	302.8	525.67	15.5	21.8	Not Sensitive

*Present potential based on estimated wet + dry SO_4^{2-} deposition (converted to alkalinity equivalents) - net wet HCO_3^-.

**Future potential based on the imposition of Crossfield East scenario on each region.

trations of sulphate in surface waters. Erickson and Trew (1987), in a similar study, surveyed 875 lakes throughout Alberta. In their study, based on surface water alkalinity values alone, approximately 9.7% of the lakes reviewed were within their classification range of sensitivity to acidification. A number of the lakes included in their study occurred in emission Regions 1, 2, and 5 all of which were outside the geographic area of this study. In each of these areas, Birch Hills, Swan Hills, and Caribou Hills, as well as in the Canadian Shield north of Lake Athabasca, pH and alkalinity values of lakes were generally lower than those in the regions examined in this study. If these areas are excluded and the results of the two studies are examined, a number of similarities and differences become apparent.

In general, results from the ADRP study agree with these and other Alberta studies conducted by Erickson (1987) and Palmer and Trew (1987). Using an expanded data base of approximately 1000 lakes, these investigators found approximately 17% of Alberta's lakes to be potentially sensitive to acidification. Erickson's (1987) study was based on sensitivity analysis on ranges of calcium and alkalinity concentrations, and pH values. The only valid comparison between the current study and that of Erickson (1987) would be to compare the results of Erikson's findings based on alkalinity ranges. In Erickson's study, the alkalinity range for highly sensitive lakes was 0 to 10 mg L^{-1}, moderately sensitive lakes between 11 to 20 mg L^{-1}, and low sensitivity lakes had alkalinities that exceeded 20 mg L^{-1} (as $CaCO_3$). A standard factor for sulphate equivalency to alkalinity loss as was used in the present study, indicates that deposition rates of between 0 to 9.62 mg L^{-1} as SO_4^{2-} (assuming all SO_4^{2-} is H_2SO_4) would be required to acidify lakes in the highly sensitive range. A review of the regional deposition estimates (Tables 9.37-9.39) indicates that this level of potential acidic deposition occurs in Emission Regions 3, 6, 7, and 9 at the present time, and this level of deposition is almost reached in Region 4 as well. If, however, Sub-regions 7b and 9b are separated from their main regions, the percentage of lakes considered to be potentially sensitive to acidification drops substantially. Since, many mountain area lakes were predicted to be highly sensitive (on the basis of Erickson's studies), the conclusion is that alkalinity ranges used for classifying acidification sensitivity for such lakes were set too high, even higher than estimates based on worst case assumptions. As indicated in the ADRP studies, under current estimated deposition rates, it is predicted that 1.6% and 1.0% of Mountain Region lakes in Sub-regions 7b and 9b, respectively, could lose all alkalinity. In fact, based on regional assessments it would seem that far fewer lakes in the Province are sensitive to acidification than were predicted by Erickson and Trew (1987) and by Erickson (1987).

The major reasons for these apparent discrepancies lie in the inclusion of highly sensitive Canadian Shield lakes in the north-eastern portion of Alberta, and a broader sample in the Erickson (1987) studies than was possible in the present work. In addition, the regional groupings used by Erickson (1987) were physiographically based rather than being based on emissions. This approach should be added to any future ADRP studies. In spite of the obvious biases present in both the present ADRP study and that reported by Erickson (1987), the estimates of sensitivity in the ADRP study are likely more representative of the provincial picture, particularly considering the conservative nature of the estimates of alkalinity loss produced by the modified Thompson (1982) model.

A suggestion for future studies of surface water acidification offered by Alberta Department of Environment (D. Trew pers. comm.), is to consider H^+ deposition rather than SO_4^{2-} per se. This change was suggested to be necessary since the amounts of alkaline dust present in Alberta make it unreasonable to use sulphate as a surrogate for acidic deposition. Nevertheless, uncertainties attached to the acid/base chemistry of dry deposition and acidic deposition/ watershed interactions (Section 9.4) compelled us to make risk assessment assuming conservative worst case assumptions: dry deposition is all acidic and this acidic deposition is perfectly transmitted to surface waters.

9.5.5.1 Modification of Deposition Chemistry by Watersheds

The data examined during this study indicated that not only are there significant sources of sulphate in Alberta's watersheds, but there are also significant sinks of sulphur. In fact, these sources and sinks appear to predominate over atmospheric contributions for most lakes. Occasionally, the concentration of lakewater sulphate resembles that predicted by deposition modelling. However, these occurrences may be more coincidental than real. These occurrences can not be scientifically accepted as indicating that lakewater sulphate is controlled by atmospheric deposition, although this is often suggested in the scientific literature (Cosby et al. 1985; Haines and Akielaszak 1983; Henriksen 1979; Kramer and Tessier 1982; Sullivan et al. 1988).

Two possible sources for the addition of sulphate to lake systems in Alberta are deep aquifer groundwater (as in the Cold Lake Region of the Province) and shallow coal seam aquifers that occur throughout the plains and foothills regions of the Province.

Deep aquifers such as the Durlingville and Empress Aquifers of the Cold Lake Region contain waters dominated by sodium bicarbonate rather than calcium bicarbonate systems. These aquifers also often exhibit high sulphate concentrations. Under some conditions, depending on the age of the groundwater, sulphate can be co-dominant with sodium. If these aquifers discharge into a lake they can cause its chemistry to alter to the extent that it resembles the groundwater. Evidence for such groundwater influences have been reported by Ozoray et al. (1980) for two Cold Lake area lakes, Moore and Hilda. Data on the chemistry exist to indicate that it is not an uncommon occurrence (Goodfish and Whitefish Lakes, Alberta Fish and Wildlife 1988, unpublished report).

In the plains and foothills regions of the Province, much of the water supply to lakes is derived from coal seam aquifers. Work conducted by the Geological Society of America (1971) and by Steiner et al. (1972) have defined the sulphur regions of Alberta coal and have put forward hypotheses regarding their sulphur content and its origin. In general, coal seams along the mountains and in southwestern Alberta range in sulphur content from 0.6 to 1.0%. This sulphur zone encompasses most of Emission Region 9. Outside this zone, coal sulphur content is lower but still ranges between 0.2 to 0.5%. These coal deposits are found throughout the remainder of Alberta and vary in quality from sub-bituminous to lignite. Coal mining activities occur, or have occurred, in almost all emission regions examined in this project except for possibly Region 4. However, coal fields also exist in Region 4 (Firebag River).

The sulphur content of the coal is thought to be derived from two potential sources: Cretaceous volcanic activity and marine deposition (Geological Society of America 1971; Steiner et al. 1972). Regardless of its origin, sulphur is easily accessible through groundwater recharge to lakes. This source plus deeper ground-water intrusion is probably a very significant factor in the hypertrophic status of many central Alberta lakes (Region 10) which also often exhibit extremely high sulphur concentrations. Coincidentally, extensive surface coal mining also occurs in these areas.

Surface water acidification was modelled in this project as a literal titration, i.e., acid titration of a basic solution in an unreactive beaker. But acid reacts with more than just alkalinity in the water. For example, acids interact with vegetation and therefore, land use practices change the overall effect on surface water systems. Acids also interact with exchangeable bases in soils and with weatherable mineral bases. All of these interactions consume acidity (Chapter 9.4). Similarly, acidity is lost through the interaction of acids with aquatic sediments. Acid consumption in natural systems not only includes ion exchange and mineral weathering, but also sulphate reduction by micro-organisms which is especially important in waterlogged, or reducing, sediments typical of lake and river bottoms, and wetlands. For example, the artificial acidification of Lake 233 (Kelly et al. 1982) in the Experimental Lakes Area of Ontario required approximately three times as much sulphuric acid to be applied to the lake directly than was predicted from H_2SO_4 titration of the lakewater alkalinity. The rate of sulphate reduction in aquatic systems is concentration dependent. The increased concentration of sulphate in Lake 233 resulted in an order of magnitude increase in the rate of sulphate reduction and alkalinity that is produced by this process.

If free carbonates are present in the soils of a watershed, acidic deposition may result in very little, or no, reduction in the alkalinity of surface run-off because acid input can be completely consumed by deposition-induced increase in mineral weathering. Most of the soils in Alberta have free carbonates in their profiles and tend to effectively remove any deposited acid (Chapter 9.4).

Acidic deposition was reported to increase the alkalinity of a Michigan lake in studies conducted by Kilham (1982). In this instance the mostly siliceous glacial soils of the watershed contained some free carbonates. These carbonates were found to have completely neutralized the atmospheric input of acidic materials and caused an increase in the flux of sulphate into the lake in the form of $CaSO_4$. Lake sediments reduced the sulphate, converting it into $Ca(HCO_3)_2$, thus increasing the overall alkalinity of the lake (Kilham 1982).

The interaction of acidic deposition with various types of soils in Alberta has been modelled by Turchenek et al. (1987) using a modified Bloom and Grigal (1985) model. In their investigations the upper 25 cm of soils from various classes commonly found in Alberta were subjected to acid input. Even though these types of surficial soils usually do not have free carbonate, they were found to be 100% effective in removing strong acids and essentially did not allow the passage of acid through to the receptor waters. Turchenek et al. (1987) found the model predicted that only the very acidic, organic rich soil material (pH $\leq$ 5.0) were somewhat ineffective in removing the deposited H^+.

9.5.5.2 Episodic Events of Runoff

It is expected that during events such as snowmelt, at least some surface waters (such as small streams, and lakes and ponds with little retention time) will become diluted and, as a result, will exhibit significantly lower alkalinities than indicated by our data. At the time of dilution, a fraction, or perhaps even most, of the acid deposited in a watershed over a year, such as in snowpack, is released to the now dilute receiving waters during freshet. The usual timing of such runoff events also coincides with the spring spawning runs and the most acid sensitive part of the lifecycle of fish (Baker and Schofield 1985). While snowmelt in most parts of Alberta is not expected to be acidic, exchangeable bases from the organic soils found in the more northern regions of the Province may not be capable of buffering deposited acids that are released in large volumes during acidic snowmelt events.

To date few studies have been conducted to determine what the actual effects are on solution chemistry during episodic events such as snowmelt. Most effects scientists do not conduct control experiments during runoff studies. Acidic runoff that is toxic to fish and emanates from acidic soils is taken as proof of an acidic rain effect. However, Jones et al. (1983) have shown that runoff from simulated unpolluted deposition that was passed through an acid, organic rich soil resulted in highly acidic, aluminum rich runoff that was highly toxic to fish.

As previously discussed, generally, very acidic, organic rich soils are thought to have little ability to buffer inputs of acid. However, in their studies Jones et al. (1983) demonstrated a remarkable ability of such soils to buffer the pH of receiving waters against the addition of incoming acids. James and Riha (1984, 1986) found that acidic surficial organic soil horizons in New York state naturally yielded acidic, aluminum bearing leachate similar to that described by Jones et al. (1983). In their studies, James and Riha simulated acidic deposition that was equivalent to an entire winter's accumulation typical for their region of the Adirondack Mountains. The simulated acidic deposition was reacted with a 1 cm thick layer of forest litter. In these studies, the more acidic the organic soil the lower its buffering capacity as a result of ion exchange, which is in agreement with current models (Chapter 9.4). Ion exchange accounted for only about 8 to 58% of the H^+ removal in these experiments. However, the soils were found to retain 67 to 96% of all H^+ added. Organic acid buffering, as proposed by Krug and Frink (1983) was found to remove between 38 to 79% of all acid added (James and Riha 1986). The more acidic and humified the soil, the greater was its organic acid buffering capacity. Organic acid buffering of this type is ignored by all current acidification models. This represents yet another factor that causes the overestimation of the aquatic effects of acidic deposition.

The net result of organic acid buffering in these studies was a decrease in the concentration of dissolved organic carbon and an increase in ionic aluminum. As suggested by Krug et al. (1985), replacement of organic acids by mineral acids influences speciation of aluminum so that such waters may become more toxic to aquatic organisms.

Acidic deposition/watershed interactions are modelled using processes of inorganic mineral systems. Potentially sensitive watersheds of Alberta tend to be mantled by organic matter, as are most of the potentially sensitive watersheds in the rest of North America. Acidic runoff principally occurs as near surface runoff from such organic soil materials. Acidic deposition/organic matter interactions merit further study to properly assess the potential aquatic effects of acidic deposition in the Province.

9.5.6 Concluding Remarks

Worst case scenarios were developed to assess the potential for acidification of Alberta's surface water resources. These scenarios indicated that a small fraction of the provincial lakes and ponds are potentially sensitive to acidification if point sources are located nearby. Such watersheds tend to be mantled with highly acidic surficial soil materials and peat. Similarly, such watersheds are also expected to be potentially sensitive to temporary acidification during episodic events such as snowmelt.

To date there has been little research to document and determine if the worst case scenarios generated in this study can be realized or not.

9.5.7 Literature Cited

Baker, J. P. and C.L. Schofield. 1985. Acidification impacts on fish populations: A Review. pp. 183-221. In: **Acid Deposition** (Environmental Economics and Policy Issues), eds. D.D. Adams and W.P. Page, New York: Plenum Press. 560 pp.

Bloom, P.R. and D.J. Grigal. 1985. Modelling soil response to acid deposition in non-sulphate absorbing soils. *Journal of Environmental Quality* 14: 489-495.

Charles, D.F. and S.A. Norton. 1986. Paleolimnological evidence for trends in atmospheric deposition of acids and metals. Appendix E. Physical and chemical characteristics of some lakes in North America for which sediment-diatom data exist. pp. 335-431, 482-506. In: **Acid Deposition.** Long-Term Trends, National Research Council, Washington, D.C. 506 pages.

Cosby, B.J., G.M. Hornberger, J.N. Galloway, and R.F. Wright. 1985. Modelling the effects of acid deposition: Assessment of a lumped parameter model of soil water and streamwater chemistry. *Water Resources Research* 21: 51-63.

Demayo, A. 1970. A Storage and Retrieval System for Water Quality Data. Department of Energy, Mines and Resources, Inland Waters Branch, Report Series No. 9, Ottawa.

Erickson, P.K. 1987. An Assessment of the Potential Sensitivity of Alberta Lakes to Acidic Deposition. Prep. by Water Quality Control Branch, Pollution Control Division, Environmental Protection Services, Alberta Environment. 137 pp.

Erickson, P.K., and D.O. Trew. 1987. An assessment of the potential sensitivity of Alberta lakes to acidic deposition. In: **Acid Forming Emissions in Alberta and Their Ecological Effects**. Second Symposium Workshop Proceedings, eds., H.S. Sandhu, A.H. Legge, J.I. Pringle, and S. Vance, 1986 May 12-15, Calgary, Alberta, pp. 309-335.

Garrels, R.M. and C.L. Christ. 1965. *Solutions, Minerals, and Equilibria*. New York: Harper Row. 89 pp.

Geological Society of America. 1971. Geologic aspects of sulphur distribution in coal and coal related rocks; Coal symposium, abstracts with program, Volume 3. Geological Society of America Annual Meeting, Boulder, Colorado.

Gherini, S.A., C.W. Chen, L. Mok, R.A. Goldstein, R.J.M. Hudson, and G.F. Davis. 1984. The ILWAS Model: Formulation and Application in the Integrated Lake-Watershed Acidification Study, Volume 4. Summary of Major Results. Electric Power Research Institute. EA-3221. pp. 7-1 to 7-45.

Goldstein, R.A. C.W. Chen, and S.A. Gherini. 1984. Integrated Lake-Watershed Acidification Study-Summary. In: **The Integrated Lake-Watershed Acidification Study. Volume 4. Summary of Major Results**, eds. R.A. Goldstein and S.A. Gherini. Electrical Power Research Institute EA-3221. pp. 1-1 to 1-15.

Haines, J.A. and J. Akielaszak. 1983. A regional survey of the chemistry of headwater lakes and streams in New England: Vulnerability to acidification. U.S. Fish and Wildlife Service Air Pollution, Acid Rain Report No. 15. 141 pp.

Henriksen, A. 1979. A simple approach for identifying and measuring acidification of freshwater. *Nature* 278: 542-545.

James, B.R. and S.J. Riha. 1984. Soluble aluminum in acidified organic horizons of forest soils. *Canadian Journal of Soil Science* 64: 637-646.

James, B.R. and S.J. Riha. 1986. pH buffering in forest soil organic horizons: Relevance to acid precipitation. *Journal of Environmental Quality* 15: 229-234.

Jones, H.C. J.C. Noggle, R.C. Young, J.M. Kelly, H. Olem, R.J. Ruane, R.W. Pasch, G.J. Hyfantis, and W.J. Parkhurst. 1983. Investigations of the Cause of Winter Fish Kills in Fish-Rearing Facilities in Raven Fork Watershed. Tennessee Valley Authority Rpt. TVA/ONR/WR-83/9.

Kelly, C.A., J.W.M. Rudd, R.B. Cook, and D.W. Schindler. 1982. The potential importance of bacterial processes in regulating rate of lake acifidication. *Limnology and Oceanography* 27(5): 868-882.

Kilham. P. 1982. Acid Precipitation: Its role in the alkalization of a lake in Michigan. *Limnology and Oceanography* 27(5): 856-867.

Kramer, J. and A. Tessier. 1982. Acidification of aquatic systems: A critique of chemical approaches. *Environmental Science and Technology* 16: 606A-615A.

Krug, E.C. 1987. Watershed effects on surface water chemistry. In: **Proceedings of the Fourth Annual Pittsburgh Coal Conference**, University of Pittsburgh. pp. 271-280.

Krug, E.C. and C.R. Frink. 1983. Acid rain on acid soil: A new perspective. *Science* 221: 520-525.

Krug, E.C., P.J. Isaacson, and C.R. Frink. 1985. Appraisal of some current hypotheses describing acidification of watersheds. *Journal of the Air Pollution Control Association* 35: 109-114.

Malanchuk, J.L. and R.S. Turner. 1987. Chapter 8. Effects on aquatic systems. In: **NAPAP Interim Assessment, Volume IV; Effects of Acidic Deposition.** National Acid Precipitation Assessment Program. Washington, D.C.

Ozoray, G., E.J. Wallick, and A.T. Lytviak. 1980. Hydrogeology of the Sand River Area, Alberta. Alberta Research Council, Earth Sciences Report 79-1, Edmonton, Alberta.

Palmer, C.J., and D.O. Trew. 1987. The Sensitivity of Alberta Lakes and Soils to Acidic Deposition: Overview Report. Prep. for Environment Protection Services, Alberta Environment. Draft report, June 1987. 28 pp.

Picard, D.J., D.G. Colley, and D.H. Boyd. 1987. Overview of the Emission Data: Emission Inventory of Sulphur Oxides and Nitrogen Oxides in Alberta. Prep. for Acid Deposition Research Program by Western Research, Division of Bow Valley Resource Services Ltd., Calgary, Alberta ADRP-A-1-87. 87 pp.

Schnoor, J.L., W.D. Palmer, Jr., and G.E. Glass. 1984. Modeling impacts of acid precipitation for northeastern Minnesota. In: **Modeling of Total Acid Precipitation Impacts**, Jerald L. Schnoor, ed. London: Butterworth Publishers. pp. 121-154.

Steiner, J., G.D. Williams, and E.J. Dickie. 1972. Coal deposits of the Alberta Plains. Proceedings of First Annual Plains Coal Symposium, Edmonton, Alberta. pg. 85-96.

Sullivan, T.J., J.M. Eilers, M.R. Church, D.J. Blick, K.N. Eshleman, D.H. Landers, and M.S. Dehaan. 1988. Atmospheric wet sulphate deposition and lakewater chemistry. *Nature* 331: 607-609.

Thompson, M.E. 1982. The cation denudation rate as a quantitative index of sensitivity of eastern Canadian rivers to acidic atmospheric precipitation. *Water, Air, and Soil Pollution* 18: 215-226.

Traversy, W.J. 1971. Methods for the Chemical Analysis for Waters and Wastewaters. Environment Canada, Water Quality Division, pp. 62, 71, 152, and 163.

Turchenek, L.W., S.A. Abboud, C.J. Thomas, R.J. Fessenden, and N. Holowaychuk. 1987. Effects of Acid Deposition on Soils in Alberta. Prep. for the Acid Deposition Research Program by the Alberta Research Council, Edmonton, Alberta. ADRP-B-05/87. 202 pp.

U.S. NAPAP. The National Acid Precipitation Assessment Program. 1987. Refer to Volumes 1 through 4.

APPENDIX
CRITICAL REVIEW OF WORLD LITERATURE ON THE EFFECTS OF ACIDIC AND ACIDIFYING AIR POLLUTANTS ON SELECTED ENVIRONMENTAL COMPONENTS

During 1987 the following peer evaluated literature reviews (10 volumes), an overview document of the literature reviews (Volume 11), and a four-volume emission inventory were released to the public. Microfiche or paper copies of these documents are available through (please cite microlog numbers): Micromedia Ltd., 158 Pearl Street, Toronto, Ontario, Canada M5H 1L3, (416-593-5211).

- •ADRP-B-01-87 Telang, S.A. 1987.
  Surface Water Acidification Literature Review. Prep. for the Acid Deposition Research Program by the Kananaskis Centre for Environmental Research, The University of Calgary, Calgary, Alberta, 123 pp.
  ISBN 0-921625-03-0 (Volume 1)
  (88-00571/ Microlog Number)
- •ADRP-B-02-87 Visser, S., R.M. Danielson, and J.F. Parr. 1987.
  Effects of Acid-Forming Emissions on Soil Micro-organisms and Microbially-Mediated Processes. Prep. for the Acid Deposition Research Program by the Kananaskis Centre for Environmental Research, The University of Calgary, Calgary, Alberta, and U.S. Department of Agriculture, Beltsville, Maryland, 86 pp.
  ISBN 0-921625-04-9 (Volume 2)
  (88-00572/ Microlog Number)
- •ADRP-B-03-87 Krouse, H.R. 1987.
  Environmental Sulphur Isotope Studies in Alberta: A Review. Prep. for the Acid Deposition Research Program by the Department of Physics, The University of Calgary, Calgary, Alberta, 89 pp.
  ISBN 0-921625-05-7 (Volume 3)
  (88-00573/ Microlog Number)

•ADRP-B-04-87 Laishley, E.J. and R. Bryant. 1987.
Critical Review of Inorganic Sulphur Microbiology with Particular Reference to Alberta Soils Prep. for the Acid Deposition Research Program by the Department of Biology, The University of Calgary, Calgary, Alberta, 50 pp.
ISBN 0-921625-06-5 (Volume 4)
(88-00574/ Microlog Number)

•ADRP-B-05-87 Turchenek, L.W., S.A. Abboud, C.J. Tomas, R.J. Fessenden, and N. Holowaychuk. 1987.
Effects of Acid Deposition on Soils in Alberta. Prep. for the Acid Deposition Research Program by the Alberta Research Council, Edmonton, Alberta, 202 pp.
ISBN 0-921625-07-3 (Volume 5)
(88-00575/ Microlog Number)

•ADRP-B-06-87 Jaques, D.R. 1987.
Major Biophysical Components of Alberta. Prep. for the Acid Deposition Research Program by Ecosat Geobotanical Surveys, Inc. North Vancouver, British Columbia, 108 pp. + maps.
ISBN 0-921625-03-1 (Volume 6)
(88-00676/ Microlog Number)

•ADRP-B-07-87 Campbell, K.W. 1987.
Pollutant Exposure and Response Relationships: A Literature Review, Geological and Hydrogeological Aspects. Prep. for the Acid Deposition Research Program by Sub-surface Technologies and Instrumentation Limited, Calgary, Alberta, 151 pp. + maps.
ISBN 0-921625-09-X (Volume 7)
(88-00577/ Microlog Number)

•ADRP-B-08-87 Torn, M.S., J.E. Degrange, and J.H. Shinn. 1987.
The Effects of Acidic Deposition on Alberta Agriculture: A Review. Prep. for the Acid Deposition Research Program by the Environmental Sciences Division, Lawrence Livermore National Laboratory, Livermore, California, 160 pp.
ISBN 0-921625-10-3 (Volume 8)
(88-00578/ Microlog Number)

•ADRP-B-09-87 Mayo, J.M. 1987
The Effects of Acid Deposition on Forests. Prep. for the Acid Deposition Research Program by the Department of Biology, Emporia State University, Emporia, Kansas, 74 pp.
ISBN 0-921625-11-1 (Volume 9)
(88-00579/ Microlog Number)

•ADRP-B-10-87 Krupa, S.V. and R.N. Kickert. 1987.
An Analysis of Numerical Models of Air Pollutant Exposure and Vegetation Response. Prep. for the Acid Deposition Research Program by the Department of Plant Pathology, University of Minnesota, St. Paul, Minnesota, and Consultant, Corvallis, Oregon, 113 pp.
ISBN 0-921625-12-X (Volume 10)
(88-00580/ Microlog Number)

•ADRP-B-11-87 Legge, A.H. and R.A. Crowther. 1987.
Acidic Deposition and the Environment: A Literature Overview. Prep. for the Acid Deposition Research Program by the Kananaskis Centre for Environmental Research, The University of Calgary, and Aquatic Resource Management, Calgary, Alberta, 235 pp.
ISBN 0-921625-13-8 (Volume 11)
(88-00616/ Microlog Number)

EMISSION INVENTORY

•ADRP-A-1-87 Picard, D.J., D.G. Colley, and D.H. Boyd. 1987.
Overview of the Emission Data: Emission Inventory of Sulphur Oxides and Nitrogen Oxides in Alberta. Prep. for the Acid Deposition Research Program by Western Research, Division of Bow Valley Resource Services Ltd., Calgary, Alberta, 87 pp.
ISBN 0-921625-18-9 (Volume 1)
(88-00581/ Microlog Number)

•ADRP-A-2-87 Picard, D.J., D.G. Colley, and D.H. Boyd. 1987. Design of Emission Inventory: Emission Inventory of Sulphur Oxides and Nitrogen Oxides in Alberta. Prep. for the Acid Deposition Research Program by Western Research, Division of Bow Valley Resource Services Ltd., Calgary, Alberta, 86 pp.
ISBN-0-921625-19-7 (Volume 2)
(88-00582/ Microlog Number)

•ADRP-A-3-87 Picard, D.J., D.G. Colley, and D.H. Boyd. 1987. Emission Data Base: Emission Inventory of Sulphur Oxides and Nitrogen Oxides in Alberta. Prep. for the Acid Deposition Research Program by Western Research, Division of Bow Valley Resource Services Ltd., Calgary, Alberta, 335 pp.
ISBN 0-921625-20-0 (Volume 3)
(88-00583/ Microlog Number)

•ADRP-A-4-87 Picard, D.J., D.G. Colley, and D.H. Boyd. 1987. Results of the Emission Source Surveys: Emission Inventory of Sulphur Oxides and Nitrogen Oxides in Alberta. Prep. for the Acid Deposition Research Program by Western Research, Division of Bow Valley Resource Services Ltd., Calgary, Alberta, 203 pp.
ISBN 0-921625-21-9 (Volume 4)
(88-00584/ Microlog Number)

SUBJECT INDEX